HAMMOND

ATLAS OF THE
20th Century

HAM

ATLAS

20th C

MOND

OF THE

entury

TIMES BOOKS
77-85 Fulham Palace Road
London W6 8JB

First published by Times Books 1996
Copyright © Times Books 1996

Editorial direction:
Philip Parker
Thomas Cussans

Sarah Allen
Sarah May

Cartographic direction:
Martin Brown

Tim Dobson
Neil Forrest

Design:
Ivan Dodd
Nick O'Toole
Colin Brown '
Phanos Anaxagorou

Additional cartography: Alan Grimwade, David Maltby,
Kevin Klein, Mark Eldridge, Laura Morris at
Cosmographics, Watford

Cartographic consultant: Andras Bereznay

Picture research: Anne-Marie Ehrlich

Place names consultant: P. J. M. Geelan

Index: Janet Smy

Glossary and chronology: Marcus Ormond

In addition to the contributors listed on page 5,
the publishers would also like to thank:
Margaret Allen; Michael Evans, *Defence Correspondent,*
The Times, *London;* Ralph Gibson, *Commercial Director,*
RIA-Novosti, London; Lucy Mennell; Gilbert Padey, *WHO*
Information Services, London; Dennis Taylor, *Foreign Desk,*
The Times, *London*; David Woodrup, *Information Officer,*
UNESCO, London

Colour separations by Colourscan, Singapore

Printed and bound in Italy by Rotolito Lombarda

HAMMOND
INCORPORATED
MAPLEWOOD, NEW JERSEY 07040-1396

ISBN
Hard cover: 0-8437-1148-5
Soft cover: 0-8437-1149-3

CONTRIBUTORS

GENERAL EDITOR | RICHARD OVERY
Professor of History
King's College London

CONTRIBUTORS | Anthony Best
Lecturer in International History
London School of Economics

Margaret Byron
Lecturer in Geography
King's College London

Kathleen Burk
Professor of Modern & Contemporary
History
University College London

Richard Clutterbuck
Honorary Research Fellow
University of Exeter

Lesley France
Retired Senior Lecturer
Division of Geography &
Environmental Management
University of Northumbria

James Gow
Lecturer
Department of War Studies
King's College London

Michael Hendrie
Astronomy Correspondent
The Times, *London*

Geoffrey Jones
Professor
Department of Economics
University of Reading

Ephraim Karsh
Professor of Mediterranean Studies
King's College London

Sarah Stockwell
Lecturer in Imperial & Commonwealth
History
King's College London

Richard Vinen
Lecturer in History
University of London

Geoff A. Wilson
Lecturer in Geography
King's College London

CONTENTS

Chapter 4
THE COLD WAR WORLD *106*

Chapter 5
TOWARDS THE NEW WORLD ORDER *144*

INTRODUCTION

IT IS A TRUISM that the 20th century has witnessed the most profound and wide-reaching changes of any century in human history. Yet the dimensions of that change are still worth recalling. In 1900 most people were wretchedly poor, by modern standards, and led lives of monotonous and physically demanding work, with almost no access to the services and goods enjoyed by small wealthy elites. The advances which have transformed the century – scientific and medical discoveries, new forms of production and exchange – were largely confined to the prosperous areas of Europe and North America.

Although this division between rich and poor, developed and developing, has scarcely diminished over the century, the fruits of modernity, bitter though they have sometimes been, have spread worldwide. The catch-phrase of the late century is globalization. The world is girdled by new routes of access: air travel; the computer superhighway; and the tentacles of thousands of multi-national organizations. Half of all people live in cities; most work in industry and services, generating levels of income undreamed of in 1900. The collapse of empire and the spread of mass culture have produced very different elites in the 1990s, whose horizons are global rather than parochial. In 1900 kings and emperors still mattered. Now they are merely ornaments.

This transformation was neither pre-ordained nor inevitable. By 1945, after two massively destructive wars in the developed world and an almost terminal crisis of capitalism in 1929, the century seemed to be going from bad to worse. The horrors of trench warfare, followed 25 years later by the revelations of German genocide and Stalinist terror, eroded the confidence in progress inherited from the 19th century. The English novelist, George Orwell, despaired of the future of the century as the war with Hitler drew to a close. He saw persistent "Führer-worship" replacing democracy, rigidly planned economies destroying economic freedom and an almost permanent state of war between "great super-powers unable to conquer one another". The result was the novel *1984*, published just before the mid-century, a savage indictment of the drift to state power and a regimented society.

Orwell was wrong, though not entirely. The abuse of state power persisted. The Cold War produced unimaginable dangers of global destruction. But when 1984 actually came, a large part of the world enjoyed levels of political freedom, economic prosperity and personal security that barely seemed possible in 1945. The problems that remain are more often than not the result of weak states, not the threat of Big Brother. The revival of the century's fortunes after 1950 resulted from three things: the power of the United States, which stabilized the world order; an exceptional economic boom; and the end of imperialism, an institution almost as old as human history. Empire collapsed everywhere between 1900 and the emancipation of Europe's colonies in the 1960s (or by 1989, if the Soviet system is counted as imperialist), and with it the idea that one state could physically and permanently control others. States now cluster together from shared interests, not through violence.

In 1900 little of this was predictable. The new century was welcomed with feelings of exhilaration, a real sense that a new age was dawning. As the century closes the temptation is to see the end of yet another age. It should be resisted. If the remarkable changes of our present century have taught us anything it is a healthy scepticism about prediction. We can now take stock of our last hundred years, but the 21st century must take care of itself.

Richard Overy
June 1996

EUROPE	AMERICAS	AFRICA	SOUTH ASIA AND THE MIDDLE EAST	EAST ASIA AND THE PACIFIC
	1902 US gains control over Panama Canal	1902 End of Boer War		1902 Anglo-Japanese alliance
	1903 Wright brothers' first flight			
1904 Anglo-French entente				1904–5 Japanese victory in war against Russia
1905 Failed revolution in Russia				
1906 Constitution granted in Russia	1906 US occupies Cuba (to 1908)		1906 Growth of nationalist movement in India 1906 Revolution in Persia 1906 Muslim League founded in India	
1907 Anglo-Russian entente 1907 Peasant revolt in Romania				1907 New Zealand acquires dominion status
1908 Bulgaria becomes independent; Austria annexes Bosnia and Herzegovina 1908 First powered flight in Europe		1908 Belgian state takes over Congo from King Leopold	1908 Young Turk revolution: Ottoman sultan deposed	
1909 Military coup in Greece: Venizelos becomes prime minister				
	1910 Mexican revolution begins	1910 Formation of Union of South Africa		1910 Japan annexes Korea
1911–12 Italo-Turkish war in Libya		1911 Italy conquers Libya		1911 Chinese Revolution: Sun Yat-sen first president of new republic: rise to power of Warlords (to 1926)
1912–13 Balkan wars		1912 ANC set up in South Africa		
1913 Treaty of Bucharest	1913 Woodrow Wilson becomes US president			
1914 Outbreak of First World War 1914 Battle of Tannenberg	1914 Panama Canal opens 1914 US intervenes in Mexico 1914 Canada joins First World War	1914 Britain proclaims protectorate in Egypt 1914–15 French and British conquer German colonies except German East Africa	1914 Turkey joins Central Powers in First World War	1914 German concessions in China and colonies in Pacific taken over by Japan, Australia and New Zealand
1915 Italy joins Allies	1915 German U-Boat warfare alienates US		1915 Gallipoli campaign	1915 Japan imposes "21 Demands" on China for economic exploitation
1916 Battles of Verdun and the Somme			1916–18 Arab revolts against Turkish rule	
1917 Revolution in Russia: tsar abdicates (Mar.), Bolsheviks take over (Nov.); first socialist state established 1917 First use of massed tanks (Battle of Cambrai)	1917 New constitution in Mexico 1917 US declares war on Central Powers 1917 US troops again occupy Cuba (to 1923)		1917 Balfour Declaration promises Jews a national home in Palestine	
1918 Germany and Austria-Hungary sue for armistice: end of First World War 1918 Civil war and foreign intervention in Russia (to 1921) 1918–19 German Revolution: democracy established	1918 President Wilson announces Fourteen Points	1918 First Pan-African Congress	1918 Britain and France promise freedom to Turkish subjects	
1919 Paris treaties redraw map of Europe	1919 First transatlantic flight	1919 Nationalist revolt in Egypt against British protectorate	1919 Amritsar incident; upsurge of Indian nationalism	1919 "May 4th" movement launched by Chinese student protestors
1920 League of Nations established (headquarters Geneva) 1920–1 Russian-Polish war	1920 Prohibition in US (to 1933) 1920 US refuses to ratify Paris treaties and withdraws into isolation		1920 Mustafa Kemal (Atatürk) leads resistance to partition of Turkey; Turkish Nationalist Movement 1920 Britain and France given "mandates" in the Middle East	1920 Japan given control of previously German islands in the Pacific
1921 German reparations bill set	1921 US restricts immigration	1921 Battle of Anual: Spanish army routed by Moroccans		1921 Chinese Communist Party formed 1921–2 Washington Conference attempts to regulate situation in East Asia
1922 Russian-German Rapallo Treaty 1922 Mussolini takes power in Italy 1922 Creation of Irish Free State		1922 Britain gives Egypt conditional independence	1922 Greek army expelled from Turkey; last Ottoman sultan deposed; republic proclaimed (1923)	
1923 France occupies the Ruhr; German hyper-inflation 1923 Military coup in Spain	1923 General Motors established (world's largest manufacturing company)	1923 Abyssinia admitted to the League of Nations		

EUROPE	AMERICAS	AFRICA	SOUTH ASIA AND THE MIDDLE EAST	EAST ASIA AND THE PACIFIC
1924 Death of Lenin; Stalin eventually emerges as Soviet leader (1929) **1924** Dawes Plan for German reparations	**1924** Virtual civil war in Brazil **1924** Military coup, Chile		**1924** Atatürk establishes right-wing secular state in Turkey	**1924** Kuomintang government established in Canton
1925 Hindenburg becomes German president **1925** Locarno Treaties stabilize frontiers in West		**1925** Abd-el Krim attracts French hostility	**1925–7** Syrian revolt against French rule	
1926 British General Strike **1926** Germany joins the League of Nations	**1926** US troops occupy Nicaragua (to 1933)	**1926** Revolt of Abd-el Krim crushed in Morocco		**1926** Chiang Kai-shek begins reunification of China (Northern Expedition) **1926** Hirohito becomes Japanese emperor **1926** Australia and New Zealand's "dominion" status defined **1927** Chiang Kai-shek suppresses communists at Shanghai
1928 Pact of Paris outlaws war **1928** First Five-Year Plan and (1929) collectivization of agriculture in Russia		**1928** Abyssinia signs friendship treaty with Italy	**1928** Gandhi becomes leader of Indian Congress, which calls for immediate independence	
1929 Mussolini signs Lateran Accord with Papacy **1929** Trotsky expelled from USSR	**1929** Wall Street Crash precipitates world Depression **1929** Calles establishes National Revolutionary Party, Mexico			
1930 Allied troops withdraw from Rhineland	**1930** First World Cup Finals held in Uruguay **1930** Military revolution in Brazil; Vargas becomes president		**1930** Gandhi leads protests against salt tax	**1930** Chiang launches operations against Chinese communists **1930–1** Communist revolt in Indo-China (Vietnam)
1931 Spanish Second Republic formed **1931** Collapse of Central European banks begins major recession **1931** National government in Britain	**1931** US president Hoover announces a one-year suspension of war debt payments			**1931** Japanese occupy Manchuria
1932 Nazis become largest party in German parliament	**1932** Chaco War between Bolivia and Paraguay (to 1935)		**1932** Kingdom of Saudi Arabia formed by Ibn Saud	**1932** "Manchukuo" Republic set up in China by Japan
1933 Hitler made chancellor in Germany; beginning of Nazi revolution	**1933** US president Roosevelt introduces New Deal			**1933** Japan leaves the League of Nations
1934 Foiled coup by Austrian Nazis **1934** Hitler becomes German *Führer* **1934** Stavisky scandal and riots in France	**1934** Cárdenas becomes president of Mexico: land redistribution and (1938) nationalization of oil	**1934** Italians suppress Senussi resistance in Libya		**1934** Long March of Chinese communists begins **1934** Japan declares "Amau Doctrine" (Japanese sphere of interest in East Asia)
1935 "Stresa Front" formed against German aggression **1935** Franco-Soviet Pact **1935** German re-armament publicly declared		**1935** Italy invades Abyssinia		
1936 German reoccupation of Rhineland **1936** Spanish Civil War begins (to 1939) **1936** Great Terror launched in Russia **1936** Mussolini proclaims Rome-Berlin "Axis"	**1936** Pan-American congress; US proclaims good neighbour policy	**1936** Anglo-Egyptian alliance; British garrison Suez Canal Zone **1936** Italian troops enter Addis Ababa	**1936** Arab revolt in Palestine against Jewish immigration **1936** Oil discovered in Saudi Arabia	**1936** Japanese military take over government **1936** Japan signs anti-Comintern pact with Germany
1937 Neville Chamberlain becomes British prime minister **1937-8** Soviet show trials of senior communists	**1937** US Neutrality Laws passed by Congress **1937** Vargas launches "New State" in Brazil	**1937** Tunisian rising against France		**1937** Beginning of full-scale war between Japan and China
1938 Germany occupies Austria			**1938** Death of Atatürk	

EUROPE	AMERICAS	AFRICA	SOUTH ASIA AND THE MIDDLE EAST	EAST ASIA AND THE PACIFIC
1938 Munich conference: dismemberment of Czechoslovakia				
1939 German occupation of Prague **1939** Nationalist victory ends Spanish Civil War **1939** German-Soviet Non-Aggression Pact; Germany invades Poland; Britain and France declare war on Germany		**1939** South Africa votes to fight with Britain against Germany	**1939** India declares war on Germany	**1939** Russian forces defeat Japan at Khalkin Gol (Manchuria); Russo-Japanese neutrality pact (1941) **1939** Australia/New Zealand enter war against Germany
1940 Germany overruns Norway, Denmark, Belgium, Netherlands, France; Italy invades Greece but is repulsed; Battle of Britain; Blitz **1940** Vichy regime established in France under Pétain	**1940** Roosevelt elected for record third term	**1940-1** Italians expelled from Somalia, Eritrea and Abyssinia	**1940** Indian Muslims demand separate state (Pakistan)	**1940** Japan begins occupation of French Indo-China
1941 Germany invades Yugoslavia **1941** Germany invades Russia; declares war on the US **1941** Tripartite Pact between Germany, Italy and Japan	**1941** Lend-Lease: US and Britain sign Atlantic charter **1941** US enters war against Germany and Japan	**1941** Germans and Italians conquer Cyrenaica and advance into Egypt (1942)		**1941** Japan attacks US at Pearl Harbor
1942 German offensive in southern Russia; Battle of Stalingrad **1942** German genocide of Jews begins in full (to 1945)		**1942** Britain occupies Madagascar **1942** Battle of El-Alamein; Italian/German defeat and retreat **1942** Anglo-American landings in Morocco and Algeria	**1942** Japanese advance halted on Indian/Burmese border **1942** Gandhi and Indian Congress leaders arrested	**1942** Japan overruns South East Asia **1942** Battle of Midway; US halts Japanese expansion
1943 German VI Army surrenders at Stalingrad; Italian capitulation **1943** Combined Bomber Offensive launched (to 1945)	**1943** Military coup in Argentina	**1943** Capitulation of Axis forces in Tunisia	**1943** Teheran Conference: Allies plan post-war order	
1944 Anglo-American landing in Normandy; Russian advance in Eastern Europe **1944** Rome falls to Allies (5 June)	**1944** Bretton Woods meeting in US			**1944** Gradual Japanese retreat: Battle of Leyte Gulf
1945 Yalta Conference; defeat of Germany and suicide of Hitler **1945** Labour government formed in Britain: welfare reforms **1945** Potsdam Conference on division of Germany	**1945** US tests first atomic bomb **1945** United Nations established (headquarters New York) **1945** Vargas falls from power in Brazil	**1945** Nationalist agitation in Algeria suppressed by French forces		**1945** US drops atom bombs on Japan, forcing surrender
1946 Nuremberg trials of Nazi war leaders **1946** Greek civil war (to 1949)	**1946** Peron comes to power in Argentina (to 1955) **1946** Philippines given independence by US		**1946** Syria gains independence from France **1946** Anglo-Indian negotiations on independence	**1946** New Japanese constitution adopted **1946** Civil war in China (to 1949) **1946** Creation of Philippine Republic **1946** Beginning of Vietnamese struggle against France (to 1954)
1947 Development of Cold War; Truman Doctrine enunciated **1947** Marshall Plan for economic reconstruction in Europe **1947** European communists set up Cominform	**1947** General Agreement on Tariffs and Trade (GATT) signed		**1947** India and Pakistan become independent amid widespread rioting; 500,000 die	**1947** First supersonic flight (US)

EUROPE	AMERICAS	AFRICA	SOUTH ASIA AND THE MIDDLE EAST	EAST ASIA AND THE PACIFIC
1948 Communist takeover in Czechoslovakia and Hungary; Berlin airlift	**1948** Organization of American States (OAS) established	**1948** Reunited National Party takes power in South Africa	**1948** Establishment of State of Israel; first Arab-Israeli war **1948** Gandhi assassinated: Indo-Pakistani dispute over Kashmir	**1948** Burma and Ceylon independent
1949 Formation of NATO alliance and of COMECON **1949** West and East Germany created: Adenauer becomes West German chancellor		**1949** Apartheid programme inaugurated in South Africa		**1949** Communist victory in China **1949** Indonesia independent
	1950 Vargas returns to power in Brazil (to 1954)			**1950** Korean War begins (to 1953)
1951 European Coal and Steel Authority set up		**1951** New constitution in Gold Coast	**1951** Mossadeq becomes prime minister in Iran; nationalizes oil	**1951** Australia, New Zealand and US sign ANZUS Pact
		1952 Beginning of Mau Mau rebellion in Kenya **1952** Military revolt in Egypt; proclamation of republic (1953)		**1952** US ends occupation of Japan
1953 Death of Stalin; East Berlin revolt crushed: Khrushchev emerges as new Soviet leader	**1953** Fidel Castro prominent in foiled Cuban rebellion		**1953** Military coup in Iran	
	1954 Growing persecution of "communists" in US (McCarthyism)	**1954** Beginnings of nationalist revolt in Algeria (to 1962) **1954** Egyptian King Farouk overthrown; Nasser prime minister		**1954** Vietnam partitioned after French defeat **1954** Geneva conference: Laos, Cambodia and Vietnam become independent states
1955 West Germany enters NATO **1955** Warsaw Pact signed				**1955** Bandung Conference **1955** Japan launches "Five-Year Plan" for economic growth
1956 Khrushchev criticizes Stalinist regime **1956** Polish revolt, Gomulka in power; Hungarian revolt crushed by Russians		**1956** Independence for Sudan, Morocco and Tunisia **1956** Suez crisis: Anglo-French invasion of Canal Zone	**1956** Second Arab-Israeli war and Suez crisis **1956** Pakistan declared an Islamic state	**1956** Japan joins the UN
1957 Treaty of Rome: formation of European Economic Community (EEC)		**1957** Beginning of decolonization in sub-Saharan Africa: Gold Coast (Ghana) becomes independent		
1958 Fifth Republic in France: de Gaulle first president	**1958** Eisenhower Doctrine proclaimed for Middle East **1958** Development of the silicon microchip in the US		**1958** US forces intervene in Lebanon **1958** United Arab Republic founded (Syria and Egypt) **1958** Kassem overthrows monarchy in Iraq	**1958** Great Leap Forward in China (to 1961)
1959 Britain establishes European Free Trade Association (EFTA)	**1959** Cuban Revolution			**1959** China reoccupies Tibet **1959** War between North and South Vietnam (to 1975) **1959** Lee Kuan Yew becomes Singapore's president (to 1993)
		1960 Sharpeville massacre in South Africa **1960** Independence of several African states; outbreak of civil war in Belgian Congo		**1960** Sino-Soviet dispute begins
1961 East Germans build Berlin Wall	**1961** Launch of Civil Rights movement in southern states of US **1961** US cuts links with Cuba: "Bay of Pigs" fiasco	**1961** South Africa becomes independent republic		**1961** Increasing US involvement in Vietnam
	1962 Cuban missile crisis	**1962** Algeria becomes independent		**1962** Sino-Indian war
1963 France vetos British entry into EEC **1963** Nuclear test ban treaty	**1963** US president Kennedy assassinated; Johnson succeeds (to 1968)	**1963** Organization of African Unity (OAU) set up: Kenyan independence achieved		
1964 Brezhnev becomes Soviet leader	**1964** US Civil Rights Act inaugurates President Johnson's Great Society Programme **1964** Military coup in Brazil		**1964** Palestine Liberation Organization set up by Arafat	
		1965 Rhodesia declares UDI	**1965** Indo-Pakistan war	**1965** Marcos becomes ruler of Philippines
1966 France withdraws from NATO command structure	**1966** Eruption of Black American discontent; growth of Black Power			**1966** Cultural Revolution in China (to 1976)
1967 Greek military coup		**1967** Civil war in Nigeria (secession of Biafra) (to 1970)	**1967** Third Arab-Israeli war (Six-Day War)	**1967** ASEAN formed
1968 Liberalization in Czechoslovakia halted by Russian invasion **1968** Student riots in France; spread across Western Europe	**1968** Assassination of Martin Luther King; Nixon elected US president (to 1974)		**1968** Saddam Hussein takes power in Iraq	

CHRONOLOGY 1969-84

EUROPE	AMERICAS	AFRICA	SOUTH ASIA AND THE MIDDLE EAST	EAST ASIA AND THE PACIFIC
				1969 Sino-Soviet border clashes
1969 Outbreak of violence in Northern Ireland	**1969** Growing opposition to Vietnam War in US			
1970 West Germany signs agreements with eastern neighbours **1970** Nuclear Non-Proliferation Treaty comes into force	**1970** Allende elected president of Chile (killed in 1973)	**1970** Sadat succeeds Nasser in Egypt	**1970** Assad takes power in Syria	
	1971 US initiates policy of détente with China and USSR **1971** US abandons Gold Standard and depreciates dollar	**1971** African nations call for "liberation" of South Africa (Mogadishu Declaration) **1971** Amin seizes power in Uganda	**1971** Indo-Pakistan war leads to breakaway of East Pakistan (Bangladesh)	**1971** People's Republic of China admitted to the UN
1972 Bloody Sunday in Londonderry, Northern Ireland **1972** EEC Paris Summit pledges potential union by 1980				
1973 Oil crisis ends post-war economic boom **1973** Britain, Ireland and Denmark join EEC **1973** Miners' strike in Britain brings three-day week	**1973** Major recession in US triggered by oil crisis **1973** General Pinochet becomes dictator in Chile		**1973** Fourth Arab-Israeli war (Yom Kippur); OPEC countries treble price of oil **1973** Gadaffi announces "cultural revolution" in Libya	**1973** US forces withdraw from South Vietnam
1974 IRA bombing campaign in Britain **1974** End of dictatorship in Portugal **1974** Giscard d'Estaing becomes French president (to 1981)	**1974** US Watergate scandal: President Nixon resigns (replaced by Ford)	**1974** Emperor Haile Selasse of Ethiopia deposed by Marxist junta	**1974** Turkish invasion of Cyprus	**1974** Indonesian occupation of East Timor
1975 Death of Franco; end of dictatorship in Spain **1975** Helsinki Accords on human rights		**1975** Lomé agreement between EEC and African states **1975** Portugal grants independence to Mozambique and Angola	**1975** Civil war in Lebanon: Syria invades (1976)	**1975** Communists take over Vietnam, Laos and Cambodia; Khmer Rouge under Pol Pot in Cambodia launches campaign of mass genocide
	1976 Military junta takes power in Argentina (to 1982)	**1976** Morocco and Mauritania partition Spanish Sahara **1976-7** Soweto riots in South Africa		**1976** Death of Mao Tse-tung; trial of Gang of Four
		1977 Bokassa proclaims "Central African Empire"	**1977** Egypt/Israeli peace talks (Camp David Peace Treaty, 1978) **1977** Military coup ends democratic rule in Pakistan	**1977** Deng Xiaoping re-admitted to power in China; begins modernization drive
			1978 General Zia becomes ruler of Pakistan	**1978** Sino-Japanese Treaty of Friendship
1979 Thatcher becomes British prime minister	**1979** Civil war in Nicaragua (to 1990) **1979** Civil war in El Salvador (to 1992)	**1979** Tanzanian forces invade Uganda and expel President Amin **1979** Fall of Bokassa and restoration of Central African Republic	**1979** Fall of shah of Iran; establishment of Islamic Republic under Ayatollah Khomeini (d. 1989) **1979** Afghanistan invaded by USSR (to 1989)	**1979** Vietnam invades Cambodia, expelling Khmer Rouge government **1979** Sino-Vietnamese war **1979** "Democracy Wall" movement in China
1980 Death of Marshal Tito **1980** Creation of independent Polish trade union Solidarity; martial law (1981)		**1980** Black majority rule established in Zimbabwe (Rhodesia) **1980** OAU meeting in Lagos pledges African common market by 2000	**1980** Outbreak of Iran/Iraq war (to 1988)	
1981 Greece joins EEC **1981** Mitterrand becomes French president (to 1995)	**1981** Reagan becomes US president (to 1989) **1981** AIDS epidemic officially announced	**1981** Assassination of Egyptian president Sadat		
1982 Death of USSR President Brezhnev, succession of Andropov (d.1984), then Chernenko (d.1985) as USSR leader **1982** Kohl forms government in West Germany	**1982** Argentina occupies South Georgia and Falkland Islands; surrenders to UK forces		**1982** Israel invades Lebanon; expulsion of PLO from Beirut **1982** Israel withdraws from Sinai Peninsula	
1983 Anti-nuclear protest against US	**1983** Democracy restored in Argentina **1983** Coup in Grenada; US invades			
		1984 Famine in Sahel and Ethiopia; continuing war against secession	**1984** Indira Gandhi assassinated: renewed Indo-Pakistan conflict over Kashmir	**1984** Sino-British agreement over Hong Kong (China to take over in 1997)

14

EUROPE	AMERICAS	AFRICA	SOUTH ASIA AND THE MIDDLE EAST	EAST ASIA AND THE PACIFIC
1985 Gorbachev becomes leader of USSR (to 1991)	**1985** Democracy restored in Brazil and Uruguay	**1985** Civil unrest in South Africa; state of emergency declared, civil rights and press freedom suspended	**1985** Israel withdraws from Lebanon, other than buffer zones in south	
1986 Spain and Portugal join EEC: Single European Act heralds closer integration	**1986** Reagan government hit by "Iran-Contra" arms scandal	**1986** US bomb Libya in retaliation for terrorist activities		**1986** Fall of Marcos in the Philippines; Aquino succeeds **1986** Pro-democracy agitation in China
	1987 INF treaty between USSR and US: phased elimination of their intermediate range land-based nuclear weapons **1987** US stock market crash		**1987** Massacre of Iranian pilgrims in Mecca **1987** "Intifada" revolt begins in Gaza Strip and West Bank	
1988 Gorbachev moves USSR towards freedom of information and debate *(glasnost)*, and industrial and social re-structuring *(perestroika)*	**1988** End of Pinochet's dictatorship in Chile; Aylwin becomes president (1989)		**1988** Benazir Bhuttto restores civilian rule in Pakistan **1988** PLO recognizes State of Israel	
1989 Democratic elections for People's Congress held in USSR; Yeltsin, president of Russia, first democratically elected leader; Poland and Hungary move towards political pluralism; popular protest topples communist regimes in East Germany, Czechoslovakia, Bulgaria and Romania (Ceauçescu executed); Berlin Wall demolished	**1989** Menem becomes Argentine president **1989** Bush becomes president of US (to 1993) **1989-90** US military intervention in Panama; arrest and extradition of Noriega	**1989** De Klerk replaces Botha as South African president	**1989** Death of Ayatollah Khomeini **1989** Russian troops leave Afghanistan, but civil war continues **1989** Establishment of Palestine National Council	**1989** Japanese emperor Hirohito dies; succeeded by Akihito **1989** Student pro-democracy demonstration crushed in Peking **1989** Vietnam withdraws from Cambodia
1990 Reunification of Germany **1990-1** Baltic republics declare independence	**1990** Democratic elections in Nicaragua end Sandinista rule	**1990** Namibia becomes independent **1990** South African government moves towards accommodation with ANC; frees Mandela; and (1991) announces intention to dismantle apartheid	**1990** Iraq invades Kuwait **1990** Yemen united **1990** Benazir Bhutto dismissed as Pakistan prime minister	**1990** Burma military regime refuses to recognize election victory of Aung San Suu Kyi
1991 Failed communist coup against Gorbachev: disintegration of the Soviet Union **1991** Disintegration of Yugoslavia: Slovenia and Croatia declare independence **1991** Maastricht Treaty on EC integration	**1991** Bush announces the end of the Cold War		**1991** Gulf War: UN coalition forces led by US attack Iraq and liberate Kuwait **1991** Middle East peace talks begin in Madrid **1991** Rajiv Gandhi killed by suicide bomber	
1992 Civil war in Georgia (to 1994) **1992** Civil war in Bosnia-Herzegovina: Bosnian Serbs fight Muslims and Croats	**1992** Leader of Peruvian "Shining Path" captured **1992** "Earth Summit" in Rio de Janeiro	**1992** US forces intervene to end Somalia's famine and civil war **1992** Election in Angola brings temporary end to civil war **1992** Algeria cancels election result after fundamentalist successes; bloody civil war erupts	**1992** UN-sponsored elections in Kurdistan	**1992** Aquino succeeded by General Ramos in Philippines
1993 Czech Republic and Slovakia emerge as separate states **1993** Russian general elections: neo-fascist Zhirinovsky wins 23%	**1993** Clinton becomes US president **1993** Siege at Waco, Texas; cult members killed		**1993** Oslo Accords between Israel and the PLO **1993** Cambodian democracy established **1993** Benazir Bhutto returns as Pakistan prime minister	**1993** Cambodian elections held under UN supervision; Sihanouk returns to throne
1994 Yeltsin orders Russian troops into breakaway Chechen republic	**1994** US invasion of Haiti, exiled leader Aristide returns	**1994** Ethnic strife in Rwanda **1994** ANC win South African election; Mandela president	**1994** Limited Palestinian autonomy in Gaza Strip	**1994** Death of North Korean dictator, Kim Il Sung
1995 Agreement signed to end civil war in Bosnia **1995** European Union (EU) established; Austria, Sweden and Finland join EU	**1995** Oklahoma City terrorist bombing **1995** Quebec votes to remain in Canada	**1995** Outbreak of Ebola epidemic in Zaire **1995** Nigeria expelled from Commonwealth for repeated human rights abuses	**1995** Israel and PLO sign agreement on West Bank autonomy **1995** Assassination of Israel's prime minster, Rabin	**1995** Protests against French nuclear tests in the South Pacific **1995** Kobi earthquake in Japan **1995** Religious sect accused of nerve-gas attack on Tokyo subway **1995** Aung San Suu Kyi released from house arrest
1996 Russian presidential elections **1996** Aznar becomes Spanish prime minister **1996** Prodi becomes Italian prime minister		**1996** De Klerk resigns as South African vice-president	**1996** Right-wing Likud government elected in Israel, Netanyahu prime minister	**1996** China attempts military intimidation of Taiwan as islands hold first free elections in Chinese history **1996** "East-West" summit in Bangkok

THE WORLD IN 1900 was poised on the threshold of one of the most remarkable periods of change in human history. An old order was giving way to a new. Under the impact of industrialization and the rise of mass politics, the established monarchical order, whose dynasties stretched back for centuries, began to crumble. The coming of mass urbanization and new technologies in the 19th century in Europe and the United States transformed societies traditionally based on landed power and peasant farming. In 1900 most of the world was still ruled by old empires – Manchu China, Ottoman Turkey, Romanov Russia, Habsburg Austria. In 1900 most of the world's population still earned its living from primitive farming. But change was irresistible and worldwide. The dominant theme of the 19th century was emancipation from royal autocracy, from imperial oppression, from poverty and ignorance, above all from political exclusion. The demands for national independence, democracy and a better way of life, with their roots in America and western Europe, worked like a strong acid on the old structures of power and wealth. As they dissolved, the world entered upon an era of exceptional turbulence and violence.

Stretcher-bearers at
the Battle of Ypres, 1915

THE END OF THE OLD WORLD ORDER

THE WORLD IN 1900: EMPIRES

AT THE BEGINNING of the 20th century, the political map of the world was overwhelmingly imperial. There were old empires in China and the Ottoman Middle East. There were European colonial empires which stretched back to the 16th century – Spanish, Portuguese, Dutch. There were newer colonial empires which reached their fullest extent in the half century before 1900 – British, French, German, Italian. Even the Americas, most of whose states were republics, had once been part of Europe's empires, and still shared the language and culture exported there. In Europe there were empires built by the Austrian Habsburg dynasty and the Russian Romanovs which had no overseas possessions, but ruled a conglomeration of subject peoples stretching from Italy in the west to the shores of eastern Asia.

In 1900 the colonial empires were at their zenith. Since the 1870s a new wave of imperialism had brought most of Africa and the Pacific islands under European rule. The last native independent state in Africa – Abyssinia (Ethiopia) – withstood Italian efforts at conquest, inflicting a humiliating defeat on Italian forces at Adowa in 1895. In southern Africa, the British fought the Dutch settlers in Transvaal and the Orange Free State in the Boer War (1899-1902) and brought both under direct British rule. The British empire was the world's largest, covering one quarter of the globe. At its heart lay India, where Queen Victoria was declared empress in 1874. A few thousand officials ruled an area of 350 million people.

Europeans looked out upon the wider world confident that what they offered was civilization and technical progress. They saw the world in their own image. European languages replaced native tongues as the medium of administration and commerce. European religion was exported along with the fruits of European technical and scientific development. In 1900 most of the world outside China was nominally Christian or was ruled by Christian officials. A flood of migrants left Europe – 25 million between 1880 and 1914. European trade dominated the world's markets. European armies and navies, armed with the most modern weapons, gave the new European empires the power to impose European interests. Japanese leaders were so

Shortly after George V ascended the throne in 1910 as king-emperor of the British empire, he visited Delhi for the Durbar marking the start of his reign. The spectacular occasion *(left)* marked the high point of British rule. The royal procession was flanked by soldiers of the imperial army, a genuine multi-racial force *(top)*. George V afterwards ordered the building of a magnificent imperial capital at New Delhi, "for all time a monument to British art and workmanship". The capital, designed by Sir Edwin Lutyens, took 20 years to build and cost £10.5 million. By the time it was completed, India was a little over a decade away from full independence.

Much of the world in 1900 was divided into empires *(map right)*. They were ruled, except for the French empire, by old dynastic houses. In the Far East the Chinese empire was in decline, that of the Japanese, invigorated by a modernizing revolution in 1868, was still expanding. The European empires covered half the globe, much of the area taken over during the previous century. Spain and Portugal were the exceptions. Their vast empires in Latin America, built in the 16th century, had won independence during the 19th century. During the 20th century, the other empires disappeared, to be replaced by the modern nation state.

The End of the Old World Order

The Pacific islands were the last part of the world to be colonized by the European powers. By this stage both Japan and the United States had a strong political interest in the area. In 1898 the United States declared war on Spain following an unexplained explosion aboard an American warship in Havana harbour. The war was quickly over. Manila in the Philippines was stormed by the 1st Colorado Volunteers on 13 August 1898 (*bottom right*). A year later the United States participated in the agreement between Britain and Germany over the division of the Samoan Islands. On 2 December 1899 the German governor of Samoa unveiled a monument to American and British officers killed in disturbances on the islands (*right*). Germany obtained Western Samoa, the USA the rest.

impressed with European expansion that they adopted western technology and military reforms, and set out to build a colonial empire of their own in east Asia. Formosa (Taiwan) was acquired in 1895, Korea in 1910. Japanese officials were made to read Sir John Seeley's *Expansion of England* as an example to follow.

Appearances in 1900 proved deceptive. As a form of political organization, empire was in its very final stages. Indeed the buoyant colonial empires of Europe contributed to

the decline of Ottoman Turkey and Manchu China, which Europeans wished to dominate for themselves. Europe was in the process of generating the social and political forces that were to transform empire in the 20th century. The unification of Italy in 1860 and Germany in 1871 showed the importance of nationalism as a political force. By 1900 agitation for national autonomy was widespread in Europe – in Ireland, Bohemia, Poland, the Ukraine, Finland – while nationalist opponents of the old dynasties in China and Turkey undermined the established order.

Nationalism was one component of the development of mass politics. Social and economic modernization in Europe in the 19th century threw up new social classes no longer prepared to accept traditional dynastic or aristocratic rule. European liberals succeeded in establishing constitutional parliamentary rule and civil rights in Britain, Italy and France (though not in their empires). Their demands for modern freedoms – democracy, the rule of law, respect for the individual, the right to self-determination – filtered beyond Europe to encourage political protest in the very areas that Europe now ruled. Europe was also the home of modern socialism. Taking inspiration from

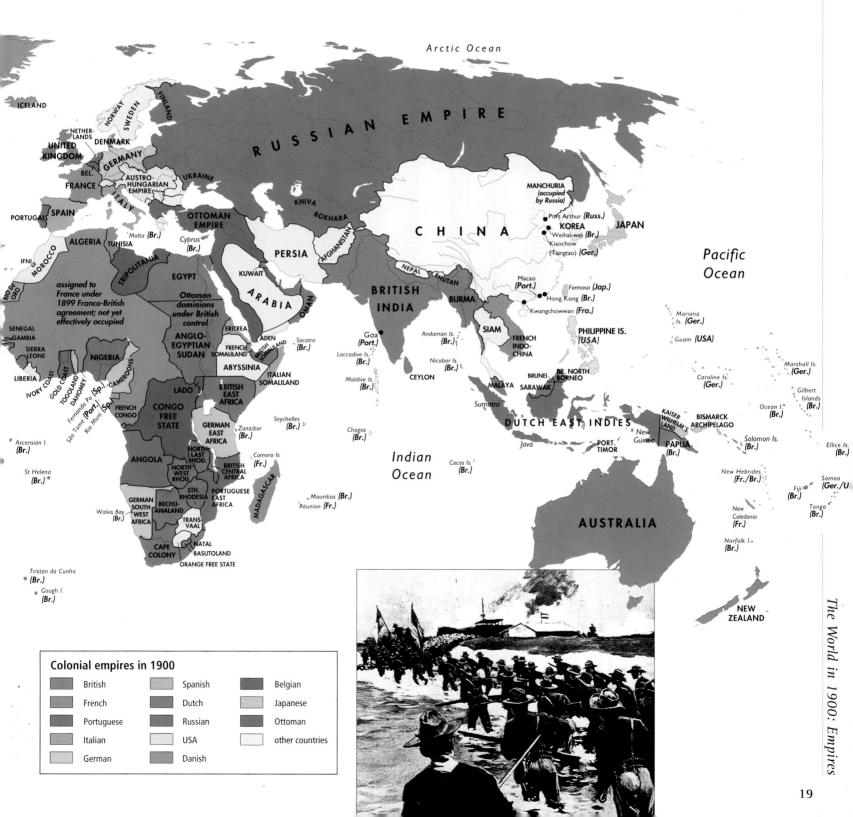

Colonial empires in 1900

- British
- French
- Portuguese
- Italian
- German
- Spanish
- Dutch
- Russian
- USA
- Danish
- Belgian
- Japanese
- Ottoman
- other countries

THE WORLD IN 1900: ECONOMY

the German philosopher Karl Marx, socialists argued for a revolutionary transformation of existing society, and the rule of the labouring masses. By 1900 there were movements for civil rights or revolution worldwide.

No single factor was as important in explaining the decline of the old world order as industrialization. Industrial growth overturned the traditional balance of power as established states failed to modernize economically or modernized only slowly, while other states – Germany, the United States, Japan – grew industrially powerful in the last third of the 19th century. The political success of Europeans overseas rested on the great wealth and technical progress brought by industrial expansion. Their appetite for empire owed much to the search for new sources of food and raw materials. In 1900 Britain was at war with the small Boer republic of Transvaal in southern Africa where, in 1886, large quantities of gold had been discovered.

Yet even in Europe the pace of industrial growth was uneven. In 1900 Europe produced over 17 million tons of steel, but two thirds of it was produced in just two countries – Britain and Germany. Britain, the oldest industrial power, produced more coal and manufactured more textiles than the whole of the rest of Europe together. But with industrial growth concentrated in particular regions, the rest of Europe's economy remained agrarian, and most Europeans, like most of the world's population, worked on the land. Outside Europe, industrialization was limited everywhere except in the United States, where abundant raw materials and an inventive and skilled workforce turned the country in the 40 years after the Civil War (1861-65) into the world's leading manufacturing nation.

Economic modernization outside Europe depended almost entirely on European investment and European or American technology. Such development as there was could be found in the processing of foodstuffs – pressed tinned beef from Latin America, cocoa from West Africa – or in the extractive industries. By 1900, the Transvaal gold mines had over 100,000 workers. Tin from Malaya or copper from Canada provided much of the world's supply of these commodities. Yet outside the small enclaves of exploited resources the rest of the world remained wedded to traditional methods of production and farming.

The process of modernization relied on the growth of commerce. By 1900 a sophisticated system of trade and currency payments was in operation. It was based on London as lender of last resort and the major centre for shipping, insurance and commodity brokerage. In 1900 Britain controlled one half of the world's merchant tonnage, while her overseas investments were greater than those of the rest of the world together. Fuelled by British credit and rising incomes in America and Europe, world trade expanded more than three-fold between 1860 and 1900. The fruits of this commerce were very unevenly spread. The highest incomes were earned in the United States and Britain, but even here the standard of living of most of the population was low and wealth was concentrated in the hands of only a few. In the less developed regions of Europe and in areas only feebly affected by economic change life was lived at, or occasionally below, the barest level of subsistence.

The transformation of material life depended on the progress of science and technology. The 19th-century industrial revolution was based on iron, coal and railways. By 1900 the components of a new wave of technical change were to hand, drawing on chemicals, electricity and the internal combustion engine. The first cars were developed in the 1880s; the first powered flight came in 1903; the electron was isolated by the British physicist Joseph Thompson in 1902; the German scientist Max Planck laid the basis for the quantum theory in 1900; Albert Einstein's theory of relativity came in 1915. The modern age of computers, jets, satellite communication and nuclear power was born in the flowering of scientific and technical development at the turn of the century. In 1900 the building blocks of a remarkable century of political, intellectual and technical change were cemented in place.

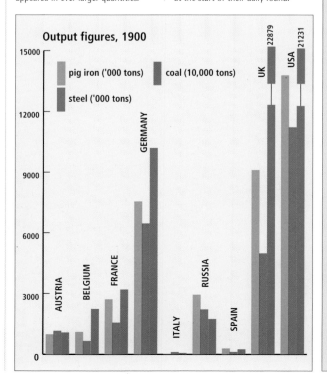

Industrialization in the 19th century was built on coal, iron and steel *(chart below)*. Coal fired steam power in factories, locomotives and ships. The iron barons were the aristocracy of the new industrial classes – Krupp in Germany, Carnegie in America. The boom in heavy industry helped to fuel a growing consumer boom from the 1890s as wages steadily rose. A new wave of consumer products, such as the Empire typewriter *(left)*, transformed everyday life. In the decade before 1914, motor cars, telephones, phonographs and electric light appeared in ever larger quantities.

Communication lay at the heart of world economic growth in the years before 1900. Shipping lanes spread worldwide and were plied by a new generation of powerful steam ships. The crowded dock at Southampton *(top right)* was typical of Europe's major ports at the turn of the century. With global shipping came a global postal service. International co-operation began with the Universal Postal Union founded in 1874, and regular postal services spread worldwide. Two postmen in German Samoa are pictured here *(above right)* c.1910, at the start of their daily round.

Output figures, 1900

- pig iron ('000 tons)
- steel ('000 tons)
- coal (10,000 tons)

AUSTRIA, BELGIUM, FRANCE, GERMANY, ITALY, RUSSIA, SPAIN, UK 22879, USA 21231

THE GOLD STANDARD

By the late 19th century most major economies based the value of their currency on a specified quantity of gold. The Gold Standard originated with Britain in 1821, when the pound sterling was fixed to a gold parity. Most other countries remained on a bi-metallic standard, gold and silver, until in 1871 Germany followed Britain's lead. France did so shortly after. A shortage of gold worldwide encouraged the survival of the use of silver. Latin America used a bi-metallic system; in the United States the silver lobby kept up pressure for bi-metallism. Although the United States adopted the Gold Standard in 1879, bi-metallism continued to operate in practice. The system relied on new gold supplies and with the discovery of gold in the Transvaal in 1886, and in the Klondike in Alaska, there came

The world economy, 1900-14

foreign investment, 1914
(in $ million)

- 535 United Kingdom
- 420 United States
- 3180 France
- 1050 Germany

— busiest shipping routes
— other major shipping routes
— international telegraph cables

By 1900 it was possible to talk about a world market. Transport, technology and capital were exported from the more developed states of western and central Europe, and from the United States, to other European states and to overseas territories (*map right*). Much of the trade and investment, however, went to other industrializing countries, notably within Europe (*map below*). German colonies provided only 0.5% of German trade, French colonies under 10%. Japan and China became major markets by 1914 for European products. Latin America, once dominated by European interests, established closer economic ties with the United States. The "north-south" divide in the global economy was already in the making.

a more general move to the Gold Standard. Japan adopted it in 1886, Austria-Hungary in 1892, India (one of the main areas of silver currency) in 1893 and tsarist Russia in 1895-7. In the United States bi-metallism produced a fierce debate in the 1890s. In 1896 the defeated US presidential candidate, William Jennings Bryan, campaigned on the slogan that "mankind shall not be crucified on a cross of gold". With the 1900 Gold Standard Act the United States finally abandoned silver. China remained the only major state committed to a silver standard. Miners at the Republic Gold Mining Company (below) in the Transvaal fuelled the economic engine which backed up the Gold Standard.

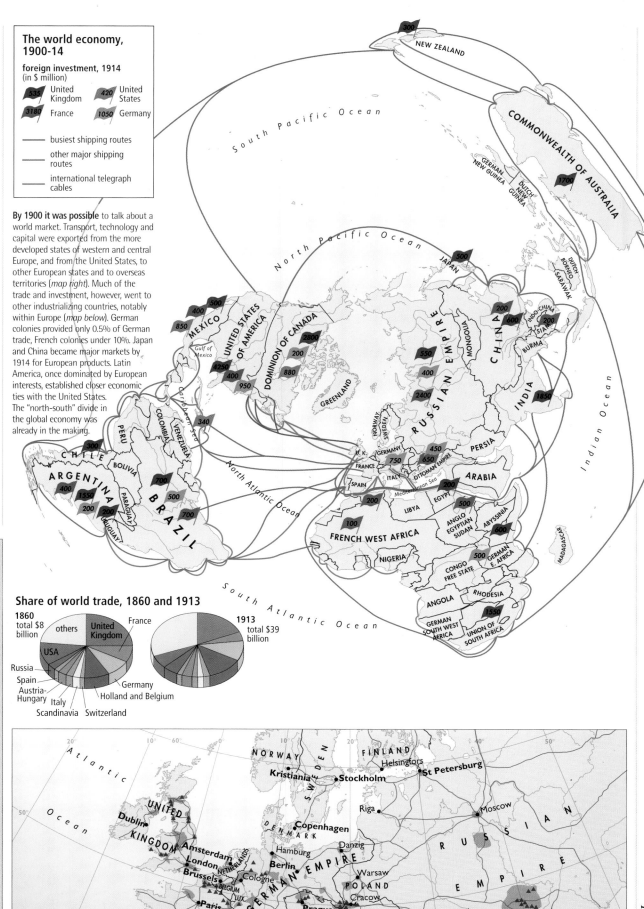

Share of world trade, 1860 and 1913

1860 total $8 billion — others, United Kingdom, France, USA, Russia, Spain, Austria-Hungary, Italy, Scandinavia, Switzerland, Holland and Belgium, Germany

1913 total $39 billion

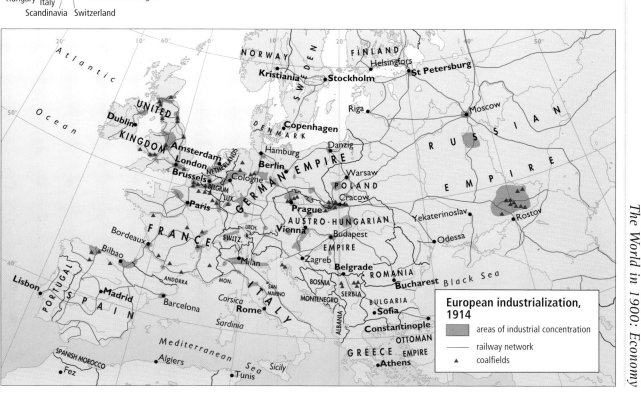

European industrialization, 1914

- areas of industrial concentration
- railway network
- ▲ coalfields

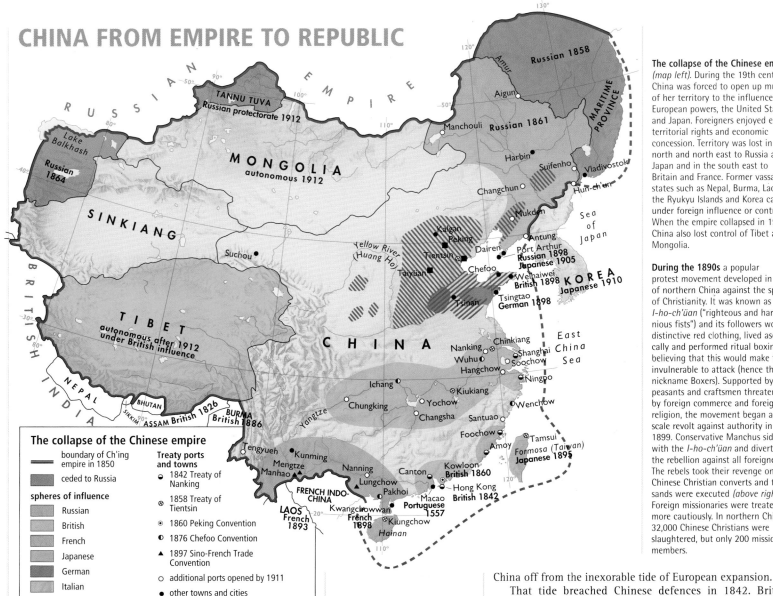

Russian 1858

Russian 1861

TANNU TUVA
Russian protectorate 1912

Lake Balkhash

Russian 1864

MONGOLIA
autonomous 1912

SINKIANG

Amur

Aigun

Manchouli

Harbin

Changchun

Suifenho

Vladivostok

Hun-ch'un

MARITIME PROVINCE

Mukden

Sea of Japan

Antung

Kalgan
Peking

Dairen
Port Arthur
Russian 1898
Japanese 1905

Yellow River
(Huang Ho)

Tientsin

Chefoo

Suchou

Taiyuan

Weihaiwei
British 1898

Tsingtao
German 1898

Tsinan

KOREA
Japanese 1910

TIBET
autonomous after 1912
under British influence

CHINA

Nanking

Chinkiang
Shanghai
Soochow
Wuhu

East China Sea

Hangchow

NEPAL

Ichang

Chungking

Kiukiang

Yochow

Ningpo

BHUTAN

SIKKIM
ASSAM British 1826

BURMA
British 1886

Yangtze

Changsha

Santuao

Foochow

Wenchow

Tengyueh

Kunming

Mengtze
Manhao

Nanning

Amoy

Formosa (Taiwan)
Japanese 1895

Tamsui

Kowloon
British 1860

Canton

Hong Kong
British 1842

Macao
Portuguese 1557

FRENCH INDO-CHINA

LAOS
French 1893

Kwangchowwan
French 1898

Pakhoi

Lungchow

Kiungchow

Hainan

The collapse of the Chinese empire

——	boundary of Ch'ing empire in 1850
▨	ceded to Russia

spheres of influence

	Russian
	British
	French
	Japanese
	German
	Italian

Treaty ports and towns

⊖	1842 Treaty of Nanking
⊗	1858 Treaty of Tientsin
⊙	1860 Peking Convention
◐	1876 Chefoo Convention
▲	1897 Sino-French Trade Convention
○	additional ports opened by 1911
●	other towns and cities

Boxer rebellion, 1899-1900

	original core, 1899
▨	additional areas affected in 1900
▨	additional areas affected in 1899
■	chief towns held by the Boxers in June-July, 1900

The collapse of the Chinese empire *(map left).* During the 19th century China was forced to open up much of her territory to the influence of European powers, the United States and Japan. Foreigners enjoyed extra-territorial rights and economic concession. Territory was lost in the north and north east to Russia and Japan and in the south east to Britain and France. Former vassal states such as Nepal, Burma, Laos, the Ryukyu Islands and Korea came under foreign influence or control. When the empire collapsed in 1911, China also lost control of Tibet and Mongolia.

During the 1890s a popular protest movement developed in parts of northern China against the spread of Christianity. It was known as the *I-ho-ch'üan* ("righteous and harmonious fists") and its followers wore distinctive red clothing, lived ascetically and performed ritual boxing, believing that this would make them invulnerable to attack (hence the nickname Boxers). Supported by poor peasants and craftsmen threatened by foreign commerce and foreign religion, the movement began a full-scale revolt against authority in 1899. Conservative Manchus sided with the *I-ho-ch'üan* and diverted the rebellion against all foreigners. The rebels took their revenge on Chinese Christian converts and thousands were executed *(above right)*. Foreign missionaries were treated more cautiously. In northern China, 32,000 Chinese Christians were slaughtered, but only 200 mission members.

No country exemplified better the tensions between the old world and the forces of change than the Chinese empire of the Manchus. For centuries China was the dominant political and economic power in Asia; the Manchu dynasty ruled over a vast area from Mongolia to Indo-China; Chinese scientific and intellectual achievements rivalled those of Europe; China's ruling classes regarded the outside world as barbarian and until the middle of the 19th century succeeded in closing China off from the inexorable tide of European expansion.

That tide breached Chinese defences in 1842. Britain defeated China in an argument over the British opium trade, and was granted a lease to Hong Kong. Five so-called "Treaty Ports" were opened to foreign merchants. Thus began a slow process of encroachment by the European colonial powers, anxious to tap what they saw as the vast potential trade with the Chinese empire. Russia seized the area of north eastern China between 1858 and 1861; defeat by France in 1885 in Indo-China finally excluded Chinese political influence from South East Asia. China also suffered from the ambitions of her Asian neighbour, Japan. Following a revolution in 1868, Japan embarked on western-style modernization. Her armed forces were reformed and re-armed, and Japanese leaders sought to imitate the west by seizing colonies. They joined the scramble for China, exerting pressure in Formosa (Taiwan), the Ryukyu Islands and in Korea and Manchuria. Finally, in 1894-5, Japan defeated China in a full-scale war and seized the island of Formosa.

The onset of European pressure provoked serious crisis in Chinese society and exposed the Manchu's military and economic weakness. The Chinese political and administrative system remained unreformed, with power in the hands of the traditional bureaucracy. Except for the European-dominated Treaty Ports, there was little modern industrial or commercial development. A burgeoning population brought growing pressure on the food supply. The Manchu defence of tradition provoked long periods of domestic unrest, just as the European version of modernity provoked waves of popular Chinese xenophobia, directed both at westerners and at the feeble regime that had been forced to admit them. Chinese leaders were aware of the need for change, but were fearful of the effects of adopting western methods, despite the evident success of Japanese emulation of the west, which was highlighted by the defeat of Chinese armies in 1895.

Gradually the Manchu regime lost effective control of the major provinces. In 1898 the emperor, Kuang-hsü, at last tried to introduce a range of radical reforms to prevent the

Tientsin lay at the heart of the area of Boxer rebellion. In June 1900, following a wave of atrocities against Chinese Christians, the Boxers seized control of the city. This early 20th-century woodblock *(below)* shows the Boxer assault. Two months later, European forces reclaimed the area amid scenes of widespread slaughter and looting.

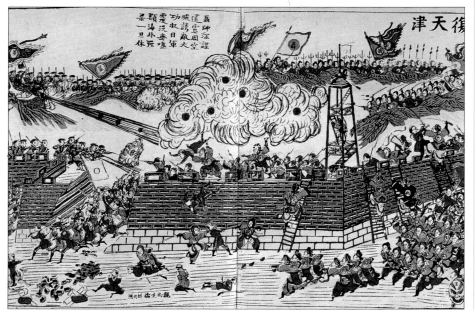

The End of the Old World Order

disintegration of the empire. He was overthrown by the dowager empress, T'zu-hsi, who rejected reform. In 1899, as the European powers closed in for what they saw as the death of the old empire, a wave of anti-western rebellion swept northern China. The "Boxer Rising" was directed at foreign missionaries and legations, and at Chinese who traded or collaborated with westerners. A European and American expeditionary force suppressed the revolt with a good deal of violence, while Russia took the opportunity to occupy most of Manchuria.

The Boxer Rising, though unsuccessful, spelt the end for the old system, as the 1905 revolution did in Russia. In 1901 the reformists won power and a programme of state modernization and economic development was launched. Military, educational and legal reforms which ended the power of the old bureaucracy were matched by a programme of railway building and the establishment of modern banks and trading houses. Progress went furthest in the areas more strongly under European influence. It was here that reformist politicians gathered, and young educated Chinese were brought into contact with western ideas and techniques.

These more radical elements were unwilling to accept the survival of the Manchu dynasty, even in its reformed guise. By 1911 the moral authority of the old regime was dead. A small army revolt in Wuchang sparked a rejection of Manchu rule across China's many provinces. The T'ung-meng-hui party (Revolutionary Alliance) set up a provisional government at Nanking. The Alliance leader, Dr Sun Yat-Sen, was proclaimed president of the new Republic of China on 1 January 1912. In ten years the world's largest and oldest empire collapsed, the weightiest victim of the European drive to modernize the world in its own image.

Ten abortive revolutionary movements had arisen since 1895, most incited by groups of radicals living abroad. By 1911 the Ch'ing Manchu dynasty was discredited, despite attempts at reform. The Manchu ruling elite was abandoned by the Chinese gentry and officials, who openly collaborated with revolutionary groups. The map (below) shows how rapidly the Chinese provinces declared for the revolution following the army mutiny in Wuchang on 10 October 1911. Four months later the new child-emperor, Pu Yi, was forced to abdicate.

In June 1900 Boxer forces seized Peking and placed the foreign legations under siege. Here (left) a headsman stands at the eastern gate of the Forbidden City, traditionally used by foreign diplomats visiting the emperor. The Boxer forces held the city for 55 days until an army of 16,000 foreign soldiers raised the siege on 14 August. In all, only 76 foreign soldiers were killed and six children. In retaliation thousands of Chinese men were murdered in the capital and an indemnity imposed on an already impoverished regime.

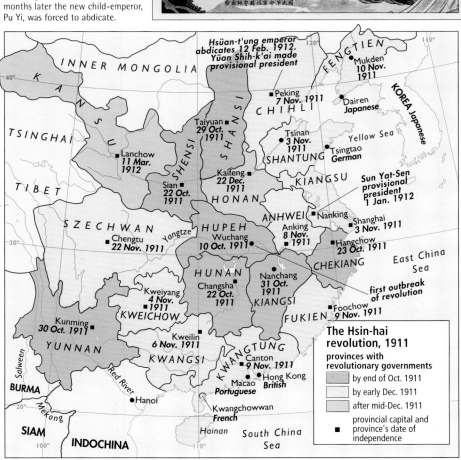

The Hsin-hai revolution, 1911
provinces with revolutionary governments

- by end of Oct. 1911
- by early Dec. 1911
- after mid-Dec. 1911
- ■ provincial capital and province's date of independence

THE DECLINE OF THE OTTOMAN EMPIRE

What China was to the Far East, the Ottoman Turkish empire was to the Middle East and North Africa. The Turkish advance ended in the 18th century with domination of the Balkans, the Middle East to the frontiers of Persia and North Africa as far as Morocco. Throughout the 19th century, however, the Ottoman empire slowly broke apart. In the 1830s, Egypt, which was only a nominal part of the empire, almost succeeded in overthrowing Ottoman rule throughout the Middle East in eight years of warfare. Greece won its independence in 1830; Algeria was conquered by the French in the 1850s; Romania, Serbia, Bulgaria and Montenegro won effective autonomy in 1878.

During the latter half of the century the European imperial powers, Britain, France and Italy, began to encroach on Ottoman interests throughout the Mediterranean and Middle East, partly to protect the interests of Christian subjects of the Ottomans, partly to extend or preserve economic interests, particularly after the opening of the Suez Canal in 1869. In 1882 the British formally occupied Egypt, in 1898 establishing Anglo-Egyptian control of the Sudan. In Palestine and Syria the French and British acted as protectors of the native Christian communities.

The presence of the Christian West in areas of former Ottoman control produced a mixed response. In the middle years of the 19th century, liberal reformers tried to imitate the West in order to strengthen the Ottoman empire, but they were resisted by reactionaries wedded to traditional Islamic values and culture, and by nationalists who, while rejecting western values, nonetheless sought secular, centralized states. In the 1870s, the reform initiative, the *Tanzimat*, succeeded in establishing a modern constitution for the empire, but when Sultan Abdul Hamid II came to the throne in 1876 he suspended the new parliament and began a 30-year reign of repressive personal rule. He encouraged those reforms which increased central power, but his rule was incompetent and corrupt. The empire's finances were a constant source of friction between the sultan and foreign creditors, and between the sultan and his long-suffering soldiers and officials, whose salaries were always in arrears.

Abdul Hamid alienated most groups in the empire during his long reign: nationalists disliked his dependence on western money and the incursions of western imperialism; traditional Islamic leaders urged a return to fundamental Islamic life. The liberals and reformers, meanwhile, many in exile in Paris or Vienna, kept up a constant cry for constitutional rule and more effective modernization. The exile movement, usually referred to as the Young Turks, established contact with disgruntled army officers and intellectuals in the empire. In 1908 widespread mutiny broke out in the Ottoman army in Macedonia and Thrace. Though the sultan hurriedly restored the constitution, his authority collapsed.

With revolution in Turkey the Christian parts of the empire began to break away. Crete joined with Greece, Austria-Hungary seized Bosnia and Herzegovina and Bulgaria declared full independence. In 1909 hard-line Islamic elements staged a coup under the dervish leader Vahdeti. It was suppressed by the Young Turk reformers led by Enver Pasha and the Paris-based Committee of Union and Progress. Abdul Hamid was forced to abdicate, and from 1913 a three-man military junta effectively ruled what remained of the empire down to 1918. The Young Turks proved anything but liberal in power. They embarked on a programme to impose the Turkish language and Turkish interests on the Arab and European parts of the empire, imprisoning and executing opponents, including the sultan's chief eunuch, Nadir Aga, who was hanged in public in Istanbul. His execution marked a symbolic break with past Ottoman practice. The Young Turks sought a centralized, Turkified and modern state. They introduced the western 24-hour clock, western modes of dress, education for women and military reforms. Yet for the Ottoman empire, the change came too late. In 1911 Italy conquered Libya, the last Ottoman outpost in North Africa, and a year later the tenuous Ottoman grip on their European possessions was torn loose by an alliance of Balkan kingdoms (*see pages 26–27*).

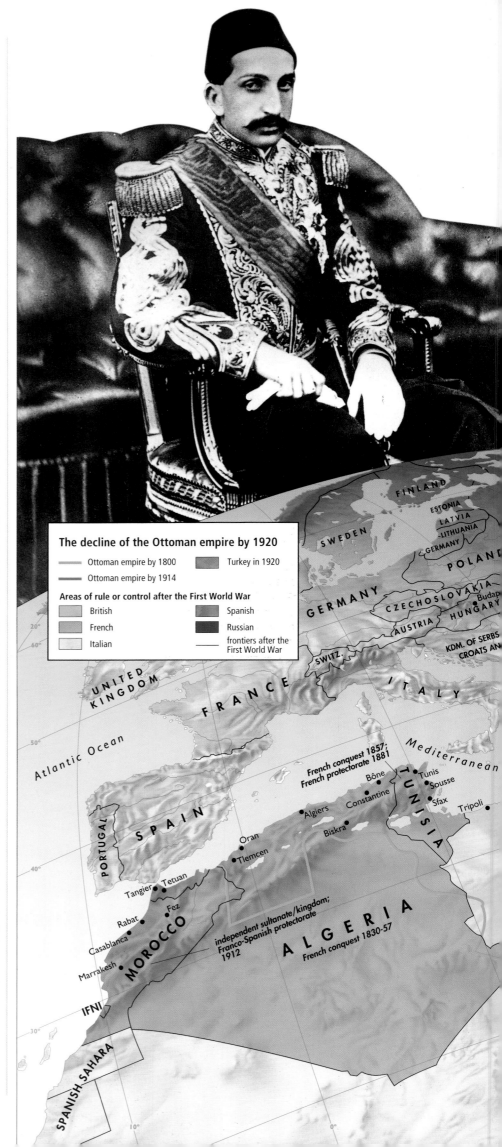

The decline of the Ottoman empire by 1920

Ottoman empire by 1800
Ottoman empire by 1914
Turkey in 1920

Areas of rule or control after the First World War

British
French
Italian
Spanish
Russian
frontiers after the First World War

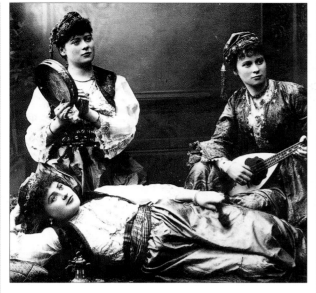

Abdul Hamid II *(left)* was the last Ottoman sultan to wield effective power over the empire. A fierce opponent of political liberalization, he nonetheless encouraged educational, legal and economic reforms. In 1897 Abdul Hamid appointed a special court martial under General Reshid Pasha to crush the freedom movement among the educated and westernized youth of the empire. His reign witnessed the early stirrings of a women's movement within the region, but polygamy, exemplified by the existence of the harem *(above)*, remained in force.

When East met West: the Ottoman capital at Constantinople (Istanbul) was a mixture of cultures from the empire's rich heritage. Trade in the old quarters was carried on as it had been for centuries, but the scene by the Galata Bridge *(above)* could be that of many European cities at the turn of the century.

The gunboat *Zafer (above)* was part of the small but increasingly modern Ottoman navy. Reorganized with British help, the navy possessed three battleships and two cruisers when war broke out in 1914.

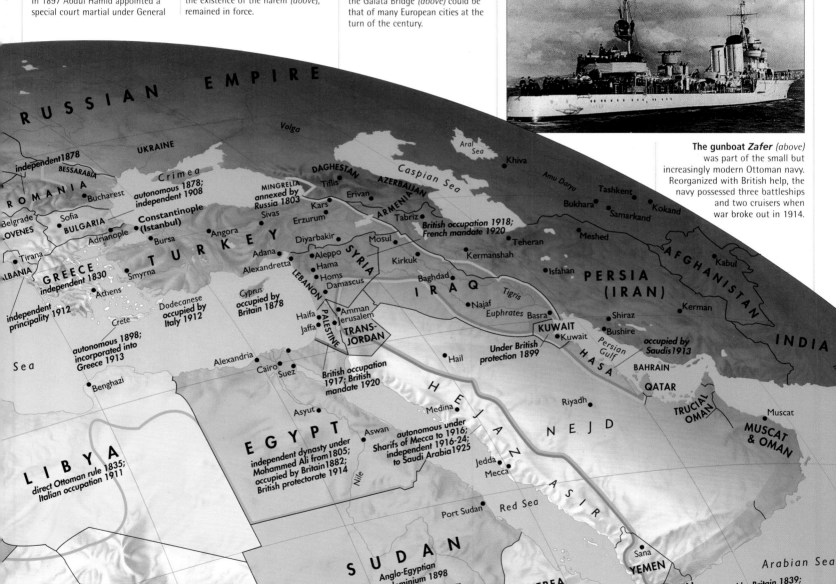

The Ottoman empire, which at its zenith in the 17th century reached as far as Budapest in eastern Europe, and included the Crimea and southern Ukraine, had shrunk by the end of the 19th century to its heartland around the eastern Mediterranean *(map above)*. Its last footholds in Africa were eliminated one by one.

Algeria was conquered by the French in 1857; Tunisia, on which Italian hopes for a north African empire were based, was taken over by France in 1881 as a protectorate. In September 1911 Italy attacked Tripoli in Libya, and the following year seized the whole area as an Italian colony.

THE BALKAN WARS 1912-13

At the beginning of the 19th century, the Balkan peninsula was ruled entirely from Constantinople, the centre of a genuinely multi-racial empire. These European provinces were sparsely populated, poor and provincial. They bordered the Christian empires of Catholic Austria and Orthodox Russia, both of which saw themselves as the natural ally of Balkan Christians. Over the course of the century Austria and Russia strove to increase their influence in the peninsula as that of the Ottomans declined, but the chief beneficiaries of Ottoman weakness were the Balkan nationalities themselves.

One by one the peoples of the Balkan area achieved their independence from Ottoman rule. At the Congress of Berlin in 1878, following a war between Russia and the Ottoman empire over the Bulgarian struggle for independence, the political map of the region was redrawn. The independence of Serbia, Romania and Greece was assured; Bulgaria became a self-governing province within the Ottoman empire, independent in all but name; the Habsburg empire took control over Bosnia, Herzegovina and the Sanjak of Novibazar. Turkish rule in the Balkans was restricted to Albania, Macedonia and Thrace and substantial parts of this legacy were ceded to Greece in 1881 and Bulgaria in 1885.

For the Balkan states, however, independence brought substantial problems. Not only were they economically backward and dominated by a numerous and impoverished peasantry, but high population growth in the second half of the 19th century led to smaller and smaller holdings of land. The average in Serbia was five acres, barely sufficient to feed a family. The only escape was emigration, most of it to the United States, or into the cities, where limited attempts were made to ape the industrial modernization of the rest of Europe. Chronic capital shortages and technical backwardness made economic progress difficult. In Romania, the most advanced Balkan economy, industry contributed only 1.5% of national wealth in 1914.

The attempt to create new national states cost the Balkan peoples dear. Taxation remained high, and governments, keen to build railways and develop their armed forces, borrowed extensively from foreign lenders. Political power was chiefly in the hands of the royal courts and a small elite of soldiers and bureaucrats, although the states were all nominally constitutional monarchies. By 1900 mass politics began to encroach more. The Radical Party in Serbia and the Agrarian Union in Bulgaria both mobilized the votes of peasants anxious for reform. In Romania a mass peasant revolt in 1909 led to 10,000 deaths. In Greece a military revolt in 1909 brought to power Eleftherios Venizelos, a leading liberal reformer who dominated Greek politics for more than a generation.

For small national states, economically weak but with pretensions to grandeur, the remaining Ottoman territories in Europe

SERBIA

Serbia, conquered by the Turks in the 14th century, began to assert a new national identity in the 19th. It was granted autonomy in 1830 and the Ottoman empire recognized its full independence in 1878. From 1817 the country was ruled by the Obrenović family whose bitter rivals, the Karadjordjević family, were in exile. Obrenović rule was corrupt and feeble. The king rigged elections and stuffed the ministries and army with his family and supporters. In 1903 the reigning Obrenović, Alexander, was overthrown. He was unpopular following his marriage to Draga Mašina, a widow with a reputation for scandal. In June an army revolt led to the brutal murder of both Draga and the king. Peter Karadjordjević, respectably married to the daughter of the king of Montenegro, was recalled from exile in Paris. His regime was more constitutional than Alexander's. Politics were dominated by the Radical Party, under the prime minister Nikola Pašić, who took the decisions that led Serbia into war in 1914. King Peter, unlike the pro-Austrian Obrenović family, was strongly pro-French and pro-Russian. Serbia became a threat to Austria. From 1905 to 1911 an economic boycott was instigated from Vienna: the so-called Pig War. Serbia looked elsewhere for economic aid and for markets, and became a rallying point for South Slav national movements.

Prince Alexander of Greece enters Janina in March 1913 during the First Balkan War *(above)*. Taking advantage of Turkish weakness, Greek forces thrust deep into Macedonia, Western Thrace and Epirus. A substantial Greek-speaking population was, however, left in the southern part of the new independent state of Albania, established by a conference of the Great Powers in London in July 1913 after the Second Balkan War.

The squabbles over territory that followed the defeat of Turkey in May 1913 finally prompted Bulgaria to attack Serbia and Greece on 29-30 June and take the Macedonian spoils for herself. Bulgaria's forces proved no match for Serbia and Greece, whose armies were soon joined by those of Montenegro, Romania and Turkey *(map below)*. During the second major battle at Tsarevo Selo on 31 July, Bulgaria sued for an armistice. Bulgarian troops, seen *(above)* retreating from the Serbian town of Krushevo, were comprehensively defeated. In the subsequent Treaty of Bucharest the victors distributed the spoils. Serbia and Greece confirmed their occupation of Macedonia; the Sanjak of Novibazar was divided between Serbia and Montenegro; Turkey regained Adrianople, and Romania received the southern Dobruja.

were an inviting asset. Encouraged by Russia, which sought to improve its diplomatic standing in the region, the Balkan states negotiated treaties of mutual assistance in the spring of 1912 directed against the Ottoman empire. It was not part of Russia's plan to create stronger Balkan states, and in October Russia and Austria warned the Balkan states to leave Turkey alone.

The warning went unheeded. On 8 October Montenegro opened the war against Ottoman forces. The other Balkan states, Serbia, Greece and Bulgaria, joined in. Their 700,000 troops were more than a match for the 320,000 Ottoman soldiers, poorly paid and with little stomach for the contest. In May 1913 the First Balkan War ended with only Constantinople and a small strip of territory left in Europe to the Turks *(map right)*. The Balkan states then squabbled over the spoils. In June Bulgaria launched a war against Serbia and Greece to increase its share. She was quickly defeated. At the Treaty of Bucharest in August 1913, Macedonia was divided between Serbia and Greece, and an independent Albania established under an International Control Commission of the Great Powers, with Dutch officials in charge of Albania's tiny security forces. The national principle triumphed over old imperialism. For Russia and Austria, presiding uneasily over creaking multi-national empires, this was an alarming precedent.

The Second Balkan War, 1913
— boundaries of areas lost by Bulgaria

territories gained according to the 1913 Treaties of London and Bucharest
- Romania
- Serbia
- Greece
- Ottoman empire

The End of the Old World Order

AUSTRO-HUNGARIAN EMPIRE

Požarevac

SERBIA
independent 1878
Nish
Novibazar
Mitrovica
Pirot
Ipek (Pec)
Morava

MONTENEGRO
Ragusa (Dubrovnik)
Podgorica
Cetinje
Lake Scutari
Scutari
siege of Scutari
Prizren
Debar

ROMANIA
Danube
Vidin
Iskur
Plevna
Ruschuk (Ruse)
DOBRUJA
Balchik
Shumla
Varna

BULGARIA
independent 1908
Trnovo
Sofia
Slivnitsa
Küstendil

siege of Adrianople;
taken by Bulgarians 1912,
restored to Turkey in
Nov. 1913 by Treaty of
Bucharest

Burgas
Black Sea

Philippopolis (Plovdiv)
Maritsa
Kumanovo
23-24 Oct. 1912
Üsküb (Skoplje)
Kocani
Krushevo
Kirdzali
Xanthi
Kirk-kilisse
22 Oct. 1912
Adrianople (Edirne)
Babaeski
Lule Burgas
29-31 Oct. 1912
Midia
Tchadalja
17-19 Nov.
San Stefano
Constantinople

ALBANIA
principality 1913
Tirana
Durazzo
Elbasan
Lake Ohrid
Lake Prespa
Monastir
15-18 Nov. 1912
Florina
Venidje Vardar
Nov. 2-3, 1912
Serrai
Kavalla
Dedeagach
Enos
Rodosto
Sea of Mamara

Valona
Kǒritsa
Salonica
8 Nov. 1912:
Salonica capitulates
to the Greeks

Thasos
30 Oct. 1912:
occupied by Greece
Samothrace
Imbros
30 Oct. 1912:
occupied by Greece
Gemlik
Bursa

Argyrokastron
Santi Quaranta
Janina
Kalabaka
Kozani
Larissa
Lemnos
Tenedos
20 Oct. 1912:
occupied by Greece
Gallipoli
Dardanelles
Balikesir

Corfu
Epirus
Preveza
Arta
Thessaly
to Greece 1881
Volos
Aegean Sea
Lesbos
21 Nov. 1912:
occupied by Greece
OTTOMAN

Lefkas
GREECE
independent 1830
Skopelos
Skyros
Chios
24 Nov. 1912:
occupied by Greece
Gediz
Manisa
Smyrna
EMPIRE

Cephalonia
Gulf of Patras
Corinth
Nikaria
17 Nov. 1912:
occupied by Greece
Aydin
Menderes
Denizli

Zakinthos
Peloponnese
Piraeus
Athens
Tripolis
Nauplia
Saronic Gulf
Andros
Tinos
Syros
Samos
Mugla

Cyclades
Naxos
Milos
Santorini
Ionian Islands
Dodecanese
Cos
Simi
Rhodes
occupied by Italy 1912
Rhodes
Fethiye
Dalaman

CRETE
independent 1898
to Greece 1913
Candia
Scarpanto
Mediterranean Sea

In October 1912 the Balkan League (Serbia, Montenegro, Bulgaria and Greece) launched what became the First Balkan War against Ottoman Turkey *(map right)*. Reinforcement for the Turks was difficult, as the Greek navy controlled the sea route to Macedonia. Bulgarian forces bore the brunt of the fighting. They besieged Adrianople (Edirne) *(left)*, which they were granted in the peace settlement in May 1913. The Greeks and Serbs attacked Macedonia and Albania, where they surrounded Turkish garrisons. These were poorly prepared for a prolonged siege, and the morale of the Turkish forces low. The Serbs and Greeks were both ambitious to capture Albania whose ports, including Durazzo (pictured *above right* during the war) offered an opening to the Adriatic Sea. The Great Powers wanted instead an independent Albania. The war ended in May 1913 with the issue unresolved. Serbia and Greece looked to the Macedonian lands captured from Turkey as compensation. They agreed secretly with each other, and with Romania, Montenegro and the recently defeated Turks, to divide up Macedonia at Bulgaria's expense.

The Balkans, 1912-13

—— western frontier of the Ottoman empire, 1912

position of armies, 18-20 Oct. 1912
◼ Bulgarian   ◼ Serbian
◼ Greek   ◼ Montenegrin
◼ Ottoman   ★ battle

areas of opposition to Ottomans at the armistice, Dec. 1912
▨ Bulgarian   ▨ Serbian
▨ Greek   ▨ Montenegrin

territory gained according to the 1913 Treaty of London by:
▨ Bulgaria   ▨ Serbia
▨ Greece   ▨ Montenegro

EUROPEAN ALLIANCES

The crisis in the Balkans proved to be more than a local conflict over the Ottoman succession. The other European powers took a keen interest in the outcome. For more than a century the so-called Eastern Question – the balance of international power in the Near East – had been a central issue in the diplomacy of the major states. The Balkans themselves were of little value beyond the insecure investments placed there, but they were long regarded as the frontier between the interests of three great empires whose preservation was thought to be in the wider interest of European security. When Ottoman power was eclipsed, the balance was rudely overturned, with neither Austria nor Russia willing to see the other fill the vacuum created by the end of Turkish rule.

The traditional solution was for the European powers to act in concert on issues that threatened to divide them. The Balkan crisis in the 1870s was resolved by the Congress of Berlin in 1878, orchestrated by the German chancellor, Otto von Bismarck, its prime motive being to maintain stability. Agreement was similarly reached on the Partition of Africa in 1884 and over the influence of the European powers in China. When the Balkan issue flared up again in 1912, the powers convened in London to do what they had done for decades: to adjudicate disputes by common agreement.

Two factors undermined the concert tradition revived at London. Since the 1870s there had developed in Europe a system of alliances between two or more of the major states which cut across efforts at multilateral co-operation. By 1913 these had solidified into two blocks. On the one hand were the Central Powers, Germany, Austria and Italy; on the other were the so-called "Entente" powers, built around the long-running military pact between France and Russia, to which Britain had finally adhered in order to settle colonial issues, first with France in 1904 and then Russia in 1907. Though the alliances were defensive in intent, they encouraged a competitive military build-up which left Europe less rather than more secure.

The second factor was the growing domestic weakness of the two empires, Russia and Austria-Hungary, whose interests were most affected by events in the Balkans. The attempts by the monarchy in both empires to maintain the old order, while encouraging economic and social modernization, produced serious tensions. Liberals and socialists wanted to scrap the old political system; nationalists demanded autonomy for the national minorities. Both empires sought to stem domestic decline through an active foreign policy. The Near East was a natural area of influence for both. Their mutual defence of the old status quo in the Balkans gave way to a growing rivalry. The area became a testing ground for the survival of the dynastic empires as great powers.

For Austria the threat was immediate. The success of Serbia against Turkey encouraged a general southern Slav movement among the Slavic peoples of the Habsburg empire. Austria, like Turkey, faced the nationalist fragmentation of its empire. When on 28 June a Bosnian nationalist, primed by Serbian military intelligence, assassinated Franz Ferdinand, heir to the

The Morocco Question, 1905-12

- progressively occupied by France 1907-12
- French protectorate from 1912
- Spanish protectorate from 1912 (partly occupied from 1909)
- notional frontier of Morocco, 1912
- frontiers, 1912

Jan. 1906: international conference; Morocco remains independent but policed by France and Spain

Mar. 1905: Kaiser pays visit to "sovereign ruler of independent Morocco" demonstrating German non-acceptance of Franco-Spanish influence. Internationalized by 1912 treaty

nominally Spanish from 1860; part of Spanish protectorate under 1912 treaty

German gunboat Panther arrives 1911; resulting crisis solved by 1912 treaty: Germany evacuates town and accepts Franco-Spanish protectorate over Morocco

Morocco provided a key flashpoint in relations between the major European states before 1914. By agreements with Italy, Britain and Spain (1902-4), France hoped to extend its influence in Morocco, with its valuable mineral deposits. German objections in 1906 led to the conference at Algeciras (*below*), where Germany was isolated diplomatically. France and Spain were permitted to police Morocco under a Swiss inspector general. Five years later a second crisis developed when the German gunboat *Panther* was sent to Agadir in a show of strength to prevent France establishing further control in Morocco. Conflict was averted in November 1911 when France granted Germany territory in the Congo in return for German recognition of French interests.

THE EUROPEAN ARMS RACE

In the 20 years before the outbreak of war in 1914 the military strength of the Great Powers was built up to unprecedented levels. Each initiative taken by one Power was matched by the others. The resulting arms race contributed to the destabilization of Europe. There were two major components to the race: a naval race between Germany and Great Britain; and a race to build up army size between France and Russia on the one side, Austria-Hungary and Germany on the other. The naval race dated from the decision, taken in 1889, to expand the British fleet. In response Germany feared for her own worldwide trade and overseas colonial position, prompting Admiral von Tirpitz to build a large modern navy. Navy laws in 1898 and 1900 laid the foundation for German

battleship building. Britain responded with a rival programme to keep her ahead, based on the most modern "Dreadnought" battleships (below). In 1914 she had 34, Germany only 20. The army race was prompted by the Franco-Russian military alliance of 1894, and the expansion and modernization of the Russian army. By contrast German army expenditure stagnated until 1912, when her army was expanded by 170,000. Meanwhile Russia planned a further growth of 500,000, while in 1913 France increased conscription from two years to three. By 1913 Britain and France both spent a higher proportion of state funds on defence than Germany. In 1914 the combined strength of the Entente Powers' armies was 2.23 million; the German and Austrian armies totalled 1.2 million.

Between 1879, when Germany and Austria-Hungary allied together, and 1907, when Britain signed an Entente agreement with Russia, Europe slowly divided into two alliance blocs. Britain remained aloof until signing an alliance with Japan in 1902, and a further agreement with France in 1904. Until then Germany had hoped to bring Britain into some kind of alliance against Russia and France. When the Balkan crises (1912-14, *see page 26*) flared up, the two alliance blocs were drawn inexorably into the conflict *(map right)*, until all the major states of Europe, save Italy, were at war.

The outbreak of World War One, 1914

- mobilizations, with date
- ultimata issued, with date
- declarations of war, with date
- Entente Powers at outbreak of war
- joined Entente Powers during the war, with date
- Central Powers at outbreak of war
- joined Central Powers during the war, with date
- frontiers, 1914

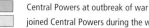

Lisbon
PORTUGAL
9 Mar. 1916
Gibraltar British

The future George V and Tsar Nicholas of Russia at a regatta in Cowes, Isle of Wight, in 1904 (right). The two were first cousins and bore a striking resemblance to each other. Both were fond of sports and the outdoor life. But George was a figure of little political significance, while Nicholas ruled Russia as an unabashed autocrat. Ten years later they were to fight on the same side in the Great War. The close relationship between the Russian and British royal families convinced the German emperor, Wilhelm II, that they were plotting the "encirclement" of Germany.

On the morning of 28 June 1914 seven Serbian nationalists mingled with the crowd in Sarajevo to greet the Archduke Franz Ferdinand and his wife Sophie. They carried bombs and handguns with the aim of assassinating the heir to the Habsburg throne. The first attempt, a bomb, missed the royal car and exploded behind it. Only Gavrilo Princip had the nerve for a second attempt. Two shots killed both the archduke and his wife. Princip tried to shoot himself but was caught, beaten and hurried into custody (right).

The military spending of all the Great Powers increased significantly between the 1890s and 1914 (chart below). However, as a proportion of GNP the growth was less marked. Germany devoted 7% of GNP to the military in 1872, but only 3% in 1913.

Austrian throne, in Sarajevo, the Austrian authorities determined to launch a third Balkan war to punish the Serbs. They expected a small war and remained blind to the wider European crisis provoked by their do-or-die gesture. Inevitably, the Serbian crisis interlocked with the wider system of alliances. Russia gave Serbia qualified support, enough for her to reject an ultimatum from Vienna that would have turned Serbia into a satellite state. Germany encouraged Austria to act quickly, but to avoid a wider war. France encouraged Russia to stand firm and mobilize. In the confusion, states believed the worst of each other. Germany mobilized and moved pre-emptively against the powers that, in her view, "encircled" her. Russia and France mobilized to avoid the German danger. Britain sided with her allies only after German troops invaded Belgium in early August on their way to fight the French. Within a week the great powers found themselves at war over a Balkan issue that a year before they had been able to resolve around the conference table.

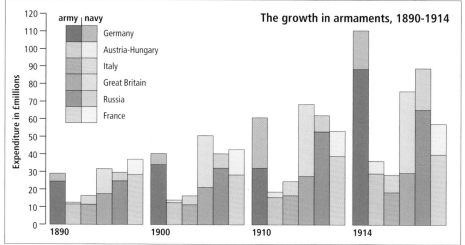

The growth in armaments, 1890-1914

Expenditure in £millions

Legend:
army | navy
- Germany
- Austria-Hungary
- Italy
- Great Britain
- Russia
- France

Years: 1890, 1900, 1910, 1914

28 June Franz Ferdinand and Archduchess Sophie assassinated in Sarajevo by Gavrilo Princip.
5 July German emperor Wilhelm II agrees to give Austria a "blank cheque" for action against Serbia.
14 July Hungarian premier, Tisza, agrees with Vienna to send ultimatum to Serbia.
20–23 July French president, Poincaré pledges support for Russia.
23 July Austrian ultimatum presented in Belgrade.
24 July Russian council of ministers recommends partial mobilization.
25 July Serb government rejects Austrian ultimatum. Austrian emperor Franz Joseph orders partial mobilization.
26 July Russian army mobilizes.
27 July French begin military preparations. Britain calls for a conference on Serb issue, excluding Austria-Hungary and Russia.
29 July Britain warns Germany that she will intervene on the side of France. Tsar Nicholas appeals to William II to avert war. Berlin warns Russia to stop mobilizing. British Royal Navy begins mobilizing.
29–30 July German chancellor, Bethmann-Hollweg urges Austria to hold back from attack on Serbia.
30 July General Russian mobilization.
31 July General Austrian mobilization. Berlin sends ultimatum to Russia to stop mobilizing.
1 August Germany declares war on Russia. Germany and France order general mobilization.
2 August German ultimatum to Belgium to allow passage of troops.
3 August Belgium rejects ultimatum. Germany declares war on France and Belgium.
4 August German troops enter Belgium. Britain declares war on Germany.

European Alliances

29

THE GREAT WAR: THE WESTERN FRONT

In the years before 1914 imperialism, economic rivalry and the rise of popular nationalism at home encouraged a widespread fatalism about the inevitability of conflict. Yet in July 1914 few Europeans expected the Balkan crisis to result in a general European war. The network of alliances and the arms race all pointed to a different conflict. In the event, it was the crisis of traditional dynasticism rather than the forces of change that produced war, the old order rather than the new.

The general assumption was that war would be over by Christmas. Military leaders prepared for a single decisive battle with the weapons to hand. There was little planning for a longer war. The German general staff exemplified this outlook. As early as 1904 the chief-of-staff, Alfred von Schlieffen, drew up a plan for a short two-front campaign. German forces were to be concentrated against France in a quick knock-out blow before wheeling eastwards to defeat the more slowly mobilizing Russian army.

It was a risky strategy, forced by necessity. When war really came in 1914 Schlieffen's successor, Helmuth von Moltke, hesitated to take the risk. He kept some forces in reserve in case the French attacked southern Germany; other forces had to be pulled back hastily to the east when Russia mobilized faster than anticipated. As a result, the German blow against France lacked sufficient strength to be decisive. French forces, reinforced by the speedily assembled British Expeditionary Force, counter-attacked the German forces 40 miles from Paris. The Battle of the Marne (5-10 September) forced a German withdrawal to the river Aisne and effectively ended the Schlieffen Plan. Both armies dug in behind a rampart of barbed wire, artillery and machine guns.

Both sides sought to break the deadlock. In the east, German forces were more successful. Together with Austro-Hungarian armies, Germany pushed Russia back hundreds of miles across Russian Poland (see page 38). The western Allies planned to circumvent the trenches by moving from southern Europe. Italy was induced to join the Entente Powers and a new front opened against Austria. When Turkey joined the German side late in 1914, attacks were launched by British Empire forces in the Middle East (see page 32). Neither new front broke the stalemate.

In 1916 the British under Field Marshal Douglas Haig prepared a more carefully planned frontal assault on German lines in the west. The Battle of the Somme began on 1 July against strongly defended German positions. Haig planned to smother the German trenches with a five-day artillery barrage and then mass divisions for a "Big Push" to break the German front. The reality was grotesquely different. On the first day of the battle the British suffered 60,000 casualties, mowed down as they advanced through barbed wire. For another four months

The Menin Road (above), vividly recalled by the British painter Paul Nash, was the scene of some of the Western Front's worst fighting. In the second half of 1917 Haig ordered a costly campaign to re-capture seven miles of Flanders towards Menin and Passchendaele. Heavy rain and mud made combat almost impossible. The British lost 265,000 men and Menin remained in German hands. The French lost even more in the desperate defence of Verdun (right), which almost broke the morale of the French army.

A German poster (left) celebrates the capture of 1,300 Allied troops near Antwerp in 1914 by a battalion of Bavarian infantry. During the war an estimated 7 million soldiers became prisoners-of-war, of whom 2.5 million were Russians and 2.2 million from the Austro-Hungarian forces. For the ordinary infantryman the arrival of the tank added a new dimension to the field of battle. They were used in the last stages of the Battle of the Somme in 1916. The clumsy and slow-moving vehicles, pictured (below right) by the British artist William Orpen, could crush barbed wire and cross trenches, and helped clear the path for Allied advances in the summer of 1918.

THE BATTLE OF VERDUN

What Stalingrad was to the Second World War, Verdun was to the First. The city of Verdun was France's most fortified strongpoint on the eastern border, with three concentric circles of forts around it. In December 1915 the German army chief, General Erich von Falkenhayn, proposed a new strategy to bleed the French army white in a major assault on Verdun, which he rightly calculated the

French would defend fiercely out of national pride. In February 1916 Crown Prince Wilhelm, with 72 battalions of specially-trained storm troops and the largest concentration of artillery yet seen, began the assault. Lightly defended, the Verdun defences crumbled. The French army sent General Philippe Pétain to hold the town. He stabilized the front, and a terrible artillery duel followed, which bled white not only the French

army but German forces too. Each side lost well over 300,000 men. On 23 June German forces reached the very last line of defence. Verdun was saved by the Brusilov offensive in the east, and the Battle of the Somme, which started two days later. The struggle for Verdun petered out in December; its survivors had experienced the most horrifying manifestation yet of modern industrialized warfare.

Haig threw forces into an unwinnable conflict. Both sides experienced terrible losses. Little was achieved.

In 1917 morale on both sides was poor. French troops mutinied rather than be pitted uselessly against machine guns. In July the German parliament passed a peace resolution calling for an end to hostilities. The mutineers were promised improved tactics and conditions; the German military leaders, Field Marshal Paul von Hindenburg and General Erich Ludendorff, rejected any thought of peace short of victory. An unrestricted submarine campaign was launched in February 1917 to blockade Britain into defeat (see page 34). The chief effect of this decision was to bring the United States into the war against Germany.

The last year of the war was fought under rapidly changing circumstances. American entry coincided with Russian withdrawal following the second revolution there in October 1917 (see page 40). Germany swung more forces to the west, and Ludendorff prepared for a final assault to break the deadlock. The March offensive of 1918 used specially-trained "storm battalions" to breach the enemy line to allow the infantry mass to follow through. Any initial success was blunted by Allied superiority in material as well as the strategic vision of the Allied supreme commander, Marshal Ferdinand Foch. In June the German effort was over, and the Allies, reinforced by American forces and money, slowly pushed the German army back towards the German frontier. Though technically undefeated in the field, Ludendorff pressed for an armistice in November 1918 to avoid an unambiguous Allied victory.

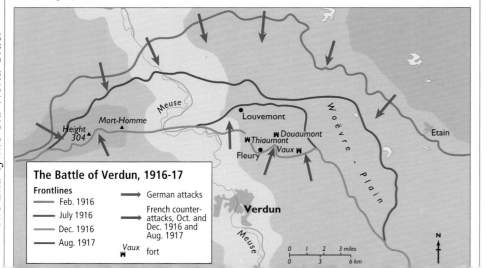

The Battle of Verdun, 1916-17

Frontlines
— Feb. 1916
— July 1916
— Dec. 1916
— Aug. 1917

→ German attacks
⇒ French counter-attacks, Oct. and Dec. 1916 and Aug. 1917
Vaux ⱳ fort

0 1 2 3 miles
0 3 6 km

N

During the Great War Europe experienced the first continent-wide conflict since the Napoleonic wars a century before. Population growth and industrialization now produced a war of extraordinary scale and destructiveness *(maps right, below and below right)*. In 1916 the Entente Allies had a combined population of almost 300 million; the Central Powers only 142 million. The balance of Europe's mineral production was more even. The Central Powers in 1913 produced 330 million tons of coal, 20 million tons of steel and 19 million tons of iron. The Allies produced more coal (392 milion tons), the same amount of steel, and four million tons more iron. By the end of the war much of this capacity had been captured by German and Austrian forces, in Belgium, northern France and the Ukraine. The Allies were saved in 1917 by the addition to their strength of the resources of the United States.

naval blockade of Germany effective end 1916

1916: Battle of Jutland

1917: Riga offensive

1914: Russian offensive in East Prussia

1914: Tannenberg counter-offensive

1916: Brusilov offensive

1918: Central Powers advance through Ukraine

1915: Gorlice campaign

1916: Asiago offensive

1917 Caporetto

1915: Isonzo

1915: conquest of Serbia

1916: Central Powers conquest of Romania

1918: Allied advance into Serbia

1915: British land at Salonica to support Serbia

1915: Gallipoli offensive

see inset below right

Kolubara 1914

The "Schlieffen Plan"

⇨	proposed route
➡	actual route

The Great War in Europe, 1914-8

▨ Entente Powers	— farthest advance by Entente Powers: (east 1914, west 1918)
▨ Central Powers	
→ major Entente Power offensives	— farthest advance by Central Powers: (west 1914, east 1918)
→ major Central Power offensives	★ battle
	✹ battle costing over 250,000 lives
	⚓ naval base

The Western Front, 1914-18

→ Entente offensives	
→ German offensives	
━ farthest German advance, 5 Sep. 1914	
━ trench line from Nov. 1914	
━ Armistice line, 11 Nov. 1918	
★ battle	
✹ battle costing over 250,000 lives	

Oct. 1914 / Apr. 1915 — Passchendaele
July 1917
June 1917
Aug. 1918 final Allied advance
Mar. 1915
Sep. 1915 — Vimy Ridge
Mar. 1915 — Arras
Apr. 1917 / Apr. 1917
July 1916
Nov. 1917
Somme offensive July 1916
Apr. 1918 — Ludendorff offensives 1918
Mar. 1918 — Cambrai
Mar. 1918
Aug. 1918 final Allied advance
Ludendorff offensives 1918
1914
Apr. 1917 / Apr. 1917
Marne 1914
Sep. 1915
Feb. 1916
Argonne 1918
Verdun offensive Feb. 1916 to Aug. 1917
see panel far left

THE GREAT WAR: SUBSIDIARY THEATRES

Victory in the First World War was decided on the Western Front, but the war was fought right across Europe and the Middle East, as well as in Germany's overseas colonies. German possessions in the Far East and the Pacific were captured in four months. German colonies in Africa fell to French, British and South African forces, except for German East Africa, where the German commander, General von Lettow-Vorbeck, fought a skilful campaign, in which he remained undefeated when the Armistice came in 1918.

Victory in the Middle East had a number of causes. The Ottoman Turks, still smarting from defeat in Europe in 1913 *(see page 26)*, were pro-Austrian and anti-Serb. The war minister, Enver Pasha, had close contacts with Berlin, and when the Central Powers offered the restoration of Turkish Macedonia in return for Turkish assistance, he persuaded his government to declare war on the Entente Powers in November 1914. The closure by Turkey of the straits to the Black Sea cut Russia off from her trade lifeline with the West, undermining the Russian war effort. The Turkish forces were sent east into the Russian Caucasus, where they were annihilated in the snow-bound mountains. In April 1915 the Turkish army prevented the British seizure of Gallipoli and the Dardanelles. More British disasters followed. A small Anglo-Indian force stationed in the Persian Gulf area to guard the oil was forced to surrender to the Turks at Kut el Amara in April 1916. A Turkish attack on the Suez Canal was repulsed by British empire forces, but the effort to dislodge the Turks from Sinai was ineffectual until General Allenby broke through in the autumn of 1917 and pushed on to take Jerusalem. In the last year of war a widespread Arab revolt helped the British cause. When Turkey sued for an armistice in October 1918 most of her remaining empire had already been occupied.

In the Balkan peninsula loyalties were divided between the two sides. Bulgaria, anxious, like Turkey, to reverse the outcome of the Balkan Wars, sided with Austria and Germany. In October 1915 Bulgaria entered the war against Serbia, whose small population had kept the Habsburg empire at bay for a year. Serbia was quickly defeated. Romania and Greece both hesitated, waiting to see which side would prevail. Romania overestimated Russian strength and when she declared for the Entente Powers in 1916 was occupied by German forces. In Greece a fierce domestic political struggle developed over intervention. British and French troops landed at Salonica in October 1915 to aid Serbia, but remained bottled up there until September 1918. Under British and French pressure Greece finally joined the war effort in June 1917. When the western powers liberated the Balkans against feeble resistance in September-November 1918, there were nine Greek and six Serbian divisions fighting alongside the British and French.

Italian belligerence was also bought by the promise of territory in the Balkans. Italy in 1914 had been formally allied to

When after a great deal of domestic argument Italy joined the Entente Powers in May 1915 in the war against Austria-Hungary, she was confined by geography to fight on a narrow stretch of her north eastern frontier *(map below)*. A group of Italian alpine troops, decked out like weekend hunters *(right)*, prepares to meet the Austrian enemy in 1915. Two years later German forces joined the battle and inflicted a devastating defeat, whose effects were only finally overcome a year later in October 1918 against a weakened enemy at Vittorio Veneto.

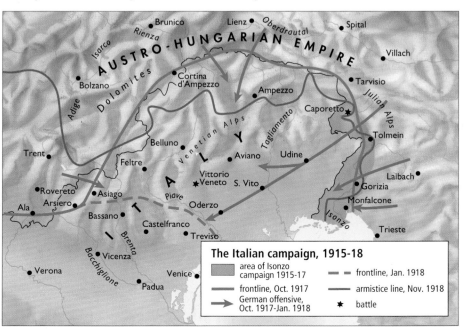

The Italian campaign, 1915-18

- area of Isonzo campaign 1915-17
- frontline, Oct. 1917
- German offensive, Oct. 1917-Jan. 1918
- frontline, Jan. 1918
- armistice line, Nov. 1918
- ★ battle

THE WAR IN THE AIR

Before the war much more was expected of the airship than the aeroplane *(left)*. Every power had them, but Germany with its fleet of ten airships designed by Count Ferdinand von Zeppelin enjoyed a considerable lead. When used against British targets in 1915 and 1916 they proved difficult to navigate and were vulnerable to air attack. They inflicted little damge and sustained high losses.

guns, began to appear. Other aircraft were modified to carry bombs, such as this Handley-Page bomber. By 1917 both sides had experimented with long-range bombing attacks.

When war broke out in 1914, aviation was in its experimental infancy; the first unassisted flight had only been made six years before. Most soldiers expected aircraft to be used for reconnaissance and artillery-spotting to supplement the cavalry. The French air force possessed 141 aircraft in 1914; the British brought 63 with the Expeditionary Force. By the end of the war the warring states had produced over 215,000 aircraft. The technology matured with extraordinary speed. Specialized fighter aircraft, armed with machine

In the summer of 1917 German Gotha heavy bombers were sent against London to try to break the British will to continue fighting. The attacks were small and ineffective – 110 tons dropped in 27 attacks – but they prompted the British to create a separate air force and to plan the bombing of Germany. In 1918 the new Royal Air Force began the systematic bombing of Germany's western cities under General Sir Hugh Trenchard. The plan to create a force of 2,000 heavy bombers to smash German industry and morale in 1919 was interrupted by the Armistice. The post-war bombing surveys convinced the RAF that morale was a more important target than industry.

The Ottoman empire entered the war at the side of Germany and Austria in October 1914, hoping to resurrect her fading fortunes in the Middle East *(map below right)*. Her actions tied down western forces in Egypt, the Persian Gulf and at Salonica in the Balkans. Turkey's major battles were fought against the Russians on the Trans-Caucasian front. The army was weakened by the confrontation with Russia, and in 1916 a widespread Arab revolt in Arabia forced Turkey to abandon much of the southern empire before an armistice was signed aboard the British warship *Agamemnon* on 30 October 1918, moored off the coast of Lemnos.

Thomas Edward Lawrence *(below)* was the most unlikely military hero. A successful scholar, he graduated from Oxford and joined a British archaeological team in the Middle East before the war. He started the war as an Arabic expert in British army intelligence, but when the Arab-Hashemite princes of the Jordan revolted against Ottoman rule in 1916, he became the liaison officer with the revolt, organizing Arab forces and planning their operations. His Arab irregulars joined forces with the British under Allenby in 1918, and in October 1918 Lawrence and Prince Faisal captured Damascus shortly before the Turkish surrender. Lawrence shunned the limelight, joined the RAF in 1922 as a simple aircraftman and died in a motorcycle accident in 1935.

Austria and Germany, but refused to honour the alliance on the grounds that Vienna had not consulted the Italian government about its plans for a Serbian war. Italian society was deeply divided over intervention in the war: the nationalists hoped to use the conflict to increase Italian power in the Adriatic and Mediterranean; the dominant liberals were split; the left was opposed to the war. In the end the government waited to see which side would offer most. It proved to be the Entente Powers. In the Treaty of London, signed in April 1915, Italy was offered the areas occupied by the Habsburgs in South Tyrol, Trieste and Istria, a slice of the Dalmatian coast, the Dodecanese Islands and a share of the German colonies.

The following month Italy duly declared war on Austria-Hungary, though not against Germany. The Italian commander, General Cadorna, mobilized his poorly-trained and ill-equipped forces on the Isonzo front in north east Italy, at the only point where Italy touched the Habsburg empire. Between 1915 and August 1917 the Italian army attacked Austrian lines 11 times, gaining only seven miles at enormous cost. By 1917 Italy was also at war with Germany, and when in October 1917 Austria at last persuaded the German Kaiser to supply German forces for Italy a counter-offensive was mounted. The deficiencies of Italian forces were fully exposed when Austro-German forces attacked at Caporetto, smashing all resistance and capturing 250,000 Italians. Cadorna withdrew to the River Piave. Famous for sacking his own generals in droves, he himself was now sacked. His successor, General Diaz, retrained and re-equipped the Italian army. Western forces and weapons appeared in larger numbers. Meanwhile, a renewed Austro-German offensive in June 1918 was repulsed. In October Diaz attacked a demoralized and disorganized Habsburg force. In the battle of Vittorio Veneto, Italy gained its first battle honours.

The war in the Middle East, 1914-18

Entente Powers
Central Powers

Advances
→ British
--→ Arab
→ French
→ Russian
→ Ottoman

Offensives
━━ area of Arab revolt against Ottomans
━━ Ottoman frontline at time of surrender, 30 Oct. 1918
━━ railways

THE GREAT WAR AT SEA

The war at sea, 1914-16

- ✳ Entente Powers minefield
- ✳ Central Powers minefield
- —— trade route
- → route of von Spee's squadron Aug.-Nov. 1914
- ⭕ area in which German merchant raiders made captures, Aug. 1914-Feb. 1915
- ⛴ naval battle (see right)

Naval battles

1 Heligoland Bight: 28 August 1914
2 Coronel: 1 November 1914
3 Falkland Islands: 8 December 1914
4 Jutland: 31 May 1916

Overseas trade with the empire and the Americas was vital to Britain and France. More than half British food and raw materials came from foreign trade. The map *(left)* shows the main trade routes bringing supplies of grain, meat, nitrates, copper and hides for the western war effort. For the first year of war individual German warships preyed on merchant vessels. Once they were eliminated the threat of minefields and submarines remained, chiefly in the approaches to British and French ports. The admiralty in London remained strongly opposed to establishing merchant convoys on the grounds that they would be large and inviting targets and were difficult to organize and escort. Armed merchantmen fought their way through individually. In 1917 convoys were tried on the Scandinavian route and losses fell from 25% to 0.24%. Convoys were then introduced on all routes. In June/July 1917, 800 ships sailed in Anglo-American convoys and only five were lost.

Submarine warfare *(maps right)*. On 4 February 1915, the German government declared the waters around Britain a war zone, and began submarine attacks on commercial shipping as a counter to the British blockade of Germany. For short spells in 1915 and 1916 German submarines (U-boats) attacked neutral shipping around Britain, but protests limited the campaign to British shipping. On 9 January 1917, unrestricted submarine warfare was finally adopted as a desperate measure to end the war. The new campaign brought the USA into the war against Germany, and it failed to undermine Britain's war effort. Countermeasures – particularly the convoying of merchant ships – reduced the number of sinkings sharply from the autumn of 1917, while U-boat losses steadily mounted *(map lower right)*.

Admiral von Spee (1861-1914) *(left)* was one of Germany's most distinguished naval commanders at the outbreak of war. Stationed in the Far East as head of the German East Asia Squadron, he led his small flotilla across the Pacific to attack British trade routes in Latin America. After beating off one attack at Coronel on 1 November 1914, his ships were caught off the Falkland Islands on 8 December and sunk. He went down in the battle cruiser *Scharnhorst* with all his men.

Britain's war effort was heavily reliant on overseas trade. The effects of submarine warfare, merchant raiders and mines, and the military demands on British empire shipping, all reduced the volume of British imports *(chart right)*. By cutting out luxuries, grain imports were well-maintained; the German plan to starve Britain into surrender in 1917 failed entirely.

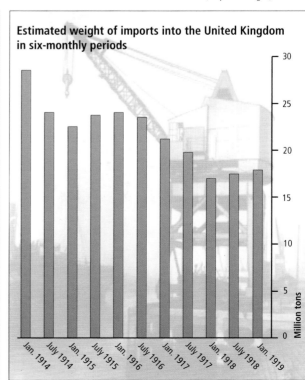

Estimated weight of imports into the United Kingdom in six-monthly periods

(bar chart, Million tons, scale 0–30; periods from Jan. 1914 to Jan. 1919)

Submarine warfare, 1916-18

- ocean convoy assembly point
- location of merchant vessel sinkings
- U-boat lost

1 September 1916 to January 1917
(restricted warfare)

Atlantic Ocean

UNITED KINGDOM

FRANCE

2 February to October 1917

Atlantic Ocean

UNITED KINGDOM

FRANCE

3 November 1917 to October 1918

Atlantic Ocean

UNITED KINGDOM

FRANCE

U-boats sunk 1914 to 1918

Atlantic Ocean

UNITED KINGDOM

FRANCE

Before the war it was widely assumed that any great power conflict would be both a land war and a naval war. The naval race before 1914 had produced great battle fleets on both sides. Germany and Austria had 56 battleships between them; Britain, France and Russia had 123, with 74 in the Royal Navy alone. In practice the naval war took second place to the land battle. There was only one serious clash between the British and German navies, the Battle of Jutland (31 May–1 June 1916), the last naval engagement with lines of large-gunned battleships.

The German High Seas Fleet was effectively blockaded in its ports by the stronger Allied force confronting it. The sea war became a conflict of blockade and counter-blockade, and its chief offensive vessel was the submarine, whose true impact few had foreseen before 1914. The target of both sides was the seaborne commerce of the other. The British began a formal blockade in March 1915, with Orders in Council which permitted the seizure of goods destined for Germany. As the stalemate continued, the blockade was tightened. Britain used her powerful trading and financial position across the world to pressure other states and private firms to limit trade with the enemy. The effects of the blockade on Germany are difficult to estimate precisely, since domestic shortages of food also had domestic causes. The loss of feedstuffs and fertilizer from abroad crippled German agriculture. By 1917 meat consumption was less than one third of the pre-war level, and grain consumption only half.

Both sides found naval inaction frustrating. In order to give the Navy a clear strategy of its own, the first lord of the admiralty, Winston Churchill, pushed for an attack on the Turkish Straits. The operation began on 19 February 1915 with five French and British battleships blasting Turkish defences. The subsequent landing was a catastrophe. The mainly Australian and New Zealand troops involved were pinned down for nine months with enormous casualties. In January 1916, they were withdrawn. Churchill resigned, and the Royal Navy developed no further independent strategy.

There were frustrations on the other side, too. In 1916 the commander of the German High Seas Fleet, Admiral Reinhard Scheer, planned to lure the Royal Navy into a major fleet battle in the North Sea, where the British ships would be sunk by a waiting U-boat trap. The plan was a disaster. Alerted by radio intelligence, the British fleet was ready for the engagement. The two met at Jutland off the Danish coast. Outnumbered, Scheer skilfully extracted his ships and retreated back to port. The British lost 14 ships, the Germans 11. The battle confirmed the powerlessness of the German navy, which was forced to sit out the war.

In the last weeks of the war German commanders at Kiel decided on a final do-or-die duel with the enemy. By then the sailors, bored and hungry, had had enough of the war and mutinied. A year later the fleet was scuttled rather than let it fall into British hands.

Souvenir OF THE VICTORY OF JUTLAND

FOR YOUR SPLENDID WORK I THANK YOU

MAY 31ST 1916.

The Battle of Jutland *(top right)* was the only major fleet engagement of the war. Though no clear winner emerged, it was hailed as a great British victory. The silk scarf *(middle right)* celebrates the triumph with portraits of George V, and Admirals Jellicoe and Beatty.

In the first year of war Allied shipping was threatened in every ocean by small numbers of German merchant raiders, German warships stranded abroad. The *Emden* and *Königsberg* entered the Indian Ocean in August 1914 from the Pacific, where they had formed part of the German East Asia Squadron under von Spee. The *Emden*'s voyage *(map right)* led to the loss of 17 merchant vessels before her surrender in November.

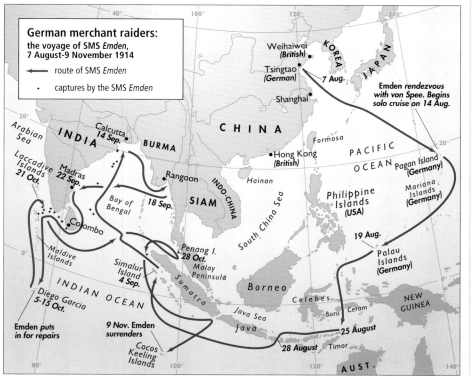

German merchant raiders:
the voyage of SMS *Emden*, 7 August–9 November 1914

→ route of SMS *Emden*
· captures by the SMS *Emden*

KOREA
Weihaiwei (British)
Tsingtao (German) 7 Aug.
Shanghai
Emden rendezvous with von Spee. Begins solo cruise on 14 Aug.
JAPAN

Arabian Sea INDIA Calcutta 14 Sep. BURMA CHINA Formosa PACIFIC OCEAN
Hong Kong (British)
Pagan Island (German)
Laccadive Islands 21 Oct. Madras 22 Sep. Rangoon Hainan INDO-CHINA
Mariana Islands (German)
Bay of Bengal 18 Sep. SIAM Philippine Islands (USA)
19 Aug.
Palau Islands (German)
Colombo
Maldive Islands Penang I. 28 Oct. Simalur Island 4 Sep. *Malay Peninsula* South China Sea Sumatra Borneo Celebes NEW GUINEA
Diego Garcia 5–15 Oct. INDIAN OCEAN Java Sea Buru Ceram
Emden puts in for repairs **9 Nov. Emden surrenders** Java 25 August
28 August Timor
Cocos Keeling Islands AUST.

THE COSTS OF WAR

When the war ended in November 1918 its costs dwarfed anything imagined four years before. The conflict took the lives of eight and a half million soldiers, and left another 21 million wounded, gassed or shell-shocked. The financial cost totalled over $186 billion worldwide and the economies of every warring state in Europe were brought close to bankruptcy as governments resorted to the printing press to fund the swelling demands of war. Civilians not only bore the financial burden, but suffered high loss rates through famine, disease or the direct effects of the war. The war also cost the lives of an estimated nine million civilians.

Conflict on this scale was without precedent. States had no experience of mobilizing and equipping forces of this size. As the war progressed the demands of the military machine forced governments to control the production of the whole economy, to ration goods, and to replace male labour with women, young workers or forced labour. By the end of the war 65 million men had been mobilized, most of them peasants or clerks. Agricultural output declined as a result. In Germany 30 million tons of grain were produced in 1913; in 1917 output was just 15 million.

General Ludendorff, the German quartermaster-general, described the conflict in 1919 as "total war". It was a new kind of war between national communities, not just between soldiers. The scientific, economic and moral resources of the nation were mobilized as ruthlessly and comprehensively as its military manpower. No state in 1914 had been prepared for such a conflict. The demands of war led to exceptional claims on domestic resources and involved a degree of state direction, even in the democracies, unheard of in peacetime. Widespread propaganda was used to maintain enthusiasm for war. Workers were placed under martial law or subjected to strict labour conscription.

The effects on the home front were often severe. Longer hours of work, declining safety standards, the difficulty of obtaining even rationed goods, the sharp fall in real income, all produced a continuous decline in the standard of life. Conditions were better in Britain, with access to the world market and a strong financial position, and worst in Germany, Austria and Russia, which were cut off from the world economy by the war. Hunger and overwork took their toll, and when a virulent "Spanish" influenza epidemic hit Europe in 1918, over six million died.

Worsening conditions provoked widespread unrest. After a period of political truce in the early stages of the war, the parties of the left became increasingly critical of the war, while the unions, their numbers swollen by new recruits to the industrial workforce, pursued improvement in conditions through strikes. In 1915 there had been 2,374 strikes in the warring states, involving 1.1 million workers; in 1917 there were 4,369 with 3.4 million taking part. In 1918 the strike movement became more radical, demanding political change as well as better conditions. There was popular resentment against businessmen who were believed to be making windfall profits out of the war, or against wealthier consumers who could buy on the black market or live life as usual.

By 1918 very few were untouched by war. An army of volunteers, many of them women, helped to run the medical services, or staff the new government offices set up to cope with administering the home front, or collect scrap and refuse to re-cycle for war production. In Germany *ersatz* or substitute materials were unavoidable. Shoes were made of cardboard, paper from potatoes, coffee from nettles. In Austria, Russia and Italy even *ersatz* could not be produced, and the supply of food and military equipment collapsed under the strain of the war, leading by 1917 to severe shortages both at home and at the front, and to widespread demoralization.

The First World War was a test of endurance – of national cohesion, of moral resilience, of economic capacity. It was also a test of the old European order, and its self-confident, morally-assured claim to be the source of peace and progress. Europe's image was irreparably tarnished by the war. Progress was shown to mask barbarism; civilization to be a veneer. The war marked the end of the Europeanization of the world, and opened the way to a new world in which Europe played just one of the parts.

In the wake of war those who survived set out to honour the almost nine million men who died. All across the towns and villages of the warring states memorials were erected, many of them, like the Canadian War Memorial at Vimy Ridge *(above)* unveiled by Edward VIII in July 1936, of monumental size. Few of the symbols of death and sacrifice were as moving as the simple tomb of the unknown soldier, captured here in a painting by the British artist, William Orpen *(bottom)*. The "eternal flame" which marked the French tomb at the Arc de Triomphe was extinguished when the Germans entered Paris in 1940.

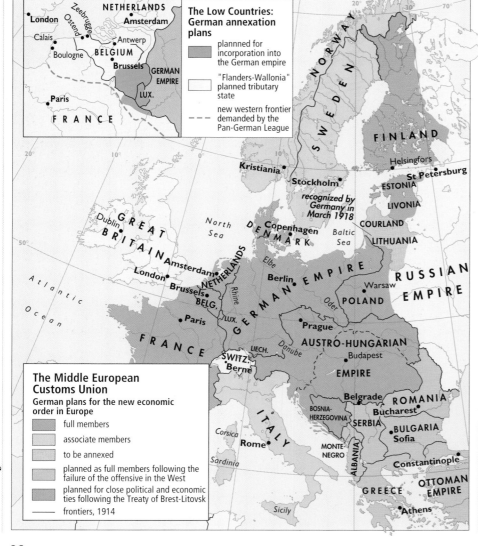

The Low Countries: German annexation plans

- plannned for incorporation into the German empire
- "Flanders-Wallonia" planned tributary state
- - - - new western frontier demanded by the Pan-German League

The Middle European Customs Union
German plans for the new economic order in Europe

- full members
- associate members
- to be annexed
- planned as full members following the failure of the offensive in the West
- planned for close political and economic ties following the Treaty of Brest-Litovsk
- —— frontiers, 1914

Towards the end of the war a number of plans circulated in Berlin about the shape of the post-war world if Germany won. A new economic order in Europe was designed to ensure German domination of central and eastern Europe *(map left)*. In Africa it was planned to create a single German colonial territory stretching across the entire continent.

During the war millions of volunteers signed up for the forces in Britain, the Dominions and the United States. Conscription was only introduced in Britain in 1916, against strong resistance. Powerful propaganda, like the American poster of 1917 *(right)*, reminded democratic youth of its duty.

I WANT YOU FOR U.S. ARMY
NEAREST RECRUITING STATION

The End of the Old World Order

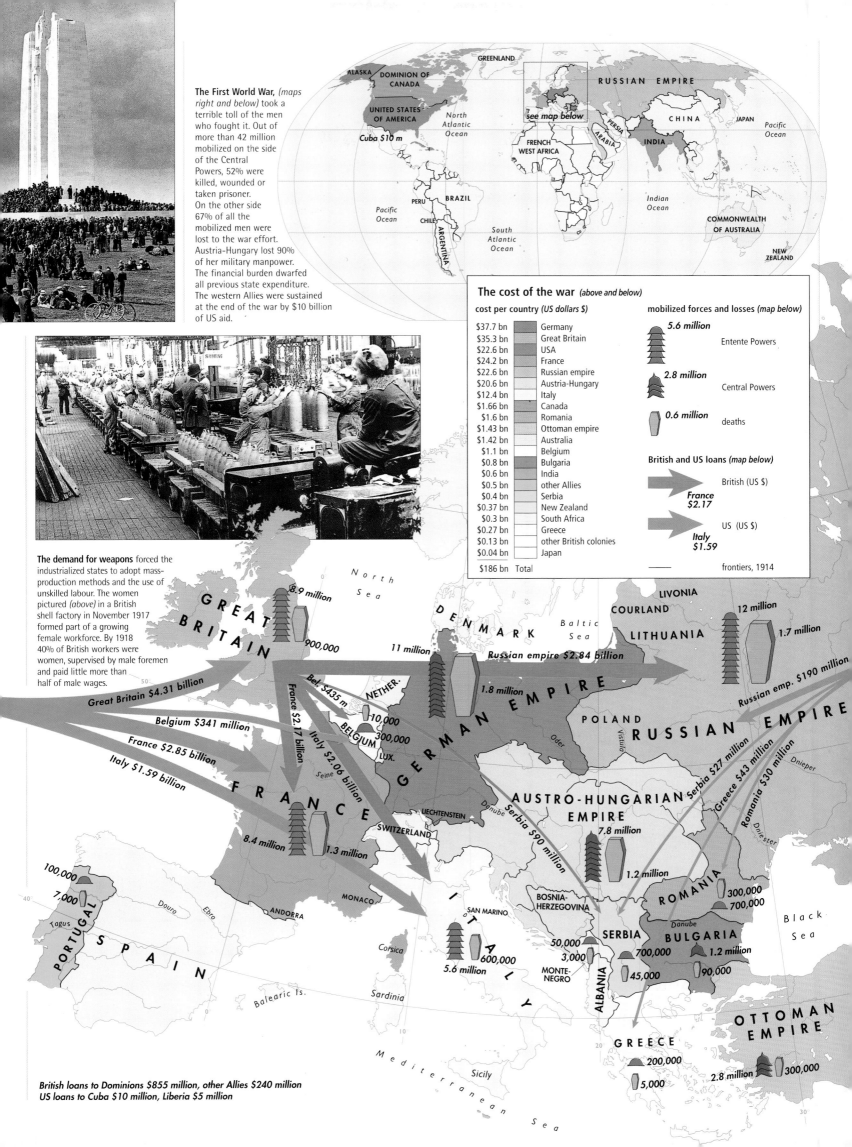

The First World War, *(maps right and below)* took a terrible toll of the men who fought it. Out of more than 42 million mobilized on the side of the Central Powers, 52% were killed, wounded or taken prisoner. On the other side 67% of all the mobilized men were lost to the war effort. Austria-Hungary lost 90% of her military manpower. The financial burden dwarfed all previous state expenditure. The western Allies were sustained at the end of the war by $10 billion of US aid.

The demand for weapons forced the industrialized states to adopt mass-production methods and the use of unskilled labour. The women pictured *(above)* in a British shell factory in November 1917 formed part of a growing female workforce. By 1918 40% of British workers were women, supervised by male foremen and paid little more than half of male wages.

British loans to Dominions $855 million, other Allies $240 million
US loans to Cuba $10 million, Liberia $5 million

The cost of the war *(above and below)*

cost per country *(US dollars $)*

$37.7 bn	Germany
$35.3 bn	Great Britain
$22.6 bn	USA
$24.2 bn	France
$22.6 bn	Russian empire
$20.6 bn	Austria-Hungary
$12.4 bn	Italy
$1.66 bn	Canada
$1.6 bn	Romania
$1.43 bn	Ottoman empire
$1.42 bn	Australia
$1.1 bn	Belgium
$0.8 bn	Bulgaria
$0.6 bn	India
$0.5 bn	other Allies
$0.4 bn	Serbia
$0.37 bn	New Zealand
$0.3 bn	South Africa
$0.27 bn	Greece
$0.13 bn	other British colonies
$0.04 bn	Japan
$186 bn	Total

mobilized forces and losses *(map below)*

5.6 million — Entente Powers

2.8 million — Central Powers

0.6 million — deaths

British and US loans *(map below)*

British (US $) — France $2.17

US (US $) — Italy $1.59

——— frontiers, 1914

RUSSIA FROM TSARDOM TO BOLSHEVISM 1905-17

The bankruptcy of the old order was most clearly evident in the Russia of the Romanovs. In the decades before 1914 Russia presented a curious blend of reform and repression. The tsars recognized that the survival of their system of personal rule depended on building a strong state. Feudalism was ended in the 1860s. The army was modernized and expanded. In the 1890s the finance minister, Sergei Witte, accelerated Russia's industrialization, so that by 1914 Russia was the world's fifth largest industrial power.

These changes helped to transform Russian society. A wave of new workers from the land moved into Russia's cities, swamping the traditional urban workforce and straining the supply of housing and food. The gentry declined as a social and political force. Modernization threw up a new business class, but it also generated a more numerous class of officials, doctors, teachers and lawyers, among whose number were many keen to maintain the pace of reform and transform Russia into a modern state. It was here that the tsarist regime refused to change. The state remained a royal autocracy, with political power imposed by the army and the bureaucracy. Nicholas II, who ascended the throne in 1894, believed that his power was granted by God and that it was his duty to exercise it undiminished.

The contradiction between old-fashioned divine-right rule and the reality of rapid social and economic change encouraged widespread political opposition. When Russia was defeated in a war with Japan over the Far Eastern frontier in 1904-5 the tsar's position weakened. Peasant unrest and growing labour protest provoked a revolutionary crisis. In October 1905 the tsar consented to a manifesto drawn up by Witte which offered civil liberties and a popularly-elected assembly. When popular protest subsided, the concessions were modified. The franchise was limited, the assembly had no real power, and civil rights – freedom

On the eve of war Russia was torn by widespread strikes, more then two thirds of them politically motivated *(chart right)*. When war broke out the strike movement subsided, but revived again by 1916-17. This was just one of the causes of Russia's economic problems during the war. On the land the loss of peasant labour, 60% of draft animals, tools and fertiliser caused a crisis of food production by 1917. Industrial production expanded slowly *(chart far right)*, but poor transport and untrained new workers reduced the quantity and quality of war material at the front.

Trench warfare was not confined to the Western Front. In the East, following the long retreat of Russian forces in 1915, both sides dug in along a more static front. German military leaders planned to use a new weapon – poison gas – against Russian armies in 1915. It was tested in April 1915 against a small sector of the Western Front, and it became a regular component of the armoury of both sides. The gas mask became regulation head gear on every front *(right)*.

Alexander Kerensky (1881-1970), pictured *(below right)* receiving a banner denoting "liberty, fraternity and equality", was a Russian social democrat and a key figure in the first revolution of 1917. He entered the Russian parliament in 1912 as a democratic socialist. In March 1917 he was the only socialist to enter the Provisional Government, as minister of justice. In May he became war minister, determined to prosecute the war more effectively. On 8 August he became prime minister, as the country lurched further to the left, and soon after, supreme commander of the armed forces. Unable to reverse defeats at the front or solve the economic crisis at home, he was overthrown by the Bolsheviks at the end of October.

of speech and assembly – never activated. Between 1906 and 1914 the tsar attempted to rule as he had always done.

By 1914 autocracy was still intact, but it co-existed with growing political movements – conservative, liberal, socialist – whose supporters expected political reform. Protest grew in 1914, and the decision for war with Austria and Germany was taken by the tsar in the midst of a general strike in St Petersburg.

With the coming of war the political tensions in Russia subsided. Two vast Russian army groups moved through East Prussia towards Berlin and then into Galicia against Austro-Hungarian forces. The Austrians suffered a crushing defeat at Lemberg, but German forces under Hindenburg, hastily deployed against a larger army, inflicted defeats at Tannenberg and the Masurian Lakes, which turned the tide in the East. In 1915 Russian forces were pressed back deep into Russian territory, suffering a million casualties and the loss of a million men captured. In the autumn the tsar insisted on taking over the high command himself, against almost universal protest. In 1916 Russian offensives, even the early successes of General Brusilov against the Austrians in June, were turned back, and another million men lost.

The effects of defeat on the home front fatally undermined the old order. The incompetence of tsarist officials and ministers undermined what voluntary efforts were made by the war industry committees set up by businessmen or by the relief and medical facilities run by the Union of Towns and *Zemstvos* (local councils). The huge losses of men and horses – two thirds of the peasant's draft animals were requisitioned – reduced the food supply. In December 1916 the army ration was cut from

Strikes in Russia, 1910-17

[Bar chart showing strikes (light bars) and working days lost ('000s, dark bars) from 1910 to 1917, with y-axis from 0 to 6000.]

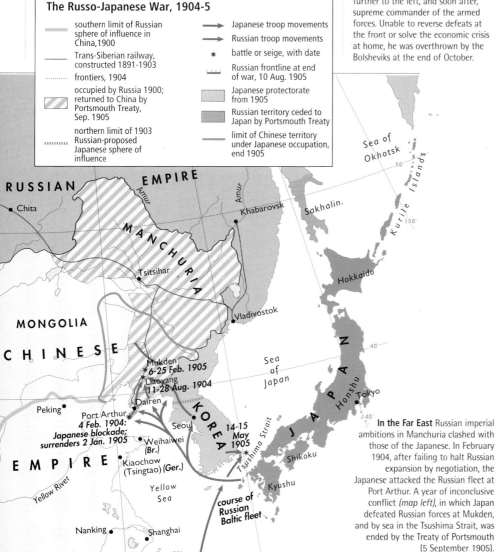

The Russo-Japanese War, 1904-5

- southern limit of Russian sphere of influence in China, 1900
- Trans-Siberian railway, constructed 1891-1903
- frontiers, 1904
- occupied by Russia 1900; returned to China by Portsmouth Treaty, Sep. 1905
- northern limit of 1903 Russian-proposed Japanese sphere of influence
- → Japanese troop movements
- → Russian troop movements
- ★ battle or seige, with date
- Russian frontline at end of war, 10 Aug. 1905
- Japanese protectorate from 1905
- Russian territory ceded to Japan by Portsmouth Treaty
- limit of Chinese territory under Japanese occupation, end 1905

In the Far East Russian imperial ambitions in Manchuria clashed with those of the Japanese. In February 1904, after failing to halt Russian expansion by negotiation, the Japanese attacked the Russian fleet at Port Arthur. A year of inconclusive conflict *(map left)*, in which Japan defeated Russian forces at Mukden, and by sea in the Tsushima Strait, was ended by the Treaty of Portsmouth (5 September 1905).

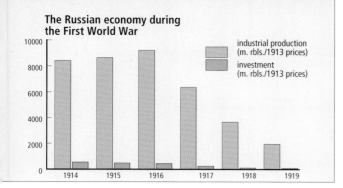

The Russian economy during the First World War

- industrial production (m. rbls./1913 prices)
- investment (m. rbls./1913 prices)

three pounds of bread a day to one. Officers and men lost confidence in the tsar. The court around the tsarina, Alexandra, and her mystic advisor, Rasputin, was isolated amidst a sea of protest across the political spectrum. In December 1916 Rasputin was murdered by a group of aristocrats. On 8 March, following a month of city-wide strikes and bread riots in St Petersburg, a demonstration for International Women's Day turned into a revolutionary protest. Soldiers in the city refused to fire on the demonstrators. The tsar's generals and the Duma political parties universally condemned the tsar. With no prospect of military support he abdicated on 15 March 1917, and the following day a provisional government was declared, led by Prince Georgii Lvov and composed largely of moderate liberals.

War on the Eastern Front *(map above right)* between 1914 and 1917 was more mobile than in the West. The field of war was much larger, and the number of troops much smaller. Early Russian victories against Austro-Hungarian armies were balanced by major defeat in the advance into Prussia in August 1914. In 1915 the Central Powers pressed Russian forces back. In 1916 Russian armies again inflicted serious reverses on Austrian forces and reached the crest of the Carpathian mountains, only to be pushed back once more by the German army.

When war broke out Russia was already heavily dependent on foreign loans, particularly from France, to fund government activity. During the war Russia accumulated $4 billion of additional debt. Patriotic Russians were encouraged to buy bonds for the war *(below right)*, but by 1916 the rouble was worth only half its value before the war.

On 22 January 1905 a demonstration for constitutional reform in St Petersburg, led by a former prison chaplain, Father Gapon, was fired on by troops and 150 killed. The massacre sparked a revolutionary crisis in Russia's cities. Workers set up local councils (soviets) to organize strike action and demand political reform *(map right)*. The government resisted until a general strike in Moscow on 7 October forced the tsar into granting a parliament and a wide franchise. Peasant and worker unrest continued, until the workers' quarters of Moscow were shelled and more than 500 killed in December.

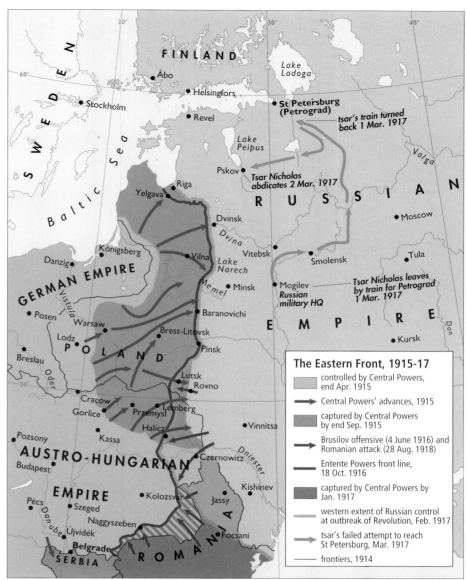

The Eastern Front, 1915-17

- controlled by Central Powers, end Apr. 1915
- Central Powers' advances, 1915
- captured by Central Powers by end Sep. 1915
- Brusilov offensive (4 June 1916) and Romanian attack (28 Aug. 1918)
- Entente Powers front line, 18 Oct. 1916
- captured by Central Powers by Jan. 1917
- western extent of Russian control at outbreak of Revolution, Feb. 1917
- tsar's failed attempt to reach St Petersburg, Mar. 1917
- frontiers, 1914

tsar's train turned back 1 Mar. 1917

Tsar Nicholas abdicates 2 Mar. 1917

Tsar Nicholas leaves by train for Petrograd 1 Mar. 1917

Russian military HQ

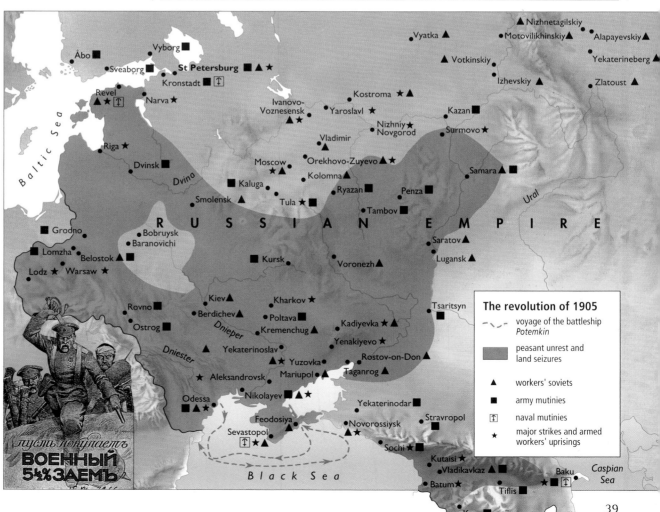

The revolution of 1905

- - - voyage of the battleship *Potemkin*
- peasant unrest and land seizures
- ▲ workers' soviets
- ■ army mutinies
- ⚓ naval mutinies
- ★ major strikes and armed workers' uprisings

39

THE RUSSIAN REVOLUTION: 1917-18

With the overthrow of the old order in Russia, there was wide-spread support for the establishment of a liberal, constitutional regime. There was hope that the war could now be prosecuted efficiently and in the people's name. The new regime promised a constituent assembly which would decide the form of the new state. The bulk of the army remained at the front to prevent a German breakthrough. The Provisional Government faced a chaotic situation. Troops at the front formed popular councils and rejected orders. In cities and villages local committees – soviets – were set up in an attempt to run local affairs in defiance of the government. The Petrograd Soviet set itself up as a rival source of authority; dominated by socialists, it called for imme-diate social reforms, economic improvement and an end to the war. But with neither the Soviets nor the Provisional Government able to compel obedience, conditions on the home front deterio-rated further.

During 1917 the crisis of food supply grew worse. By October Moscow and Petrograd were down to a few days' supply of what were already meagre rations. Real wages fell by more than a third over the summer, factories were closed down for want of materials, the transport system was strained to breaking point, and runaway inflation set in. The public mood became more radical. Peasants, who had hoped for a redistribution of land which never came, began to seize the large estates for themselves. Workers, many of whom had not initially been hostile to the regime, were alienated by further deprivation and the decision to renew the war. In April the Provisional Government took in moderate socialists; in July the socialist Alexander Kerensky became premier. As the government moved to the left, it alienated conservative and liberal support without solving popular grievances. When a renewed offensive in Galicia in June 1917 was defeated by German forces with heavy loss of life, a popular revolution was declared in Petrograd by angry workers and soldiers. The "July Days" were ended by repression, but Russia's cities were becoming ungovernable.

The main beneficiary of the radicalization of Russian society was the extreme wing of Russian social democracy, the Bolshevik Party. Support for other socialist parties – the Social Revolutionaries and the moderate Mensheviks – also increased, but it was the spectacular growth of Bolshevism, from around 22,000 party members in February 1917 to more than 200,000 eight months later, that constituted the chief threat. Bolshevik leaders refused to co-operate with the Provisional Government. They argued for an end to the war, the granting of land to poor peasants and the transfer of power to the local soviets, which

Bolshevik sympathizers entered in large numbers. Lenin, their chief spokesman, stressed the importance of propaganda and political activism. By the autumn many Russians saw Bolshevism as the only way out of the chaos of war and economic collapse, and the only way to save the revolution.

When in September the army commander-in-chief, General Kornilov, attempted a march on Petrograd to stamp out unrest and stiffen the war effort, Bolsheviks were prominent among the workers who halted his trains and persuaded his soldiers to defect. Russia began to polarize between right and left, and violence increased. On 14 September Kerensky declared a republic, hoping to appease radical opinion, but the Provisional Government had lost all credibility. Early in October the Bolshevik central committee decided to stage a coup. The Petrograd Soviet established a Military Revolutionary Committee on 29 October, controlled by the Bolshevik, Leon Trotsky. Between 6 and 8 November the Military Committee seized control of Petrograd, while the All-Russian Congress of Soviets, meeting in the city, approved an exclusively Bolshevik Council of People's Commissars as the new government, with Lenin as its chairman.

The new regime announced sweeping changes. Local power was granted to the soviets and popular committees; land was promised to the peasantry (who had already seized most of it); non-Russian nationalities were promised autonomy; the workers were offered control of the factories. Above all, Lenin urged the search for peace. This was almost the only promise he redeemed. In December the German government agreed to an armistice. In March Trotsky travelled to the Polish city of Brest-Litovsk, where he was compelled to recognize the loss of the former tsarist territories of Poland, the Baltic States, the Ukraine and Georgia. Two months before, the long-promised Constituent Assembly, called reluctantly by Lenin, returned 75% non-Bolshevik dele-gates. It was closed down immediately, and a *de facto* Bolshevik dictatorship set about the daunting task of building a new Russia.

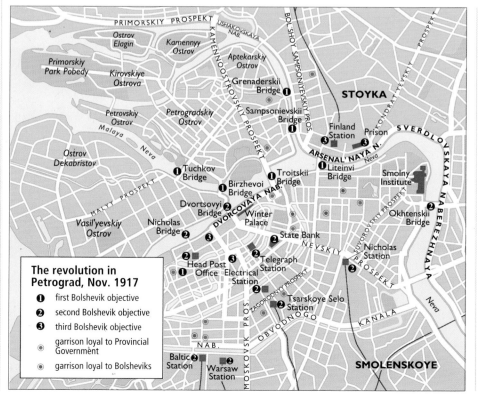

The revolution in Petrograd, Nov. 1917

❶ first Bolshevik objective
❷ second Bolshevik objective
❸ third Bolshevik objective
⊙ garrison loyal to Provisional Government
⊙ garrison loyal to Bolsheviks

The Bolshevik Revolution in November was a carefully planned seizure of power in the capital, Petrograd *(map left).* On 6-7 November the bridges and the main railway stations were seized by armed workers and soldiers. On 25 October the cruiser *Aurora* fired blank shells at the Winter Palace, which housed the Provisional Government. On 8 November the palace was occupied and the government disbanded. The Bolsheviks set great store by the use of revolutionary force. Here *(top)* Lenin is seen at an early military review in Red Square in 1919, inspecting young trainees for the civil war that followed the Bolshevik coup.

By the winter of 1916-17 Russia was in crisis *(map right).* When hostility to the regime reached boiling point in February 1917, there were spontaneous protests in many Russian cities. When the tsar abdicated, he was succeeded by a provisional government whose authority was difficult to establish in the country and in the army. Popular local councils (soviets) sprang up in the cities in the countryside. By October domestic order and military discipline had collapsed to such an extent that the radical socialist Bolshevik movement was able to seize power.

By early **1917** the growing opposition to the tsarist regime could no longer be controlled. The situation in Petrograd prompted the tsar, who was at his military headquarters in Mogilev, to return to his capital. His train was stopped by radical railway-men and diverted to Pskov, where the tsar abdicated on 15 March. In Petrograd the streets were packed with soldiers and workers eager for news *(right)*. Revolutionary propaganda turned street demonstrations into a widespread movement for social and political change *(left)*.

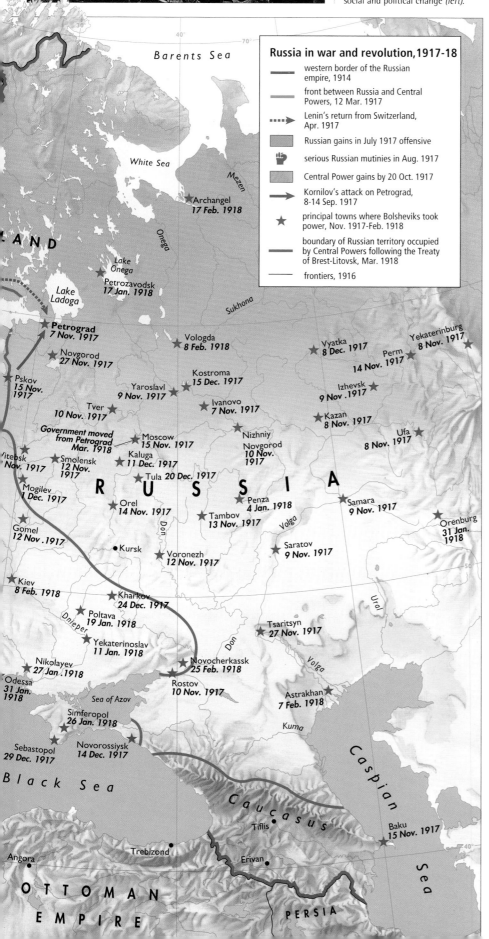

Russia in war and revolution, 1917-18

— western border of the Russian empire, 1914

— front between Russia and Central Powers, 12 Mar. 1917

····▶ Lenin's return from Switzerland, Apr. 1917

▨ Russian gains in July 1917 offensive

✊ serious Russian mutinies in Aug. 1917

▨ Central Power gains by 20 Oct. 1917

→ Kornilov's attack on Petrograd, 8-14 Sep. 1917

★ principal towns where Bolsheviks took power, Nov. 1917-Feb. 1918

— boundary of Russian territory occupied by Central Powers following the Treaty of Brest-Litovsk, Mar. 1918

— frontiers, 1916

The revolution saw a sharp shift to the left in Russian politics. The last pre-war Duma of 1912 was overwhelmingly centre-right, dominated by liberals and conservatives. A distinct change took place in 1917: the first pro-visional government of March was largely liberal, while that of July had mainly moderate left ministers. The Constituent Assembly in 1918 was overwhelmingly moderate left. Lenin's Bolsheviks had only 12% of delegates to the Congress of Soviets in June 1917, and only 24% of the Constituent Assembly, two months after seizing power.

Composition of the Russian assemblies and governments, 1912–18

■ hard left   ■ right
■ left   ■ hard right
□ liberal   ■ nationalist

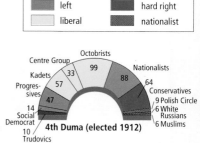

4th Duma (elected 1912)
Centre Group · Octobrists 99 · Nationalists · Kadets 57 · Progres-sives 47 · Social Democrat 14 · Trudovics 10 · Nationalists 88 · Conservatives 64 · 9 Polish Circle · 6 White Russians · 6 Muslims · 33

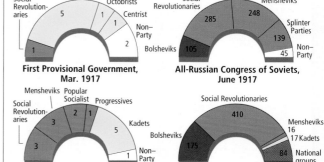

First Provisional Government, Mar. 1917
Social Revolution-aries 5 · Kadets · Octobrists 1 · Centrist 1 · Non–Party 2 · 1

All-Russian Congress of Soviets, June 1917
Social Revolutionaries 285 · Mensheviks 248 · Splinter Parties 139 · Bolsheviks 105 · Non–Party 45

Provisional Government, 24 July 1917
Mensheviks 3 · Popular Socialist 2 · Social Revolution-aries 3 · Progressives 1 · Kadets 5 · Non–Party 1

Constituent Assembly, Jan. 1918
Social Revolutionaries 410 · Mensheviks 16 · 17 Kadets · Bolsheviks 175 · National groups 84

LENIN

Vladimir Ilyich Ulyanov (1870-1924), better known by the name he adopted as a revolutionary, Lenin, was the central figure in the transformation of Russia from a royal autocracy to a socialist state. The son of a school inspector from the Middle Volga, Lenin graduated in law from St Petersburg University. A student rebel, he became a convinced Marxist by 1889. Lenin became a leading spokesmen of social democracy in the Russian capital and in 1895 he was imprisoned. Exile in Siberia followed, and in 1900 Lenin moved abroad. Convinced of the need for a revolutionary voice, he founded the paper Iskra (The Spark) in 1900. In his early writing he argued that the workers were incapable on their own of seeing beyond bread-and-butter issues: they needed a tightly organized revolutionary party to act on their behalf. In 1903 the Russian social democrats split over these issues, Lenin carrying the majority (the Bolsheviks) with him. In 1904 he returned to St Petersburg briefly to help organize the revolution. He then found himself once again in exile in western Europe, where he remained until the German high command conveyed him in a sealed train back to Russia in April 1917

in the hope that his agitation would destabilize the Russian war effort. Lenin's April Theses, published on his return, laid the theoretical foundations for the growing Bolshevik movement. By the October Revolution, Lenin was the undisputed leader of the radical socialist movement. His idea of the "dictatorship of the proletariat" was ruthlessly imposed during the civil war. In 1921 he introduced the New Economic Policy to reverse the moves to a communist economy, but he died in 1924, too soon to see the full consequences of proletarian dictatorship.

The Russian Revolution: 1917-18

THE COLLAPSE OF THE HABSBURG AND GERMAN EMPIRES

The collapse of the Central Powers, 1918

- controlled by Entente Powers, 30 September
- limit of Central Powers' control, 29 September
- lost by the Central Powers before the Armistices
- evacuated by the Central Powers under the Armistices of 3-11 Nov.
- ceded to former nationalities of Austro-Hungarian empire by 6 Dec.
- German-Austria, declared independent 12 Nov.
- Hungary, declared independent 16 Nov.
- controlled by Czechoslovakia by 31 Dec.
- ✗ social/political unrest
- ◐ troop mutiny
- ◑ naval mutiny
- ◆ Armistice
- ★ declaration of independence
- ▲ overthrow of monarchy

independent from 6 Dec. 1917 (Finland)

250,000 POWs returning to Germany and 2 million to Austria, bring with them Bolshevik and revolutionary ideas

as determined by the Treaty of Brest-Litovsk, Mar. 1918

occupied by Romania, 11 Nov.

Romania as agreed by the Treaty of Bucharest

14 Sep; beginning of Entente offensive

The Treaty of Brest-Litovsk marked the high watermark of the German and Austrian war effort. With the Central Powers now controlling the supplies of much of Eurasia, victory was regarded as a real possibility not some distant hope. Plans were drawn up for a German-dominated European order, and for German imperial supremacy in Africa once the war in the west was won.

Triumph over a weakened Russia disguised weaknesses in the Central Powers. The Habsburg empire now faced the very nationalist crisis she had gone to war in 1914 to prevent. In January 1918 the American president, Woodrow Wilson, announced his 14 Points for the post-war settlement of Europe. These included an independent Poland and self-determination for other peoples. This was considerably more than the nationalities had been demanding. Whether or not they may have accepted a federal monarchial state, Wilson's promise of genuine independence encouraged the subject races to break away from Habsburg rule.

In April 1918 a Congress of the Oppressed Peoples was called in Rome at which Poles, Czechs, Slovaks and south Slavs called for the fragmentation of the empire on national lines. Even the Austrian German social democrats called on the emperor to liberate the nationalities. When in October Emperor Karl granted a manifesto giving autonomy to his ethnic minorities, it was already too late.

In Poland a national council was formed with western backing. Polish soldiers stopped fighting for Germany and Austria, and Polish officials resigned their posts. A Czech-Slovak national council in Paris under Edvard Benes was recognized by the West as the *de facto* Czech government, and on 28 October

an independent Czech state was declared in Prague. A few weeks earlier a council of Serbs, Croats and Slovenes set itself up in Zagreb to found a Yugoslav state. Everywhere the authority of the empire crumbled. Local councils sprang up in defiance of the central authority. Soldiers deserted, no longer willing to die for a bankrupt order. In Hungary the prime minister, Count Istvan Tisza, was murdered and amidst mounting violence in the capital his successor, Mihály Károlyi, declared an independent Hungary. This was the final blow to the monarchy. On 3 November the Austrian forces signed the Padua Armistice and stopped

The Habsburg empire was a melting pot of nationalities. From 1867 the two chief "peoples of state" were the Germans and the Magyars, to whom the interests of the other smaller nationalities were subordinate. In the early years of the century a programme of Magyarization directed against the subject nationalities drove the different Slav peoples to demand self-determination.

Austria-Hungary: ethnic divisions

- Germans
- Italians
- Czechs
- Slovenes
- Croats and Serbs
- Poles
- Magyars
- Slovaks
- Ruthenes
- Romanians

The collapse of the Central Powers when it came in 1918 was sudden and complete (map left). The three allies, Germany, Austria and Hungary, recognizing that the war was effectively lost in September 1918, broke away from each other in the hope of securing a separate peace and better treatment. The old ruling classes conceded political reform and peace. In Hungary Mihály Károlyi called for a separate peace on 16 October and broke completely with Austria two weeks later. On 21 October German deputies in the Vienna parliament voted for a "Greater Germany", hoping to persuade the Allies to accept their self-determination. The Allies refused a German-Austrian state and treated each separately.

During the last days of the war in Germany, the sailors of the Kiel fleet mutinied and signalled a wave of revolutionary violence throughout the country. Soliders, sailors and workers formed councils (Räte) to take over local administration (bottom). Some workers and intellectuals wanted a communist revolution. The Spartakists, as they were known, seen here in the streets of Berlin in January 1919 (right), were brutally crushed.

Captured by the Russians during the war, Béla Kun (1886-1937), seen here addressing a crowd in Budapest in 1918 (below), returned to Hungary as a Bolshevik agitator and, in the confusion after the Armistice, drove Károlyi from office in March 1919. For four months he ruled a communist state until ousted by Romanian, Czech and nationalist Hungarian forces.

THE FATE OF THE MONARCHS

There was a point in central Europe where the three empires of Russia, Germany and Austria-Hungary met at a common frontier. Near the town of Myslowitz lay the Dreikaiserreichsecke (corner of the three empires) where postcards, such as the one pictured (right), could be franked with the stamps of all three empires. With the collapse of imperial rule in 1918, Myslowitz ended up in modern Poland. The end of empire left the three imperial dynasties in a vulnerable and uncertain position. Tsar Nicholas and his family were sent in March 1917 to the palace of Tsarskoye Selo, where they lived as prisoners under increasingly harsh conditions. In April 1918 they were sent to Yekaterinburg in the Urals, where, as anti-Bolshevik forces drew near in the civil war, they were murdered on the night of 16 July 1918. Kaiser Wilhelm was more kindly treated. On 10 November, following his formal abdication, he set off in a car for the Dutch border near Maastricht. Unrecognized, he crossed the frontier and was granted sanctuary by Queen Wilhelmina.

He was settled in a country house at Amerongen. The victorious powers hoped to extradite him to face war crime trials, but the Dutch authorities refused to reverse the decision to offer the Kaiser political asylum. In 1920 he moved to a house at Doorn. Within a year his wife died and his youngest son committed suicide. Wilhelm led the life of a country squire. There was no real effort in Germany to restore the dynasty, but when German

forces occupied the Netherlands in 1940 he was left in peace. His Austrian ally was less fortunate. Emperor Karl fled to Switzerland without formally abdicating. Twice in 1921 he tried to reclaim the throne of Hungary until his loyalist forces were defeated. The Hungarian Diet then passed a law which ended the Habsburg monarchy, and the British sent Karl into exile in Madeira, where shortly afterwards he died.

fighting. On 11 November the emperor Karl withdrew unconditionally from state affairs. A day later a republic was declared in Vienna. At this final stage of the war the Austrian Germans looked to Germany for salvation. They hoped that the self-determination of peoples promised in the 14 Points would apply to them, and they could build a Pan-German state.

For much of 1918 Germany appeared less crisis-ridden than the Habsburg empire, but there already existed a strong undercurrent of political tension. Hostility to the Kaiser and to military rule was widespread and serious strikes broke out in Berlin in the spring. Strikers began to add political demands to the call for higher wages and more food. When news of German reverses in August and September reached the home front, morale declined sharply.

A day later, in the hope that it would satisfy the demands of the Allies and pave the way for an armistice before they reached German soil, Ludendorff recommended to the Kaiser the establishment of a parliamentary government. On 3 October the autocracy came to an end. While negotiations continued, the home front reached crisis point. When the sailors of the Kiel fleet refused to sail on 29 October there was talk of revolution. On 9 November the Kaiser fled to the Netherlands. The same day a republic was declared and the social democrat leader, Friedrich Ebert, became chancellor.

Within the space of a year the three empires – Romanov, Habsburg and Hohenzollern – had become republics, with all of them dominated by the popular parties of the left. Of the major pre-war empires, the only one to survive the turmoil of the war and to retain its king-emperor into the 1920s was that of the British.

The Collapse of the Habsburg and German Empires

IN THE AFTERMATH of the First World War the victors hoped to build a better world based on democratic principles and collective efforts for peace. It proved difficult to heal the wounds of war. The world economy grew unevenly in the 1920s and then collapsed in 1929, throwing more than 20 million out of work. The Russian Revolution set the stage for bitter ideological divisions between communist and conservative forces worldwide. The communist threat provoked a new political force, Fascism, committed to destroying Marxism and building modern authoritarian regimes based on mass nationalism. Finally, the legacy of the war's political settlement left a whole number of unsettled scores. When the 1929 slump undermined collective action, there followed a nationalist backlash and escalating international tension. The post-war dreams turned sour. The powers that imposed peace in 1919 found themselves 20 years later facing war once again.

A young Italian fascist makes the salute

THE WORLD BETWEEN THE WARS

THE POST-WAR SETTLEMENT IN EUROPE

ON 12 JANUARY 1919 the victorious Allies met at the palace of Versailles, on the outskirts of Paris, to draw up a peace settlement. The conference was dominated by the great powers: Britain, France, the United States. Italy and Japan were both full participants, but their political weight was never sufficient to force the hand of the other three powers. The lesser Allies – Greece, Romania, Serbia – were allowed to send representatives to Paris, but had no say in the final settlement except in matters that affected them directly.

The American president, Woodrow Wilson, represented his country in person at the conference. He came expecting to impose a lasting peace, based on the liberal principles he outlined in the Fourteen Points. Chief of these was the right to national self-determination, a right Wilson thought would encourage popular democratic regimes in Europe. His vision of a new liberal Europe was shared by few of his allies. France came to the conference led by the fiery veteran politician Georges Clemenceau, whose chief concern was to guarantee French security and to make the Germans pay for the physical destruction of much of north eastern France. The British representative, Prime Minister David Lloyd-George, shared some of Wilson's hopes for a liberal Europe, but was not willing to put principle before national interest. Britain, too, wanted to repair the economic damage of war.

The resulting settlement was an uneasy compromise between enlightened principle and raison d'état. It was agreed that Germany should be disarmed, but no other power was similarly obliged. There was a vague commitment in the League Covenant to the goal of disarmament, but a formal conference to address the issue did not convene until 1932, and broke up two years later with little achieved. Self-determination was applied only loosely, as the confused ethnic pattern of eastern Europe made a neat solution almost impossible. Austrian Germans were denied the right to join with their fellow Germans, while many other Germans were forced to live under Czech rule in the Sudetenland. The American idea to create a peace-keeping League of Nations was enfeebled by the failure to agree on a multi-national army to enforce the peace; nor could the architect of the League, Wilson, sell the idea to Congress. The United States remained outside the League, leaving the settlement to be dominated by the interests of France, the only major armed power left on the continent.

Nor did it prove possible to impose a settlement in any coherent way. Peace was signed with Germany on 28 June 1919, but the other Central Powers were treated separately. Agreement was reached slowly and only after a great deal of bickering between the Allies and between victors and vanquished. In much of eastern Europe, fighting continued for several years and the settlement in the east was only completed in 1923. A treaty with Austria was signed at St Germain on 3 November 1918. The loss of the non-German areas of the Habsburg empire was confirmed, together with the loss of about one third of the German-speaking part of the old kingdom. The Austrian army was limited to 30,000 and reparations imposed on the rump state. In the Trianon Treaty, signed by Hungary on 4 June 1920, two thirds of the old Hungarian state was lost, principally to Yugoslavia and Romania. Hungary, too, was made liable for reparations, and her armed forces restricted to 35,000.

The gainers were the new republics of eastern Europe. Poland became a sovereign state again after years of partition, and was able to expand her territory at the expense of the weak Soviet state on her eastern borders and a disarmed Germany in the west. Czechoslovakia was carved out of the northern territories of the Habsburg empire. Both contained substantial minorities. Three million Hungarians, eight million Germans and five million Ukrainians lived under the rule of other races. The other indirect beneficiary of the settlement was Ireland. Granting self-government to Poles and Czechs made it difficult to deny it elsewhere, and in 1921 Lloyd-George finally conceded autonomy to all of Ireland save the northern province of Ulster. The national question here, as in much of eastern and central Europe, remained unresolved.

The post-war settlement *(map right)* was arrived at in a series of treaties devised by the major powers in sessions in and around Paris between 1919 and 1920. Heavy territorial penalties were imposed on Germany, Austria, Hungary, Bulgaria and Turkey. Though not party to the settlement, the new Soviet state also lost extensive pre-war Russian territory in Poland, the Baltic states and Romania. The gainers were the eight new national states created in central and eastern Europe. Despite the Allies' desire to satisfy demands for self-determination by the peoples of the old empires, they left large minority groups living under the rule of other nationalities, creating an unstable foundation for the new post-war order.

The Versailles Conference was the largest in Europe since Vienna in 1815. Some 32 states were invited to participate, though not the defeated powers. There were 70 plenipotentiaries with the power to negotiate and some 1,037 delegates in all. They met in full-scale plenary sessions in the Hall of Mirrors in the Louis XIV palace at Versailles *(above)*.

At the end of the war Hungary was occupied by Romanian, Serb and Czech forces in two thirds of its territory *(map left)*. The peacemakers in Paris regarded the small Hungarian state that remained as the core of Magyar settlement, and drafted a treaty which approximated most of the unoccupied area with the new Hungary. Following the collapse of the Communist Bela Kun government in August 1919, Romanian forces occupied almost the whole of Hungary until forced back by Allied intervention. A new national army under Admiral Horthy entered Budapest in November 1919, and six months later his government signed the Trianon Treaty confirming territorial losses.

THE FORMATION OF YUGOSLAVIA

The disintegration of the Habsburg empire created the conditions for the creation of a south Slav state. The Serb government, in exile on Corfu since Serbian defeat in 1915, was torn between ideas of a Greater Serbia and a federation with other southern Slav peoples. The Croats and Serbs of the empire formed a Yugoslav National Committee in 1917, which took the lead a year later in establishing a Yugoslav state in collaboration with the exiled Serbs. On 1 December 1918 the Kingdom of the Serbs, Croats and Slovenes was established in Belgrade, under the Serbian Karadjordjević dynasty. Fear of Italian ambitions drove the Montenegrins and other minorities into the new Slav state. Over the next two years there developed strong arguments between the federalists – mainly Croats – and the Serbian leadership, which favoured a unitary state based around Serb institutions. The constitution of June 1921 was a victory for the Serb idea. Serbs, who constituted 43% of the population,

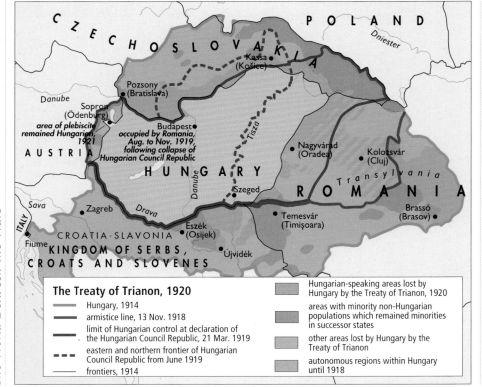

The Treaty of Trianon, 1920

— Hungary, 1914

— armistice line, 13 Nov. 1918

— limit of Hungarian control at declaration of the Hungarian Council Republic, 21 Mar. 1919

--- eastern and northern frontier of Hungarian Council Republic from June 1919

— frontiers, 1914

- Hungarian-speaking areas lost by Hungary by the Treaty of Trianon, 1920
- areas with minority non-Hungarian populations which remained minorities in successor states
- other areas lost by Hungary by the Treaty of Trianon
- autonomous regions within Hungary until 1918

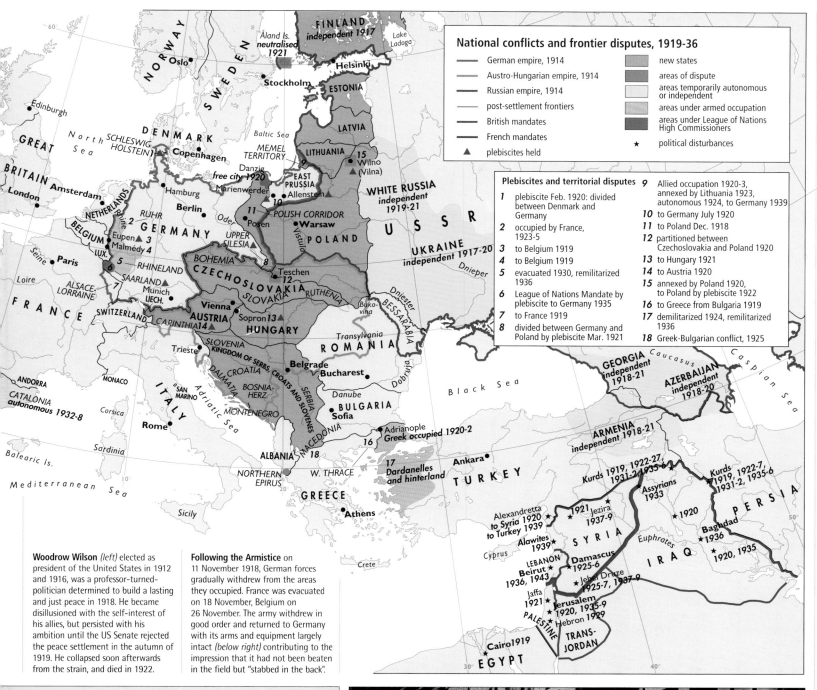

National conflicts and frontier disputes, 1919-36

— German empire, 1914
— Austro-Hungarian empire, 1914
— Russian empire, 1914
— post-settlement frontiers
— British mandates
— French mandates
▲ plebiscites held

■ new states
■ areas of dispute
■ areas temporarily autonomous or independent
■ areas under armed occupation
■ areas under League of Nations High Commissioners
★ political disturbances

Plebiscites and territorial disputes

1 plebiscite Feb. 1920: divided between Denmark and Germany
2 occupied by France, 1923-5
3 to Belgium 1919
4 to Belgium 1919
5 evacuated 1930, remilitarized 1936
6 League of Nations Mandate by plebiscite to Germany 1935
7 to France 1919
8 divided between Germany and Poland by plebiscite Mar. 1921
9 Allied occupation 1920-3, annexed by Lithuania 1923, autonomous 1924, to Germany 1939
10 to Germany July 1920
11 to Poland Dec. 1918
12 partitioned between Czechoslovakia and Poland 1920
13 to Hungary 1921
14 to Austria 1920
15 annexed by Poland 1920, to Poland by plebiscite 1922
16 to Greece from Bulgaria 1919
17 demilitarized 1924, remilitarized 1936
18 Greek-Bulgarian conflict, 1925

Woodrow Wilson (*left*) elected as president of the United States in 1912 and 1916, was a professor-turned-politician determined to build a lasting and just peace in 1918. He became disillusioned with the self-interest of his allies, but persisted with his ambition until the US Senate rejected the peace settlement in the autumn of 1919. He collapsed soon afterwards from the strain, and died in 1922.

Following the Armistice on 11 November 1918, German forces gradually withdrew from the areas they occupied. France was evacuated on 18 November, Belgium on 26 November. The army withdrew in good order and returned to Germany with its arms and equipment largely intact (*below right*) contributing to the impression that it had not been beaten in the field but "stabbed in the back".

The formation of Yugoslavia, 1919

■ Serbia and Montenegro, 1913
■ annexed by Serbia and Montenegro from Ottoman empire, 1913
— frontiers, 1914
■ annexed from Bulgaria, 1919
■ Austro-Hun. territory united with Serbia and Montenegro, 1920, to create the Kingdom of Serbs, Croats and Slovenes
■ remained Austrian by plebiscite, 1920
— Kingdom of Serbs, Croats and Slovenes in 1929, when renamed Yugoslavia

against the Croats' 23%, set up a centralized state, dominated by the Serbian political elite and a national army officered mainly by Serbs.

THE POST-WAR SETTLEMENT IN GERMANY

Of all the states that had sued for peace at the end of the war, Germany was the only one to do so on the basis of Wilson's Fourteen Points, rather than surrender unconditionally. As a result many Germans assumed that Germany would be treated as a participant in the peace settlement, able to negotiate the terms on her own behalf. The achievement of democratic government, confirmed with the election of a Constituent Assembly in January 1919, appeared to fulfil the wishes of the Allied powers. When the German delegation arrived at Versailles, however, they found that the terms were dictated to them, and that far from reflecting any spirit of democratic goodwill the terms were punitive and non-negotiable.

The settlement provoked strong resentment inside Germany at a time when the fragile democratic government was trying to damp down popular revolutionary movements and cope with military threats from newly independent Poland. A heated debate within the new parliament over acceptance or defiance was finally resolved in June in favour of signing because of Germany's feeble military situation and the Allied decision to maintain the blockade, which left millions of Germans close to starvation. Extreme nationalists were never reconciled to the humiliation, and dubbed those who signed the "November criminals" for seeking an armistice in the first place. The main proponent of acceptance, Matthias Erzberger, was murdered in 1921 by a right-wing extremist in the Black Forest.

The Allied powers had two major objectives in imposing the settlement: they wished to weaken Germany so that she would no longer impose a military threat to the other powers; and they hoped to limit Germany's economic revival by stripping her of assets and resources and forcing her to pay reparations. They also wanted Germany to accept responsibility for the war as a moral basis for their own claims against her. In Article 231 of the Treaty, Germany was forced to accept "war guilt". No other provision provoked greater resentment, for the German public believed that the war had been the product of a collective crisis of the Powers.

The rest of the settlement was bad enough from the German viewpoint. One eighth of the pre-war territory was lost, and all German colonies. East Prussia was divided from the rest of Germany by a corridor of territory designed to give Poland access to the sea. German assets abroad were seized and her merchant fleet confiscated. Germany's armed forces were emasculated. The General Staff was disbanded, training schools closed down, fortifications and munitions works destroyed, while the right to possess or develop any weapons of an offensive character was refused. An Allied Control Commission was established to secure verification of German compliance. Similar bodies were set up for the permanent monitoring of the conditions of the Treaty. Even German missionaries abroad were to be supervised by Allied commissioners of the same Christian denomination.

At the heart of many of the arguments between the Allies lay the issue of reparations. The Treaty laid down in precise detail what the Allied powers wanted to repair the economic losses of war. The total sum was only finally agreed at a conference in London in 1921, when Germany was asked to pay 132 billion gold marks in annuities down to the year 1988. But well before then large deliveries in kind were made according to the terms of the Treaty, which specified everything from schedules of coal deliveries to the supply of 500 stallions, 2,000 bulls and 1,000 rams to replenish the stock on French farms caught up in the fighting on the Western Front.

The reparations demands came on top of Germany's own vast war debts, which totalled more than 150 billion marks. The strain of paying for the war and demobilization produced serious inflation, which was exacerbated in 1923 when French and Belgian troops were sent to the Ruhr in January to force the delivery of coal reparations. By November 1923 the mark was worth one trillionth of its pre-war value. The currency collapse wiped out the cost of the war, but it also wiped out the savings of millions of ordinary Germans. In 1924 the currency was stabilized with Allied help, and a new reparation schedule drawn up, geared to Germany's ability to pay. The German public blamed the Treaty for the currency collapse and for Germany's economic weakness. The foreign minister, Gustav Stresemann, argued that fulfilment of the Treaty was the only way to achieve German rehabilitation in the international arena. While Germany settled down to work within the framework of the Versailles system, a legacy of bitterness and a profound sense of injustice lived on in the German mind.

For many Germans the final insult was the invasion of the Ruhr industrial area by French and Belgian troops on 11 January 1923 to enforce reparation deliveries. Troops were posted in factories *(bottom left)*. Several hundred Germans and Allies died in the fighting and 150,000 Germans were forcibly expelled from the occupied zone. The occupation ended in September 1923.

On 7 May 1919 a draft copy of the Treaty of Versailles was handed to the German delegation at the Trianon Palace with 22 days for comment. The German delegation rejected most of the proposals, but the Allies agreed only to a modification of the rate of German demobilization and to a plebiscite in Upper Silesia. The main settlement remained unchanged *(map above)*: Germany lost territory in Poland, Belgium, Denmark and France, and its western frontier was demilitarized. Its rivers were freed to international traffic and Danzig was made a free city under the League of Nations.

After the loss of German colonies in the peace settlement a lively propaganda campaign was mounted to keep alive German commitment to an overseas empire. Mourning stamps *(below left)* were issued for the colonies to be used as labels on ordinary mail.

Sailors from the German cruiser *Nürnberg* surrender *(below)* in Scapa Flow after scuttling their vessel just hours before the German fleet was due to be handed over to the British. The German submarine fleet was surrendered at Harwich on 20 November.

One of the most bitterly contested decisions of the peace conference concerned the fate of Upper Silesia – a coal-rich area, which Poland wanted to detach from Germany. The Allies agreed to a plebiscite for the area, and 60% voted to stay with Germany, including some of the Slav population. The poster from 1920 *(right)* was part of the propaganda campaign to keep Silesia German. In the end the Allies insisted on giving part to Poland, including the bulk of the rich coalfield. This was among the largest single economic losses from the settlement. The Allies also took industrial equipment and agricultural resources, most of Germany's merchant marine, and significant quantities of natural resources *(chart bottom right)* . The money reparations proved impossible for Germany's economy to cope with and in 1924 and again in 1929 the schedule of payment was adjusted *(chart bottom)*. The picture *(below)* shows the two Americans who led the re-scheduling commissions: General Charles G. Dawes *(centre)*, the architect of the 1924 plan; and Owen D. Young *(left)*, whose 1929 plan caused a nation-wide protest in Germany, and first brought Adolf Hitler into the national political limelight.

Herrgott, laß meine Heimaterd.
deutsch bleiben

WALTER RIEMER ~ PLAKAT 1920
WIEDERGABE DINSE & ECKERT · JNH.OTTO DINSE · BERLIN · SO·

Memel Territory
Entente occupation, 1920; to Lithuania, 1923
141,000

LITHUANIA

Baltic Sea

Königsberg

EAST

Danzig
free city, protected by League of Nations
356,740

WEST PRUSSIA

PRUSSIA

Stettin

to Poland
2,700,000

Vistula

Berlin

Oder

Posen

Warsaw

N Y

Dresden

P O L A N D

Breslau

Lodz

UPPER SILESIA

to Poland
980,926

Prague

Cracow

CZECHOSLOVAKIA

Hultschin to Czech. 45,000

The post-war settlement in Germany, 1918-35

— German empire, 1918

areas lost and retained by Germany

- lost, 1919-20
- lost by plebiscite, 1920-1
- retained by plebiscite, 1920-1
- under League of Nations control, plebiscite stipulated for 1935

61,000 population of lost territories

— frontiers, 1923

zones occupied by Allied troops

- for 5 years
- for 10 years
- for 15 years

- - - - eastern frontier of demilitarized zone
- - - - - southern frontier of defortified area
——— line beyond which no repair or fortification was allowed
——— rivers under international control

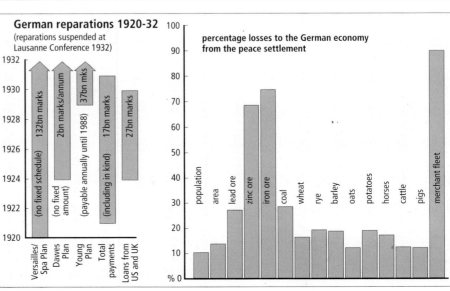

German reparations 1920-32
(reparations suspended at Lausanne Conference 1932)

1932		
1930		
1928		
1926		
1924		
1922		
1920		

Versailles/Spa Plan — 132bn marks (no fixed schedule)
Dawes Plan — 2bn marks/annum (no fixed amount)
Young Plan — 37bn mks (payable annually until 1988)
Total payments — 17bn marks (including in kind)
Loans from US and UK — 27bn marks

percentage losses to the German economy from the peace settlement

population, area, lead ore, zinc ore, iron ore, coal, wheat, rye, barley, oats, potatoes, horses, cattle, pigs, merchant fleet

% 0 10 20 30 40 50 60 70 80 90 100

THE POST-WAR SETTLEMENT: THE LEAGUE OF NATIONS

One of the fruits of Stresemann's policy of realism was the admission of Germany to the League of Nations in 1926. The League was approved at the Versailles Conference on 28 April 1919, and it met in formal session for the first time in Geneva in 1920. Its purpose was to preserve the peace through the collective action of its members. Though the League had no armed force of its own, economic sanctions and the imposition of a kind of quarantine on the offending state were considered sufficient deterrent against aggression. In practice the League spoke with anything but a collective voice. Germany and Communist Russia were both excluded. The United States Senate refused to ratify the Versailles Treaty and never joined the League.

The League Council consisted of four permanent powers – Britain, France, Italy and Japan – and four others chosen at intervals from the remaining member states. The first four were Belgium, Brazil, Spain and Greece, but Brazil became the first state to leave the League, in 1926, because of the inferior status enjoyed by the co-opted states. The first real success for the League, when it forced Italy to withdraw from its unilateral occupation of Corfu in 1923, was scored against one of the organization's own principal council members.

The League assumed responsibility for important parts of the post-war settlement. German colonies and Ottoman provinces were distributed as territories mandated by the League to Britain, France and Japan. The city of Danzig in Prussia was placed under a League commissioner as a free city, to allow the Poles to have a port on the Baltic. The most serious test faced by the League came in the effort to impose a settlement on the Balkans and the Near East, the area whose instability had helped provoke the Great War in the first place.

Here there was an issue which was difficult to resolve. Italy had joined the war in 1915 after signing a secret convention in London promising her substantial territorial spoils in Dalmatia and Slovenia. At the peace conference Wilson rejected secret agreements, and the London agreement was shelved. The Italian representative stormed out of the conference, but nothing could persuade the other Allies to concede all Italy wanted. In Italy Versailles was christened the "mutilated victory" by angry nationalists. One of their number, the poet Gabriele D'Annunzio, seized the city of Fiume (Rijeka), which had been promised to Yugoslavia. He was driven out by force after a year, though in 1924 the city was finally given to Italy in return for other concessions to the Yugoslavs.

The second problem was Greece. A minor Allied power, Greece's ambitions were fired by the power vacuum which the defeat of Bulgaria and Turkey opened up in the Near East. The Greek premier, Venizelos, looked for compensation in mainland Turkey, where there were large Greek minorities, and in Thrace. Under the Treaty of Neuilly, which Bulgaria signed on 29 November 1919, her gains in the Balkan Wars were largely lost. Serbia retained a large share of Macedonia, while Greece took western Thrace. The Greeks reached a secret agreement with Italy, granting them a free hand in western Turkey in return for Greek support for an Italian mandate over Albania. Here again the League intervened. Albania's independence was guaranteed and the 1913 frontiers restored with minor adjustments in November 1921. Italy's loss in Albania was followed by Greek disaster in Turkey. Venizelos's delusions of grandeur were exposed when Turkish forces crushingly defeated the Greek armies in August 1922. The conflict was settled by the League, which secured a final peace treaty with Turkey at Lausanne on 23 August 1923, and supervised the exchange of national minorities between the two warring states.

By 1923 the post-war settlement was complete. Six new national states had been created, but the principle of national self-determination could not be reconciled with the ambitions of the victors, large and small. Central and Eastern Europe represented an untidy ethnic map, in which sizeable and resentful minorities lived in uneasy partnership with the dominant race who ruled over them.

The League faced its most serious test in the settlement with Turkey. The terms of the treaty to be imposed on the Turks were handed over in May 1920, and produced a nationalist backlash. Kemal Atatürk set up a national republic in Ankara and his forces began the re-conquest of Turkey, much of which was controlled by foreign forces, mainly Greek, and by nationalist rebels, Kurds and Armenians. By August 1922 he had defeated the Greek armies, and a year later the League powers signed a final treaty with the Turks, giving full sovereignty to Turkey itself, but removing the remaining Ottoman provinces under League Mandate. The League supervised the exchange of refugees, such as those pictured here *(below right)*, between the two sides. More than one million Greeks were returned to Greece from Asia Minor, and some 350,000 Muslims returned from the Balkans to Turkey.

The League of Nations was founded by the delegates at the Paris Peace Conference. It was composed mainly of states in Europe and Latin America. During the inter-war years most other independent states joined, except the US *(map right)*. By 1939, 18 states had left or been expelled, including Germany, Japan, Italy and the USSR. The League was one of a number of associated global organizations set up in Geneva, where the magnificent Palais des Nations was built to house it. The International Labour Office and the Bank of International Settlements, a forerunner of the World Bank, were also established in Switzerland.

The founding committee of the League *(below)*, set up at the Paris Peace Conference, included representatives from France, the UK, the US, Italy, Japan, Greece, Serbia, South Africa, Belgium and China.

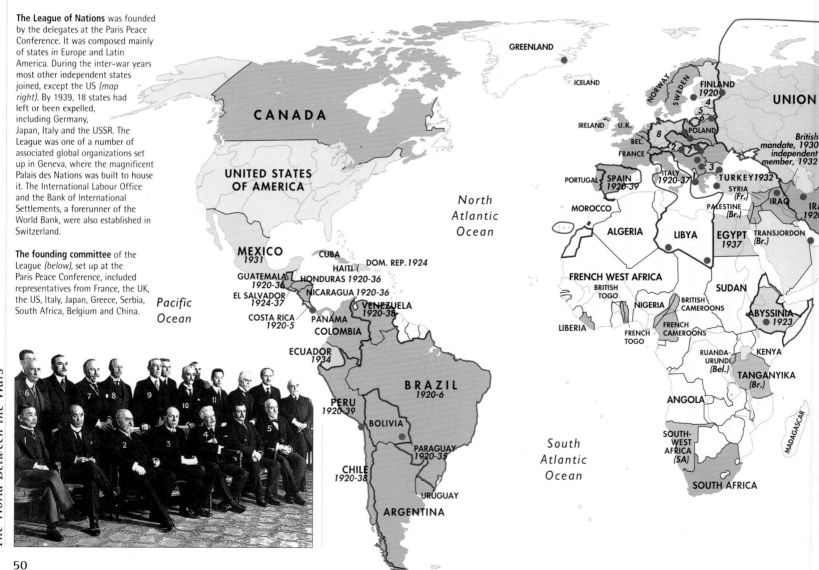

A hotly contested area at the peace conference was the port of Fiume, which had been Hungary's outlet to the sea in 1914. In defiance of the peacemakers, the port was occupied in September 1919 by a small Italian volunteer force led by the flamboyant nationalist poet, Gabriele D'Annunzio (below). He was finally expelled in December 1920, but the issue was not resolved until 1924 when Italy and Yugoslavia agreed on a division.

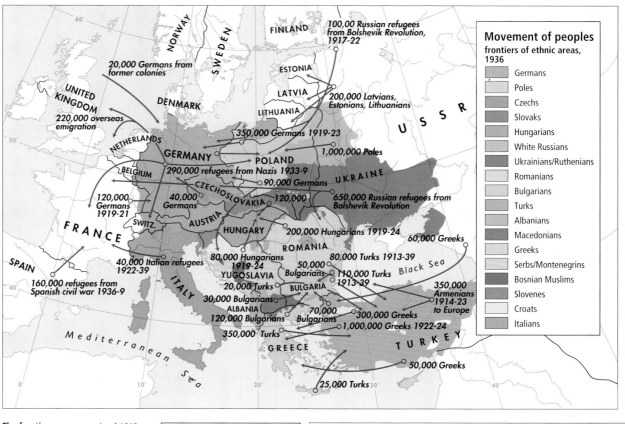

Movement of peoples
frontiers of ethnic areas, 1936

- Germans
- Poles
- Czechs
- Slovaks
- Hungarians
- White Russians
- Ukrainians/Ruthenians
- Romanians
- Bulgarians
- Turks
- Albanians
- Macedonians
- Greeks
- Serbs/Montenegrins
- Bosnian Muslims
- Slovenes
- Croats
- Italians

Map labels: 100,00 Russian refugees from Bolshevik Revolution, 1917-22 · 20,000 Germans from former colonies · 220,000 overseas emigration · 200,000 Latvians, Estonians, Lithuanians · 350,000 Germans 1919-23 · 1,000,000 Poles · 290,000 refugees from Nazis 1933-9 · 90,000 Germans · 120,000 Germans 1919-21 · 40,000 Germans · 120,000 · 650,000 Russian refugees from Bolshevik Revolution · 200,000 Hungarians 1919-24 · 60,000 Greeks · 40,000 Italian refugees 1922-39 · 80,000 Hungarians 1919-24 · 50,000 Bulgarians · 80,000 Turks 1913-39 · 110,000 Turks 1913-39 · 160,000 refugees from Spanish civil war 1936-9 · 20,000 Turks · 350,000 Armenians 1914-23 to Europe · 30,000 Bulgarians · 70,000 Bulgarians · 300,000 Greeks · 120,000 Bulgarians · 350,000 Turks · 1,000,000 Greeks 1922-24 · 50,000 Greeks · 25,000 Turks

The frontier arrangements of 1919-24 (map above) left large numbers of Europeans under the rule of a different nationality. Though some were protected by clauses in the peace treaties respecting minority rights, hundreds of thousands chose to leave their homes and return to their national homelands. Thousands of others fled communist rule, or fascist rule in Italy and Nazi rule in Germany.

League of Nations, 1920-1939

- founder members and states invited to join at foundation, 1920
- subsequent members, with dates of membership
- mandated territories
- possessions of member states
- non-member states
- states, including their possessions, which were withdrawn or expelled
- ● territorial conflicts over which the League made decisions
- frontiers, 1930

Subsequent members within Europe

1 Albania, 1920
2 Austria, 1920
3 Bulgaria, 1920
4 Estonia, 1921
5 Latvia, 1920
6 Lithuania, 1921
7 Hungary, 1922-37
8 Germany, 1926-33

Map labels: OF SOVIET SOCIALIST REPUBLICS 1934-9 · AFGHANISTAN 1934 · CHINA · INDIA · SIAM · JAPAN 1920-33 · Indian Ocean · Pacific Islands (Jap.) · NEW GUINEA (Aus.) · NAURU (Br.) · AUSTRALIA · NEW ZEALAND

THE TURKISH SETTLEMENT

The process of post-war nation building was completed with the emergence of the modern Turkish state from the ruins of the Ottoman empire. In 1918 the Young Turks who had led the empire into the war were overthrown, and the power of the sultan briefly revived. In practice Turkish affairs were dominated by the Allied powers, who occupied Constantinople and attempted to dismember the Ottoman state. The Treaty of Sèvres forced on the sultan in July 1920 stripped the empire of its Arab and European provinces, gave sections of Anatolia to Greek, French and British control, and authorized the establishment of an independent Armenia and Kurdistan. The harsh Allied occupation and the Greek threat mobilized a new nationalism throughout the Turkish-speaking area. Led by one of the heroes of Gallipoli, Mustafa Kemal, an alternative government was set up in Ankara, backed by nationalist forces. Unwilling to risk the costs of war, the Allied powers agreed to restore the integrity of the Turkish state and to end the occupation. In November 1922 a National Assembly in Ankara ended the sultan's rule and a year later declared a republic with Mustafa Kemal (Kemal Atatürk, pictured below) as president. In July 1923 the Treaty of Lausanne confirmed the loss of the Arab lands, but endorsed a new national Turkish state.

THE RISE OF THE US AS A GREAT POWER TO 1929

The central role played in the Versailles Settlement by Woodrow Wilson, the US president, highlighted the emergence of the United States as a major player on the world stage. Although European diplomats regarded the New World as a marginal factor in the balance of power, the United States had, since the 1860s, transcended her geographical isolation and come to play a fuller part in the international order.

The basis of the United States' new power was economic. Like Germany, the US was able to use rapid and large-scale industrialization as an entry to the club of great powers. By 1914 the United States was the world's largest industrial producer with vast natural resources, a large population swollen by mass immigration from Europe and a tradition of technical and scientific innovation. Farming remained an important activity, with one third of the working population still engaged in agriculture in 1910, but heavy investment in transport and in scientific farming methods turned the United States into a major supplier of world foodstuffs.

US statesmen prided themselves on their nation's republican and democratic foundations, so different from the Europe many had recently left. But in the late nineteenth-century climate of colonization and imperial rivalry even the United States was tempted into an expansion of its territorial claims and political influence, particularly in the Caribbean and the Pacific. The acquisition of Hawaii was a model of colonial expansion – a trade treaty with its native sovereign in 1875, the establishment of a coaling and naval base at Pearl Harbor in 1887, and a final decision to annex the island group in 1898 following the overthrow of the king and the formation of a pro-American republic.

That same year, 1898, the United States fought a war with Spain, one of the oldest and most decrepit of the European empires. The United States' growing naval power made victory a formality. The outcome turned the US into a power with extensive overseas interests. In the Pacific, the Philippines and the island of Guam were acquired. In the Caribbean, Puerto Rico and Cuba were brought under US protection, while the other Caribbean imperialists, Britain and France, agreed by the Hay-Pauncefote Treaty of 1901 to give the United States a virtually free hand throughout the region.

Some American imperialists began to dream of turning eastern Asia into an American sphere. Japan was opened up to

western influence by US pressure. Manchuria was regarded as an area ripe for economic penetration, and the whole of China was viewed by American businessmen and politicians as a potential sphere of influence. US insistence produced the so-called "open door" policy in China to ensure that no one power pre-empted the others by gaining special economic privileges, but the principle of "open door" trade was soon applied wherever American merchants had strong interests. This was particularly so in Latin America, which was regarded from Washington as the United States' back yard. In 1903 the United States forced Colombia to abandon its claims to Panama; a virtual US protectorate was established and work begun on a US canal across the isthmus. The canal was completed in 1914, linking America's Atlantic and Pacific interests.

Increasingly before the Great War, the US came to see itself as an arbiter between the warring monarchies of Europe and Asia. In 1906 the United States hosted the peace conference between Japan and Russia. The US was also represented at the Algeciras conference in Morocco in 1906. The eventual decision to enter the war in 1917 was backed by a growing belief that the US was destined to transform the old balance of power. Woodrow Wilson represented a powerful strand of American opinion which wished to produce a new global order based on open diplomacy, open trade and liberal values.

Though Wilson failed to carry Congress with him in his effort to re-make the world order in 1919, the decade that followed was dominated by American culture and economic power. The Washington Conference of 1921-2 set a new balance of global naval power; the United States intervened regularly over the issue of war debts and reparations; the world economy began to orientate itself away from London and Paris and towards New York. Though formally isolated from international commitments, the aggressive modernity of American life – jazz, cars, the cinema – brought the American Dream to millions of non-Americans worldwide.

In 1906 US president Theodore Roosevelt won the Nobel Peace Prize for his part in bringing together the two warring parties in the Russo-Japanese conflict. The two sides met at Portsmouth, New Hampshire, where they signed a formal treaty on 5 September 1906 (above left). The next major international conference hosted in the US was the brainchild of a formidable Republican, Charles Evans Hughes (1862-1948) (below). Described by Roosevelt as a "bearded iceberg", Hughes was secretary of state from 1920 to 1925, and later chief justice during the New Deal. In 1921 he got Britain, Japan, China and the US to agree in Washington to a Pacific settlement which guaranteed Chinese sovereignty and produced an arms limitation deal on naval strengths (chart below right).

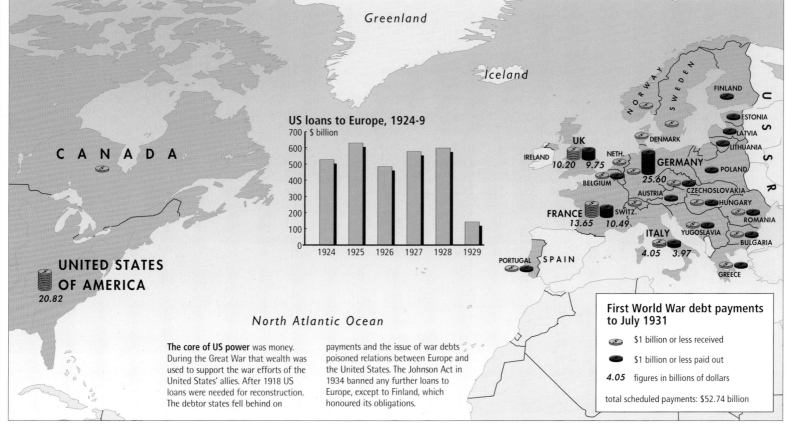

US loans to Europe, 1924-9

The core of US power was money. During the Great War that wealth was used to support the war efforts of the United States' allies. After 1918 US loans were needed for reconstruction. The debtor states fell behind on payments and the issue of war debts poisoned relations between Europe and the United States. The Johnson Act in 1934 banned any further loans to Europe, except to Finland, which honoured its obligations.

First World War debt payments to July 1931

- $1 billion or less received
- $1 billion or less paid out
- **4.05** figures in billions of dollars

total scheduled payments: $52.74 billion

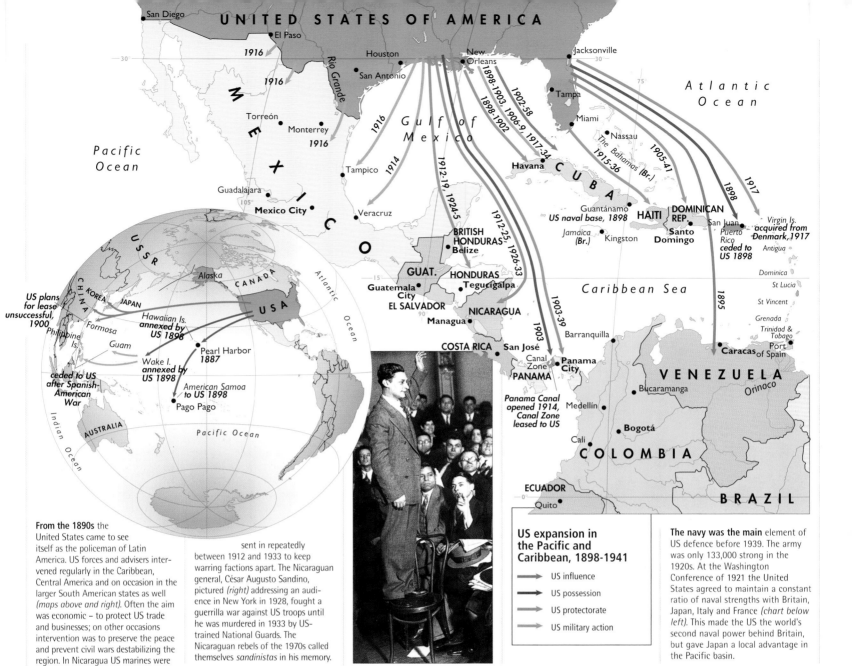

US naval base, 1898 (Guantánamo)

Virgin Is. acquired from Denmark, 1917

Puerto Rico ceded to US 1898

US plans for lease unsuccessful, 1900

Hawaiian Is. annexed by US 1898

Wake I. annexed by US 1898

Pearl Harbor 1887

American Samoa to US 1898
Pago Pago

ceded to US after Spanish-American War

Panama Canal opened 1914, Canal Zone leased to US

From the 1890s the United States came to see itself as the policeman of Latin America. US forces and advisers intervened regularly in the Caribbean, Central America and on occasion in the larger South American states as well *(maps above and right)*. Often the aim was economic – to protect US trade and businesses; on other occasions intervention was to preserve the peace and prevent civil wars destabilizing the region. In Nicaragua US marines were

sent in repeatedly between 1912 and 1933 to keep warring factions apart. The Nicaraguan general, César Augusto Sandino, pictured *(right)* addressing an audience in New York in 1928, fought a guerrilla war against US troops until he was murdered in 1933 by US-trained National Guards. The Nicaraguan rebels of the 1970s called themselves *sandinistas* in his memory.

US expansion in the Pacific and Caribbean, 1898-1941

→ US influence
→ US possession
→ US protectorate
→ US military action

The navy was the main element of US defence before 1939. The army was only 133,000 strong in the 1920s. At the Washington Conference of 1921 the United States agreed to maintain a constant ratio of naval strengths with Britain, Japan, Italy and France *(chart below left)*. This made the US the world's second naval power behind Britain, but gave Japan a local advantage in the Pacific basin.

Naval strengths agreed at Washington Conference, 1921-2

Bar chart showing battleships and cruisers (shaded) and aircraft carriers (white), in Tons:

- Great Britain: battleships and cruisers ~580,000; aircraft carriers ~135,000
- United States: battleships and cruisers ~525,000; aircraft carriers ~135,000
- Japan: battleships and cruisers ~305,000; aircraft carriers ~80,000
- France: battleships and cruisers ~220,000; aircraft carriers ~65,000
- Italy: battleships and cruisers ~185,000; aircraft carriers ~60,000

Henry Ford (1863-1947) became the symbol of American modernity and industrial power. Seen *(below)* with his son Edsel beside the 27-millionth Ford car, exhibited at the New York World Fair in 1939, Ford typified the "rags to riches" mythology of American enterprise. An apprentice machinist in the 1890s, he built his first car in 1893. Ten years later he set up a factory in Detroit, and was soon selling millions of cheap, standard cars, including the famous Model-T. Ford exported his methods abroad. "Fordism" came to stand for modern mass production and scientific management the world over.

THE US AT WAR

The onset of war in Europe in August 1914 was met in the United States by a formal declaration of neutrality on 4 August. In 1917 the US was provoked into declaring war first by the onset of unrestricted German submarine warfare, then by the "Zimmerman Telegram", a confidential letter from the German foreign minister to the Mexican government proposing to return Texas and California to Mexico if she joined the German side. On 2 April 1917 Congress approved a declaration of war. The US had virtually no air force and a tiny army, but the navy began at once to assist in the convoying of

merchant ships. The greatest assistance given to the Allied cause was economic. Already from 1914 a stream of steel, machinery and explosives had fed the British and French war efforts, though at a price. After 1917 the United States began to supply generous credits to her new allies as well. By October 1917, the US expeditionary force numbered fewer than 100,000 men. But by August 1918 there were 1.5 million Americans in Europe, fresh and well-equipped. Only in 1919 were US ground and air forces intended to take a decisive role, and that threat contributed to the German decision to abandon the fight in 1918.

LATIN AMERICA TO 1939

US influence was widespread in Latin America, but Latin America was never simply the United States' back yard. Emancipation from the Spanish and Portuguese empires early in the 19th century did not end Latin America's close economic and cultural links with Europe. A stream of migrants brought with them new skills and new political ideas, which competed with the liberal republicanism on which the newly independent states had been based.

The traditional social order was sustained by the new export economies. The old land-owning class, based on the large estates (*haciendas*), monopolized the export trade and dominated the mulatto and native Indian populations who worked on the land. The landed elite controlled politics by a complicated system of patronage and through rigged elections. Though nominally liberal, the political systems were in reality oligarchies, working for the interests of the rural elites who profited from the exceptional boom in export earnings from the 1870s to the end of the First World War.

The war proved a watershed for Latin American politics and the export economy. In the 1920s and 1930s export growth declined sharply. Overseas protection and chronic oversupply of food and raw materials on world markets ended the decades of prosperity and eroded the economic power base of the old elites. The social balance was also changing. The export economies had produced large new cities with a new educated middle-class to service the trade and to run bureaucracies and an urban proletariat whose ties with the land were cut. These groups had no allegiance to the old *hacienda* system and resented the power of the rural elites. Their political hostility to the old system was fuelled by European immigrants bringing socialist, anarchist and nationalist ideas with them. In 1917-20 there occurred a wave of social protest across the continent against low wages and poor conditions. The crisis was brutally suppressed, but the consensus on "order and progress" on which the old system relied collapsed in the years that followed.

There were powerful signs of change before the war. In Uruguay José y Ordónez began a programme of democratic and social reform in 1911. In 1912 the urban radicals succeeded in forcing electoral reform, and in 1916 won their first election. The most far-reaching transformation occurred in Mexico, where the reform programme of Francisco Madero, begun in 1910, turned into a full-scale revolution against the old order. In the 1920s, with increased urbanization and the export crisis, the grip of the old elites slackened almost everywhere in Latin America. New, predominantly middle-class parties emerged preaching a new nationalism. They had a quasi-fascist outlook, with their emphasis on authoritarian politics, corporatist social policy and state-led economic modernization. They were based in cities and won support among army officers anxious to create some kind of new order out of the crisis of the declining liberal states. Nationalists all shared a growing resentment against the United States, which had steadily increased its economic presence in Latin America by taking over commodity production (sugar in Cuba, copper in Chile, oil in Mexico etc.) and repatriating the profits.

The nationalist revolt produced a period of confused and violent politics between landowners, soldiers and middle-class radicals. In Chile in 1924 the army established a dictatorship for Carlos Ibáñez, but seven years later he was overthrown and one of his successors, Marmaduke Grove, briefly established a socialist republic. In Argentina urban nationalists and the army took control in 1930, but an effective corporatist dictatorship was only finally established by General Juan Péron in 1943. In Brazil, too, the military and the urban radicals ended the old order in 1930, and Getúlio Vargas set up a single-party dictatorship to establish a "New State". The new regimes attempted to create a consensus by making concessions to labour and by encouraging *indigenismo*, a movement to revive native culture and values and to reject US and European influence. But the rift between rich and poor, rural and urban proved difficult to bridge and more and more Latin American regimes relied on crude authoritarianism to survive.

San Diego

1916 ⊗ Carranza 1913

Pancho Villa 1914

M E X

1910 Zapata revolt
1911 Díaz overthrown
1913 Madero deposed by Huerta
1913 Carranza, Villa and Zapata fight Huerta
1914 Carranza seizes power;
Villa turns against Carranza
1917 new constitution adopted
1920 Carranza overthrown by rebel generals and killed; Villa surrenders;
Obregón becomes president
1927 and 1929 insurrections;
1936 General Cárdenas seizes power

1898-1920 Cabrera dictatorship
1921 popular revolt
1930-44 Ubico dictatorship
1944 first free elections

1913-27 Meléndez dictatorship
1931-44 Martínez dictatorship

1909 Zelaya dictatorship overthrown
1932-6 limited democracy
1936 General Somoza military dictatorship

At the beginning of the 20th century Latin America was mainly ruled by large landowners and rich exporters who relied on money and orders from abroad to keep the traditional elites in power *(map far right)*. In the 1920s the decline of the export economies and the rise of a class of educated urban officials and soldiers keen to create vigorous new nation-states less dependent on the developed world produced a period of political instability across the continent. By the mid-1930s the old elites were in retreat and a generation of army-backed authoritarian politicians seized power in a confused series of coups and counter-coups *(map right)*.

THE MEXICAN REVOLUTION

For the quarter century until 1910 Mexico had been dominated by the dictatorship of Porfirio Díaz. Following an economic crisis in 1907 and popular demands for democracy and land reform for the rural poor, political resistance to Díaz hardened into a challenge to his re-election in 1910, led by a northern landowner-turned-radical, Francisco Madero. Although Díaz was re-elected, the northern states rallied to Madero, while a peasant civil war was summoned up by the ex-bandit "Pancho" Villa in Chihuahua and by Emiliano Zapata in the central state of Morelos. The unlikely alliance of northern landowners and peasant radicals led to the fall of Díaz in 1911 and Madero's election as president. Over the following nine years Mexico descended into political confusion and civil war. On one side the old

elites of the Díaz years (urban modernizers and rural bosses) rallied together; against them were ranged the constitutionalists of the northern states in a loose often hostile alliance with forces of popular democracy and social reform. The old forces were finally defeated in 1914, but their opponents split into warring factions. The northern elites, led by Venustiano Carranza and Alvaro Obregón, defeated the radical Zapatist forces in Mexico City in August 1916. The following year a new constitution was drawn up to establish a democratic Mexico, but not until 1920, with the election of Plutarco Calles as president, was the period of revolutionary crisis finally ended. Under Calles' National Revolutionary Party, a single movement designed to embrace all the elements of Mexican revolutionary politics, the revolution was institutionalized.

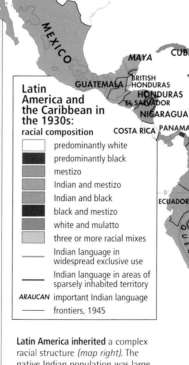

Latin America and the Caribbean in the 1930s: racial composition

- predominantly white
- predominantly black
- mestizo
- Indian and mestizo
- Indian and black
- black and mestizo
- white and mulatto
- three or more racial mixes
- —— Indian language in widespread exclusive use
- —— Indian language in areas of sparsely inhabited territory
- ARAUCAN important Indian language
- —— frontiers, 1945

MEXICO

MAYA   CUBA   DOMINICAN REPUBLIC   PUERTO RICO
GUATEMALA   BRITISH HONDURAS   HAITI
HONDURAS
EL SALVADOR
NICARAGUA
COSTA RICA   PANAMA

VENEZUELA   BRITISH   DUTCH   FRENCH
CHIBCHA   GUIANA
COLOMBIA
ECUADOR

QUECHUA   PERU   AYMARA   B R A Z I L
50% white
36% black and mulatto

BOLIVIA   QUECHUA   GUARANI

PARAGUAY

CHILE   ARGENTINA   URUGUAY

ARAUCAN

Latin America inherited a complex racial structure *(map right)*. The native Indian population was large in Mexico, Peru and Brazil, where it was dominated by white settlers. Further south white communities comprised a larger proportion. Across Latin America there developed mixed race communities: mestizos and mulattos. The black population in the Caribbean were ex-slaves. Like those of mixed race, they were dominated by the European settlers except in Haiti, where a successful slave revolt put power in the hands of the black and mulatto population.

In the early 1920s in Brazil a series of provincial revolts, aimed to end the rule of the coffee-exporting elite, gave rise to a political crisis centred in the province of Rio Grande do Sul. Here Luís Carlos Prestes launched a rebellion by junior army officers and led a column of guerrillas on a gruelling march through the Brazilian interior *(map right)*. Prestes went on to lead the Brazilian Communist Party for 30 years.

Latin America, 1900-45

Acre, independent 1899-1903

frontiers of Brazil secured 1900-9

areas affected by Mexican revolution, from 1910

Federation of Central America, 1921-2

disputed territory added to Colombia, 1922 and 1934

gained by Paraguay in Chaco war, 1938

disputed territory added to Peru by 1942

★ war and frontier dispute, with date

★ civil war, with date

⊗ US military action, with date

frontiers, 1945

Latin America economies to 1913

→ commodities exported

→ major US company

→ major UK company

• United Fruit Company plantations

Guggenheim (mining)

Standard Oil

tobacco

sugar

oil

J. P. Morgan (railways)

oil

Weetman Pearson (contracting/oil)

oil

MEXICO

CUBA

coffee

tobacco

coffee

cacao

cattle

United Fruit Company (plantations)

VENEZUELA

COLOMBIA

bananas

ECUADOR

cacao

sugar

PERU

copper

guano

silver

tin

nitrates

John Thomas North (nitrates)

copper

Peruvian Corporation (railways)

rubber

BRAZIL

cotton

sugar

coffee

British investment in government bonds

BOLIVIA

PARAGUAY

CHILE

Antony Gibbs and Sons (finance/trading)

Williamson Balfour (finance/trading)

cattle

ARGENTINA

URUGUAY

hides

beef

wool

beef

Liebigs (beef products)

wheat

wool

British railways, public utilities, shipping, land

1915 popular revolt, US intervention
1915-34 US occupation
1934 Vincent dictatorship

Bahamas (Br.)

1916-24 US control
1930 Trujillo military dictatorship

1902 self-government from US
1906 liberal revolt
1925 Machado dictatorship
1940 Batista dictatorship

CUBA (US)

⊗ 1906, 1917

1916

HAITI
1915

DOM. REP.

JAMAICA (Br.)

Puerto Rico (US)

Windward Islands

1910-20

Mexico City

Veracruz

⊗ 1914

Zapata 1910

Madero 1911

1920

C

O

BRITISH HONDURAS
1912, 1919

GUATEMALA

EL SALVADOR

HONDURAS
1909 -11

★ 1907

NICARAGUA

1916, 1926
1916, 1926-7

COSTA RICA

⊗ 1919
★ 1921

1916-37 insurrections
1933-48 Carias dictatorship

1917 military seize power
1919 democracy restored

1903 seceded from Colombia, dominated by US
1931 president overthrown by coup

PANAMA
Canal Zone (US)

⊗ 1903

Panama

Cartagena
Mompos

Maracaibo

Caracas

Trinidad (Br.)

Angostura

VENEZUELA

Georgetown

Cayenne

(Br.) (Dutch) (Fr.)

GUIANA

1899-1908 Castro dictatorship
1908-35 Gómez dictatorship
1935 limited democracy restored

1904 military rule
1909 limited democracy
1936-40 social disturbances

Antioquia

1900-3

Bogotá

COLOMBIA

1925 military coup
1929 democracy restored
1931-5 period of turmoil
1935 military rule
1938 democracy restored again

Quito
Guayaquil

ECUADOR

★ 1938

Iquitos

Leticia
★ 1932-4

Amazon

B R A Z I L

1900-30 limited democracy
1924 military revolt
1930 military revolt
1934-45 dictatorship of Vargas
1935 communist uprising
Nov. 1937 Vargas declares "New State"

1914 military revolt
1919-30 Leguia dictatorship
1930 popular military revolt
1931 revolts
1936-9 dictatorship

PERU

Lima

Callao

Cuzco

BOLIVIA

La Paz

Potosí

Tacna
to Peru, 1929

Arica

Santa Cruz

Corumbá

Campo Grande

1928, 1932-6
★

Concepción

Ciudad Real

PARAGUAY

Asunción

1920 liberal regime overthrown
1926-30 Siles dictatorship
1931 democracy restored
1934-46 military dictatorship

1912-36 limited democracy
1940-8 Morinigo dictatorship
1936-7 military rule

Rio de Janeiro

The Long March, 1924-7

→ line of march

★ battle

São Luis

winter 1925: joined by new rebel recruits

MARANHÃO

Natal

Recife

battles (with bandits)

Aug. 1925: defeated federal forces

BAHIA

Aracaju

Salvador

Córdoba

Rosario

Alegrete

URUGUAY

Montevideo

1903-33 democratic rule
1933-4 Terra dictatorship

B R A Z I L

MATO GROSSO

Tocantins

GOIAS

June 1925

Cuiabá

Brazilian Highlands

BOLIVIA

Santa Cruz

Feb. 1927: 620 rebels enter Bolivia

PARAGUAY

MINAS GERAIS

Belo Horizonte

Vitória

SÃO PAULO
Apr. 1925: Long March begins 1,500 rebels

Campos

Rio de Janeiro

Curitiba

25 Mar. 1925: rebel forces unite in Parana

RIO GRANDE DO SUL

Alegrete

Pôrto Alegre

24 Oct. 1924: Alegrete Rebellion

URUGUAY

ARGENTINA

South Atlantic Ocean

Valparaíso

Mendoza

Santiago

San Luis

Constitución

Valdivia

1902

Jujuy

Buenos Aires

A R G E N T I N A

C H I L E

1890-1912 limited democracy
1912 electoral reform
1930-2 Uriburo dictatorship
1932 democracy restored
1933-4 revolt in north eastern provinces
1933 economic recovery programme launched
1943 military revolt

Falkland Is. (Br.)

1891-1925 limited democracy
1924-31 Ibáñez military dictatorship
1931 unrest
1932 democracy restored
1938 aborted fascist uprising
1938 Popular Front government

In 1930 the military in Brazil brought to power Getúlio Vargas (cartoon below). He established a dictatorship, and in 1937 enshrined it in the so-called Estado Nôvo (New State), a quasi-fascist authoritarian system which sought to reduce dependence on the export economy and build up Brazil's own economic and military resources.

Nosso Século

A crise da República Velha. Elites descontentes de São Paulo lançam o Partido Democrático. Outubro de 1929: o colapso do café. Getúlio Vargas: uma estrela sobe.

48

1910/1930

COLONIAL EMPIRES: CRISIS AND CONFLICT

The challenge to European influence was not confined in the inter-war years solely to the independent states of Latin America. Though the European empires reached their fullest extent during the period, with the acquisition of Turkey's Arab provinces at the end of the war and with Italy's later conquest of Abyssinia, nationalist forces in the empires began to challenge the whole basis of imperial power. The main beneficiaries of the post-war settlement were Britain and France. The League gave Britain trusteeship for Palestine, Iraq and Transjordan, as well as former German colonies in Samoa, New Guinea, Togoland and Tanganyika. France gained Syria, Lebanon and German Cameroon. These areas were integrated into the global economic interests of the two metropolitan powers. Oil discoveries in the Persian Gulf area gave a new strategic significance to the Middle East.

The expansion of territory disguised a great number of weaknesses in the whole imperial structure. There was growing criticism of colonialism from the United States and the Soviet Union. The empires were expensive to maintain, despite the very real economic rewards which they brought with them. Above all the European empires faced growing opposition from within their territories, led by educated elites who sought a role in local administration or even national autonomy. There were serious, sometimes violent, challenges to colonial rule.

Nowhere was opposition to British rule more apparent than in India, where the nationalist Indian Congress party gathered widespread popular support. In 1935 the British conceded the India Act, which gave self-government to the 11 provinces and set up an All-India Federation of provinces and the remaining principalities. The Act satisfied neither old elites nor new nationalists, and by 1939 British rule rested on fragile foundations. Congress refused to assist the war effort between 1939 and 1945 and began a "quit India" campaign under the radical nationalist, Jawaharlal Nehru.

Nationalist opposition was less developed elsewhere, but

During the inter-war years European overseas empires remained an important economic asset against a background of economic stagnation and recession (map below). But the material advantages were offset by growing political and social conflict. Some of this was the result of impoverishment and trade decline, but much of the protest came from nationalists hostile to colonial rule. In the Middle East, India and the British Dominions concessions were made. By 1939 the long-term prospects for the survival of colonial empires were bleak.

A poster for the Orient Steam Navigation Company from 1920 (bottom). The empire was held together by British shipping and naval power, but the rise of the Italian and Japanese navies threatened empire security in the 1930s. For all the effort to present the empire as a traveller's delight, it was a more dangerous place to live than in 1914.

Jewish population of mainly European origin in 1939: 429,605 (28%). By 1948, Jewish population own 14% of cultivable land. Iraq Petroleum Company pipeline from Iraq to Haifa.

European population in 1936: 213,000 (8%) own one tenth of cultivated land.

European population in 1931: 881,600 (15.7%) own one third of cultivated land.

European population in 1936: c.202,000 (3.4%)

The Rif 1921-6

European population in 1935: 50,000 and 100,000 Greeks with Turkish nationality (less than 1%). Indigenous banks and beginnings of coal mining and iron and steel industry.

French investment in public utilities. Iraq Petroleum Company pipeline from Iraq to Tripoli.

Cyrenaica 1914-32

European (or European-protected) population in 1937: 225,000 (1.5%). Considerable French, British and Belgian investment mostly in mortage banks and land companies. Indigenous industry beginning with Banque Misr group in 1920s.

Some French investment; major agricultural development of Jezira (north-east) after 1938.

Condominium shared between Britain and Egypt. Cotton produced by partnership between government, tenants and British-owned Sudan Plantations Syndicate.

Petroleum beginning to be produced by Arabian-American Oil Company. 0.7 million tons produced in 1940.

Resistance by Sayyid Muhammed, "the Mad Mullah", 1891-1920

conquered by Italy 1936

Ashanti rebellion 1900

Anyang revolt, 1904

anti-Portuguese risings 1913

Nandi resistance 1895-1905

Maji-Maji revolt 1905-7

Herero and Hottentot revolts 1904-6

Zulu revolt, 1906

Autonomous Dominion within British empire from 1926

1898-1904

Map labels

U S, TURKEY, Istanbul, Ankara, Teheran, CYPRUS, LEBANON, SYRIA, Baghdad, P E R S I A, IRAQ, Abadan, PALESTINE, TRANSJORDAN, KUWAIT, BAHRAIN, TRUCIAL OMAN, Alexandria, Cairo, EGYPT, 1906, 1919, Aswan, Mecca, SAUDI ARABIA, OMAN, FRANCE, ITALY, Tunis, SPAIN, Algiers, TUNISIA, Tripoli, LIBYA, Red Sea, YEMEN, Sana, ADEN PROTECTORATE, PORTUGAL, Ceuta, Melilla, Tangier, Rabat, Casablanca, SPANISH MOROCCO, MOROCCO, ALGERIA, ANGLO-EGYPTIAN SUDAN, Khartoum, ERITREA, FRENCH SOMALILAND, Aden, Gulf of Aden, BRITISH SOMALILAND, RIO DE ORO, FRENCH WEST AFRICA, FRENCH EQUATORIAL AFRICA, Addis Ababa, ABYSSINIA (ETHIOPIA), ITALIAN SOMALILAND, THE GAMBIA, PORT. GUINEA, NIGERIA, BRITISH CAMEROONS, FRENCH CAMEROONS, BRITISH EAST AFRICA, UGANDA, KENYA, SIERRA LEONE, GOLD COAST, TOGOLAND, Lagos, FERNANDO PO, SPANISH GUINEA, Mombasa, LIBERIA, SAO TOME & PRINCIPE, CABINDA, BELGIAN CONGO, RUANDA URUNDI, TANGANYIKA, ZANZIBAR, Dar-es-Salaam, ANGOLA, NYASALAND, MOZAMBIQUE, MADAGASCAR, NORTHERN RHODESIA, SOUTHERN RHODESIA, SOUTH WEST AFRICA, BECHUANALAND, SWAZILAND, BASUTOLAND, UNION OF SOUTH AFRICA, Durban, Cape Town, Indian Ocean

was still significant. In the Dutch East Indies and French Indo-China nationalism was sustained by communist opposition. In the Dutch East Indies a communist revolt was mounted in 1926, and communist resistance reached a climax in Vietnam in the early 1930s, led by, among others, the young revolutionary Ho Chi Minh, who had picked up his radical politics in Paris. In North Africa the European powers faced attacks not only from popular native uprisings – the Rif revolt in Morocco led by Abd el-Krim, the Sanussi resistance to Italian conquest in Cyrenaica – but also from more organized political movements, such as the Etoile Nord-Africaine in Algeria or the Destour in Tunisia, which drew on the example of European radicalism.

In the former Ottoman territories of the Middle East European rule was difficult to establish in the face of Arab nationalism which had been mobilized to throw off Turkish rule. Britain conceded independence to Egypt in 1922 and to Iraq in 1932, though both had to accept a continued British military presence. A Syrian revolt against French rule (1925-7) was suppressed and a harsh authority imposed. In Palestine

the British faced a prolonged crisis, made worse by the promise to grant the Jews a homeland. When Jewish emigration from Europe expanded following Hitler's achievement of power in 1933, a virtual guerrilla war between Arabs, Jews and the British tied down more British troops than were stationed in mainland Britain. In sub-Saharan Africa resistance to Europeanization, which had produced decades of violent conflict, developed into more formal political protest movements: the Young Kikuyu Association set up in Kenya in 1921; the African Nation Congress founded in 1923 in South Africa; the National Congress of British West Africa established in 1920.

All of these movements, and those in the Caribbean and South Asia, were fuelled by the economic distress following the slump of 1929. Falling prices, high unemployment and restricted rights to the land produced serious unrest among the sugar islands of the West Indies, in the Gold Coast (Ghana) and the Rhodesian copper belt. The gulf between the metropolitan powers, mainly democracies, and their undemocratic colonies became harder to justify or sustain.

At the end of April 1924 King George V opened the British Empire Exhibition at Wembley outside London. It was a celebration of Britain's imperial greatness. Every corner of the empire was represented. In the vast Palace of Engineering *(above left)* were displayed 15 avenues of British technological achievements.

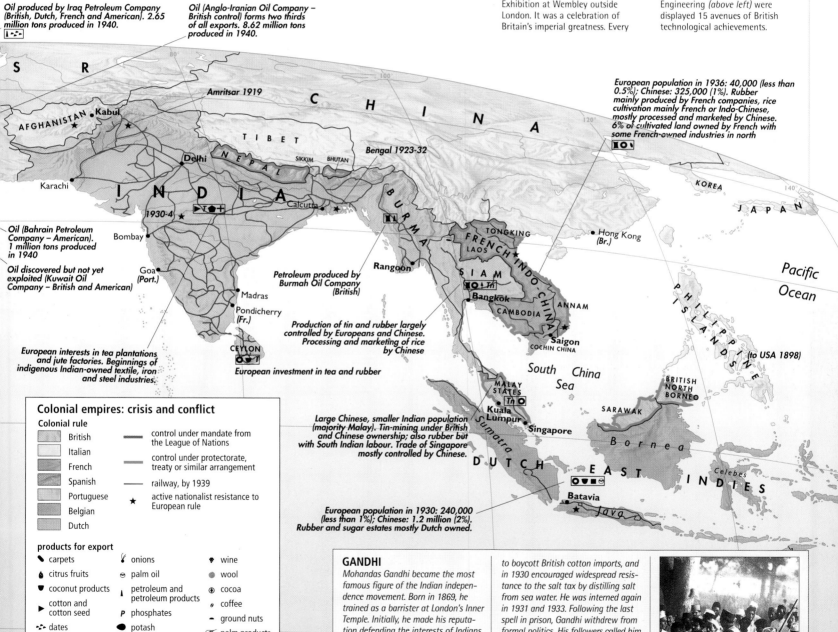

Oil produced by Iraq Petroleum Company (British, Dutch, French and American). 2.65 million tons produced in 1940.

Oil (Anglo-Iranian Oil Company – British control) forms two thirds of all exports. 8.62 million tons produced in 1940.

European population in 1936: 40,000 (less than 0.5%); Chinese: 325,000 (1%). Rubber mainly produced by French companies, rice cultivation mainly French or Indo-Chinese, mostly processed and marketed by Chinese. 6% of cultivated land owned by French with some French-owned industries in north

Oil (Bahrain Petroleum Company – American). 1 million tons produced in 1940

Oil discovered but not yet exploited (Kuwait Oil Company – British and American)

European interests in tea plantations and jute factories. Beginnings of indigenous Indian-owned textile, iron and steel industries.

European investment in tea and rubber

Petroleum produced by Burmah Oil Company (British)

Production of tin and rubber largely controlled by Europeans and Chinese. Processing and marketing of rice by Chinese

Large Chinese, smaller Indian population (majority Malay). Tin-mining under British and Chinese ownership; also rubber but with South Indian labour. Trade of Singapore mostly controlled by Chinese.

European population in 1930: 240,000 (less than 1%); Chinese: 1.2 million (2%). Rubber and sugar estates mostly Dutch owned.

Amritsar 1919
Bengal 1923-32

Colonial empires: crisis and conflict

Colonial rule

British	control under mandate from the League of Nations
Italian	control under protectorate, treaty or similar arrangement
French	railway, by 1939
Spanish	★ active nationalist resistance to European rule
Portuguese	
Belgian	
Dutch	

products for export

carpets	onions	wine
citrus fruits	palm oil	wool
coconut products	petroleum and petroleum products	cocoa
cotton and cotton seed	P phosphates	coffee
dates	potash	ground nuts
foodstuffs	rice	palm products
fruit	rubber	gold
gum arabic	sugar	M manganese
jute	T tea	diamonds
maize	teak	iron ore
nuts	Tn tin	C copper
oil seeds	tobacco	sisal
olive oil	wheat	

GANDHI
Mohandas Gandhi became the most famous figure of the Indian independence movement. Born in 1869, he trained as a barrister at London's Inner Temple. Initially, he made his reputation defending the interests of Indians in South Africa. During the Boer War and the First World War he raised and ran an Indian ambulance unit. In 1918 he became leader of the Indian nationalist movement, preaching non-violent resistance (satyaghra). In 1922 he was imprisoned for six years by the British authorities, but was released in 1924 on becoming president of the Indian National Congress. He led the campaign

to boycott British cotton imports, and in 1930 encouraged widespread resistance to the salt tax by distilling salt from sea water. He was interned again in 1931 and 1933. Following the last spell in prison, Gandhi withdrew from formal politics. His followers called him the "Mahatma", the "Great Soul", and even in retirement he cast a great influence on the nationalist movement. During the Second World War he again advocated non-cooperation, and was a key figure in the eventual negotiations for Indian independence and partition. On 30 January 1948 he was assassinated by a Hindu fanatic for his part in the religious break-up of India.

THE USSR FROM LENIN TO STALIN: WAR AND REVOLUTION

For many anti-imperialist politicians the beacon they followed was the Soviet Union. The triumph of communism in the Russian empire demonstrated the power of popular politics directed to a clear revolutionary ambition. The new communist state was regarded as a rallying point for all those struggling against exploitation and imperial rule. To Sidney Webb, the veteran British socialist, the Soviet Union was the "New Civilization", an island of progress amidst a sea of reaction.

The reality was very different. The infant revolutionary movement became isolated internationally and was almost stifled in Russia itself. Russian communists were a minority in the new state and had to fight to establish their political survival. The first priority was to end the war. At Brest-Litovsk in March 1918 the new Russian leaders had to concede extensive territorial losses, including the Baltic states and the Ukraine. By July 1918 full-scale civil war had broken out. Anti-Bolshevik forces were supported by contingents of foreign troops, some of them, like the Czech Legion, prisoners of war fighting to get back to Europe, others sent from abroad to get Russia back into the war and to overthrow communism. British, French, American and Japanese forces, together with the so-called "White" Russian armies, succeeded in controlling large areas of the old Russian empire, leaving the Bolsheviks with the Russian heartland around Moscow and Petrograd. By 1919 the survival of the new state was in the balance.

The Bolsheviks won the civil war in 1921 only by imposing a brutal dictatorship on the areas they controlled, and by militarizing both the Party and society. Five million were called into the Red Army set up at the start of 1918. Over 50,000 former tsarist officers, under the command of Leon Trotsky, commissar for defence, fought for the Reds. Under a system of "War Communism" all Russian businesses, except the very smallest, were nationalized and the "money economy" widely suspended. Grain was requisitioned, workers regimented. Bolshevik Russia became an armed camp, and the new Party members who joined after 1918 did so in most cases via the army. The civil war also forced Lenin's government to run the state from above. All potential sources of opposition, from separatist movements in the non-Russian areas to workers' groups demanding greater democracy, were ruthlessly crushed. The soviets, instead of becoming the instruments of democratic participation, were sidelined. The regime set up a new secret police force, the Cheka, in November 1917, which murdered and imprisoned anyone accused of counter-revolutionary activity. When the civil war finally ended in 1921, the political system had crystallized into a virtual one-party state, ruled from above by the Council of People's

Commissars in tandem with the Central Committee of the Communist Party.

When the dust of the civil war settled the re-named Soviet Union was in chaos. Much of the area ceded at Brest-Litovsk had been recaptured. But the economy was close to collapse, with raging inflation, falling grain production and a shrinking urban and industrial population. By 1920 only 1.5 million factory workers remained of the 3.6 million in 1917. The peasantry had retreated into subsistence agriculture, seizing the estates and converting 99 per cent of the land area into old-fashioned communes. Among the peasantry fewer than one per cent of households had a Communist Party member.

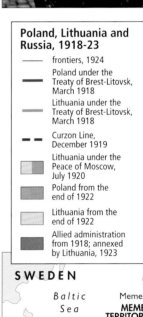

The Transcaucasus, 1918-23

—— frontiers, 1914	controlled by Armenia, Azerbaijan, Georgia and British forces, May 1919
to the Ottoman empire under Treaty of Brest-Litovsk, Mar. 1918	—— disputed between Armenia and Azerbaijan, Jul. 1919- Aug. 1920
Transcaucasian SSR at declaration of independence, Apr. 1918	----- northern frontier of Turkey under Treaty of Kars, Oct. 1921
to the Ottoman empire under Treaty of Batum, June 1918	—— Transcaucasian SSR, Mar. 1922
Batum-Baku railway; under British control, Nov. 1918-Aug. 1919	→ Ottoman advances, with dates
	→ Red Army advances, with dates

Poland, Lithuania and Russia, 1918-23

—— frontiers, 1924
—— Poland under the Treaty of Brest-Litovsk, March 1918
—— Lithuania under the Treaty of Brest-Litovsk, March 1918
– – Curzon Line, December 1919
Lithuania under the Peace of Moscow, July 1920
Poland from the end of 1922
Lithuania from the end of 1922
Allied administration from 1918; annexed by Lithuania, 1923

The Polish national state created in 1918 had at first very ill-defined borders (map right). The commander of the Polish armies, General Josef Pilsudski (above right), pictured playing patience, had ambitions to re-create the Polish state of the 18th century, including much of White Russia and the Ukraine. War broke out between Polish and Bolshevik forces in the spring of 1919 and continued until the decisive defeat of the Red Army before the gates of Warsaw in August 1920.

After the fall of the tsar a Transcaucasian Federal Republic was set up independent of Russia (map left). It was invaded and occupied by the Turks until May 1918, when Georgia, Armenia and Azerbaijan set up separate independent states under British and French protection. When the British left in December 1919, the Bolshevik armies began the slow reconquest of the region.

The World Between the Wars

For the urban radicals who led the revolution, the prospects for turning a peasant society into a modern workers' state were bleak. Revolution failed everywhere else outside Russia. In 1921 Lenin reluctantly proposed economic reforms designed to restore private trade and production and to create an orthodox central banking system. The reforms became known as the New Economic Policy (NEP). Lenin saw them as a necessary retreat in order to rebuild Russian industry and stabilize the social order after six years of war. When Lenin died in 1924, the Party was left with the unhappy compromise of a radical socialist political system trying to rule a deeply conservative, peasant-dominated society.

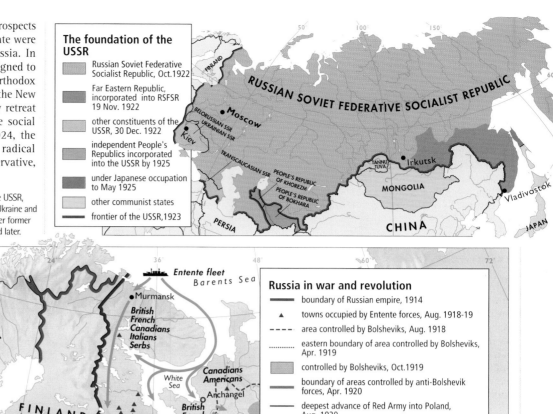

The foundation of the USSR

- Russian Soviet Federative Socialist Republic, Oct.1922
- Far Eastern Republic, incorporated into RSFSR 19 Nov. 1922
- other constituents of the USSR, 30 Dec. 1922
- independent People's Republics incorporated into the USSR by 1925
- under Japanese occupation to May 1925
- other communist states
- frontier of the USSR, 1923

A 1922 treaty established the USSR, linking Russia, Belorussia, the Ukraine and Transcaucasia *(map right).* Other former tsarist territories were absorbed later.

Following the Bolshevik revolution in November 1918, the former tsarist empire broke down into civil war *(map right).* The battle lines remained exceptionally confused. The communist Red Armies fought against the "White" armies, which were made up of a mixture of monarchists, anti-communists and nationalists, who wanted the independence of the non-Russian peoples. Foreign armies also intervened from north, east and south, while former Czech POWs fought the communists along the Trans-Siberian railway in their effort to return to Europe. A fragile peace came in 1921 with the final defeat of anti-communist forces.

Russia in war and revolution

- boundary of Russian empire, 1914
- ▲ towns occupied by Entente forces, Aug. 1918-19
- area controlled by Bolsheviks, Aug. 1918
- eastern boundary of area controlled by Bolsheviks, Apr. 1919
- controlled by Bolsheviks, Oct.1919
- boundary of areas controlled by anti-Bolshevik forces, Apr. 1920
- deepest advance of Red Army into Poland, Aug. 1920
- final anti-Bolshevik advance, Oct. 1920
- boundary of Soviet territory, Mar. 1921
- movement of White Russian armies
- movement of non-Russian anti-Bolshevik forces
- anarchist military activities
- ⊙ centres of Confederation of Anarchist Organizations
- O Makhno's headquarters, 1918-20
- frontiers, 1923

THE USSR FROM LENIN TO STALIN: MODERNIZATION AND TERROR

In the period after Lenin's death, the Communist Party in the Soviet Union searched for ways to modernize the new state while retaining the momentum of revolutionary progress. There was some hope that a new socialist international organization, Comintern, set up in Moscow in 1920, would encourage revolution outside Russia, but that prospect seemed a distant one by 1924. The Party divided into those who believed that NEP would gradually produce a successful urban economy which would evolve through Party guidance into a socialist system, and those who urged rapid modernization from above before communism was swamped by peasant capitalism.

This "Great Debate" was resolved by the drift of internal Party politics. The moderates around Bukharin, who probably represented the majority in the Party and in the country, became the enemies of the Party General Secretary, Joseph Stalin (pictured *far right*). He was appointed in 1922 after years of distinguished revolutionary service. He used his powers of patronage to build up a power base in local Party branches, from where he then strengthened his political position in the national leadership following Lenin's death. He thought there was little prospect of revolution abroad and argued for a strategy of "socialism in one country". This strategy, in Stalin's view, was to push through a second revolution from above, stamping out any vestiges of political pluralism or cultural diversity in the name of a rigorous communism, while at the same time

The maps of the Russian village of Sof'ino near Moscow *(below)* show the pattern of agricultural change from pre-war large estates to the establishment of communal peasant ownership in the 1920s and finally the state collective farm of the 1930s. The return to old-fashioned strip farming is evident in the 1920s. The Soviet regime saw the collectives as a means to modernize agriculture. Propaganda lorries, carrying party enthusiasts, toured the farm areas, explaining the reform.

The collectivization drive took a terrible toll on the rural population and its livestock. In the Ukraine – the Soviet "breadbasket" – a deliberate famine was induced in 1932-3 *(far right centre)*, designed to break the resistance of the Ukrainian peasantry and punish Ukrainian nationalism. Millions died and agricultural output went into sharp decline.

transforming Soviet society and the economy to match the communist model. This meant rapid industrial and urban growth, and an end to free-market peasant agriculture.

The trigger for the second revolutionary wave came in the winter of 1927-8 when falling grain supplies to the cities coincided with a series of war scares prompted by a breach in Anglo-Soviet diplomatic relations. Stalin threw his weight behind those who argued for rapid industrialization, the enforced modernization of agriculture and a build-up of military power to defend the revolutionary achievement. The Bukharin moderates were defeated and Stalin's vision of a communist reform imposed. In 1929 the first Five-Year Plan for industrial development was launched. In the summer of the same year, following an unsuccessful campaign to persuade the peasantry to adopt large-scale "collective" farming, a revolutionary wave was unleashed against the countryside. The chief victims of collectivization were the so-called "kulaks", the richer and more successful farmers, but all peasants suffered the widespread destruction of their traditional way of life. Communal farming was replaced by labour for the state; millions of peasants were shipped to work in cities or on the

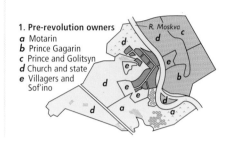

1. Pre-revolution owners
a Motarin
b Prince Gagarin
c Prince and Golitsyn
d Church and state
e Villagers and Sof'ino

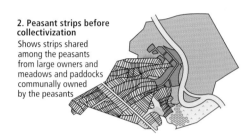

2. Peasant strips before collectivization
Shows strips shared among the peasants from large owners and meadows and paddocks communally owned by the peasants

In the late 1920s the Soviet regime embarked on a colossal experiment in social engineering and forced industrial growth. The object was to turn the Soviet Union from a mainly peasant society into a mainly urban-industrial one within ten years *(map below)*. In 1926 81% of Russians worked on the land. In 1939 the figure was only 52%. The city population expanded by 30

million from 1926 to 1939. For those who opposed the reforms or who were deemed enemies of the people, forced labour camps were set up in the far north and in Siberia. Up to ten million suffered imprisonment and exile, working on vast building projects in terrible conditions, while possibly as many as three million were executed.

During the 1930s every effort was made to encourage the participation of workers and peasants in building the new Soviet society. Model workers, known as Stakhanovites after the miner Stakhanov, who exceeded his work norms by exceptional efforts, were the new heroes of the Soviet state *(above)*. In the drive to modernize, almost 1.5 million new members were brought into the Communist Party.

By 1933 pressure had grown to root out elements regarded as corrupt or incompetent in a nationwide purge. The committee sitting in judgement here in May 1933 *(left)* was one of thousands which interviewed party members. Some 792,000 were expelled in the 1933 purge. In all, over one million were purged during the 1930s (chart *far right*).

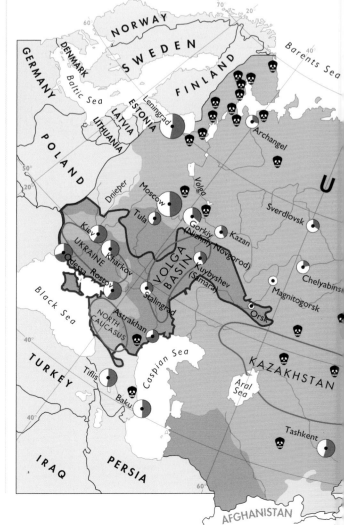

new construction sites; religion was suppressed as contrary to communist progress. By 1937 as many as 93 per cent of peasant households were collectivized.

The revolution from above had many consequences. The Soviet Union did become a major industrial power by the late 1930s. But above all the modernization drive allowed Stalin to complete his progress towards personal dictatorship. To push through the reforms it became necessary to rely on increased state power. The police forces of the interior ministry, the NKVD, were used indiscriminately against anyone, even those in the Party, deemed an enemy of Stalinist modernization. The regime adopted a propaganda of frantic revolutionary endeavour and isolated those who stood outside it as saboteurs and capitalist spies. The "cult of personality" built Stalin up as the supreme hero of the revolution. Those who opposed him, such as Bukharin, were forced to confess to outlandish crimes in a series of staged show trials in 1937 and 1938. Stalin arranged the death of almost all the old cadre of Bolsheviks. The tally of those who died was in excess of 12 million. By 1940 the Soviet Union was unarguably more modern, but the New Civilization was sustained by a vicious despotism.

SOVIET RE-ARMAMENT

In the early 1930s the Soviet Union embarked on a large-scale re-armament drive. Soviet leaders were aware that the security of the socialist state depended on defence in a world of hostile capitalist powers. In 1927 a major war scare, prompted by Britain's breaking of diplomatic relations, influenced Stalin's decision to go for all-out industrial growth. Priority was given to armaments industries. Between 1930 and 1940 the Soviet Union was transformed from a minor military power to the world's largest, at least on paper, with production of over 10,000 aircraft, 2,700 tanks and 15,000 artillery pieces. In the process, the foundation for the post-war Soviet super-power was laid. The price was high. Not only were Soviet living standards in the 1930s appalling, the rush for re-armament created a serious regional imbalance with over 90% of the Soviet aircraft industry in western Russia in areas near to any likely war zone. The military expansion also brought conflict between army and state. In 1937 Stalin, worried about a military coup, arrested the senior generals and put them on trial, including the architect of military revival, Marshal Mikhail Tukhachevsky. About 30,000 of the 75,000 officers in the Red forces were purged and executed or imprisoned, fatally weakening the readiness of the Red Army when war came in 1941.

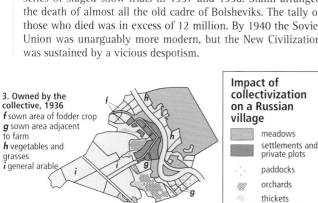

3. Owned by the collective, 1936
f sown area of fodder crop
g sown area adjacent to farm
h vegetables and grasses
i general arable

Impact of collectivization on a Russian village

- meadows
- settlements and private plots
- paddocks
- orchards
- thickets

Collectivization and population movements, 1923-39

principal areas of collectivization
- 2-10% of all farms collectivized by 1928
- 25-50% of all farms collectivized by 1933
- 50-70% of all farms collectivized by 1933
- 70-85% of all farms collectivized by 1933
- principal famine areas, 1932
- *Kraslag* labour camp administration zone
- ☠ corrective labour camps, 1932

city populations by census of Jan. 1939
(in thousands)
over 1 million | 500-1,000 | 250-500 | 100-250 | 50-100

○ population by census of Dec. 1926
◑ growth of population Dec. 1926 to Jan. 1939

— frontiers, 1930

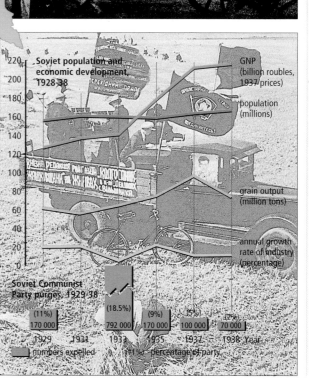

Soviet population and economic development, 1928-38

GNP (billion roubles, 1937 prices)
population (millions)
grain output (million tons)
annual growth rate of industry (percentage)

Soviet Communist Party purges, 1929-38

Year	1929	1931	1933	1935	1937	1938
percentage of Party	(11%)	(18.5%)	(9%)	(5%)	(7%)	
numbers expelled	170 000	792 000	170 000	100 000	70 000	

BOOM AND SLUMP

The growth of the Soviet economy in the 1930s contrasted sharply with the rest of the capitalist world. The depression that gripped the world economy in 1929 produced the biggest business slump of modern times. Economic crisis undermined confidence in the survival of capitalism as an economic system, just as the Russian Revolution had hit confidence in the survival of the liberal political order.

The seeds of the economic crisis can be found buried in the more prosperous decade which followed the end of the war. For many the 1920s were years of boom, though not for all. After the initial disruption caused by massive demobilization in 1919, the world economy began to expand again on the lines of the pre-war years. The core of that expansion was the American industrial economy. The war turned the United States into the world's banker and the world's largest trader. During the 1920s the United States continued to boom. Industrial production increased by 45 per cent between 1922 and 1929 in a country that was already the largest industrial producer. Industrialization also made strides in those areas cut off from European imports during the war. The flow of US funds helped to drag the more backward parts of the world economy towards the modern age.

The boom years were sustained by a whole range of new products – cars, radio, cinema, chemicals – and by the development of large industrial corporations, like Ford or Dupont, which used the most up-to-date techniques in management and production. Scientific management, first developed by the American engineer Frederick Taylor, transformed the productive performance of industry and introduced time-and-motion studies to the shop floor. Improvements in efficiency boosted profits and lowered prices, releasing more money for investment and encouraging a boom in consumer durables, such as cars.

The American-led boom disguised many surviving weaknesses. The pre-war trading economy, based on the gold standard and free convertibility, proved impossible to resurrect fully in the 1920s. Free trade, the hallmark of the prosperity of the 19th century, gave way to widespread protectionism as states sought to shield the living standards of their own populations. Three of the major pre-war industrial economies were, for different reasons, unable to play the part in the world economy that they once did. Germany revived slowly from the war; even by 1928 her trade was a third smaller than in 1913. The Soviet Union played only the smallest part in the world economy. Finally Britain, the hub of the old global economy, was weakened by the financial cost of the war and the rise of overseas competitors. British trade never recovered to the level of 1913 and Britain pumped far less capital into the world economy than she did before 1914.

The United States replaced Britain as the primary lender. Over $6 billion were invested overseas between 1925 and 1929. Unlike in Britain, however, the US economy had a weak international finance sector. Many of the loans were speculative or short-term, subject to sudden recall. Moreover the US economy was self-sufficient in many commodities, unlike the British, and so imported less, blunting the trade growth of other states. If the American boom was good for Americans, the impact abroad was more mixed.

The fragility of the new boom was glaringly exposed when in October 1929 a sustained speculative investment bubble finally burst. American creditors began to call in overseas loans, causing panic among debtors who had re-loaned the money for long-term projects, as had been the case in Germany. As the arteries of world finance began to silt up, the US economy reacted by putting up prohibitive tariffs in the 1930 Hawley-Smoot Act, thereby cutting off imports from economies desperate to pay back dollar debts. Competitive protection followed, cutting world trade by almost two thirds of its value between 1929 and 1932. Prices and profits collapsed, and in the United States, Germany and half a dozen other developed states industrial output sank by more than a third in three years. In a general atmosphere of *sauve qui peut* international economic collaboration broke down, and the major capitalist economies withdrew increasingly into economic nationalism.

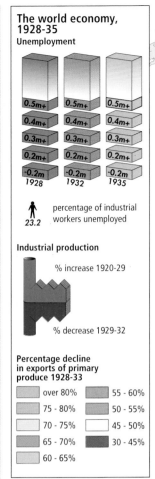

The world economy, 1928-35

Unemployment

	1928	1932	1935
0.5m+			
0.4m+			
0.3m+			
0.2m+			
-0.2m			

23.2 percentage of industrial workers unemployed

Industrial production

% increase 1920-29

% decrease 1929-32

Percentage decline in exports of primary produce 1928-33

over 80%	55 - 60%
75 - 80%	50 - 55%
70 - 75%	45 - 50%
65 - 70%	30 - 45%
60 - 65%	

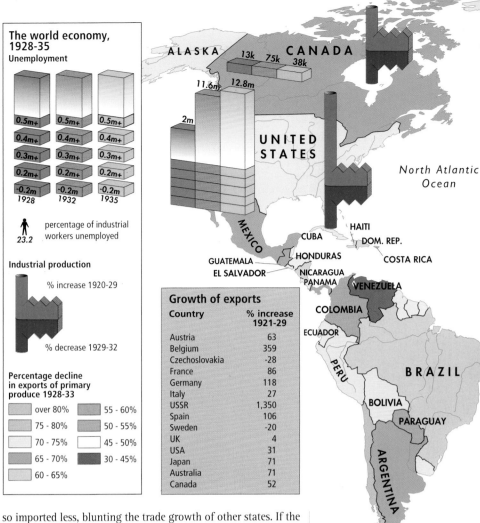

Growth of exports

Country	% increase 1921-29
Austria	63
Belgium	359
Czechoslovakia	-28
France	86
Germany	118
Italy	27
USSR	1,350
Spain	106
Sweden	-20
UK	4
USA	31
Japan	71
Australia	71
Canada	52

Decline in exports

Country	% decrease 1929-32
Austria	-65
Belgium	-53
Czechoslovakia	-64
France	-60
Germany	-57
Italy	-54
USSR	-37
Spain	-65
Sweden	-47
UK	-50
USA	-70
Japan	-62
Australia	-54
Canada	-60

THE WALL STREET CRASH

No event symbolized better the world's slide into depression in the 1930s than the spectacular collapse of the American share market in October 1929. The Wall Street Crash was a heart attack for the world economy. Its roots lay in the remarkable growth of American industry in the boom years of the mid-1920s. The insatiable demand for cars, radios, aviation and entertainment fuelled a largely unsupervised rush for shares in the new profit boom. The number of shares listed on the New York Stock Exchange rose between 1925 and 1929 from 113 million to more than one billion. Unscrupulous traders and middlemen lured small investors with promises of riches; but even respectable corporations speculated irresponsibly in a future that seemed limitless and assured. The country was gripped by speculation fever. Then, in September 1929, came unmistakable signs that the economy was approaching a downturn. The smart investors began to pull out before the fall. What began as a trickle became in the last week in October a flood, as banks, corporations and brokers realized that the bubble was bursting. On Black Tuesday, 29 October, the collapse came, suddenly and completely. In a day of frantic, sometimes violent trading over 16 million shares were sold off at a loss of $10 billion – twice the value of money in circulation in the US. When the stock market slide finally ended in July 1933, $74 billion had been wiped off share values and millions of Americans faced ruin.

The Dow Jones Index 1926–36

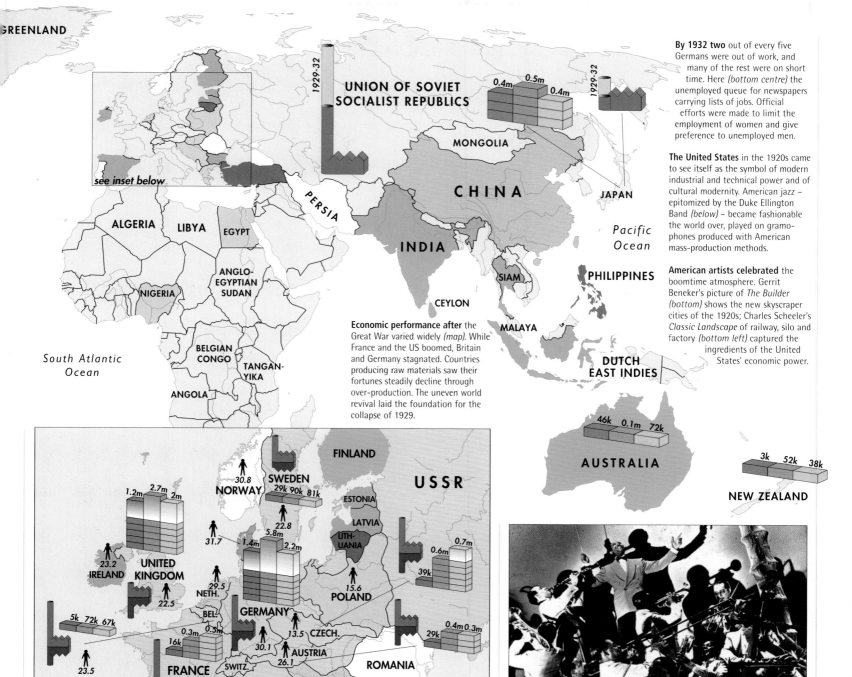

GREENLAND

1929-32

UNION OF SOVIET
SOCIALIST REPUBLICS

0.4m 0.5m
0.4m

1929-32

MONGOLIA

CHINA

JAPAN

ALGERIA LIBYA
EGYPT

PERSIA

Pacific Ocean

INDIA

ANGLO-
EGYPTIAN
SUDAN

NIGERIA

SIAM

PHILIPPINES

CEYLON

MALAYA

BELGIAN
CONGO

South Atlantic Ocean

TANGAN-
YIKA

ANGOLA

DUTCH
EAST INDIES

Economic performance after the Great War varied widely *(map)*. While France and the US boomed, Britain and Germany stagnated. Countries producing raw materials saw their fortunes steadily decline through over-production. The uneven world revival laid the foundation for the collapse of 1929.

By 1932 two out of every five Germans were out of work, and many of the rest were on short time. Here *(bottom centre)* the unemployed queue for newspapers carrying lists of jobs. Official efforts were made to limit the employment of women and give preference to unemployed men.

The United States in the 1920s came to see itself as the symbol of modern industrial and technical power and of cultural modernity. American jazz – epitomized by the Duke Ellington Band *(below)* – became fashionable the world over, played on gramophones produced with American mass-production methods.

American artists celebrated the boomtime atmosphere. Gerrit Beneker's picture of *The Builder (bottom)* shows the new skyscraper cities of the 1920s; Charles Scheeler's *Classic Landscape* of railway, silo and factory *(bottom left)* captured the ingredients of the United States' economic power.

46k 0.1m 72k

AUSTRALIA

3k 52k 38k

NEW ZEALAND

see inset below

— INSET MAP —

FINLAND

1.2m 2.7m 2m

30.8
NORWAY

SWEDEN
29k 90k 81k

USSR

ESTONIA

31.7

22.8

LATVIA

23.2
IRELAND

UNITED
KINGDOM

1.4m 5.8m 2.2m

LITH-
UANIA

0.6m 0.7m

22.5

29.5
NETH.

15.6
POLAND

39k

5k 72k 67k

BEL.

GERMANY

13.5 CZECH.

0.4m 0.3m

23.5

0.3m 0.5m

30.1

29k

16k

AUSTRIA

FRANCE

SWITZ.

26.1

ROMANIA

24.3

YUGOSLAVIA

BULGARIA

1m 1m

SPAIN

21.3

ITALY
0.3m

TURKEY

GREECE

UNEMPLOYMENT WORLDWIDE

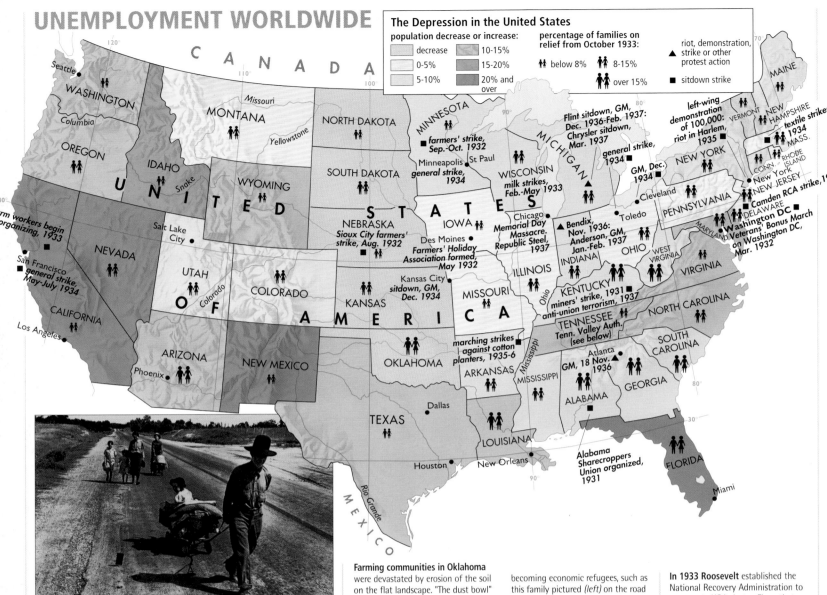

The Depression in the United States

population decrease or increase:
- decrease
- 0-5%
- 5-10%
- 10-15%
- 15-20%
- 20% and over

percentage of families on relief from October 1933:
- below 8%
- 8-15%
- over 15%

- ▲ riot, demonstration, strike or other protest action
- ■ sitdown strike

farm workers begin organizing, 1933

San Francisco general strike, May-July 1934

farmers' strike, Sep.-Oct. 1932

Minneapolis general strike, 1934

Sioux City farmers' strike, Aug. 1932

Des Moines Farmers' Holiday Association formed, May 1932

Wisconsin milk strikes, Feb.-May 1933

Chicago Memorial Day Massacre, Republic Steel, 1937

Kansas City sitdown, GM, Dec. 1934

Bendix, Nov. 1936; Anderson, GM, Jan.-Feb. 1937

Flint sitdown, GM, Dec. 1936-Feb. 1937; Chrysler sitdown, Mar. 1937

GM, Dec. 1934

left-wing demonstration of 100,000; riot in Harlem, 1935

general strike, 1934

textile strikes, 1934

New York

Camden RCA strike, 1936

Washington DC Veterans' Bonus March on Washington DC, Mar. 1932

miners' strike, 1931 anti-union terrorism, 1937

Tenn. Valley Auth. (see below)

marching strikes against cotton planters, 1935-6

GM, 18 Nov. 1936

Atlanta

Alabama Sharecroppers Union organized, 1931

Farming communities in Oklahoma were devastated by erosion of the soil on the flat landscape. "The dust bowl" which resulted forced many into becoming economic refugees, such as this family pictured *(left)* on the road in Pittsburg County in 1938.

The most conspicuous element of the slump between 1929 and 1933 was unemployment and the miserable poverty that marched with it. In the developed world alone by 1932 there were over 24 million out of work. At its peak unemployment affected one in four of the American workforce; in Germany the figure by 1932 was two in every five. At its worst almost nine million Germans were thrown out of work in a workforce of 20 million. For many unemployment lasted for years; millions of young workers spent the first years of their working lives on the dole.

The figures for the registered unemployed told only part of the story. Many of the long-term unemployed disappeared from the registers as their entitlement to unemployment relief lapsed. Many women left work without being registered. Those in work found themselves on short-time with reduced earnings. In the less developed regions of the world unemployment was disguised by the return of peasant workers from mines and plantations to their villages, where households survived on the edge of subsistence. The direct result of unemployment was a desperate poverty. Even in those states with welfare systems the scale of unemployment soon exhausted the relief budgets. As state revenues fell it proved impossible to give adequate welfare. In cities across Europe and the United States there sprang up hundreds of unofficial charitable projects giving out a hot meal or bread and fuel for starving families. Working class communities fell back on their resources, those in work helping those without.

The Depression was a social catastrophe for which governments were ill-prepared. In a pre-Keynesian age it was assumed that the state could do little to alleviate the crisis until the market revived of its own accord. Not until the very end of the slump, when the damage had been done, did governments recognize that recovery was only possible with more state

intervention. In the United States the election of Franklin Roosevelt as president in 1932 brought to office a man committed to the idea of a state-led revival. A package of welfare and economic reforms, known collectively as the New Deal, was introduced into Congress. Under the National Recovery Act of 1933 a programme of public works was established to restore output and employment. By the end of the decade over $10 billion had been spent and 122,000 public buildings, 664,000 miles of road, 77,000 bridges and 285 airports constructed.

In Europe recovery was patchy. It went furthest in Germany following Hitler's appointment in 1933 as chancellor. A package of state-backed programmes soaked up many of the unemployed, including a new system of multi-lane motorways,

In 1933 Roosevelt established the National Recovery Administration to regenerate US industry. Firms that co-operated with its labour code and minimum wage rates were allowed to display the Blue Eagle symbol *(right)*. One of the largest projects of the revival authority was the Tennessee Valley Authority, set up in 1933 to build state-sponsored hydro-electric power *(below)*, and to revive the agriculture of the area. The Norris Dam *(below right)* was the first dam built, completed in 1936 and named after Senator George Norris, who had championed state utility development in the 1920s. Although there was strong opposition to the TVA from other private utility operators, all the projects were completed in ten years.

The Tennessee Valley Authority: dam-building projects
➤ *Norris* dam

In four years German unemployment was almost eliminated *(maps right)*. Re-armament and conscription soaked up some of the jobless, but the main cause lay with state efforts to revive employment through work creation projects, the building of the new motorways and investment in house-building, canals and other major construction projects. Unemployment fell fastest among construction workers and farm labourers. Marriage loans, introduced in 1933, were designed to remove women from the job market. The compulsory state labour service, for men and women, took more than 250,000 18-year-olds off the unemployment totals. By 1939 unemployment was down to just 33,000 – most of whom were largely unemployable.

The Depression in the United States created poverty on a massive scale. In 1934 some 17 million Americans lived on relief, many in the poorer southern states *(map left)*. Even by the late 1930s there were still over eight million unemployed. The Depression produced mass migration from the impoverished small farm states of the Mid West and South to new industrial regions in the Far West, Florida and to established northern industrial cities. Others obtained temporary work from the Works Project Administration, which provided 500,000 temporary jobs from public funds.

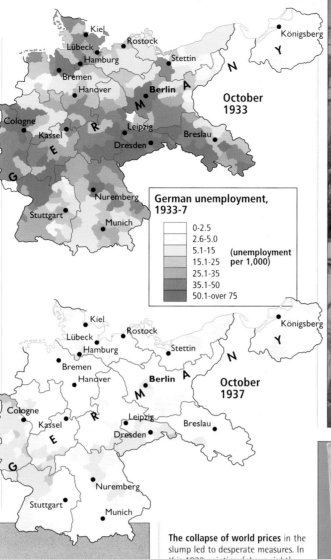

German unemployment, 1933-7

☐	0-2.5
☐	2.6-5.0
☐	5.1-15
☐	15.1-25 (unemployment per 1,000)
☐	25.1-35
☐	35.1-50
■	50.1-over 75

October 1933

October 1937

NRA MEMBER — U.S.

WE DO OUR PART

The collapse of world prices in the slump led to desperate measures. In this 1932 painting *(above right)*, Brazilian coffee is being thrown into the sea in order to try to keep up its price. The coffee crisis hit the rich elite of São Paulo, who lost their grip on political power in Brazil in 1930 to the cattle ranchers of the south.

DEPRESSION IN BRITAIN
During the 1930s, the effects of the slump in Britain were most damaging in the older industrial areas of northern England, Scotland and Wales. In the south and the Midlands the revival was led by new industries: motor-vehicles, radio, aviation. A wave of housebuilding in the south allowed higher earners to move from the crowded city centres to tree-lined suburbs – symbolized by the poster (left) for suburban north London, served by the newly built Metropolitan railway.

METRO-LAND
PRICE TWO-PENCE

the *Autobahnen*. Investment was pumped into house-building, agriculture and rearmament using state deficits to jerk the industrial economy into life. By 1936 the pre-Depression position was restored; by 1939 the German economy was one third larger than in 1929. In Britain, home of free trade, the state began to institute a higher level of economic management, backed by a new tariff structure established at the Commonwealth Conference in Ottawa in August 1932. Home demand expanded rapidly as the British consumer benefitted from cheap food imports. For those still employed, the 1930s brought a remarkable consumer boom.

The gradual recovery of national economies did little to stimulate the world market. The 1930s saw a wave of competitive protectionism designed to stimulate domestic production and to avoid reliance on an uncertain export economy. The term "autarky" was coined to describe a policy of economic self-sufficiency. In Germany a Four-Year Plan launched in October 1936 aimed to reduce Germany's dependence on overseas supply by producing synthetic oil, rubber and textiles and by mining low-grade German ores. In 1938 a large programme of import substitution was set up in Japan, centred around synthetic oil. Much of world trade was reduced to a simple barter system. International efforts to combat the recession were confined to a World Economic Conference in London in 1933 whose failure highlighted the changed outlook of post-Depression governments.

There existed a widespread belief that the days of liberal economics and global trade were over, to be replaced by state-regulated economic development and small self-contained trading blocs. Even in the United States, where the idea of an open world economy still had powerful advocates, state intervention and tariff protection were the key features of the Depression decade.

Unemployment Worldwide

STATES OF THE "NEW ORDER": JAPAN

The political consequences of the slump were profound. The breakdown of international collaboration and the collapse of world trade undermined willingness to sustain the post-war settlement. The search for a new economic order, based on economic nationalism, was soon followed by efforts to construct a new international political order to replace that established by Britain, France and the United States in the early 1920s. The driving force behind this New Order was a triumvirate of states, Germany, Italy and Japan, where militaristic nationalism was in power, trading on resentment towards other races and harbouring ambitions to launch a new wave of imperial conquests.

Japan exemplified this profound dissatisfaction with the existing order. In the 19th century, following a revolution in 1868 against the old feudal system, Japanese leaders made a sustained effort to imitate the West in order to build a strong and prosperous state. Modern industry was adopted; the armed forces were reformed along European lines; European styles of dress and European culture were imitated. In 1902 Japan consolidated her new international role when she signed a treaty of friendship with Britain. Two years later she defeated imperial Russia at sea and on land and became the major power in East Asia (see page 38). Japan saw herself as an imperial power, like Britain or France. By 1910 Japan had acquired control of Korea and Taiwan, and a string of islands and mainland bases, which were treated like colonial possessions. When war came to Europe in 1914, Japan joined the Allies and sat at the Versailles Conference as one of the five major powers. At the Washington Conference, which opened in November 1921, she signed a Four-Power Treaty with Britain, France and the US confirming the existing territorial settlement in the Pacific region, and in 1922 was party to a Nine-Power Pact guaranteeing the sovereignty and independence of the Chinese Republic.

The policy of integration with the West had brought Japan great gains, but it also provoked widespread criticism at home

When Emperor Hirohito *(below, at his coronation)* was crowned in 1926, he chose the word *Showa* – Enlightened Peace – as the emblem of his reign. A studious man, who loved marine biology and ballroom dancing, he found himself ruling a Japan that plunged into domestic political violence and military expansion abroad.

During the 1930s Japan embarked on a programme of imperial expansion into China *(map right)*. Encouraged by the army and by patriotic associations in Japan, the Japanese government approved the army's initiative in seizing the Chinese province of Manchuria in 1931. A year later Japanese forces threatened the port of Shanghai, with its large European population. After a restless peace, full-scale war came in 1937, which brought Japanese forces down the eastern seaboard, threatening European colonies in the Far East.

The World Between the Wars

66

Japanese expansion from 1914, (left)

▨	Japanese territory, 1914
▨	spheres of Japanese influence in 1918
▨	expansion to 1933
▨	expansion to Dec. 1941
▨	spheres of Japanese influence in 1941
✳	Japanese conflict with USSR
→	Japanese attacks
■	Chinese capitals
▪	Allied bases

Pu Yi Hsüan-t'ung (1906-67) was the last emperor of the Chinese Manchu or Ch'ing dynasty *(above)*. When the Japanese captured Manchuria they renamed the state Manchukuo, and in 1934 made Pu Yi the puppet emperor of the new state. He ruled formally through Chinese-run "self-government committees", but in practice power lay with the Japanese army.

Silk was among the most important of Japan's exports in the 1920s. The silk-workers *(above)* are separating the cocoons from the leaves. When recession came in 1929 the silk trade declined sharply and millions of small farmers faced ruin. The crisis provoked a wave of anti-Western nationalism.

On 18 September 1931 Japanese soldiers of the Manchurian Kwantung army blew up a short stretch of line on the South Manchurian railway near Mukden, run by Japan. The "incident" was taken as the opportunity to extend Japanese control *(map right)* over an area rich in mineral resources, particularly coal and shale oil. Chinese resistance was vigorous but sporadic. By 1 March 1932 the area was under Japanese control.

Japanese military spending *(chart below)* increased slowly in the 1930s until 1936-7 when it more than quadrupled in a single year.

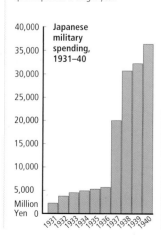

Japanese military spending, 1931–40 (Million Yen)

among a new generation of young nationalists, who wanted to reassert traditional Japanese culture and values, and who rejected what they saw as Japan's humiliating dependence on the West. When the slump of 1929 shattered the Japanese silk industry and closed the door to overseas trade, the nationalists, prominent in Japan's armed forces, agitated for a new direction in Japan. The military leadership began to dominate the cabinet and override parliament; military radicals assassinated hundreds of politicians and businessmen with Western links; radical nationalists mobilized popular support for Japanese imperialism in Asia.

The obvious area for Japanese expansion was mainland China. In September 1931, in defiance of the Washington agreements, the Japanese army in Manchuria, stationed there to protect Japan's economic interests in the region, seized the whole province for the Japanese empire. Though Japan had long enjoyed an extensive economic presence in Manchuria, its seizure was the first serious challenge to the League of Nations and the post-war settlement. Japan was censured by the League, and left it in 1933. In 1934 the nationalist politician, Prince Konoye Fujimaro, declared the Amau Doctrine, which amounted to a rejection of Western influence in China and the establishment of a new Asian order, centred on Japan.

The Western states, anxious about their own economic interests, did nothing to obstruct Japan. Following Manchuria, which was turned into a puppet kingdom under the Manchu Pu Yi Hsüan-t'ung – who had been the last Ch'ing emperor of China – Japan continued to put pressure on China for further concessions, while embarking on an extensive programme of rearmament. Elaborate schemes were drawn up to create a new regional economy, the "Greater East Asia Co-prosperity Sphere", with Japan at the core and a circle of other Asian and Pacific states tied to her economically and politically. In July 1937, following a clash between Chinese and Japanese forces in Peking, full-scale war was launched against China. Japan seized much of northern and eastern China, including the capital, Nanking, captured with huge loss of life in December 1937. In 1938 Japan announced that a New Order was in the making, which would restore Asia to the Asians.

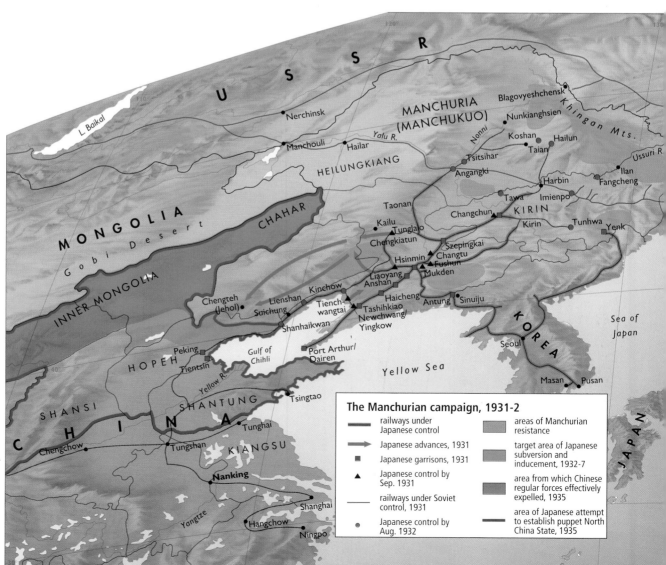

The Manchurian campaign, 1931-2

— railways under Japanese control
→ Japanese advances, 1931
■ Japanese garrisons, 1931
▲ Japanese control by Sep. 1931
— railways under Soviet control, 1931
● Japanese control by Aug. 1932

areas of Manchurian resistance
target area of Japanese subversion and inducement, 1932-7
area from which Chinese regular forces effectively expelled, 1935
area of Japanese attempt to establish puppet North China State, 1935

STATES OF THE "NEW ORDER": ITALY

Italy was too late to join the ranks of Europe's colonial powers. By 1900 she possessed only small territories in the Horn of Africa. Over the next 40 years the empire was extended in Africa and the Mediterranean *(map right)*. Libya was seized from Turkey in 1911. The Dodecanese Islands were taken over from Turkey the next year. A second wave of imperialism began in the 1930s with the conquest of Abyssinia in 1935-6 and the occupation of Albania, which the Italians regarded as a virtual protectorate, in April 1939. Mussolini had visions of turning the Mediterranean and North Africa into a new Roman empire.

Italy, like Japan, was a new power. Unified between 1859 and 1870, the Italian state, weak economically and militarily, remained marginal to the European power balance. In the two decades before 1914, Italy acquired a colonial empire in Africa and, like Japan, began to dream about becoming a major regional power. When in 1911 Italy defeated Ottoman forces in Libya, her ambitions turned towards the Middle East and the Mediterranean basin as a natural area of Italian dominance. The Great War exposed the feebleness of this ambition. Italy made few gains from the conflict, despite the promises made by Britain and France when Italy intervened on their side in 1915. Like the Japanese, Italian leaders found themselves compelled to work within an international framework manipulated by Britain and France.

Italy's weak international standing was one of the factors that fuelled the rise after the war of a radical nationalist movement, which found its chief expression in a small party of veterans led by an ex-socialist agitator, Benito Mussolini. His Fascist Party, founded in 1921, (named after the *fasces* – the bundle of rods and axes carried by those who executed the law in ancient Rome) attracted support from those Italians – veterans, students, intellectuals – frustrated with the existing parliamentary system and from businessmen hostile to Labour. The fascists developed a reputation for brutal anti-Marxist

THE ITALIAN COLONIAL EMPIRE

When Fascism came to power in Italy in 1922, it sought to revive Italy's weak colonial spirit. Italy had gained only minor colonies from the scramble for Africa: Eritrea and Southern Somaliland. The Libyan provinces of Tripolitania and Cyrenaica were conquered from Turkey in 1911, but by 1922 Italy controlled only a small coastal strip; the rest was lost to native Arab forces. In the whole Italian empire there were only 50,000 colonists (over 100,000 lived in the French colony of Tunisia). In the 1920s Italian colonial troops began a vicious programme of reconquest. In 1929 the new governor, Marshal Badoglio, had

the nomad people of Cyrenaica herded into concentration camps, while the Senussi rebels were hunted down and slaughtered. 20,000 died in the camps, and the bedouin population of Libya was halved. Fascist leaders planned to send ten million settlers to the colonies. With the conquest of Abyssinia in 1935-6. the whole state was to be turned into a rich source of food sustained by Italian peasants. Yet in 1940 only 854 Italian farmers worked in the East African colonies and their economies absorbed from Italy ten times what was sent in return. In 1937 and 1938 race laws were introduced to create an apartheid in the African colonies. By 1942 British troops had taken over Italy's whole African empire, ending a brutal chapter of imperialism.

Map: Italian empire, 1910-39

Legend:
- Italy and possessions, 1910
- acquired 1911-12
- acquired 1919
- acquired 1924-34
- acquired 1935-9
- temporary occupation
- → Italian invasion of Ethiopia, 1935
- → British sea route to India
- ⚓ Italian naval port
- ⚓ British naval port
- — frontiers, 1939

Map labels: PORTUGAL · SPAIN · FRANCE · SWITZ. · GERMANY · Trieste · Marseilles · Genoa · Fiume 1924 · Barcelona · Corsica · Rome · Zara 1919 · YUGOSLAVIA · Balearic Islands 1936 · Sardinia · ITALY · Naples · Adriatic Sea · ALBANIA 1939 · Gibraltar (Br.) · Sp. MOROCCO · MOROCCO (Fr.) · Algiers · Tunis · Taranto · Sicily · Corfu 1923 · GREECE · Athens · TURKEY · ALGERIA (Fr.) · TUNISIA (Fr.) · Malta · Mediterranean Sea · Crete · Dodecanese Islands 1912 · Cyprus · 1919-21 · Antalya · Tripoli · TRIPOLITANIA · CYRENAICA · Benghazi · Alexandria · PALESTINE · Port Said · LIBYA 1911 · 1919 · 1926 · Cairo · Suez · Sanussi resistance to Italian rule to 1934 · Nile · SAUDI ARABIA · FRENCH WEST AFRICA · FRENCH EQ. AFRICA · Kufra · 1919 · EGYPT · Aswan · Aozou Strip 1935 · 1934 · Port Sudan · Red Sea · Suakin · ANGLO-EGYPTIAN SUDAN · Khartoum · Asmara · ERITREA · Massawa · Gondar · Aksum · Makale · Mai Chio · Dessie · ABYSSINIA (ETHIOPIA) · ITALIAN EAST AFRICA · Addis Ababa · Diredawa · Harar · Segag · 1936 · Wabera · Negheli · KENYA (Br.) · JUBALAND ceded by Kenya to Italy 1925

The World Between the Wars

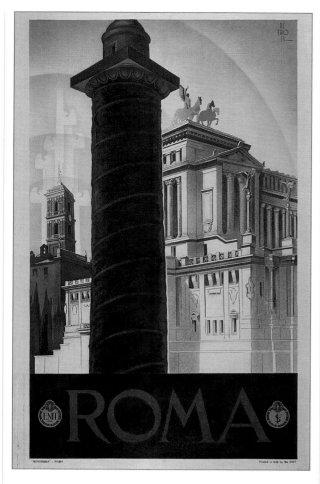

"NOVISSIMA" – ROMA   Printed in Italy by the ENIT

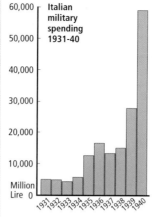

The conquest of Abyssinia was carried out with considerable brutality. Villages were bombed and Abyssinian soldiers and civilians were subjected to gas attack. It took seven months to defeat Abyssinia's ill-equipped forces with the loss of 1,537 Italian lives. On 9 May 1936 Mussolini officially declared the establishment of the new Italian empire. Abyssinia was forced into the colonial model. The Habsburg thaler currency was replaced by the lira. Peasants became labourers on Italian plantations and construction sites. In 1937 a guerrilla campaign was launched against the occupying power, which was savagely suppressed. The country was flooded with propaganda posters of Mussolini *(below)*, but few Abyssinians accepted their new rulers. In 1941 Italian forces were defeated in a matter of weeks by British empire troops and Emperor Haile Selassie restored to the Abyssinian throne.

within the League system in the 1920s. Mussolini was concerned with the establishment of domestic power. In 1926 Italy became a one-party state, with Mussolini as its undisputed leader, or *Duce*. By mobilizing business support, the economy was stabilized and labour unrest died down. In 1929 the Lateran Accords signed with the Papacy ended the conflict between church and state that dated back to 1870. In foreign policy Mussolini sought to have Italy taken seriously as a great power. Italian foreign policy was respectable. Italy worked within the League and supported the Locarno Pact of 1925, which re-affirmed the territorial settlement in western Europe from which Italy had profited little. In April 1935 at Stresa, Mussolini hosted a meeting of British, French and Italian representatives, which was the last occasion on which the three European victor powers publicly endorsed their commitment to the survival of the Versailles Settlement.

By 1935, however, Italian foreign policy had taken a new turn. In the aftermath of the slump, Mussolini saw an opportunity from the domestic preoccupation of the other powers and the collapse of international collaboration. His nationalist supporters were keen for Italy to overturn her reliance on the West, and to develop an independent and imperial policy in the areas of historic Italian interest – North Africa, the Middle East and the Balkans. Mussolini was also anxious to keep the Fascist revolution on the boil; he saw himself as a new Caesar, called to build a second Roman empire.

For nationalists the natural area for empire-building was independent Abyssinia, where Italy had been humiliatingly defeated in 1896 and where there were extensive Italian economic interests. On 3 October 1935 Italian forces invaded Abyssinia, and by May had brought it under Italian rule. The League again faced a challenge from one of its most prominent members. Sanctions were imposed, which had little effect, but so alienated Italian opinion that Italy left the League in 1937. Mussolini now burned his boats. In July 1936 he committed Italian forces to help nationalist rebels in Spain overturn the republican regime in defiance of Britain and France. In 1937 Italy aligned herself publicly with Hitler's Germany when she signed the Anti-Comintern Pact (directed against international communism) to which Japan had already subscribed the previous year. From an attitude of co-operative internationalism, Italy had become by 1938 a power committed to challenging the status quo.

activism and for tub-thumping patriotism. Though with only 35 seats in the national assembly, Mussolini was invited by the king, on the advice of leading conservatives, to become prime minister in October 1922. His success was sealed by a mass demonstration of uniformed Fascist militia whose black-shirted members converged on Rome on 30 October, the day of Mussolini's appointment.

For all the nationalist rhetoric, Mussolini's Italy remained

Mussolini planned to turn Rome into the heart of the new empire *(poster above)*. Elaborate arrangements were made from the late 1930s for a magnificent Fascist Exhibition to mark the 20th anniversary of the Fascist achievement of power in 1922. Mussolini wished Rome to host permanently this monument to Fascist imperialism. The war put an end to the scheme.

General Italo Balbo (1896-1940) *(above)* was Italy's star aviator in the 1920s. He was made minister of aviation by Mussolini in 1926, and later became famous for flying the Atlantic. He was sacked in 1933 because of his popularity, and sent to be governor of Libya in 1934. In June 1940 he was shot down by Italian anti-aircraft fire while flying over Tobruk.

STATES OF THE "NEW ORDER": GERMANY

Italy's decision to move closer to Hitler's Germany in 1937 tied her to the most dangerous and powerful of the states seeking political revision in the 1930s. Hitler and his Nazi party colleagues made no secret before 1933 of their hostility to the Versailles Settlement. When Hitler was appointed chancellor on 30 January 1933 as part of a conservative scheme to stabilize the German political system following the political turmoil of the years of depression, Germany was ruled by a man convinced that she would rise again as a great power, and that he was the chosen instrument of destiny to achieve it.

For Hitler there were two sides to the idea of a New Order – a political and social revolution in Germany itself; and a revolution in the international order established at the end of the Great War. Within 18 months Germany was turned into a one-party state, dominated at every level by the Nazi Party and its numerous affiliated organizations and, following the death in 1934 of the president, Field Marshal von Hindenburg, a one-man dictatorship. Hitler merged the offices of chancellor and president, and declared himself to be simply the leader – *der Führer*. Trade unions were abolished, political prisons and a political police force, the Gestapo, set up. The latter was run by Heinrich Himmler who, by 1936, was in control of all the country's police and security services. In 1935 the first active steps were taken to remodel Germany racially with the notorious Nuremberg Laws denying Jews full civil rights. This was part of a more general programme of "racial hygiene", which included racial teaching in schools and the compulsory sterilization of the hereditarily ill and mental patients.

On the international front Nazi goals were less clear. There was general agreement in German society on the justice of overturning Versailles. Hitler wanted to create a pan-German state in central Europe and to remilitarize Germany. For many in the Nazi movement this was the limit of German ambition. But from the early 1920s Hitler had harboured the desire for a war of revenge which would turn Germany into a world power.

He had no fixed plan or blueprint, but his long-term goal was to build up German power to the point where Germany could carve out a large territorial empire in Eurasia and become an imperial super-power. Hitler regarded war as an integral feature of relations between states, to be welcomed rather than avoided. Nevertheless the early years of the Nazi regime saw a cautious approach to foreign policy, from fear of provoking other states while Germany was still relatively powerless. Rearmament began slowly, and was not publicly declared until 16 March 1935, when Hitler announced a new 36 division army, five times larger than the 100,000 force allowed in the Versailles Treaty. In March 1935 the Saarland returned to German sovereignty when 90.8 per cent of its citizens voted for union. In March 1936 German forces reoccupied the demilitarized zone of the Rhineland, which Allied forces had vacated in 1930. The Allies did nothing to maintain the post-war settlement on these issues, partly because of hostility to the risk of war among their home populations, partly because they privately recognized the futility of maintaining a punitive peace on questions that did not constitute a serious threat to western interests.

In 1936 Hitler stepped up the pace. New military programmes were authorized for the modernization and expansion of the armed forces, which were intended to make Germany the foremost military power in Europe by the early 1940s. In November 1937 Hitler announced a new course in German foreign policy: union with Austria and possible war with Czechoslovakia to return three million Sudeten Germans to German rule. In March 1938 German forces entered Austria and an *Anschluss*, or annexation, was imposed. In May 1938 Hitler ordered plans for war with the Czechs in the autumn. This did alarm the West. Although Britain and France pressured the Czechs to concede German occupation of the Sudetenland in October 1938, both states prevented outright conquest. At the Munich Conference in September 1938 – the first occasion on which Versailles was revised through discussions on German soil – Hitler backed away from full-scale war. By the end of 1938 he had achieved much of his domestic and international ambition. Germany was poised to begin a more radical revision of the international order.

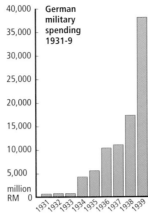

German military spending 1931-9

40,000 / 35,000 / 30,000 / 25,000 / 20,000 / 15,000 / 10,000 / 5,000 / million RM 0

1931 1932 1933 1934 1935 1936 1937 1938 1939

Germany's military spending *(chart above)* increased rapidly after 1935, buttressing Hitler's foreign policy ambitions. In spring 1938 he turned his attentions to Czechoslovakia *(maps below left)*, where the large German population in the west gave him a pretext for intervention. At Munich in September 1938, Britain and France acquiesced to German occupation of the Sudetenland. Six months later, Hitler seized the rest of the Czech lands, leaving a rump Slovakia as a Nazi client state, while Hungary and Poland each took a share of the spoils of what had been central Europe's most democratic and stable country.

Austrian forces were integrated with the German Wehrmacht and conscription was introduced following the *Anschluss*. In the town of Bregenz near Austria's border with Switzerland, German motorized troops can be seen *(left)* replacing the mounted Austrian guards in March 1938. Under Hitler German society was thoroughly militarized. Parties of soldiers or Nazi stormtroopers became part of everyday life *(below)*. Militarism helped to unite Nazism with the German military tradition.

The ethnic divisions of Czechoslovakia, 1938

- German
- Czech
- Polish
- Slovakian
- Hungarian
- Ukrainian

The Munich Agreement, 1938

- to Germany, Oct. 1938
- to Hungary, Nov. 1938
- to Poland, Dec. 1938
- frontiers, Dec. 1938

The dissolution of Czechoslovakia, 1939

- to Germany, Mar. 1939
- to Hungary, Mar. 1939
- independent, Mar. 1939
- frontiers, Apr. 1939

Austria had its own Nazi Party campaigning for a pan-German state until it was banned in 1934 following the Nazi murder of Chancellor Engelbert Dollfuss. Working in secret it established firm contacts with the Hitler government and agitated for union. In the spring of 1938 Hitler used the agitation as an excuse to send in German forces. Austrian Hitler Youth, pictured *(above)* at the time of the *Anschluss*, helped to Nazify Austria almost overnight.

HITLER

Adolf Hitler (1889-1945) was born in a small Austrian town, Braunau-am-Inn, the third son of an Austrian customs official. When he left school he drifted to Vienna in pursuit of a career in architecture or painting. He had sufficient talent for neither. When he was due for conscription he fled to Munich. Here in 1914 he joined the German army on the outbreak of war. He was wounded and gassed on the Western Front but, unlike almost all his early companions, he survived the whole conflict. He was recruited by the army to act as a political educator during demobilization, and quickly developed a reputation as an extraordinary speaker. In 1919 he joined a small radical nationalist party in Munich,

which he soon came to dominate under its new title, the National Socialist German Workers' Party. In 1923 he led an unsuccessful coup in Munich against the Weimar authorities. During the two-year jail sentence which followed he wrote his major work, Mein Kampf, in which he blamed the Jews and Marxists for Germany's problems and called for revenge for the defeat of 1918 by building a racially strong Germany and overturning the Versailles Settlement. He projected an image of the German messiah. In the Depression after 1929 this proved attractive to Germany's conservative masses. In January 1933 he assumed the chancellorship shortly after taking on German citizenship. The messiah became the "Leader" (Führer).

In the 1930s Germany, under Hitler's leadership, began to reverse the conditions imposed under the Versailles Settlement *(map above).* The Saarland returned to German control in 1935; the demilitarized Rhineland was occupied by German forces in March 1936 and, by 1939, a wall of fortifications (the Siegfried Line) had been built on the western border. In 1938 Austria was united with Germany, the dream of Austrian nationalists in 1919, achieved by the ex-patriate Austrian, Hitler. Three million Germans who had been Austrian subjects in 1918 lived under Czech rule. In 1938 Hitler planned to seize these German areas by force, but they were achieved by negotiation with Britain and France. The Sudeten Germans, who had organized themselves in the Sudeten German Party, with more than 1.3 million members in 1938, welcomed the unification with Germany with overwhelming enthusiam. Hitler entered the town of Asch shortly after the German occupation *(right).*

German expansion, 1935-9

Germany, 1935

Saarland, incorporated by plebiscite, 1935

Rhineland (demilitarized zone) reoccupied by Germany, 1936

— frontiers, 1937

Slovakia (client state of Nazi Germany), nominally independent from Mar. 1939

German annexations:

Mar. 1938

Oct. 1938

Mar. 1939

THE SPANISH CIVIL WAR

The struggle between an old and new political order in Europe came to be exemplified in the 1930s by the civil war which tore Spain apart between 1936 and 1939. It was a conflict widely regarded as a struggle between the forces of light – socialism and liberalism – and the forces of darkness – fascism and reactionary nationalism. In reality the issues at stake in Spain were far more complex.

In many respects Spanish politics and society in the early part of the century resembled Latin America. Spain, too, was dominated by a traditional rural elite, with a large mass of impoverished and landless rural workers. The Catholic Church was a major force, and the military had a tradition of intervention in Spanish politics. Spanish modernization was slower than in other parts of western Europe. Agriculture was the main source of livelihood even in the 1930s, while Spain relied on exporting food and raw materials. But, as in Latin America, the cities expanded from the late 19th century, particularly Barcelona, centre of Spain's limited industrial development. In the cities flourished a liberal middle class, keen to shift political power to the urban populations, anti-clerical in outlook, radical in their desire for social reform and effective democracy. There also developed a new urban proletariat, entirely hostile to the old order and attracted to the more radical wing of European socialism, anarchism and syndicalism.

The old political system was based on a parliament – the *Cortes* – which was dominated by the traditional elites, who rigged elections and stifled popular political participation. It was a system in decay. In 1918-20 there was widespread and violent political unrest – the *Trieno Bolshevista* – fuelled by the rural and urban poor. In 1923 the army overthrew the feeble constitutional monarchy of Alfonso XIII, and General Primo de Rivera became military dictator, committed to a programme of social and economic modernization. Unable to cope with the effects of the slump, de Rivera was overthrown in turn, and in April 1931 the urban radicals, in alliance with Labour, established a republic under the radical intellectual, Miguel Azaña.

The republic embarked on a thorough programme of reform directed against the Church, which was disestablished; the army, which was much reduced in size; and the landlords. The rural issue was the most bitterly contested. The September Law of 1932 aimed to transform Spanish agriculture by giving rural workers minimum wages, regular all-year employment and the chance to own land seized from absentee landlords. Within two years conservative opinion in Spain was mobilized in mass nationalist movements: the CEDA (Spanish Confederation of Autonomous Rights), the Nationalist Party and the fascist Falange. This bloc achieved power in the elections of November 1933 and set about reversing the reforms and enforcing landlord power with savage violence. The republican forces divided between the moderate liberals and socialists, who sought reform through parliament, and the radical socialists, anarcho-syndicalists and communists, who saw the conflict in revolutionary terms and met violence with violence.

Spain was hit hard by the recession and rural and urban poverty sharpened the political conflicts. A wave of political murders, church burnings and land seizures made Spain all but ungovernable by 1936. In the elections of February the forces of the republic, moderate and revolutionary, combined in a "Popular Front" to defeat the right at the polls. In July 1936, Nationalist army officers, fearful of the prospect of Bolshevik revolution, launched an abortive coup d'etat. Republican forces organized for a military show of strength, and for three years the two sides fought a bitter and vicious war, which left 600,000 dead.

The Nationalists were supported with arms and men by Italy and Germany; the Republicans obtained volunteer support from International Brigades organized overseas to fight "fascism". The Nationalists were not so much fascists as an alliance of conservative, clerical and nationalist forces with some fascist support. Their leader, General Franco, came from the tradition of *caudillismo*, or military dictatorship. He succeeded in welding Nationalist forces into a modern armed force. After early defeats the Nationalists captured Barcelona on 26 January 1939 and Madrid on 28 March, bringing the war to an end. Franco became head of an authoritarian regime committed to a strategy of modernization within a conservative framework.

The political map of Spain *(right)* roughly followed the pattern of land-holding. In the more prosperous and independent farming communities of the north was found the heartland of Spanish support for Catholicism and the conservative order. In the tenant farming areas of the south were found the supporters of anarcho-syndicalism, hostile to the landowners and the state. In the central area of large estates the labourers were predominantly socialist. In Catalonia there existed an independent movement for Catalan autonomy, dominated by the left-wing *Esquerra*. A statute of autonomy was granted to the Catalans in September 1932, but then rescinded when the right came to power. There was a separate Basque nationalist movement in the north, but this was dominated by clerical and nationalist movement groups hostile to the Republic.

Spanish political affiliations, 1931-6

anarchist, 1931	catholic-conservative, 1931
socialist, 1931	Popular Front, Feb. 1936

Spain between the wars was faced with a serious crisis on the land *(map right)*. Over-population, soil erosion and the slump in food prices added to the traditional tensions between landlords and the rural poor. The areas of greatest poverty in the south and south east were also the areas of greatest aridity. In the north of the country agriculture was more prosperous, and small peasant-owned farms were the norm. In the south small tenant farmers and sharecroppers prevailed. In the central regions large landlord estates had survived, worked by an army of impoverished labourers. In 1932 the Republic pushed through an agrarian law to improve wages and working conditions against fierce resistance. Against a background of world recession, the law achieved little. The Basque peasant family *(below)* watching Franco's troops in 1936 were still part of a traditional rural way of life.

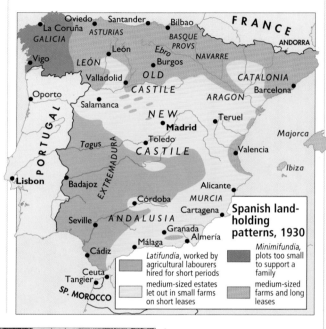

Spanish land-holding patterns, 1930

Latifundia, worked by agricultural labourers hired for short periods	*Minimifundia*, plots too small to support a family
medium-sized estates let out in small farms on short leases	medium-sized farms and long leases

The Church dominated much of Spanish life: by 1914 the Jesuit communities were said to own around one third of the country's wealth. As a result, there was a long tradition of violent anti-clericalism. During the Civil War churches were burned and desecrated and churchmen murdered. In the Carmelite church in Barcelona *(below)*, skeletons have been disinterred and stacked on the church steps.

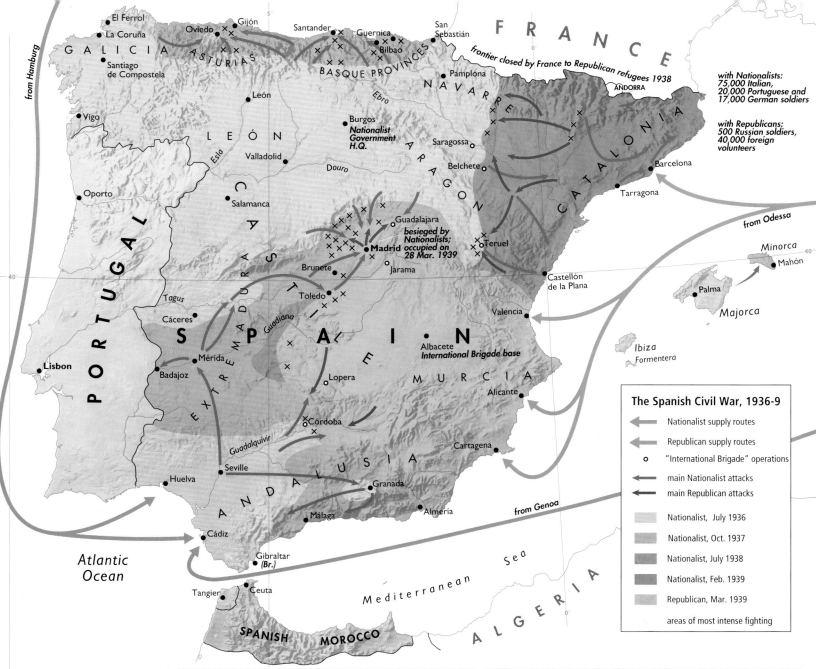

F R A N C E

frontier closed by France to Republican refugees 1938

ANDORRA

with Nationalists:
75,000 Italian,
20,000 Portuguese and
17,000 German soldiers

with Republicans:
500 Russian soldiers,
40,000 foreign
volunteers

from Hamburg

GALICIA

El Ferrol
La Coruña
Santiago de Compostela
Vigo
Oporto

ASTURIAS

Oviedo
Gijón
Santander
Guernica
Bilbao
San Sebastián
Pamplona

BASQUE PROVINCES

NAVARRE

LEÓN

León
Burgos
Nationalist Government H.Q.
Valladolid
Esla
Douro
Salamanca

ARAGON

Saragossa
Belchete

CATALONIA

Barcelona
Tarragona

from Odessa

Minorca
Mahón

CASTILE

Guadalajara
besieged by Nationalists; occupied on 28 Mar. 1939
Madrid
Brunete
Jarama
Toledo

Teruel

Castellón de la Plana

Valencia

Palma
Majorca

PORTUGAL

Lisbon

EXTREMADURA

Tagus
Cáceres
Mérida
Badajoz
Guadiana

S P A I N

Albacete
International Brigade base

Lopera
MURCIA

Ibiza
Formentera

Alicante

Guadalquivir

ANDALUSIA

Córdoba

Cartagena

Huelva
Seville
Granada
Málaga
Almería

from Genoa

Atlantic Ocean

Cádiz
Gibraltar (Br.)

Tangier
Ceuta

SPANISH MOROCCO

M e d i t e r r a n e a n   S e a

A L G E R I A

The Spanish Civil War, 1936-9

- Nationalist supply routes
- Republican supply routes
- ○ "International Brigade" operations
- main Nationalist attacks
- main Republican attacks

Nationalist, July 1936
Nationalist, Oct. 1937
Nationalist, July 1938
Nationalist, Feb. 1939
Republican, Mar. 1939
areas of most intense fighting

The bombing of Guernica by the German "Condor" Legion on 26 April 1937 became the symbol of Nationalist atrocity in the Civil War and of the horror of modern war. This French anti-fascist poster by Pierre Mail *(right)* captures the sense of outrage at the bombing of defenceless civilians. The German air force learnt the lesson that bombing cities did not achieve much militarily, and concentrated on air-army co-operation thereafter.

FRANCO

Francisco Franco (1892-1976), the victor of the Spanish Civil War, was the dominant figure in Spanish history from the 1930s to the 1970s. Born in Galicia, Franco became a career soldier, serving in Spanish Morocco in the wars of the 1920s. He rose rapidly under the New Republic after 1931, becoming army chief-of-staff in 1935, and then governor of the Canary Islands. He was one of a group of army officers hostile to the Republican efforts to secularize and modernize Spanish society. Fearful of a communist coup in the summer of 1936, Franco played a key part in the conspiracy which resulted in the attempted military coup of 17 July. Franco raised the revolt among the Moroccan garrison and flew the troops to mainland Spain. The coup became a prolonged war. In September, with the death of the more senior General Sanjurjo, the rebel generals chose Franco as their leader. He was declared head of a new Nationalist State on 21 September, and on 1 October proclaimed himself the Caudillo – military chief of the new Spain. In spring 1939 he became ruler of all Spain, and imposed a harsh dictatorship which cost the lives of 200,000 Spaniards, who were executed between 1939 and 1943.

The Spanish Civil War

CHINA BETWEEN THE WARS

Civil war between nationalists and communists was not confined to Europe. In the 1930s China was plunged into a war between rival political factions until the common threat of Japanese aggression in 1937 brought an uneasy truce. The roots of the Chinese civil war went back to the revolution of 1911 (*see page 22*). The overthrow of the Manchu emperors led to a period of turmoil throughout China as Chinese political forces struggled to find a post-imperial state which could command widespread allegiance.

The first Chinese president, Yüan Shih-k'ai, elected in 1912, though nominally in favour of a constitutional democracy, by 1915 turned his office into a virtual dictatorship, based on the military force of the northern Chinese generals. When he tried to make himself emperor in 1916 his allies deserted him. His death a few months later ushered in a period of warlord rule which lasted until 1927. China fragmented into a number of military dictatorships whose forces fought among themselves for regional advantage. In the absence of a settled central government the other powers maintained the privileged position they had enjoyed under the emperors, dominating Chinese trade, customs, railways, even the post office, while enjoying extraterritorial rights on Chinese soil.

The rise of warlordism and the continued presence of foreigners prompted a nationalist revolt in the 1920s. The call for national unity and sovereignty was loudest in China's universities, where students demanded social reform. The May 4th Movement, named after a demonstration by Peking students in May 1919, sparked a wave of strikes and boycotts which were crudely suppressed. The "New Culture Movement" that followed produced a period of intense intellectual debate on the path of modernization China should follow.

Two major political groups emerged from the debate. The first was based on Sun Yat-sen's National People's Party (Kuomintang), first founded in 1912 and revived by Sun in 1924; the second was the Chinese Communist Party, set up in July 1921. The two co-operated on a shared anti-imperialism, the communists winning support among the working classes of the main ports and the Kuomintang recruiting from among the educated urban classes and native Chinese businessmen of

the south. Following Sun's death in 1924, a Kuomintang government was set up in Canton as a rival to the government in Peking dominated by the northern warlords. Sun had learned from the warlord era and the Kuomintang had its own trained army by the mid-1920s, run by a young officer, Chiang Kai-shek, who by 1925 was the leading figure in the movement.

In July 1926 Chiang began a year-long war against the north – the so-called Northern Expedition – which led a year later to the consolidation of much of China under one regime, based at Chiang's new capital at Nanking. Up to this point Kuomintang and communists co-operated, but Chiang's fear of a broader social revolution turned him against communism. In 1927 his forces destroyed the communists in the major cities. However, one young communist leader, Mao Tse-tung, kept resistance alive in the province of Kiangsi and when Chiang attacked his group in 1934, the fragments of the Chinese movement trekked 6,000 miles to the northern province of Shensi. In the 1930s Chiang became undisputed leader of the new national China; sovereignty was largely restored, though China remained reliant on western help. The social issue of China's millions of poor peasants and workers remained unresolved.

In China in the 1920s the secret of political success was armed power. The Christian general Wu P'ei-fu trained his men in traditional martial arts *(below)*, which were useless against modern weapons. Chiang Kai-shek, seen *(below left)* with two other warlords, Feng Yü-hsiang and Yen Hsi-shan, was the most successful of the provincial generals.

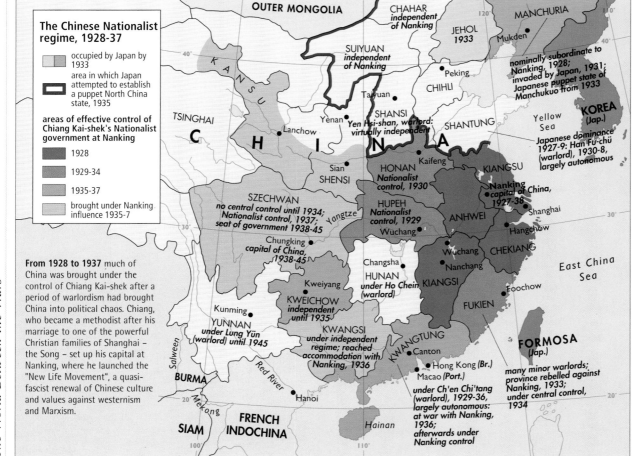

The Chinese Nationalist regime, 1928-37

- occupied by Japan by 1933
- area in which Japan attempted to establish a puppet North China state, 1935

areas of effective control of Chiang Kai-shek's Nationalist government at Nanking

- 1928
- 1929-34
- 1935-37
- brought under Nanking influence 1935-7

From 1928 to 1937 much of China was brought under the control of Chiang Kai-shek after a period of warlordism had brought China into political chaos. Chiang, who became a methodist after his marriage to one of the powerful Christian families of Shanghai – the Song – set up his capital at Nanking, where he launched the "New Life Movement", a quasi-fascist renewal of Chinese culture and values against westernism and Marxism.

MAO

Mao Tse-tung (1893-1976) emerged during the warlord era as the most prominent Chinese communist. Born to wealthy peasant parents in Hunan, Mao was largely a self-taught student. In 1913 he entered Changsha Normal College to train as a teacher. Here he came into contact with the political debates going on about the future of Chinese society after the 1911 revolution. His early views were liberal, based on European individualist philosophy. Even when in 1919 he first discovered socialism, he read almost no Marx or Lenin. His socialism derived from sources which emphasized popular democracy and small communities. As an active communist in the 1920s, Mao found himself at odds with other leaders, who wanted to reproduce the

Shanghai, pictured (above) in 1927, was the centre of the crisis between Chiang's Kuomintang movement and the popular union and communist movements. In April 1927 foreign trade concessions in the city were threatened by angry striking workers. A right-wing secret society, the Green Gang, was armed by Europeans and unleashed on the strikers, ending radicalism in Shanghai until Mao's revolution.

Russian revolutionary experience in China. Maoism emphasized much more reliance on the popular masses and the importance of cultural change in bringing about revolutionary transformation. When, in the late 1920s, communism was destroyed in China's cities, Mao set up a rural-based Chinese soviet state in Kiangsi province, but was driven out by Nationalist forces in 1934. His supporters followed him in a long and destructive march to the north, where in Yan'an a new communist stronghold was established. Here Mao finally outmanoeuvred many of his opponents. In the Rectification Movement (1941-5) Mao imposed his ideas about organizing the peasantry and modernizing China. Two years later, following the civil war, Mao became communist China's first leader.

On 4 May 1919 3,000 students from Peking University protested against China's subjection to Japanese demands at the Versailles Conference (map right). The protests became nationwide, with boycotts, strikes and a run on the banks. The Chinese government of Tuan Ch'i-jui bowed to the protests and refused to sign the treaty.

China, 1920

▨	Japanese empire, 1920
▨	international zones, 1920
○	treaty ports by 1920

foreign ports, occupied areas and spheres of influence:

▨ ●	British
▨ ●	Japanese
▨ ●	French
●	Portuguese

the May 4th Movement

■	student demonstrations and strikes
▲	boycott movement and commercial strikes
♦	worker demonstrations
✕	worker strikes

In 1926 the Kuomintang and the communists allied to launch a military campaign against the northen warlords to unify the country (map right). Despite their numerical inferiority, Chiang's forces defeated the divided warlord armies and brought much of central China under nationalist rule.

The Northern Expedition, 1926-8

▨	area controlled by Fengtien faction (Chang Tso-lin)
▨	area controlled by Kuo-min-chün (Feng Yü-hsiang)
▨	area controlled by Chihli faction (Sun Ch'uan-fang)
▨	area controlled by Chihli faction (Wu P'ei-fu)
▨	Kwangsi clique (warlords)
▨	T'ang Chi-yao, warlord of Yunnan and Kweichow
▨	area controlled by Kuomintang
→	main Kuomintang forces
⇢	minor Kuomintang forces
→	Yen Hsi-shan (warlord of Shansi 1912 onwards)
→	Kuo-min-chün

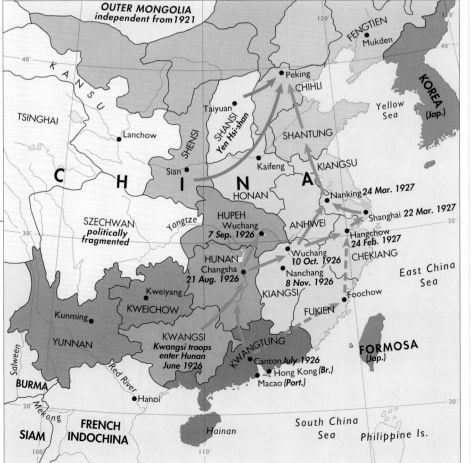

DEMOCRACY AND DICTATORSHIP

The history of China between the two wars exemplified the problem of adapting to modern politics in the vacuum left by the collapse of an old imperial order. The assumption held by most radical modernizers was that some form of parliamentary democracy was the natural successor to old-fashioned authoritarianism. In reality most democratic experiments soon collapsed. China was democratic for a brief moment in 1912 when a president was popularly elected; Russia's short taste of democracy in the elections of 1917 was crushed in the civil war; Turkey held free elections in 1920, but from 1924 Kemal Atatürk's Republican People's Party operated a one-party system.

Democracy survived little longer in post-war Europe. In 1920 most European states were parliamentary democracies in imitation of the victorious democratic powers. But one by one the new democratic regimes gave way to dictatorships. The first transformation was in Italy, where the Fascist Party leader, Benito Mussolini, became prime minister in a right-wing coalition, then in 1926 head of a one-party state. Spain followed suit in 1923 when the military seized power. A brief democratic interlude between 1931 and 1936 ended in civil war and the reassertion of military dictatorship under Franco (see pages 72-3). In Poland the military seized power in 1926 under Marshal Pilsudski, and was run by the so-called "Colonels' Group".

Democracy collapsed in the Baltic States between 1926, when Antanas Smetona seized power in Lithuania, and 1934 when Konstantin Päts imposed one-party rule in Estonia. In Portugal a weak parliamentary state was overturned in 1926, and in 1932 Antonio de Salazar, leader of the National Union Party, became dictator. Hungary was ruled by Admiral Horthy's National Union Party from 1919, and Austria was turned into an authoritarian state in 1933 under Engelbert Dollfuss until it was absorbed by Hitler's Germany five years later. Greece became a military dictatorship in 1935, Bulgaria in 1936. Romania became a royal dictatorship under King Carol in 1938, and a military dictatorship three years later. The only one of the new post-war democracies to survive as such was Czechoslovakia, and even this fell victim to German expansion in 1939. By the time Germany became a single-party dictatorship in the summer of 1933, democracy was already deep in crisis. By 1939 it survived as a political form only in Britain, France, the United States, the Low Countries, Scandinavia and a handful of British Dominions.

The failure of democracy had many causes. There developed in the 1920s a powerful movement against liberal politics, which were seen as serving the interests of the wealthy western elites rather than meeting the needs of the masses. Democracy gave the masses the chance to express their hostility to the old elites, but neither the new mass right nor left was particularly democratic in outlook except where parliamentary government was well entrenched, as in Britain and France. The new authoritarian parties made a striking contrast with liberal organizations. They were militaristic, violent, active: the endless rallies, marches and rituals gave them an appeal which staid parliamentary politics lacked in a period of crisis and transition. They also provided a source of status and power to those who lacked wealth and social position. They were genuinely populist movements, led by men such as Stalin and Mussolini, the sons of craft workers or, like Hitler, the son of a clerk. Dictators, left or right, imposed consensus and persecuted opponents in order to build their version of the modern state.

At the beginning of the 1920s most European states were democracies. By 1939 most were dictatorships, some fascist, some nationalist, some royalist (map below). The new states created at Versailles were faced with numerous difficulties in establishing a modern parliamentary system. The slump exacerbated domestic political tensions. Europe's population began to move towards radical extremes – communist, nationalist or fascist. Democracy survived in Britain and France, the Low Countries and Scandinavia, but in these too there developed native fascist and communist movements which threatened democratic stability. The most disruptive change came in Germany. The electoral triumph of Nazism (maps centre and below right), which secured 44% of the vote in March 1933, encouraged radical right-wing movements in the rest of Europe. Hitler and his closest associates, seen here greeting Nazi Reichstag deputies in August 1932 (bottom right), used the means of democracy in order to subvert it. Elsewhere in the 1930s democracy was overthrown by coup d'etat or military violence rather than by mass political mobilization.

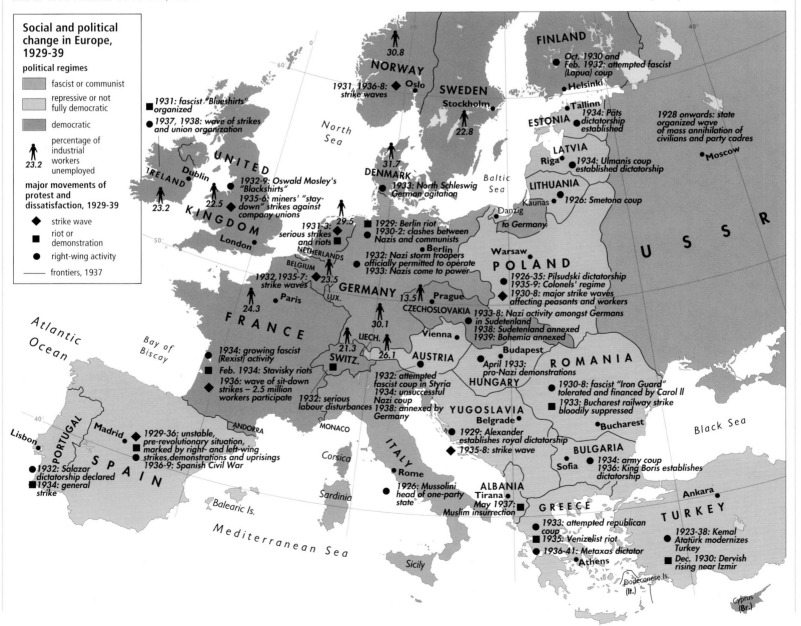

Social and political change in Europe, 1929-39

political regimes

- fascist or communist
- repressive or not fully democratic
- democratic

23.2 — percentage of industrial workers unemployed

major movements of protest and dissatisfaction, 1929-39

- ◆ strike wave
- ■ riot or demonstration
- ● right-wing activity
- — frontiers, 1937

1931: fascist "Blueshirts" organized
1937, 1938: wave of strikes and union organization
1932-9: Oswald Mosley's "Blackshirts"
1935-6: miners' "stay-down" strikes against company unions
1931-3: serious strikes and riots
1932, 1935-7: strike waves
1934: growing fascist (Rexist) activity
Feb. 1934: Stavisky riots
1936: wave of sit-down strikes – 2.5 million workers participate
1932: serious labour disturbances
1929-36: unstable, pre-revolutionary situation, marked by right- and left-wing strikes, demonstrations and uprisings
1936-9: Spanish Civil War
1932: Salazar dictatorship declared
1934: general strike

1931, 1936-8: strike waves ◆ Oslo
1930.8
Oct. 1930 and Feb. 1932: attempted fascist (Lapua) coup ● Helsinki
1934: Päts dictatorship established ● Tallinn
1928 onwards: state organized wave of mass annihilation of civilians and party cadres ● Moscow
1934: Ulmanis coup established dictatorship Riga
1926: Smetona coup Kaunas
1933: North Schleswig German agitation
1929: Berlin riot
1930-2: clashes between Nazis and communists
1932: Nazi storm troopers officially permitted to operate
1933: Nazis come to power ● Berlin
1926-35: Pilsudski dictatorship
1935-9: Colonels' regime
1930-8: major strike waves affecting peasants and workers Warsaw
1933-8: Nazi activity amongst Germans in Sudetenland
1938: Sudetenland annexed
1939: Bohemia annexed
1932: attempted fascist coup in Styria
1934: unsuccessful Nazi coup
1938: annexed by Germany
April 1933: pro-Nazi demonstrations
1930-8: fascist "Iron Guard" tolerated and financed by Carol II
1933: Bucharest railway strike bloodily suppressed
1929: Alexander establishes royal dictatorship
1935-8: strike wave
1934: army coup
1936: King Boris establishes dictatorship
1926: Mussolini head of one-party state
May 1937: Muslim insurrection
1933: attempted republican coup
1935: Venizelist riot
1936-41: Metaxas dictator
1923-38: Kemal Atatürk modernizes Turkey
Dec. 1930: Dervish rising near Izmir

30.8 · 31.7 · 22.8 · 23.2 · 22.5 · 29.5 · 24.3 · 23.5 · 13.5 · 30.1 · 21.3 · 26.1

FINLAND · NORWAY · SWEDEN · Stockholm · ESTONIA · LATVIA · LITHUANIA · Danzig to Germany · North Sea · Baltic Sea · IRELAND · Dublin · UNITED KINGDOM · London · NETHERLANDS · BELGIUM · LUX. · Paris · FRANCE · DENMARK · GERMANY · Prague · CZECHOSLOVAKIA · Vienna · LIECH. · SWITZ. · AUSTRIA · Budapest · HUNGARY · POLAND · U S S R · Atlantic Ocean · Bay of Biscay · ANDORRA · MONACO · Corsica · Madrid · Lisbon · PORTUGAL · SPAIN · Balearic Is. · Mediterranean Sea · Sardinia · ITALY · Rome · Sicily · YUGOSLAVIA · Belgrade · ROMANIA · Bucharest · BULGARIA · Sofia · ALBANIA · Tirana · GREECE · Athens · Black Sea · Ankara · TURKEY · Cyprus (Br.) · Dodecanese Is. (It.)

The World Between the Wars

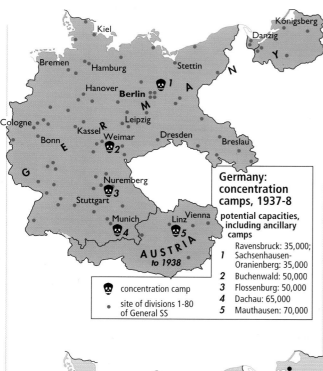

Germany: concentration camps, 1937-8

potential capacities, including ancillary camps

1 Ravensbruck: 35,000;
  Sachsenhausen-Oranienberg: 35,000
2 Buchenwald: 50,000
3 Flossenburg: 50,000
4 Dachau: 65,000
5 Mauthausen: 70,000

☠ concentration camp

• site of divisions 1-80 of General SS

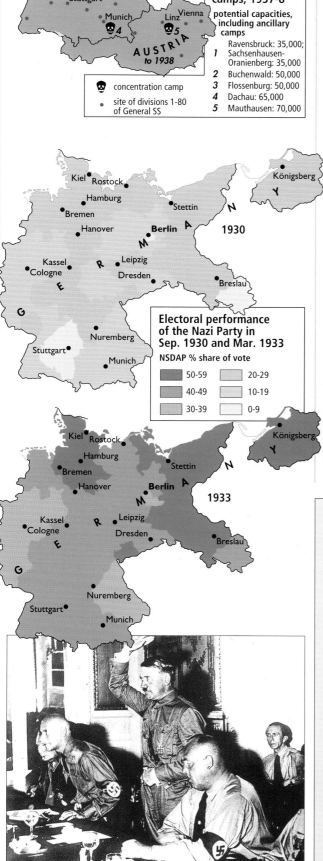

1930

Electoral performance of the Nazi Party in Sep. 1930 and Mar. 1933

NSDAP % share of vote

50-59	20-29
40-49	10-19
30-39	0-9

1933

The secret of the Nazi Party success in the 1930s was to organize a parallel state based on the party regions or *Gaue* and to create institutions – youth groups, labour service, the Labour Front – which forced ordinary Germans to collaborate or face party hostility or worse. By 1945 1.3 million Germans had been imprisoned at one time or another for political crimes *(map above left)*.

All over the world political movements became militarized between the wars. Here *(below left)* members of the Japanese "Black Dragon" secret society, responsible for political assassinations in the 1930s, march past the Imperial Palace. In Britain Sir Oswald Mosley's fascist Blackshirts imitated Europe's fascists. Mosley is seen here inspecting female members in July 1935 *(below right)*.

In the mid-1930s left-wing political forces began to collaborate in so-called "Popular Fronts" uniting social-democrats, liberals and communists in the fight against fascism. This poster *(above)* was part of the French communist campaign for the Popular Front election of 1936, which the left narrowly won. The victory of the Spanish Popular Front the same year provoked the Civil War.

FRANCE IN CRISIS, 1934

In the early 1930s the French parliamentary system faced a serious crisis. The economic recession bit deep. Workers' wages fell by up to one third. The extremes on right and left began to attract more support. The French Communist Party's membership rose from 30,000 in 1931 to 285,000 five years later. An array of small nationalist and quasi-fascist groups emerged, known collectively as the "Leagues". Few were truly fascist and the largest, the Croix de Feu, was an association of war veterans. In 1934 the parliamentary regime was rocked by a serious scandal, the Stavisky Affair. The suicide of a small-time swindler exposed corruption in high places. On 6 February the Leagues called for anti-government demonstrations. Thousands gathered in the Place de la Concorde and a pitched battle followed (right) with police and soldiers in front of the Chamber of Deputies. In all 15 were killed and 1,326 injured. The government resigned and a government of National Solidarity was installed to rally democratic forces. Stability was slowly restored, but French politics was poisoned for the rest of the decade by bitter conflicts between extremes of right and left.

OVERTURE TO THE SECOND WORLD WAR

The rise of authoritarian dictatorships in states with a vested interest in revising the international order established in the early 1920s posed a severe challenge to the democratic powers. The sharp polarization between liberal democracy, communism and fascism produced a period of intense crisis in the 1930s, a form of European civil war. By 1939 the optimism of the peacemakers of 1919 gave way to a growing fatalism about the inevitability of another great war.

The international order which the democratic states set up after the Great War contained fundamental flaws. Two major powers, the US and the USSR, effectively stood outside it, though both possessed great potential weight in the balance of power. The United States refused to join the League of Nations or to ratify the Versailles Treaty, and in 1937 isolationist opinion at home forced neutrality legislation through Congress. The Soviet Union was never fully integrated within the post-war order, even during a brief period as a League power between 1934 and 1939. Germany also stood outside the system, whose treaty structure most Germans wished to overturn. In the 1920s when Germany was disarmed these ambitions did not matter, but when Germany became a real military force again in the 1930s under Hitler, she was a permanent threat to the status quo.

In effect the post-war order was dominated by Britain and France. Neither power was in a position to afford the military and economic effort required to defend the status quo, particularly in the 1930s when their priority was to rebuild their economies after the recession and to avoid domestic social conflict. The system always depended on the collective goodwill of other states, and when that evaporated in the 1930s Britain and France found themselves overstretched defending global empires with shrunken resources. The weakness of their position encouraged them not to run risks in foreign policy and to try to meet the revisionist states halfway. The policy of "appeasement", as it became known, was a recognition of realities by statesmen who knew that their electorates were hostile to war and their forces too weak to restrain aggression by force.

Nonetheless, in 1936 both Britain and France began to rearm in order to deter potential aggressors and to be able to defend their empire if it came to war. The British prime minister, Neville Chamberlain, and his French counterpart, Edouard Daladier, both knew that to defend the status quo they must argue from a position of strength. By 1939 the combined output of British and French tanks and aircraft exceeded that of Germany. Germany was seen in the West as the key to the international crisis, but Italy and Japan, if allied to her, constituted a formidable threat in the Mediterranean and Far East. Between 1936 and 1939 neither Britain nor France was yet in a position to face these multiple threats with any real force.

In early 1939 both the British and French governments came to realize that unless they could curb German expansion, both the post-war settlement (and the balance of power and

The Prussian city of Danzig at the mouth of the River Vistula became the cause of war in 1939. In 1919 under the terms of the Versailles Settlement, Danzig was created a free city under the League of Nations. The object was to allow Poland access to the Baltic here for trade. A corridor of Prussian territory between Danzig and the rest of Germany was given to the Poles for the same reason. The Corridor contained substantial German minorities. However, instead of using Danzig, the Poles developed the port of Gdynia as their outlet to the sea, while the Danzig enclave, populated largely by Germans, came to be dominated in 1933 by a Nazi government. In 1939 Hitler's Germany asked for the return of the city *(map below)*. Poland refused to revise the League settlement. Hitler prepared for war with Poland and, to prevent outside intervention, sent Foreign Minister Ribbentrop to sign a pact with Stalin *(bottom far right)* in August 1939.

The Polish government in 1939 was the first to stand up to Hitler *(map above right)*. Foreign Minister Josef Beck argued that Germany was bluffing and its military strength exaggerated. Boosted by a guarantee of support from Britain and France, the Polish army drew up defence plans. Poland's 37 divisions and 200 aircraft would fight a brief holding action and then withdraw to central Poland to fight in front of Warsaw. They had few tanks, relying on cavalry for mobility. During August 1939 troops and arms were smuggled into Danzig by German forces in East Prussia. On the night of 31 August SS units captured the city, and the following day German soldiers, seen here *(below right)* crossing the border with Poland at Steinfleiss, began the invasion of Poland.

WESTERN INTELLIGENCE

On 25 July 1939 a meeting took place at a secret location in the Pyry forest near Warsaw between representatives of the Polish, French and British secret services. The meeting was to discuss progress in deciphering German communications made on the Enigma machine *(right)*. The machine was first sold commercially by a German firm as a way to keep business secrets, but it was bought by the German armed forces as their basic equipment for sending messages in cipher form. German forces believed the messages to be unbreakable since the cipher could be changed every day and the machinery allowed endless permutations. Much of the pioneering work in

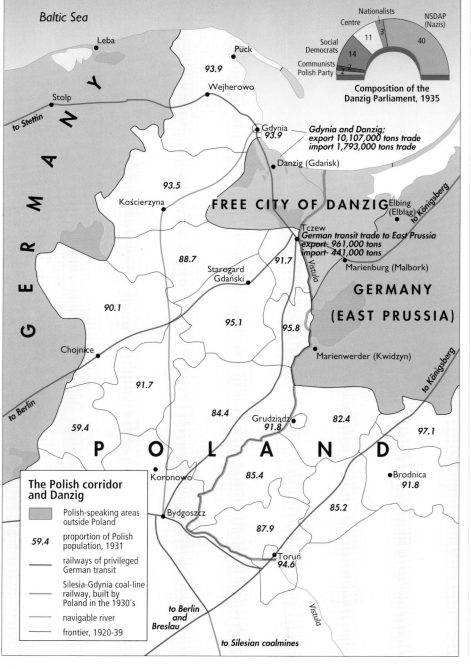

Baltic Sea

Leba

Puck

93.9

Stolp

to Stettin

Wejherowo

Gdynia
93.9

Gdynia and Danzig;
export 10,107,000 tons trade
import 1,793,000 tons trade

Danzig (Gdańsk)

93.5

Kościerzyna

FREE CITY OF DANZIG Elbing (Elbląg)

to Königsberg

Tczew
German transit trade to East Prussia
export 961,000 tons
import 441,000 tons

Marienburg (Malbork)

88.7

91.7

Starogard
Gdański

GERMANY
(EAST PRUSSIA)

90.1

95.1

95.8

Chojnice

Marienwerder (Kwidzyn)

to Königsberg

91.7

84.4

Grudziądz
91.8

82.4

97.1

59.4

P O L A N D

Koronowo

85.4

Brodnica
91.8

to Berlin

Bydgoszcz

85.2

The Polish corridor and Danzig

- Polish-speaking areas outside Poland
- 59.4 proportion of Polish population, 1931
- railways of privileged German transit
- Silesia-Gdynia coal-line railway, built by Poland in the 1930's
- navigable river
- frontier, 1920-39

to Berlin and Breslau

87.9

Toruń
94.6

Vistula

to Silesian coalmines

Nationalists / Composition of the Danzig Parliament, 1935

Nationalists
Centre 3
NSDAP (Nazis) 40
Social Democrats 14
11
Communists Polish Party 2 2

Composition of the Danzig Parliament, 1935

North Sea

Amsterdam

Osnabrück

Rotterdam

HOLLAND

Kassel

Ostend

Antwerp

BELGIUM

Brussels

Aachen

Cologne

GERMANY

Calais

Namur

Liège

Koblenz

Lille

Ardennes

Maubeuge

Frankfurt

LUX.
Lux.

Châlons-sur-Marne

Alsace

Strasbourg

French defence lines, 1940

- Westwall line
- Maginot line
- Grebbe line (Fortress Holland)
- projected "Zone of Defence"
- lightly defended zone
- French strongpoints
- Belgian strongpoints
- railway

Lorraine

F R A N C E

Dijon

Geneva

SWITZERLAND

A L P S

The Polish campaign plans, 1939

Polish forces
German forces

their democratic way of life) were dangerously threatened. In February Britain committed herself to fighting alongside France, and the two military staffs began to plan for a possible war in 1939. On 15 March German troops occupied the rest of Czechoslovakia. On 31 March Chamberlain gave an unconditional guarantee to Poland of British help if Germany violated Polish sovereignty. Poland, with its large German minority, seemed likely to be Hitler's next ambition. In the event, the guarantee was given before Hitler had decided to attack Poland and tear up the last shreds of the Versailles Settlement.

On 6 April Hitler instructed his armed forces to prepare a brief campaign against Poland in the autumn. He expected the conflict to be localized: he was convinced that Britain and France were too militarily weak and too politically spineless to oppose him seriously, despite the growing evidence to the contrary. To ensure western non-intervention Hitler made overtures to the Soviet Union for an agreement. On 23 August the Nazi-Soviet Non-Aggression Pact was sealed in Moscow. Three days later Britain entered a formal military alliance with Poland. Throughout the summer Britain and France had given Hitler clear warnings of their intention to fight. Preparations for war were well advanced by August, though both Chamberlain and Daladier expected Hitler to stand back from war when he realised the risk. Deterrence failed. On 1 September German forces attacked on a broad front. Mussolini's last-minute attempt to mediate between the powers on 2 September failed, and Britain and France declared war on Germany the following day.

cracking the German system was done in Poland, where from 1932 Polish experts established the nature of the machine, read some of the early messages and created a special mechanical device (known as a "bombe") for speedy calculation of the possible settings. By 1939 all of the German ciphers were still unreadable and the Allies were denied clear intelligence on what the German forces were doing. In August 1939 the Polish army sent two copies of the Enigma machine to Paris, and one reached London two weeks before the outbreak of war. On the basis of the Poles' early work, the British cipher school at Bletchley Park broke the Luftwaffe cipher in May 1940 and Enigma intelligence became a key source of Allied information throughout the war.

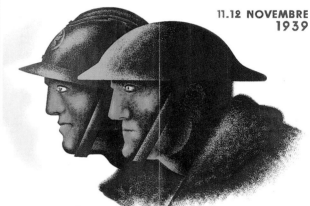

During the 1930s French governments spent six billion francs building a complex defensive wall to protect France from German and Italian invasion *(map below left)*. The defensive system took its name from the minister of war, André Maginot, whose inspiration it was. The main fortifications in Alsace-Lorraine were almost impregnable, but a gap was left opposite the Ardennes Forest, which was regarded as a natural obstacle, and in the north on the Belgian frontier, where low-lying land prevented deep fortifications. These were the two places where German forces attacked in 1940. The Maginot Line was a recognition of French weakness. France needed allies. Not until February 1939 did Britain give a guarantee of military assistance if Germany attacked. The Franco-British alliance, illustrated in the poster for "soldiers' day" in November 1939 *(above right)*, was essential to make the defensive wall work.

11.12 NOVEMBRE 1939

JOURNÉE FRANCO-BRITANNIQUE
AU BÉNÉFICE DE CEUX QUI COMBATTENT ET DE LEURS FAMILLES

Overture to the Second World War

THE EUROPEAN WAR which began in September 1939 became, two years later, a world war involving all the treaty powers. It was the most destructive war in history. This was "total war", fought against civilians as well as soldiers and waged across the globe, in which the mobilization of national resources reached unprecedented levels and at least 55 million people were killed. The war marked a watershed in world history. Both sides fought to promote their version of a new world order to replace that established in 1919. The involvement of the Soviet Union and the United States from 1941 ensured that the final outcome would no longer be decided by the European powers. The post-war world, dominated by these two new super-powers, left little room for traditional European imperialism. With the defeat of the Italian, Japanese and German empires in 1945, the age of imperialism was dead.

Fighting on the Eastern Front, June 1943

AXIS CONQUESTS, 1939-41

AT 4.45 ON THE MORNING of 1 September 1939, the German training ship *Schleswig-Holstein*, on a visit to Danzig, opened up its guns on Polish installations. SS units, smuggled into Danzig, overpowered Polish officials in the city. Within hours it was in German hands, the first stage in a campaign of conquest which in two years took German forces across Europe: to the Atlantic coast in the west and to Moscow and the Crimea in the east.

The seizure of Danzig prefaced a rapid assault on the whole of Poland. Against Poland's 30 divisions and 11 brigades of cavalry and her tiny air force, Germany mustered 55 divisions – including the six so-called "Panzer" divisions with large numbers of tanks and vehicles – and 1,929 aircraft. Using the tanks and aircraft together to outflank and demoralize the enemy, German forces quickly overran western Poland. Although the French high command had promised a campaign in the west after mobilization was complete, nothing was done to ease the pressure on Poland's armies. On 17 September Soviet forces began to occupy the almost undefended eastern areas of Poland, and on 27 September, following a fierce aerial bombardment, Warsaw surrendered to German forces. Poland was partitioned between Germany and the Soviet Union on lines agreed in Moscow on 28 September 1939.

Hitler's instinct was to strike at Britain and France while the iron was hot. But disagreements over the plan of operations and deteriorating weather conditions led to 29 postponements. The final date was fixed for 10 May 1940. The plan of campaign, devised by General von Manstein and approved by Hitler, was to attack with three army groups, one of which would contain most of the tanks and heavy vehicles designed to penetrate the enemy line north of the Maginot Line fortifications and to encircle Allied armies in north east France and Belgium. While the final preparations were put in place, Hitler became increasingly anxious about possible British plans to occupy Scandinavia in order to threaten Germany's northern flank. Two operations were hastily improvized against Denmark and Norway. The former was occupied with little resistance on 9-10 April. The attack on Norway on 9 April was resisted with British support, and Norwegian forces did not surrender until 7 June.

On 10 May the long-awaited German attack in the west began. Against a total of 144 Allied divisions (French, British, Dutch, Belgian) the Germans mustered 141. The German air force had 4,020 operational aircraft, the Allies a little over 3,000. The gap in tank strength favoured the Allies: 3,383 against 2,335. Yet in six weeks, and at a cost of only 30,000 dead, German forces conquered the Netherlands, Belgium and Luxembourg, and on 21 June forced the capitulation of France. Spurred on by German success, Italy declared war on Britain and France on 10 June. The secret of German success was not vast numerical superiority, but effective operational planning and the fighting skills of its army. British forces were divided – much of the air force stayed at home to avert the bombing threat – while the French forces, tanks included, were parcelled out along the Maginot Line. The Germans were concentrated for a short, sharp blow against the Allies in north eastern France.

While the French struggled to cope with Germany's mobile armies, British forces on the continent were evacuated. Some 370,000 troops (139,000 of them French) were shipped back to Britain in May and June 1940, with almost none of their equipment. For several weeks Hitler hesitated between offering terms to Britain and invading. In the end he decided to invade in the autumn (Operation Sealion), as long as British air power could be neutralized. During August and September the "Battle of Britain" was fought across the skies of southern England. The failure to neutralize the British air force, whose number and organization had been seriously underestimated, dented Hitler's enthusiasm. In July he began to talk of possible war with the Soviet Union, his current ally. By December 1940 he had made up his mind to defeat Stalin first, before finishing Britain off at his leisure.

German plans for the invasion of Britain, 1940

→ planned German advances
— operational objectives

The German advance, 1939-41

- Axis territory, 1 Sep. 1939
- Axis satellites
- Axis occupied
- Soviet occupied territory, 1939-40
- neutral powers
- → Axis advances
- Axis airborne landings
- → Allied forces
- → Soviet forces
- Allied retreat and withdrawal
- ✳ cities severely damaged by bombing

On 28 June 1940 Hitler landed at Le Bourget Airport in Paris to begin a tour of the French capital. For three hours in the early morning he toured the city, visiting the Eiffel Tower *(left)* among other sites, and then left. He later abandoned the idea of a victory parade and never saw Paris again.

Though the elite divisions in the German army were motorized, the great bulk of the German army used horses, bicycles or legpower to move around *(below)*. The horses in the invasion of France were part of the army's stock of over one million. During the Polish campaign, both sides made use of cavalry for scouting.

The division of Poland, 1939-40

- Greater Germany, Aug. 1939
- USSR, Aug. 1939
- Poland, Aug. 1939
- frontier of German-Soviet spheres of interest as agreed, 23 Aug. 1939
- frontier of German-Soviet spheres of interest as modified, 28 Sep. 1939
- annexed by German empire, Oct. 1939
- annexed by USSR, Oct. 1939
- annexed by USSR, 1940
- frontiers, end 1939

Under the Nazi-Soviet Pact of August 1939, Poland was divided between Germany and the USSR *(map left)*. Western Poland was absorbed into "Greater Germany" and a rump client state, the General Government, was established.

THE BATTLE OF BRITAIN

In the summer of 1940 Britain found herself alone against an Axis-dominated Europe. Her small and ill-equipped army was no match for German forces, but she was protected by a much larger navy, and an air defence system built up during the late 1930s to counter the German air threat. When on 16 July 1940 Hitler finally decided to invade Britain, the air force commander, Hermann Göring, promised to defeat the Royal Air Force in a couple of weeks as a prelude to the invasion of the English south coast. From July to September the German Air Force attacked British airfields, radar installations (below), ports and military bases. Despite inflicting high losses, the campaign failed. German aircraft casualties were considerably higher (from July to October they totalled 1,733 against British aircraft casualties of 915), while German aircraft production failed to make good the deteriorating strength of German units. By 1 October there were only 275 serviceable fighter aircraft left, against a British figure of 732. Frustrated at the lack of success, Hitler switched to a strategy of terror bombing of British cities. The subsequent "Blitz" reduced the pressure on the RAF and brought the Battle of Britain to an end.

On 16 July Hitler gave his blessing to Operation Sealion for the invasion of Britain *(map far left)*. It was planned to move 100,000 men across the Channel in a first wave of attacks on the coast of Kent and Sussex. The main objective was London, which was to be seized by two German armies. Shortages of shipping and of adequate air cover led to the plan's abandonment in October 1940.

The air battle fought over England between the RAF and the German air force was vital for keeping Britain in the war *(map below)*. British forces had the advantage of radar which gave accurate readings of the approaching German aircraft. German forces took high losses of pilots and aircraft between July 1940, when German attacks began, and October 1940, when the battle ended.

Between September 1939 and April 1941, German and Italian forces conquered nine European states *(map left)*. Germany dominated Europe, while Italy extended its power in the Mediterranean basin. While Germany was engaged in the war in the west against France and Britain, the Soviet Union extended its political sphere in eastern Europe. A short war with Finland (November to March 1940) gave the USSR control of the Karelian peninsula; in June 1940 Lithuania, Latvia and Estonia were annexed; in the same month Romania was compelled to give up Bessarabia.

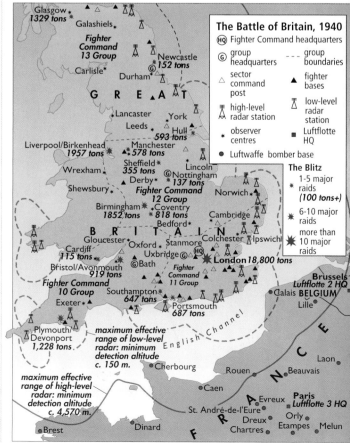

The Battle of Britain, 1940

- HQ Fighter Command headquarters
- G group headquarters
- △ sector command post
- high-level radar station
- * observer centres
- ● Luftwaffe bomber base
- - - group boundaries
- ▲ fighter bases
- low-level radar station
- ■ Luftflotte HQ

The Blitz
- * 1-5 major raids (100 tons+)
- ✳ 6-10 major raids
- ✸ more than 10 major raids

maximum effective range of low-level radar: minimum detection altitude c. 150 m.

maximum effective range of high-level radar: minimum detection altitude c. 4,570 m.

83

THE NAZI-SOVIET CONFLICT, 1941-3

The operation against the Soviet Union was codenamed "Barbarossa" after the medieval German emperor, Frederick I, who led the Third Crusade. Hitler viewed the attack on the Soviet Union as a modern crusade – the forces of European culture against the heathen Slavs and their Marxist masters. From the 1920s onwards Hitler had looked to the east as an area for German conquest and colonization. The rich resources of western Russia were to give him the means to turn Germany into a super-power.

The preparations were conducted in the strictest secrecy. The attack was scheduled for May to give a summer of good fighting weather for German tanks and aircraft. The pretence was maintained that Britain was still the object of the new campaign, but by the late spring there was growing intelligence evidence that German forces were swinging to the east. Efforts to persuade Stalin that his ally was about to betray him were brushed aside by the Soviet leader as Western propaganda. Then in April 1941 Hitler was diverted to the Balkans in response to the failure of the Italian attack on Greece and an anti-German coup in Yugoslavia. On 6 April Belgrade was bombed and after a brief campaign both Yugoslavia and Greece were defeated and occupied by a mixed Italian-German force.

The Balkan conflict meant the postponement of Barbarossa until 22 June. Stalin considered the date too late for the onset of hostilities that year, and Soviet defences were not alerted to any threat. Though Soviet forces outnumbered those of Germany – 20,000 tanks to 3,350 and 10,000 aircraft to the Luftwaffe's 3,400 – they were poorly organized and short of modern equipment. The German army of 146 divisions, organized into three army groups, North, Centre and South, and spearheaded by 29 Panzer and motorized divisions, achieved complete surprise when the orders were given to roll across the Soviet frontier on 22 June.

The German forces achieved a series of spectacular victories against a poorly prepared and demoralized Red Army. Huge pincer movements by mobile forces enveloped Soviet armies: two million prisoners were taken in the first three months. By the autumn almost all Soviet tanks and aircraft in the western areas had been destroyed, Leningrad and Moscow were threatened and German armies in the south had penetrated deep into the Ukraine, where food supplies and industrial production were concentrated.

In October 1941 Hitler returned from his headquarters behind the battle lines to Berlin, where he announced to an ecstatic crowd that the Soviet Union was on the point of complete defeat and that the time had now come for Germany to begin the construction of a New Order in Europe. In the east

On 22 June 1941 Germany launched an attack against the Soviet Union on a front of almost 1,000 miles with three million troops *(map below)*. In three months German armies, supported by troops from Romania, Finland, Hungary and Italy, almost reached Moscow and Leningrad. By December the advance was halted and the Red Army began the counter-offensive which saved the Soviet capital.

German forces practised a policy of scorched earth in the east. Food and livestock were seized, villages routinely burned to the ground, such as the one pictured here *(bottom)*, caught in the German retreat before Moscow. By 1945 70,000 Soviet villages had been destroyed.

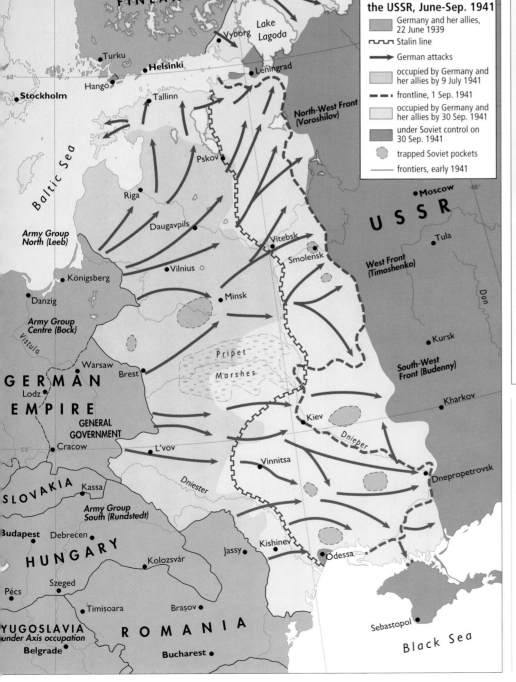

The German assault on the USSR, June-Sep. 1941

- Germany and her allies, 22 June 1939
- Stalin line
- German attacks
- occupied by Germany and her allies by 9 July 1941
- frontline, 1 Sep. 1941
- occupied by Germany and her allies by 30 Sep. 1941
- under Soviet control on 30 Sep. 1941
- trapped Soviet pockets
- frontiers, early 1941

THE SIEGE OF LENINGRAD

Leningrad (St Petersburg), Russia's second city, was the target of the German Army Group North in the summer of 1941. By July it was cut off from the Soviet interior by encircling German and Finnish forces. For 900 days it was subjected to a blockade which took a terrible toll of the civilian population. An estimated 900,000 people starved or froze to death or were killed by the constant shelling and bombing. Hitler wanted the city obliterated and its inhabitants wiped out. The only source of supply for the beleaguered population was a thin trickle of supplies across Lake Ladoga. The rations were quite inadequate. Some resorted to cannibalism. Soups were made from paper, glue or leather. Priority was given to feeding the defending soldiers and militia and the factory workers who turned out improvised weapons. In January 1943 the blockade was eased when a Soviet counter-offensive restored a limited rail link. Not until January 1944 was the blockade lifted.

A poster highlighting the importance of the Urals area (right) for the Soviet war effort (Ural-Front). In 1941 the Soviet Union moved 1,523 factories away from the danger zone to the Urals and Siberia. Here miracles of production were performed in harsh conditions.

Germany and Italy both had ambitions in the 1930s to construct a new European order, dominated by the two fascist powers (map below). By 1942 most of Europe was under Axis control or was allied to or dependent on the Axis bloc. Hitler gave Mussolini a free hand in the Mediterranean area. The rest of Europe was the German sphere. The Versailles Settlement was turned on its head. Germany annexed neighbouring areas to create "Greater Germany". The rest of occupied Europe was controlled by German plenipotentiaries or German puppet governments.

the plan was to destroy the Soviet state and raze its major cities to the ground. The bulk of the Slavic population would be pushed back beyond the Urals. The rest of the Soviet Union was to be broken into colonial regions ruled by Nazi commissars and permanently settled and garrisoned by Germans. The rest of Europe was to be organized hierarchically: the more developed and racially superior areas would hold privileged positions within the German empire; those less developed and inhabited by Slav or Latin peoples would form a lower tier of poorer, rural states. At the centre was to be the rich and industrialized Germany, dominating the continent as Rome had once done.

The declaration of the New Order proved premature. The German campaign in the east slowed with the onset of autumn rains and high losses of equipment and men. Confident of a quick victory, little effort had gone into supplying equipment and clothing for winter fighting. Soviet forces fought fiercely when they stood their ground. In front of Moscow, whose outskirts were reached in December 1941, a young Soviet general, Georgiy Zhukov, organized a frantic but effective defence. Supported by fresh troops pouring in from the Soviet eastern provinces, equipped with winter clothing and weapons newly produced in the east, the Red Army inflicted the first reverses on German forces. Mobile warfare was replaced by two defensive lines that stretched for more than 1,000 miles deep into Soviet territory.

The Axis New Order, 1939-43

——	pre-war borders
– – –	frontiers, Nov. 1942
▨	Grossdeutsches Reich (Greater Germany)
▨	occupied by Germany
RADOM	German administrative areas
■	German civil administration
▲	German military administration

▨	occupied by Italy
▨	Axis satellites
▨	neutral
▨	Allied territory

civil administrative areas
1 GENERAL GOVERNMENT
2 PROTECTORATE OF BOHEMIA-MORAVIA
3 DANZIG-WEST PRUSSIA
4 KATTOWITZ
5 WARTHELAND
6 CARINTHIA AND CARNIOLA
7 LOWER STYRIA
8 BIAŁYSTOK

regions
9 TRANSNISTRIA (to Romania 1941)
10 BESSARABIA (returned to Romania)
11 N. BUKOVINA (returned to Romania)
12 TRANSYLVANIA (to Hungary 1940)
13 SPIŠ (to Slovakia 1939)
14 BAČKA (to Hungary 1941)
15 MEDJIMURJE (to Hungary 1941)
16 PREKMURJE (to Hungary 1941)
17 LJUBLJANA (LUBIANA) (to Italy 1941)
18 ZICHENAU (to E. Prussia)
19 SUWALKI (Sudauen) (to E. Prussia)
20 KOSOVO (to Albania 1941)
21 THRACE (to Bulgaria 1941)

THE NAZI-SOVIET CONFLICT, 1943-5

In the spring of 1942 Hitler was confident that he could complete the defeat of the Soviet Union which had eluded him in 1941. Against the advice of his generals, who were keen to capture Moscow, Hitler opted for a drive on the southern flank to secure the Ukraine's rich resources and to seize the oil of the Caucasus region. He hoped that the Soviet front would be unhinged and that German forces could then wheel from the south to the rear of Moscow, encircling what remained of the Red Army.

On 28 June the southern "Operation Blue" was launched. Again German forces made remarkable gains. The Red Army was weaker in the south and withdrew in disorder towards the Volga and the Caucasus mountains. It stopped to fight on the mountain passes, which proved the limit of the southern advance towards the oil. Only the oil town of Maikop was captured, though following Soviet demolitions fewer than 70 tons of oil a day could be delivered. The Soviet forces also halted the German advance on the banks of the Volga at Stalingrad.

The Red October Factory in Stalingrad *(below top)* was the scene of bitter fighting in the attack on the city. Small groups of Soviet soldiers fought a hit-and-run campaign, wearing down the stronger attacking force.

On 6 June 1945 Marshal Zhukov, deputy commander-in-chief of Soviet forces, signed a pact with the other Allies on the defeat of Germany *(below bottom)*. Within months a jealous Stalin had demoted him.

From early 1943 the Soviet army experienced an almost unbroken run of successes *(map below)*, pushing the Germans back as far as Kiev in late 1943. 1944 saw the frontline move as far west as Warsaw, Romania and Bulgaria. In 1945 the Red Army swept through eastern Germany. Berlin surrendered on 2 May.

The city of Stalingrad was a major industrial centre and the key to the flow of oil northwards to the Soviet armies. Hitler ordered General Friedrich Paulus and his Sixth Army to seize the city from the retreating Soviet forces. A bitter battle ensued as the city was demolished street by street. The Soviet 62nd Army under General Chuikov was pressed back to the very edge of the river, where it fought with a fanatical tenacity. Both armies reached the very limits of endurance, driven on by Stalin and Hitler, who saw the contest as a symbol of the struggle between them. In November a Soviet counter-offensive, "Operation Uranus", broke the German front around Stalingrad and left Paulus and his army trapped. On 31 January he surrendered.

The Red Army drove the Germans back across the territory they had conquered in 1942 until, by the spring, poor weather and exhaustion brought a halt. In 1943 both sides prepared for what they saw as the decisive confrontation. Around the steppe city of Kursk lay a large Soviet salient in the German frontline. Here the German army concentrated almost one million men, 2,700 tanks and 2,000 aircraft. Zhukov prepared a series of defensive lines to absorb the German attack, while he built up large reserves behind the battlefield for a massive blow against the German front. Soviet forces were transformed from the demoralized, ill-prepared troops of 1941. Improved technology and training, a clearer command structure and the reorganization

The defeat of Germany, 1942-5

- "Grossdeutches Reich" 1942
- maximum extent of Axis control, 1942
- → Axis attacks
- ⇠ Axis withdrawals
- ← Allied attacks
- cities under heavy air attack
- partisan/resistance movements
- ✳ major battle with date

The German offensive in the Caucasus, 1942

→ German advances
━━ 28 June ━━ 11 July
━━ 6 July ━━ 22 July
━━━ 18 Nov.

Bryansk
Bryansk front
Orel
2nd Army (Weichs)
Yelets
Tambov
Kursk
4th Panzer Army (Hoth)
Voronezh
Army Group B
6th Army (Paulus)
South-West front
Don
Kletskaya
Kachalinskaya
6th Army
Stalingrad
Morozovsk
Stalingrad front
Donets Corridor
Donets
1st Panzer Army (Kleist)
Army Group A
17th Army
Kotelnikova
Volga
Taganrog
Rostov
4th Panzer Army
Astrakhan
North Caucasus front
Sea of Azov
17th Army
31 July
Elista
1st Panzer Army
Kerch
Army Group A
Stavropol 5 Aug.
Pyatigorsk
Trans-Caucasus front
11th Army
Krasnodar
Novorossiysk
Maikop (oilfields) 9 Aug.
Mozdok
Grozny (oilfields)
Black Sea
Tuapse
▲ Mt. Elbruz
Ordzhonikidze
Sochi
South front

In the summer of 1942 Hitler ordered his forces to seize the rich oilfields of the Caucasus. After three months they reached the banks of the Volga and as far as the Caucasus mountains. This was the limit of German expansion (map left). Overstretched at the end of long supply lines, German forces were too weak to withstand the Soviet winter counter-offensive.

In Stalingrad German armies drove Soviet defenders to the very edge of the river Volga in the factory district and the city centre (map below). In November 1942, German forces were encircled, and two months later capitulated with a total loss of 200,000 men.

In common with the inhabitants of Leningrad in the winter of 1941-2, those trapped in Stalingrad during the ferocious German assault on the city in the autumn of 19420 endured appalling privations (left).

of mechanized and air forces in imitation of German practice all contributed to a narrowing of the gap in military effectiveness between the two sides. When the German assault began on 5 July against the 1.3 million men and 3,400 tanks of the Red Army it was blunted within a week. Zhukov then hurled his reserves into the battle. The German front broke, and German armies began the long gruelling retreat back to the Reich.

By 6 November Soviet armies had reached Kiev. Early in January 1944 they crossed the old Polish border. Over the next six months German forces were cleared from the southern areas of the Soviet Union and from eastern Belorussia. On 22 June 1944, timed to coincide with the invasion of France by the western Allies, Stalin ordered a massive offensive – "Operation Bagration" – to clear German armies from Belorussia, the Baltic States and western Poland. By July the Red Army reached the Vistula opposite Warsaw. They inflicted 850,000 casualties on the defending German forces. The rapid Soviet advance brought the collapse of Germany's allies, Finland, Romania and Bulgaria. Massively outnumbered in men and equipment, German armies fought a desperate rearguard defence until, in January 1945, their frontline protecting Germany finally broke. Within four months the Soviet armies swept to Berlin and Vienna. Rather than be captured, Hitler shot himself on 30 April 1945 as Soviet forces were storming the last streets of the German capital.

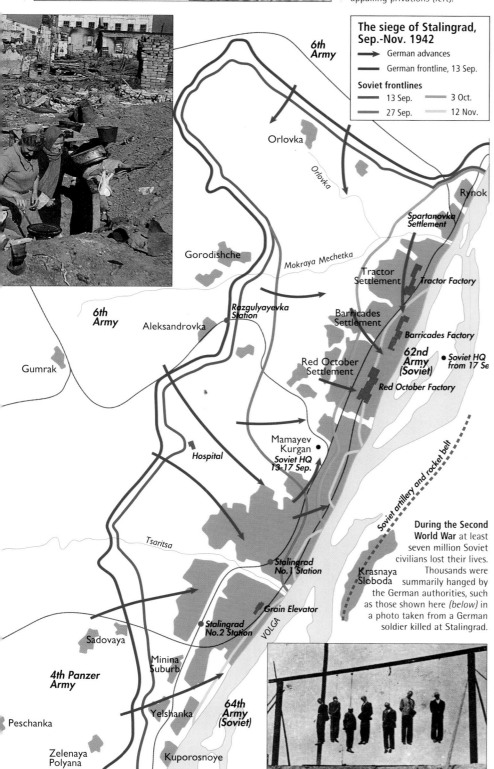

The siege of Stalingrad, Sep.-Nov. 1942

6th Army

→ German advances
━━ German frontline, 13 Sep.

Soviet frontlines
━━ 13 Sep. ━━ 3 Oct.
━━ 27 Sep. ━━ 12 Nov.

6th Army
Orlovka
Orlovka
Rynok
Spartanovka Settlement
Gorodishche
Mokraya Mechetka
Tractor Settlement
Tractor Factory
6th Army
Razgulyayevka Station
Barricades Settlement
Aleksandrovka
Barricades Factory
62nd Army (Soviet)
Soviet HQ from 17 Se
Gumrak
Red October Settlement
Red October Factory
Mamayev Kurgan
Soviet HQ 13-17 Sep.
Hospital
Soviet artillery and rocket belt
Tsaritsa
Stalingrad No.1 Station
Krasnaya Sloboda
VOLGA
Stalingrad No.2 Station
Sadovaya
Grain Elevator
Minina Suburb
4th Panzer Army
64th Army (Soviet)
Yelshanka
Peschanka
Kuporosnoye
Zelenaya Polyana

During the Second World War at least seven million Soviet civilians lost their lives. Thousands were summarily hanged by the German authorities, such as those shown here (below) in a photo taken from a German soldier killed at Stalingrad.

In July 1943 around the city of Kursk, Soviet forces inflicted a decisive defeat on the German army. From deep defensive positions they absorbed the German punch, then delivered an annihilating counter-attack that broke the German line (map bottom).

Commanders of the First Bulgarian Army at the front in Hungary are pictured here (below). Germany was assisted in the Soviet campaign by her allies, Finland, Hungary, Romania and Bulgaria, who hoped to profit territorially from Soviet defeat.

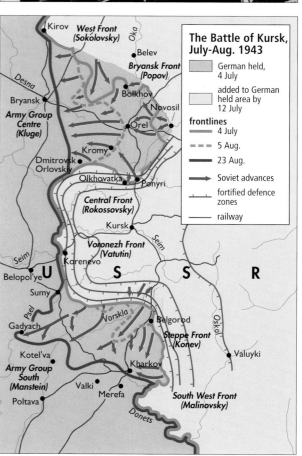

The Battle of Kursk, July-Aug. 1943

Kirov
West Front (Sokolovsky)
Oka
Belev
Desna
Bryansk Front (Popov)
Bolkhov
Bryansk
Novosil
Army Group Centre (Kluge)
Orel
Kromy
Dmitrovsk Orlovskiy
Olkhovatka
Ponyri
Central Front (Rokossovsky)
Kursk
Seim
Voronezh Front (Vatutin)
Korenevo
Seim
Belopol'ye
U   S   S   R
Sumy
Psel
Gadyach
Vorskla
Belgorod
Oskol
Kotel'va
Kharkov
Valuyki
Army Group South (Manstein)
Valki
Merefa
Steppe Front (Konev)
Poltava
Donets
South West Front (Malinovsky)

German held, 4 July
added to German held area by 12 July
frontlines
━━ 4 July
┅┅ 5 Aug.
━━ 23 Aug.
→ Soviet advances
⊣⊢ fortified defence zones
── railway

JAPANESE CONQUESTS, 1941-2

While Hitler's Germany was fighting to conquer Asia from the west, Japan was carving out an empire in the east. The conquest of eastern China, begun in July 1937, had sucked Japan into a long war of attrition against Chinese forces, both nationalist and communist. The Chinese war was seen in Tokyo as the key to the establishment of the Japanese New Order in Asia, but the military threat posed by an imperialist Japan involved her in increasing conflict with the Soviet Union and the United States.

Japanese leaders faced a dilemma. The army wanted to concentrate its efforts on the conquest of China and face the threat from a heavily armed Soviet Union on the Manchurian frontier: two major battles were fought with the Red Army in 1938 and 1939 along the border, and Japan was defeated on both occasions. The navy meanwhile looked south to the rich resources of oil and other materials on which the future of any Japanese war effort depended. However, southern advance would bring Japan into conflict with the US, which was already giving aid to China and preparing to reinforce its possessions in the Pacific basin.

The war in Europe opened up new possibilities. The defeat of France encouraged Japan to look towards the vulnerable European colonies of South East Asia. In September 1940 Japan signed a Tripartite Pact with Germany and Italy on the redivision of the world into a New Order. The same month Japan occupied the northern part of French Indo-China, prompting the

An American poster for distribution in China (opposite right) shows an American airman as a god of war, crushing the Japanese. Under the leadership of a retired US army officer, Claire Chennault, an American volunteer group of airmen (known as the Flying Tigers) fought for the Chinese from 1940.

Oil was Japan's weak spot. With few domestic resources and limited quantities of oil shale from Manchuria, Japan relied on oil from south west Asia (map right). A synthetic production programme achieved little. Supplies were secured by conquering Burma, the Dutch East Indies and Borneo.

Pilots were the elite of Japanese forces, heirs to the Japanese samurai tradition (poster bottom right). Yet though there was no shortage of volunteers, the early Japanese victories at sea were secured by just 600 highly trained naval aviators.

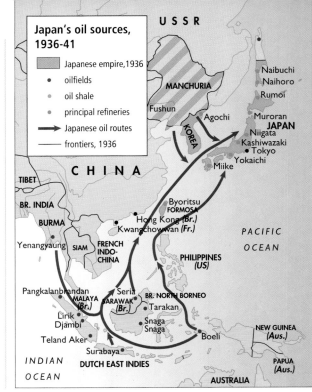

Japan's oil sources, 1936-41

- Japanese empire, 1936
- • oilfields
- • oil shale
- • principal refineries
- → Japanese oil routes
- — frontiers, 1936

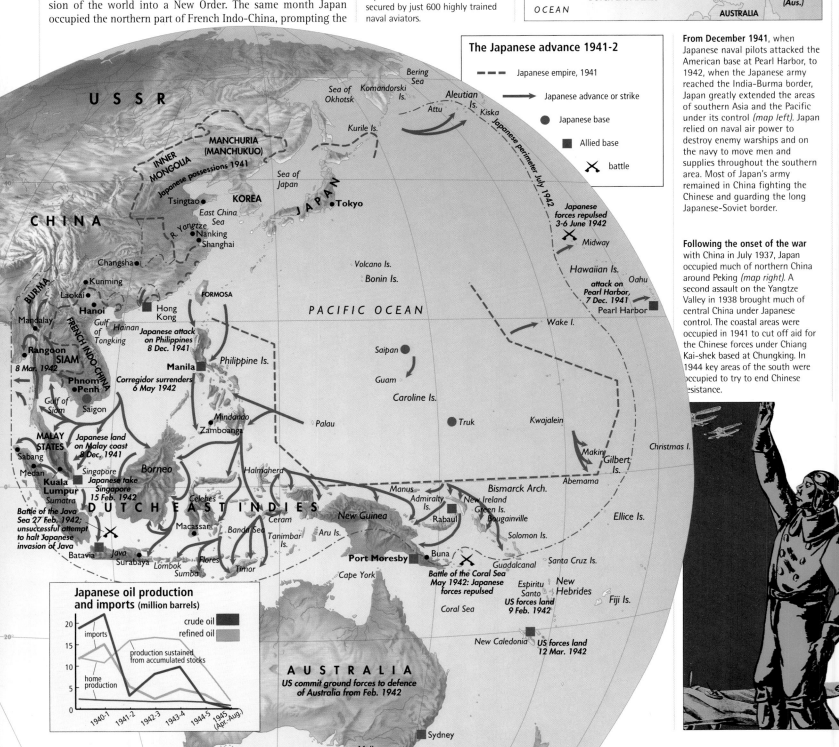

The Japanese advance 1941-2

- – – – Japanese empire, 1941
- → Japanese advance or strike
- ● Japanese base
- ■ Allied base
- ✕ battle

From December 1941, when Japanese naval pilots attacked the American base at Pearl Harbor, to 1942, when the Japanese army reached the India-Burma border, Japan greatly extended the areas of southern Asia and the Pacific under its control (map left). Japan relied on naval air power to destroy enemy warships and on the navy to move men and supplies throughout the southern area. Most of Japan's army remained in China fighting the Chinese and guarding the long Japanese-Soviet border.

Following the onset of the war with China in July 1937, Japan occupied much of northern China around Peking (map right). A second assault on the Yangtze Valley in 1938 brought much of central China under Japanese control. The coastal areas were occupied in 1941 to cut off aid for the Chinese forces under Chiang Kai-shek based at Chungking. In 1944 key areas of the south were occupied to try to end Chinese resistance.

Japanese oil production and imports (million barrels)

- crude oil
- refined oil
- production sustained from accumulated stocks
- imports
- home production

US to impose oil and steel sanctions. When the German-Soviet conflict in 1941 ended the threat from the north, the advance south was accelerated. In July 1941 the rest of Indo-China was occupied by 40,000 Japanese soldiers. In retaliation sanctions were tightened, depriving Japan of 80 per cent of her overseas oil supplies. Rather than retreat, the expansionists argued for a campaign to secure a perimeter from Burma to Australia and the Pacific islands, defensible by Japanese naval and air power.

The plan was approved in September after much argument. The prime minister appointed in October, General Tojo, tried one more diplomatic offensive while Japanese forces moved into position for attack. While negotiations continued in Washington, Japanese aircraft launched an attack on the US base at Pearl Harbor in Hawaii on the morning of 7 December 1941. The same day Japanese forces attacked Hong Kong, which surrendered on Christmas Day, and swept down the Malayan peninsula to Singapore, which surrendered on 15 February 1942. The US possessions – Guam, Wake Island, the Philippines – all fell one after the other. The heroic resistance of the US garrison at Corregidor ended on 6 May 1942. In the west Japanese forces reached the Indian border at the end of May, when the offensive finally paused for breath.

The southward advance exceeded the wildest Japanese expectations. Well-prepared operations against weaker, poorly armed forces, who had greatly underestimated Japanese fighting skills, brought rich dividends. Japan's military leaders decided to consolidate their position by seizing a further ring of islands, including Midway on the approaches to Hawaii. This time, forewarned by their radio intelligence, US naval forces were deployed to intercept. The first wave of renewed attacks around the Coral Sea were repulsed on 5-7 May. Against Midway the Japanese naval commander, Admiral Yamamoto, sent the bulk of the fleet and all four major aircraft carriers to destroy US naval power in the Pacific. Vastly outnumbered, US naval forces this time enjoyed the element of surprise. On 3-6 June US naval aviators destroyed all the Japanese carriers. The Battle of Midway decisively halted Japanese expansion and shifted the initiative to the United States.

Japanese soldiers advance in the oil-rich Dutch East Indies in early 1942 (above right). Despite plans to demolish the oil installations, most fell into Japanese hands intact and were producing oil again within weeks.

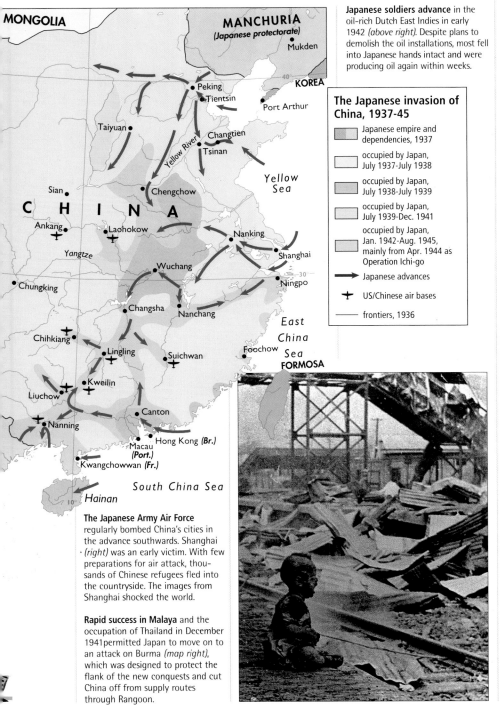

The Japanese invasion of China, 1937-45

- Japanese empire and dependencies, 1937
- occupied by Japan, July 1937-July 1938
- occupied by Japan, July 1938-July 1939
- occupied by Japan, July 1939-Dec. 1941
- occupied by Japan, Jan. 1942-Aug. 1945, mainly from Apr. 1944 as Operation Ichi-go
- Japanese advances
- + US/Chinese air bases
- frontiers, 1936

The Japanese Army Air Force regularly bombed China's cities in the advance southwards. Shanghai (right) was an early victim. With few preparations for air attack, thousands of Chinese refugees fled into the countryside. The images from Shanghai shocked the world.

Rapid success in Malaya and the occupation of Thailand in December 1941 permitted Japan to move on to an attack on Burma (map right), which was designed to protect the flank of the new conquests and cut China off from supply routes through Rangoon.

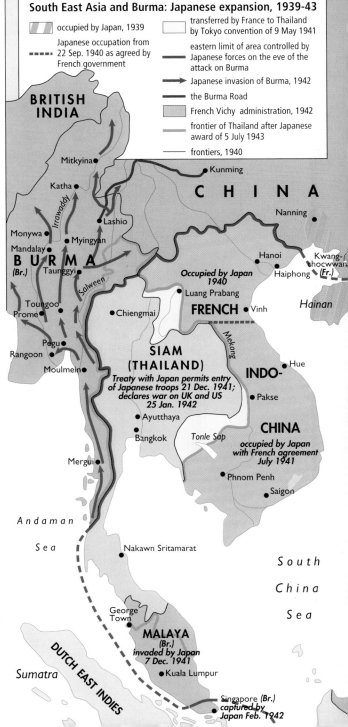

South East Asia and Burma: Japanese expansion, 1939-43

- occupied by Japan, 1939
- Japanese occupation from 22 Sep. 1940 as agreed by French government
- transferred by France to Thailand by Tokyo convention of 9 May 1941
- eastern limit of area controlled by Japanese forces on the eve of the attack on Burma
- Japanese invasion of Burma, 1942
- the Burma Road
- French Vichy administration, 1942
- frontier of Thailand after Japanese award of 5 July 1943
- frontiers, 1940

SIAM (THAILAND) Treaty with Japan permits entry of Japanese troops 21 Dec. 1941; declares war on UK and US 25 Jan. 1942

INDO-CHINA occupied by Japan with French agreement July 1941

Occupied by Japan 1940

MALAYA (Br.) invaded by Japan 7 Dec. 1941

Singapore (Br.) captured by Japan Feb. 1942

THE DEFEAT OF JAPAN

The Battle of Midway may have ended Japanese expansion in the Pacific, but how to defeat Japan posed serious problems. The brunt of the war against Japan was borne by the United States, which had to supply its forces across 6,000 miles of ocean and balance these demands with other commitments in the Middle East and Europe. Furthermore the war with Japan was fought against an enemy that had no concept of surrender. Every island and outpost was defended with a fanatical determination which made progress slow, even when the material balance between the two sides so clearly favoured the Allies.

The Pacific campaign was begun in August 1942 when American forces invaded Tulagi and Guadalcanal in the Solomon Islands. Responsibility was divided between the US navy, under Admiral Chester Nimitz, and the US army in the Pacific, under General Douglas Macarthur. The British had responsibility for the Indian Ocean and the Burma front, but with the defence of India their main priority they were only able to play a modest part in the campaign against Japan until enough forces could be spared from other theatres to renew the offensive in 1944 with US and Chinese assistance. Burma was reconquered in 1945.

US forces advanced across the Pacific under a strong air umbrella and a screen of fast aircraft carriers. Japanese defence was stubborn, but the island garrisons were gradually isolated by American air and submarine attacks on Japanese

The Allied bombing campaign against Japan, 1944-5

— Allied air attack
☁ atomic bomb target
✴ the "Big Six" fire raid targets
∗ secondary fire raid targets
. coastal bombardment by US forces

For the bombing of the Japanese home islands *(map left)* the US air force developed the B-29 Superfortress bomber. Against weak defences it was able to bomb in daylight using incendiaries to burn Japan's wooden cities.

In the last stages of the war Japan sent *kamikaze* suicide aircraft against Allied shipping *(above far right)*. Some 3,000 attacks were made from October 1944 to August 1945, and 402 ships were sunk or damaged.

MAGIC INTELLIGENCE
During the Pacific war, US forces were greatly aided by the successful interception and decipherment of Japanese radio communications. This intelligence was known as MAGIC and military information as ULTRA. The Japanese naval code JN-25 was broken sufficiently by the efforts of Joseph Roquefort (right) and his team to discover the timing and direction of the attack on Midway and to achieve complete surprise in the battle. The army codes proved harder, but captured codebooks allowed a breakthrough in January 1944, and when the codes were changed the new books were also captured. By 1945 the collapse of the Japanese signals system made it more difficult to get reliable radio intelligence on

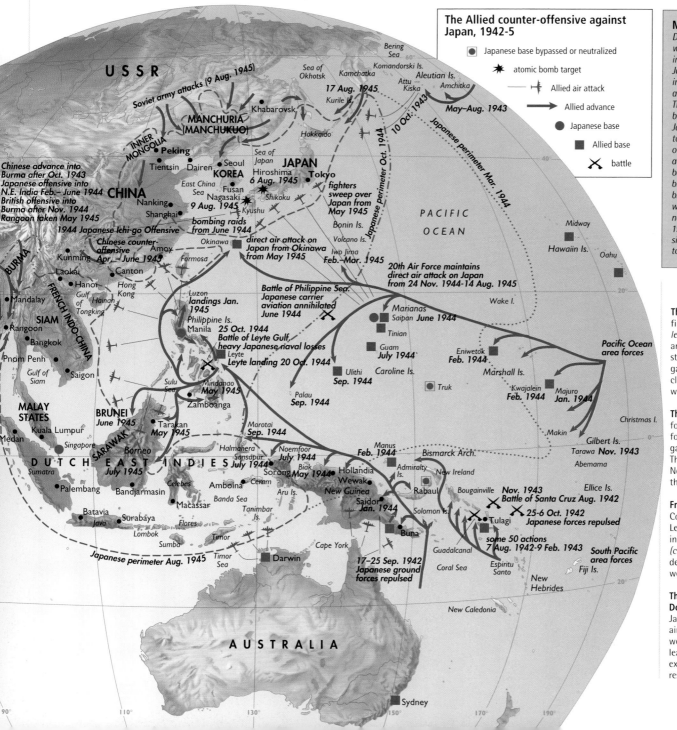

The Allied counter-offensive against Japan, 1942-5

⊙ Japanese base bypassed or neutralized
✴ atomic bomb target
⊢✈ Allied air attack
→ Allied advance
● Japanese base
■ Allied base
✕ battle

The tide of Japanese advance was finally halted in May 1942 *(map left)*. The Allies stabilized the front and then began limited step-by-step offensives against Japanese garrisons until in 1945 they were close enough to Japan to finish the war from the air.

The US army battled its way forward island by island. Here US forces *(above right)* attack Bougainville in the Solomon Islands. This island campaign lasted from November 1943 until the end of the war.

From January 1945 the XXI Bomber Command, under General Curtis LeMay, dropped 98,000 tons of incendiaries on Japan's major cities *(chart above right)* causing massive destruction. Altogether 58 cities were attacked and burnt down.

The Allies planned "Operation Downfall" for the invasion of Japan *(map right)*, and 14,000 aircraft and 100 aircraft carriers were assigned to the campaign. At least one million casualties were expected from fanatical Japanese resistance.

shipping, which made the supply of munitions and oil inter-mittent at best. The US forces had radar, which most Japanese units did not; their intelligence supplied regular interceptions of Japanese intentions; and their aircraft were sturdier, more heavily armed and gave their pilots armoured protection. Japanese aircraft were light and long-range, but only because they lacked adequate protection. Japanese pilots regarded armour as incompatible with the samurai tradition, but as a result lost more than 50 per cent of their number each month by 1944.

In 1944 the US was poised to retake the Philippines and to attack the Mariana Islands as a potential jumping-off point for the offensive on the Japanese home islands. The attack on Saipan in June 1944 was treated by Japanese leaders as the decisive engagement of the war, and strong naval and air forces were sent to intercept. In the Battle of the Philippine Sea (19 June 1944) the Japanese force was thoroughly defeated. In October 1944 the American invasion of Leyte Gulf provoked one final attempt by Japan to defend her new empire. A powerful three-pronged attack was directed at the superior American force, which again ended in complete disaster.

The way was now open to secure bases for a final attack on Japan. In February 1945 US marines landed on Iwo Jima, where Japanese troops fought almost literally to the last man – 20,700 were killed, and only 216 taken prisoner. The next target was Okinawa in the Ryukyu islands. The island took three months to conquer, from April to June 1945. By May 50,000 Japanese troops had been killed, and only 227 taken

prisoner. Suicidal defence made the prospect of invasion of the Japanese home islands a bleak one. Instead the new bases allowed the latest American heavy bomber, the B–29 Superfortress, to reach targets within Japan. Japanese cities were reduced one by one to ash, and then on 6 August 1945 the first atomic bomb was exploded on Hiroshima. On 15 August Japan surrendered, after months of bitter argument in the cabinet. Many Japanese leaders had accepted Japan's failure long beforehand, but military domination of Japanese politics made peace negotiations impossible. The emperor's decision to surrender was taken in the face of military opposition to anything other than a fight to the death for the honour of Japan.

Japanese soldiers were not taught how to surrender, but were expected to die fighting. The Allied leaflet *(bottom far right)* was dropped on Japanese lines to explain to them that surrender was possible. The difficulty of facing surrender certainly postponed Japan's effort to end the war in 1945. On 2 September Japanese representatives finally signed the act of surrender *(above)* on the battleship USS *Missouri* in Tokyo Bay.

Japanese movements and strengths. Intercepted intelligence gave US and British Commonwealth forces a vital advantage over the enemy.

Fire raids on the Big Six, 10 March-15 June 1945			
target	date	US attack force	area of target destroyed (sq. miles)
TOKYO	10 March	334	15.8
	13 April	327	11.4
	15 April	109	6.0
	23 May	520	5.3
	25 May	502	16.8
area of city: 110.8 sq. miles, area destroyed 56.3 sq. miles = 50.8% of target			
NAGOYA	12 March	285	2.1
	20 March	290	3.0
	14 May	472	3.2
	16 May	457	3.8
area of city: 39.7 sq. miles, area destroyed 16.0 sq. miles = 31.2% of target			
KOBE	14 March	307	2.9
	5 June	473	4.4
area of city: 15.7 sq. miles, area destroyed 8.8 sq. miles = 56.1% of target			
OSAKA	14 March	301	8.1
	June	521	3.2
	7 June	458	2.2
	15 June	516	2.5
area of city: 59.8 sq. miles, area destroyed 15.6 sq. miles = 26.1% of target			
YOKOHAMA	15 April	–	1.5
	29 May	454	6.9
area of city: 20.2 sq. miles, area destroyed 8.9 sq. miles = 44.1% of target			
KAWASAKI	15 April	194	3.6
area of city: 11.0 sq. miles, area destroyed 3.6 sq. miles = 32.7% of target			

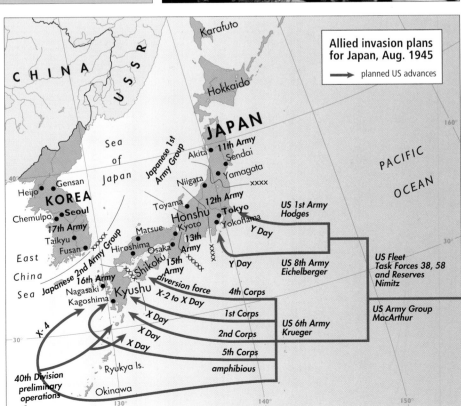

Allied invasion plans for Japan, Aug. 1945
→ planned US advances

ATTENTION AMERICAN SOLDIERS!

I CEASE RESISTANCE

THIS LEAFLET GUARANTEES HUMANE TREATMENT TO ANY JAPANESE DESIRING TO CEASE RESISTANCE. TAKE HIM IMMEDIATELY TO YOUR NEAREST COMMISSIONED OFFICER.

By Direction of the Commander in Chief.

上の英文の内容は「こゝの人は最早敵く國際條約により生命衣食住は六小意味が書かれて居る左圖は既に當方に来て居る諸戰友の一部

医療等が完全に保証さるべき者なり

THE BATTLE OF THE ATLANTIC

While the Japanese fought to control the sea routes of the Pacific, the Western Allies fought a prolonged struggle to maintain the sea links between the Old World and the New against the threat of German submarines. Victory in the Battle of the Atlantic was critical to the Western war effort, for only Allied naval power could secure the lifeline of military supplies, food and materials for the campaigns in Europe.

In the early stages of the war this threat was not immediately apparent. The British and French navies vastly exceeded the German navy in weight and number of ships: the 22 battleships and 83 cruisers of the Western navies faced only three small "pocket" battleships and eight cruisers. But after the defeat of France the balance began to change. In the summer of 1940 Hitler ordered an air and submarine assault on British shipping to cut off British imports and starve Britain into submission.

The air attacks were immediately successful. In 1940 aircraft sank 580,000 tons of shipping; the following year they sank over one million tons, more than Britain could make good from her dockyards. The submarine attack took longer to achieve results, but as a larger number of new submarines came into service, their commander, Admiral Karl Dönitz, ordered them to hunt in "wolf-packs" at night, where they would be undetected by current anti-submarine technology. In the first four months of 1941, two million tons of shipping were sunk. By the beginning of 1942 the number of German submarines had increased to 300 and German interception of British naval cyphers led the wolf-packs accurately to the convoy routes.

The loss of US merchant ships brought the United States into the Atlantic battle even before the onset of hostilities with Germany on 11 December 1941. US forces occupied Greenland and Iceland, where aircraft were stationed to give shore-based air cover; US warships patrolled the western Atlantic. When hostilities broke out between the United States and Germany,

The Battle of the Atlantic, 1939-45

· · · areas of merchant ship sinkings
- - - maximum extent of air cover

SEP. 1939-DEC. 1941

JAN. 1942-JULY 1942

The Atlantic was a battlefield between Western navies and aircraft and the German submarine force throughout the war *(maps left and below)*. The battle was fought in a number of phases. In the early part of the war most losses were in and around the British Isles. From January 1942, with American entry, submarines moved to the Caribbean and the American east coast. Through the winter of 1942-3 most victims were sunk in the "Atlantic Gap", outside Allied air cover. In May 1943 the submarine was defeated and only isolated sinkings continued away from the main convoy routes.

By 1941 the first of a new generation of heavy German battleships was ready to attack convoy routes in the Atlantic. The *Bismarck*, 42,000 tons and with eight 15-inch guns, set out in May 1941 with the cruiser *Prinz Eugen (map below)*. After sinking HMS *Hood* in a battle in the Denmark Strait, the German vessel was damaged by a torpedo aircraft. Sighted making for the port of Brest, she was immobilized by naval planes and then sunk by British warships on 27 May with the loss of all but 115 of her 2,222 crew.

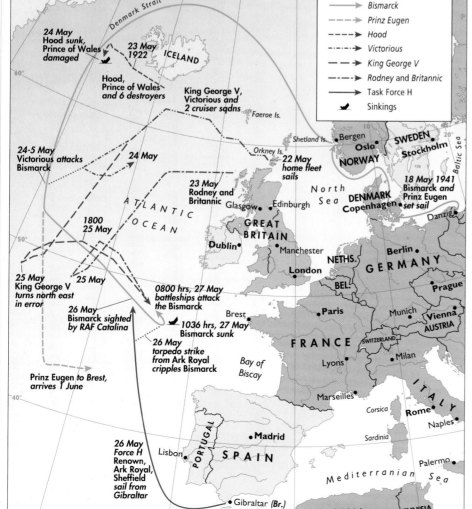

The pursuit of the *Bismarck*, May 1941

Battleship routes
→ *Bismarck*
→ *Prinz Eugen*
→ *Hood*
→ *Victorious*
→ *King George V*
→ *Rodney and Britannic*
→ Task Force H
⚓ Sinkings

24 May Hood *sunk*, Prince of Wales *damaged*

23 May 1922 ICELAND

Hood, Prince of Wales and 6 destroyers

King George V, Victorious and 2 cruiser sqdns

24-5 May Victorious *attacks* Bismarck

24 May

Faeroe Is.

Shetland Is. • Bergen

Orkney Is. SWEDEN
Oslo Stockholm
22 May home fleet sails NORWAY

23 May Rodney and Britannic Glasgow • Edinburgh

1800 25 May

North Sea DENMARK Copenhagen

18 May 1941 Bismarck and Prinz Eugen set sail

Danzig

25 May King George V *turns north east in error*

25 May

0800 hrs, 27 May battleships attack the Bismarck

26 May Bismarck *sighted by RAF Catalina*

26 May torpedo strike from Ark Royal cripples Bismarck

1036 hrs, 27 May Bismarck sunk

GREAT BRITAIN
Dublin • Manchester
NETHS. Berlin •
London GERMANY
BEL. Prague •
Brest • Paris • Munich Vienna •
FRANCE SWITZERLAND AUSTRIA
Bay of Biscay Lyons • Milan •
ITALY
Marseilles •
Corsica Rome •
Naples •
Sardinia
Palermo •
Mediterranean Sea

Prinz Eugen to Brest, arrives 1 June

26 May Force H Renown, Ark Royal, Sheffield *sail from Gibraltar*

Lisbon PORTUGAL SPAIN • Madrid
• Gibraltar *(Br.)*
ALGERIA TUNISIA

U-boat losses by cause, 1939-45

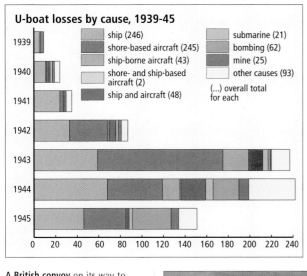

Legend:
- ship (246)
- shore-based aircraft (245)
- ship-borne aircraft (43)
- shore- and ship-based aircraft (2)
- ship and aircraft (48)
- submarine (21)
- bombing (62)
- mine (25)
- other causes (93)

(...) overall total for each

Years: 1939, 1940, 1941, 1942, 1943, 1944, 1945
Scale: 0 20 40 60 80 100 120 140 160 180 200 220 240

German submarine losses *(chart left)* were low until the advent of improved radar on Allied ships and aircraft. In 1943 aircraft destroyed 149 out of 237 submarines sunk and turned the tide of the ocean battle.

The loss of merchant ships *(chart below right)* peaked in 1942 with 1,662 sinkings. Many were made good from US shipyards, which turned out 21 million tons of shipping in three years.

A British convoy on its way to Russia in 1942 *(right)*, painted by Charles Pears. The route to Murmansk and Archangel was the most hazardous in the war, subject to air attack from Norway, as well as atrocious weather.

A Liberator B-24 *(below left)*, converted for use by RAF Coastal Command in the Battle of the Atlantic. These very-long-range aircraft closed the "Atlantic Gap" and hastened the defeat of the submarine.

**AUG. 1942-
MAY 1943**

The captain and crew of a German destroyer *(below right)*. There was little action for German surface vessels. When war broke out the naval commander, Admiral Erich Raeder, thought his forces would only discover "how to die gallantly". Many did, but most big ships remained bottled up in port, where they became sitting targets for Allied bombing.

**JUNE 1943-
MAY 1945**

The German submarine U-123 *(left)* in 1942. Submariners ran a great risk. Out of 39,000 some 28,000 were killed during the war, more than three quarters of the force.

Dönitz turned his attention to the American eastern seaboard, where ships still sailed with lights burning and radios on and without the use of a convoy system. Within months the US navy learned its lesson and the area became too dangerous for the U-boats to operate in strength. But there remained the "Atlantic Gap", the wide stretch of water in mid-ocean out of range of aircraft. It was here that Dönitz concentrated his submarines in 1942 and early 1943. In 1942, 5.4 million tons of shipping were sunk. By 1943 the Royal Navy was down to just two months' supply of oil.

The Western navies turned to technology in their efforts to fight the submarine. Aircraft were converted for long-range patrols and fitted with a new centimetric radar and powerful searchlights. They began to impose heavy loss rates on submarines returning to their bases on the French coast across the Bay of Biscay, where they could be attacked by day or night when they surfaced. Ship-borne radar was much improved, and the escorts for convoys strengthened. In 1943 British intelligence began to break German signal ciphers regularly. In November 1942 the submariner, Admiral Max Horton, was appointed to command the Atlantic battle. It was his insistence on the use of very long-range aircraft over the Atlantic Gap and on the use of quick-response support groups made up of a powerful flotilla of anti-submarine warships that turned the tide. In March 1943 two large convoys, HX229 and SC122, were severely mauled by the waiting submarines. But in April and May the new tactics reduced sinkings sharply, and increased the destruction of submarines. In May 41 were sunk; in June and July 54 more. On 31 May Dönitz recalled his boats from the Atlantic and the battle was over.

From May 1943 until the end of the war the submarine threat disappeared. In 1944 only 31 Allied ships were sunk in the Atlantic against a figure of 1,006 in 1942. Though German industry developed new long-range submarines capable of avoiding detection from the surface, they were not brought into service in time. Victory over the submarine made possible the build-up of forces for a land attack on Europe and ended the threat of blockade against Britain.

Allied merchant ship losses, 1939-45

Years: 1939, 1940, 1941, 1942, 1943, 1944, 1945
Scale: 0 200 400 600 800 1000 1200 1400 1600

Legend:
- submarine (2,828)
- aircraft (820)
- mine (534)
- warship raider (104)
- merchant raider (133)
- E-boat (99)
- unknown/other causes (632)

(...) overall total for each

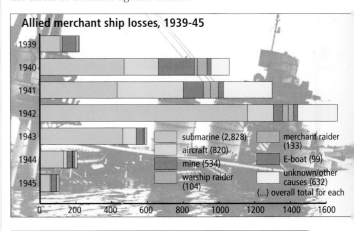

THE WAR IN THE MEDITERRANEAN

During the first half of the war, Britain divided her efforts between the Battle of the Atlantic and the conflict in the Mediterranean, which had been opened up by Mussolini's decision to enter the war in June 1940. Britain had long regarded the Mediterranean as a lifeline to her eastern empire and a key element in her global strategic security and, despite fears of a German invasion, diverted troops to the Mediterranean immediately Mussolini declared war. There could be no more certain indication of how vital to her interests Britain viewed the Mediterranean theatre.

British Commonwealth forces were initially successful, defeating the Italians in Abyssinia and, the following year, in Eritrea. The bulk of the Italian army stationed on the Egyptian border in Libya was driven back by British Commonwealth forces in 1940 and early 1941, and hundreds of thousands of Italian prisoners taken. But in 1941 Hitler was reluctantly forced to intervene to save his Axis ally from defeat. In February 1941 a small German army and air force was sent to Libya. Two months later German troops were also sent to Greece, where an ill-judged Italian invasion was stoutly resisted by the Greek army, assisted by a small British expeditionary force. However, Greece was conquered in April, and the British driven out. Crete fell to a costly paratroop assault in May 1941. General Erwin Rommel, a tank commander and hero of the Battle of France, was sent to lead the assault in North Africa, where his Afrika Korps took Axis forces to the Egyptian frontier by June 1942.

In 1942 Britain faced a crisis. The sea lanes through the Mediterranean were subject to air and submarine attack; Malta and Gibraltar were at risk; and Axis forces in North Africa and the Caucasus threatened to sweep through the Middle East and secure the valuable oil resources and the key supply route to Britain's Far Eastern empire. All this contributed to Churchill's pressure on his American ally to consider an attack in 1942 by Anglo-American forces somewhere in the Mediterranean theatre, rather than the strategy preferred by Roosevelt's chief of staff, General Marshall, of a direct attack on Hitler's Europe.

The lack of US preparation for such an attack persuaded Roosevelt that the Mediterranean was the most expedient option. US forces would be seen to be in action in Europe, and something decisive might be achieved with slim forces. The two allies planned an assault on North Africa codenamed Torch. In November heavy reinforcement of the Egyptian front produced a clear victory over Rommel and the Italian army at El Alamein. In the same month Allied forces landed in Morocco and Algeria. Axis forces were squeezed into a defensive pocket in Tunisia, where over 230,000 surrendered on 13 May 1943.

Despite American misgiving, the argument for continuing the pursuit into Italy was strong. The Torch landings had made a cross-Channel invasion in 1943 impossible. In July 1943 "Operation Husky" led to the capture of Sicily, and in early September Anglo-American forces crossed to Italy. On the evening of 8 September Italy surrendered. Allied forces expected the landings at Salerno the following day to be unopposed, but German forces almost drove the assault back into the sea. The Italian war turned into a long one of attrition between an Anglo-American force and a stubborn German defence. German forces retreated up the peninsula to prepared defensive lines, which proved difficult to penetrate despite overwhelming Allied air power. Churchill's vision of a hard blow at the soft Axis underbelly failed to materialize.

Allied forces under Field Marshal Alexander gradually found themselves taking a back seat to the preparations for the invasion of occupied France in 1944. Progress was painfully slow and losses high. Rome finally fell to the US Fifth Army on 5 June 1944. German forces fell back on the so-called Gothic Line, running across the peninsula just north of Florence, which was not breached until September 1944. The Allied front stalemated over the winter of 1944-5. Only in April 1945 did the final Allied assault break German resistance and lead to capitulation on 2 May. The campaign tied down more than 20 German divisions, and 35 more were caught up in the anti-partisan war in the Balkans. Nevertheless the Mediterranean campaign failed to achieve any decisive results. Victory in Europe was won only through direct attack on Germany.

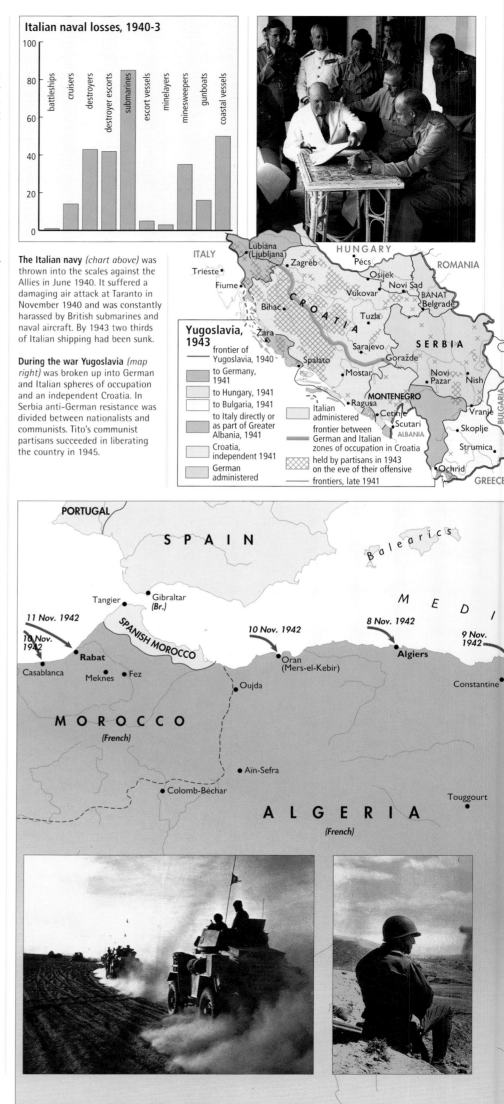

Italian naval losses, 1940-3

The Italian navy (chart above) was thrown into the scales against the Allies in June 1940. It suffered a damaging air attack at Taranto in November 1940 and was constantly harassed by British submarines and naval aircraft. By 1943 two thirds of Italian shipping had been sunk.

During the war Yugoslavia (map right) was broken up into German and Italian spheres of occupation and an independent Croatia. In Serbia anti-German resistance was divided between nationalists and communists. Tito's communist partisans succeeded in liberating the country in 1945.

Yugoslavia, 1943

frontier of Yugoslavia, 1940
to Germany, 1941
to Hungary, 1941
to Bulgaria, 1941
to Italy directly or as part of Greater Albania, 1941
Croatia, independent 1941
German administered
Italian administered
frontier between German and Italian zones of occupation in Croatia
held by partisans in 1943 on the eve of their offensive
frontiers, late 1941

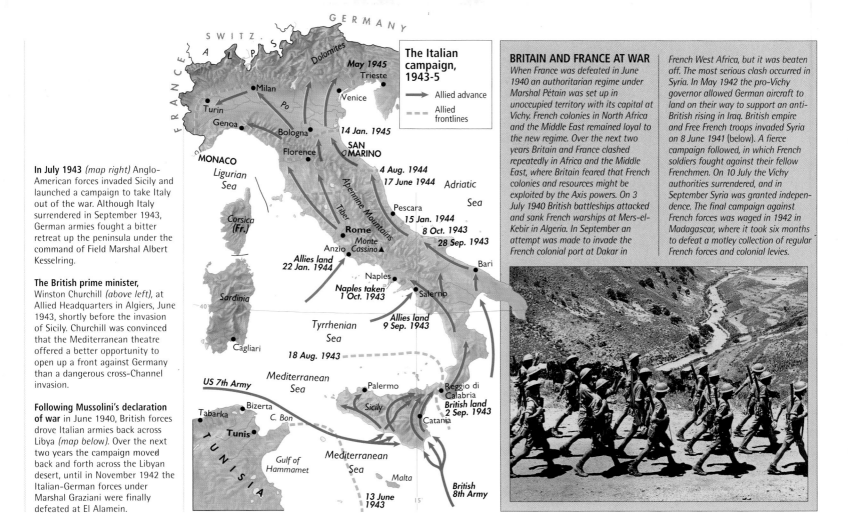

The Italian campaign, 1943-5

Allied advance →
Allied frontlines (dashed)

In July 1943 *(map right)* Anglo-American forces invaded Sicily and launched a campaign to take Italy out of the war. Although Italy surrendered in September 1943, German armies fought a bitter retreat up the peninsula under the command of Field Marshal Albert Kesselring.

The British prime minister, Winston Churchill *(above left)*, at Allied Headquarters in Algiers, June 1943, shortly before the invasion of Sicily. Churchill was convinced that the Mediterranean theatre offered a better opportunity to open up a front against Germany than a dangerous cross-Channel invasion.

Following Mussolini's declaration of war in June 1940, British forces drove Italian armies back across Libya *(map below)*. Over the next two years the campaign moved back and forth across the Libyan desert, until in November 1942 the Italian-German forces under Marshal Graziani were finally defeated at El Alamein.

Map labels: GERMANY, SWITZ., FRANCE, ALPS, Dolomites, May 1945 Trieste, Milan, Turin, Po, Venice, Genoa, Bologna, 14 Jan. 1945, Florence, SAN MARINO, MONACO, Ligurian Sea, Apennine Mountains, 4 Aug. 1944, 17 June 1944, Corsica (Fr.), Tiber, Pescara, 15 Jan. 1944, Rome, 8 Oct. 1943, 28 Sep. 1943, Monte Cassino, Adriatic Sea, Bari, Allies land 22 Jan. 1944, Anzio, Sardinia, Naples, Naples taken 1 Oct. 1943, Salerno, Allies land 9 Sep. 1943, Cagliari, Tyrrhenian Sea, Reggio di Calabria, British land 2 Sep. 1943, 18 Aug. 1943, Mediterranean Sea, Palermo, US 7th Army, Bizerta, Tabarka, C. Bon, Tunis, Sicily, Catania, TUNISIA, Gulf of Hammamet, Mediterranean Sea, Malta, 13 June 1943, British 8th Army

BRITAIN AND FRANCE AT WAR

When France was defeated in June 1940 an authoritarian regime under Marshal Pétain was set up in unoccupied territory with its capital at Vichy. French colonies in North Africa and the Middle East remained loyal to the new regime. Over the next two years Britain and France clashed repeatedly in Africa and the Middle East, where Britain feared that French colonies and resources might be exploited by the Axis powers. On 3 July 1940 British battleships attacked and sank French warships at Mers-el-Kebir in Algeria. In September an attempt was made to invade the French colonial port at Dakar in French West Africa, but it was beaten off. The most serious clash occurred in Syria. In May 1942 the pro-Vichy governor allowed German aircraft to land on their way to support an anti-British rising in Iraq. British empire and Free French troops invaded Syria on 8 June 1941 (below). A fierce campaign followed, in which French soldiers fought against their fellow Frenchmen. On 10 July the Vichy authorities surrendered, and in September Syria was granted independence. The final campaign against French forces was waged in 1942 in Madagascar, where it took six months to defeat a motley collection of regular French forces and colonial levies.

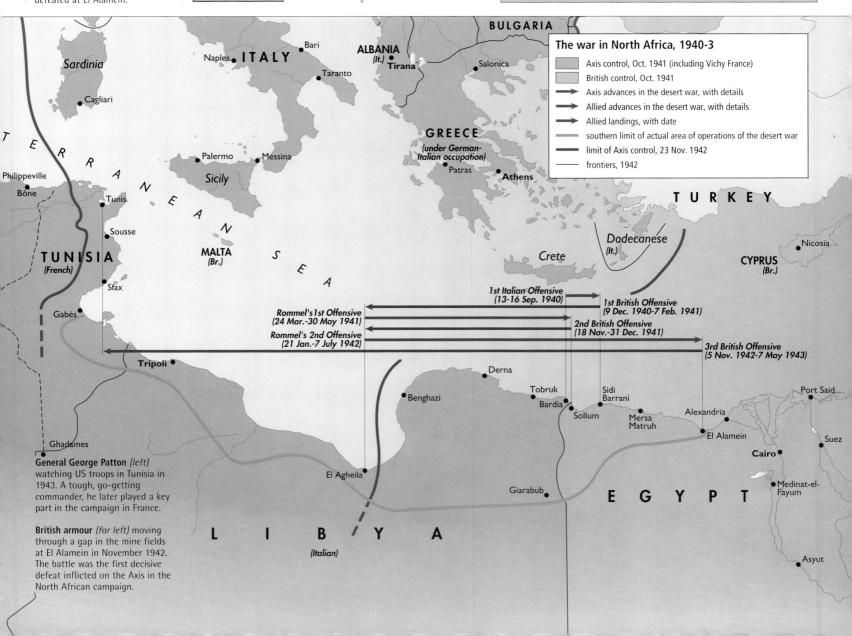

The war in North Africa, 1940-3

- Axis control, Oct. 1941 (including Vichy France)
- British control, Oct. 1941
- → Axis advances in the desert war, with details
- → Allied advances in the desert war, with details
- → Allied landings, with date
- southern limit of actual area of operations of the desert war
- limit of Axis control, 23 Nov. 1942
- frontiers, 1942

Map labels: BULGARIA, Sardinia, Cagliari, ITALY, Naples, Bari, Taranto, ALBANIA (It.), Tirana, Salonica, GREECE (under German-Italian occupation), Patras, Athens, TURKEY, Philippeville, Bône, Palermo, Messina, Sicily, MALTA (Br.), Dodecanese (It.), Crete, CYPRUS (Br.), Nicosia, Tunis, Sousse, TUNISIA (French), MEDITERRANEAN SEA, Sfax, Gabès, Ghadames, Tripoli, 1st Italian Offensive (13-16 Sep. 1940), 1st British Offensive (9 Dec. 1940-7 Feb. 1941), Rommel's 1st Offensive (24 Mar.-30 May 1941), 2nd British Offensive (18 Nov.-31 Dec. 1941), Rommel's 2nd Offensive (21 Jan.-7 July 1942), 3rd British Offensive (5 Nov. 1942-7 May 1943), Derna, Tobruk, Bardia, Sidi Barrani, Sollum, Mersa Matruh, Alexandria, Port Said, Benghazi, El Alamein, Suez, Cairo, El Agheila, Giarabub, EGYPT, Medinat-el-Fayum, LIBYA (Italian), Asyut

General George Patton *(left)* watching US troops in Tunisia in 1943. A tough, go-getting commander, he later played a key part in the campaign in France.

British armour *(far left)* moving through a gap in the mine fields at El Alamein in November 1942. The battle was the first decisive defeat inflicted on the Axis in the North African campaign.

THE INVASION FROM THE WEST

The arguments over whether to concentrate on defeating Germany by pushing north from the Mediterranean or invading across the English Channel were not finally resolved until the Teheran Conference in November 1943. Stalin, with wholehearted support from the US, insisted that an invasion of France was the only way to defeat Germany. In January 1944 serious preparation began for "Operation Overlord".

Eisenhower was named supreme commander, with British deputies for army, air and naval forces. The plan was to attack on a narrow front in Normandy, initially deploying five divisions. Artificial "mulberry" harbours were to be towed into place on the invasion day so that the invasion beaches could be supplied rapidly with forces and equipment.

The critical issue was to keep the destination secret from the Germans. By keeping the enemy guessing, the German forces would be stretched out along the entire coast, rather than concentrated at the invasion point. A deception plan, "Operation Bodyguard", was mounted and succeeded against all reasonable expectation in persuading Hitler that the major target was the Pas de Calais. Although Germany had 58 divisions in France, there were only 14 facing the Normandy beaches. The second imperative was to use Allied air superiority to neutralize the German air force and to cut off the communications net in northern France to prevent German reinforcement. Both campaigns were successful. On invasion

The D-Day landings relied heavily on Allied seapower. The naval operation, codenamed Neptune *(map right)*, required meticulous planning. The naval commander, Admiral Bertram Ramsay, had to move 7,000 vessels around the coasts of Britain to rendezvous areas where they crossed the Channel in great convoys. Off the Normandy coast warships gave essential protective fire for the first waves of invasion.

On 25 August 1944 Paris was liberated from four years of German occupation. The first units that entered the capital came from the Second French Armoured Division. Ever since French defeat in June 1940, hostilities were maintained by exiled French forces under the leadership of General Charles de Gaulle. With a reputation for arrogance and inflexibility, he proved a difficult ally. On 26 August he entered the city in triumph *(below left)*. He rekindled the flame at the Tomb of the Unknown Soldier, and then marched down the Champs Elysées until gunfire from surrounding rooftops interrrupted the procession.

day – popularly known as D-Day – there were 12,000 Allied aircraft against 170 German aircraft, while reinforcements from Germany to the front took weeks to reach the fighting, so poor had communications become.

The invasion was set for 5 June. In the event bad weather meant it had to be postponed until the following morning. Paratroopers were sent in overnight to secure the flanks and initial positions established successfully. German forces were divided and reinforcements were slow to arrive. Air power gave Allied forces – British, American, French and Canadian – great advantages. The German commander, Field Marshal Rommel, hero of the fast-moving desert campaigns, found himself forced to fight a tough defensive campaign in hedgerows and fields.

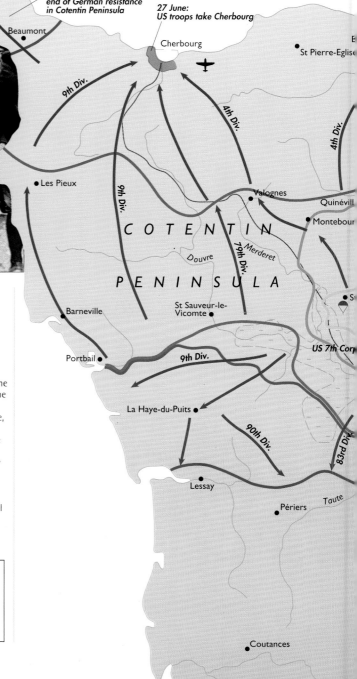

After the collapse of German resistance in France, the Western Allies stopped to regroup on Germany's borders and to allow the supply system to catch up with the armies *(map left)*. In December Hitler ordered a counter-offensive, the Ardennes Offensive, which failed to break the Allied line. The effort severely weakened forces already outnumbered. In February the Allies breached the West Wall fortifications and drove into the Rhineland. In three months they pushed across western and central Germany until they met with Soviet forces on the bridge over the Elbe at Torgau.

The Allied advance into Germany, Jan.-May 1945
- Allied front 28 Jan.
- Allied advance
- Allied front 8 May

On 25 July 1944 the American forces in Normandy launched "Operation Cobra" to force a breakout (map below right). American air power and armour made remarkable strides. General Patton's Third Army pushed into Brittany and then wheeled east towards Paris as the Germans retreated. By September the Allies had crossed France, but the attempt to drive into Germany was defeated. British paratroopers, pictured here (below) at Oesterbeek near Arnhem, were part of an Allied force which tried, unsuccessfully, to open the way into Germany in September 1944.

EISENHOWER AND MONTGOMERY

The Western war effort in 1944 was dominated by two men of very different temperament. The US general Dwight D. Eisenhower (above left) became supreme commander of Allied forces in the Mediterranean in November 1942, having never seen combat before. In January 1944 he was made supreme commander for the Normandy invasion. As his army commander-in-chief he chose the British field marshal, Bernard Law Montgomery (above right), a Great War veteran and the hero of El Alamein.

He was a man of strong views, with great confidence in his strategic capabilities. Eisenhower was the more conciliatory. They worked together well enough in planning, but in Normandy Eisenhower became irritated with Montgomery's slow and methodical approach. Their relationship rapidly deteriorated, and was saved only by the remarkable success of the US breakout. They clashed again over the timing of the invasion of Germany in autumn 1944. Montgomery became chief of the imperial general staff after the war, while Eisenhower was elected US president in 1952.

The Allied plan was to force the German armies to concentrate on the eastern wing of the invasion front, while in the west the Cotentin Peninsula was cleared and American forces broke out into France towards Paris. Progress was slow, and in the east the city of Caen, fiercely defended by the Germans, took over a month to capture. But the constant attrition wore down the under-strength German forces. On 25 July General Omar Bradley broke out into the west and within days Bradley had swept up resistance and his forces were wheeling towards the Seine. A desperate counter-attack at Mortain by remaining German armoured forces was repulsed and a headlong retreat followed. By late August the Allies were across the Seine and by September stood on the German frontier.

While the Allied supply system caught up with the rapidly advancing air forces and armies, German forces were given time to regroup to defend the Reich. Montgomery, the British army commander, wanted to press on into Germany, but the assault by paratroopers at Arnhem on 17 September was bloodily repulsed. Allied forces paused during the winter, but on Christmas Day were surprised by a strong German counter-thrust towards Antwerp. The "Battle of the Bulge" was won through rapid redeployment of Allied forces. By January the Western Allies had 25 armoured divisions and 6,000 tanks massed against 1,000 German tanks. Montgomery pushed north into the Ruhr and towards Berlin, while American generals swung into southern Germany. Despite strong local resistance the final defeat of Germany was achieved by early May when German forces capitulated to Montgomery on Lüneburg Heath.

The D-Day Landings, June-July 1944

- airborne landings
- **GOLD** beach heads

Allied advances
- British & Commonwealth
- United States
- airfields
- mulberry harbours
- German advance
- built-up areas
- flooded areas
- woodland

frontlines
- midnight 6 June
- 10 June
- 18 June
- 1 July
- 24 July

The invasion of Normandy on 6 June 1944 (map below) was the largest amphibious assault ever launched. 12,000 aircraft and over 7,000 ships of all sizes supported the operation. Five divisions were landed on a series of five invasion beaches. Paratroopers were sent in on both flanks during the night of 5-6 June. The initial beach heads were secured by 7 June, and positions established by 11 June. Progress thereafter was slow. Marshy ground, thick hedgerows and determined German defence prevented a breakout and threatened to re-create the trench stalemate of the First World War.

The Allied advance through France, July-Dec. 1944

- Allied advances
- Allied frontlines and date
- planned Allied frontlines
- West Wall

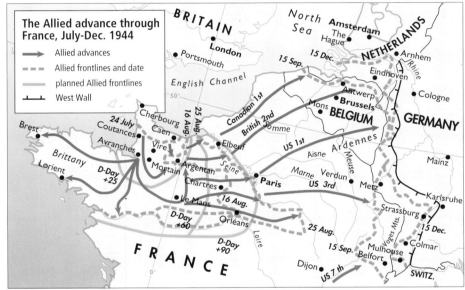

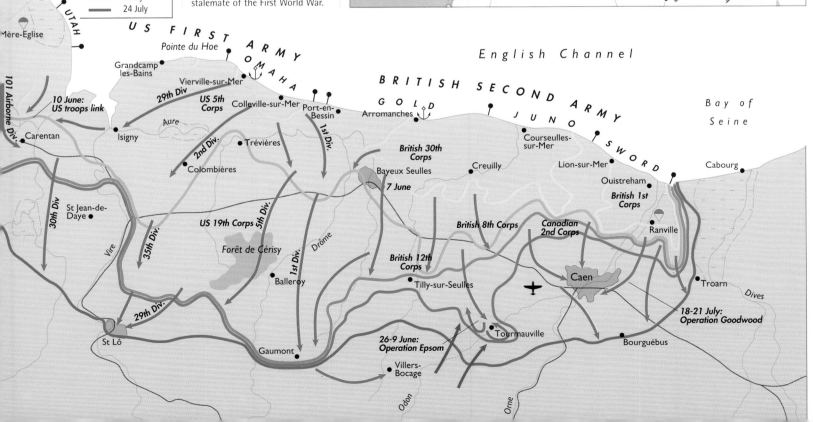

The Invasion from the West

THE BOMBING CAMPAIGN

The bombing of enemy cities and industry – or strategic bombing as it became known – began in the First World War but became a serious technical possibility only by the early 1940s. There were widespread fears in 1939 that bomb attack might end the war on its own, even before a shot had been fired on the ground. But strategic bombing did not become a significant force until 1944, and only in the war with Japan did bombing play a decisive part.

Strategic bombing began in 1940 with the first British attacks on German targets, authorized by Churchill in May in retaliation for the German destruction of Rotterdam during the Battle of France. In September 1940 Hitler ordered the Luftwaffe to attack British cities in an attempt to drive Britain out of the war without a costly invasion. Although German attacks killed 42,000 in the winter of 1940-1, they had little effect on British war production and failed to bring the British government to the negotiating table. The Blitz demonstrated to Hitler that bombing could not achieve war-winning results and from 1941 the German air force concentrated on tactical air support over the battlefield.

Britain maintained the bombing offensive for want of any other way of retaliating once her armies had been expelled from Europe. The early attacks achieved little and were shown by photo reconnaissance to be wildly inaccurate. British bombers attacked by night to avoid German fighters, which made it difficult to attack anything much smaller than a city.

In February 1942 Bomber Command was ordered by the British chiefs of staff to concentrate on area attacks against German industrial cities. On 31 May 1942 the first 1,000-bomber raid was staged, against Cologne.

From February 1942 Bomber Command was led by Air Marshal Arthur Harris. He introduced tactical changes – including a specialized Pathfinder Force to increase bombing accuracy – and benefited from the introduction of new technology: the Lancaster heavy bomber and improved navigational aids. In 1942 Bomber Command was joined by the US Eighth Air Force, which bombed by day, attacking specific industrial and military targets rather than whole cities. In 1943 the two forces were directed by the British and American combined chiefs of staff to undertake a combined bomber offensive against key targets – oil, aviation, submarines – as well as against the war-willingness of the German population.

During 1943 the combined offensive began to have real effects on German strategy. Extensive resources were put into air defence with the construction of the Kammhuber Line of

German oil supplies had been a priority target from the start of the war *(map below right)*, but only in 1944 did it prove possible to launch devastating attacks. Bomber Command and the US Eighth Air Force hit German synthetic oil plants, while the US 15th Air Force struck at the Romanian oilfields around Ploesti.

In 1943 Hitler gave the go-ahead to attack Britain with pilotless bombs and rockets *(map bottom right)*, the V (for vengeance) weapons. On 22 June 1944 the first ten V1 flying bombs were fired at London. On 8 September the first V2 rockets hit London. 2,420 V1s and 517 V2s reached the capital, killing over 9,000 people. They failed to deter Britain from further bombing.

A B-24 Liberator bomber *(below)* of the US 15th Air Force flies low over the Romanian oil centre of Ploesti. The first raid was mounted on 5 April 1944 and attacks continued until 19 August, when the Red Army overran Romania. Romanian supplies to Germany dropped by half in 1944. Oil transports were also damaged by mines laid by the RAF in the Danube.

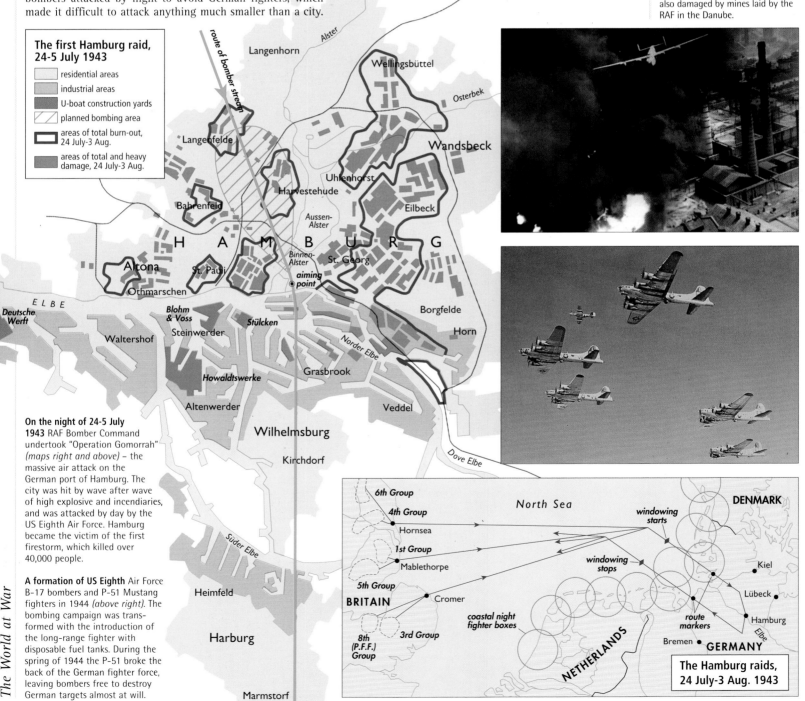

The first Hamburg raid, 24-5 July 1943

- residential areas
- industrial areas
- U-boat construction yards
- planned bombing area
- areas of total burn-out, 24 July-3 Aug.
- areas of total and heavy damage, 24 July-3 Aug.

route of bomber stream

Langenhorn · Alster · Wellingsbüttel · Osterbek · Wandsbeck · Uhlenhorst · Harvestehude · Eilbeck · Langenfelde · Bahrenfeld · Aussen-Alster · H A M B U R G · Binnen-Alster · St. Georg · aiming point · Altona · St. Pauli · Othmarschen · Borgfelde · Horn · ELBE · Deutsche Werft · Blohm & Voss · Stülcken · Steinwerder · Norder Elbe · Waltershof · Howaldtswerke · Grasbrook · Altenwerder · Veddel · Wilhelmsburg · Dove Elbe · Kirchdorf · Süder Elbe · Heimfeld · Harburg · Marmstorf

On the night of 24-5 July 1943 RAF Bomber Command undertook "Operation Gomorrah" *(maps right and above)* – the massive air attack on the German port of Hamburg. The city was hit by wave after wave of high explosive and incendiaries, and was attacked by day by the US Eighth Air Force. Hamburg became the victim of the first firestorm, which killed over 40,000 people.

A formation of US Eighth Air Force B-17 bombers and P-51 Mustang fighters in 1944 *(above right)*. The bombing campaign was transformed with the introduction of the long-range fighter with disposable fuel tanks. During the spring of 1944 the P-51 broke the back of the German fighter force, leaving bombers free to destroy German targets almost at will.

North Sea · DENMARK · 6th Group · 4th Group · windowing starts · Hornsea · 1st Group · windowing stops · Kiel · Mablethorpe · 5th Group · Lübeck · Cromer · Hamburg · BRITAIN · coastal night fighter boxes · route markers · Elbe · 8th (P.F.F.) Group · 3rd Group · Bremen · GERMANY · NETHERLANDS

The Hamburg raids, 24 July-3 Aug. 1943

ATOMIC WEAPONS

By 1939 German physicists had demonstrated the physical possibility of developing a weapon of exceptional force by the nuclear fission of uranium. The United States was the only state to capitalize on this discovery and to produce a usable bomb during the war. The German nuclear programme was given low priority once the cost and complexity of production was clear. Hitler distrusted modern physics, dismissing it as a "Jewish science". German émigré scientists contributed instead to the nuclear research of the Allies. British scientists organized under the Maud Committee made further progress, but in 1942 their project, which was beyond the technical and financial capabilities of the British war effort, was taken over by the United States.

The American "Manhattan Project", led by a team of international scientists, involved 150,000 people over four years. By 1944 enough fissionable material had been produced to make two bombs. On 16 July 1945 the bomb was tested by the Los Alamos research team at Alamogordo Air Base. Its horrifying success prompted President Truman to use the weapon against Japan to shorten the war and save American lives. On 6 August 1945 a single bomb, nicknamed Little Boy (below), was loaded onto a B-29 Superfortress bomber, the Enola Gay, and dropped on Hiroshima. On 9 August a second bomb was dropped on Nagasaki. Over 100,000 people were killed instantly in the attacks. These last days of the war were the first to witness the birth of the nuclear age.

From 1941 the RAF began to attack German industrial cities systematically *(map below right)*. The object was to reduce industrial output and to "de-house" the workforce. By 1945 45 per cent of the housing in Germany's major cities was destroyed and eight million Germans had been evacuated.

A German family *(bottom right)* in the ruins of Berlin in 1946. The German capital was attacked repeatedly from 1940 onwards. By 1945 over one third of its housing stock was completely destroyed, a total of 556,500 dwellings. The population of the capital fell from 4.3 million to 2.6 million.

anti-aircraft guns, radar and fighters. But by the winter of 1943-4 the bombing offensive was grinding to a halt because of mounting losses of men and planes. The campaign was saved by the introduction of long-range "strategic" fighters – in the main the P-51 Mustang, fitted with disposable fuel tanks, which won air superiority over Germany. Better navigational aids, improved training and much higher numbers of heavy bombers turned bombing into a real threat. The defeat of German air power in 1944 left the Reich open to attack. By 1945 both forces could roam over German air space without serious resistance. Some 83 per cent of the 1.9 million tons of bombs dropped fell in 1944 and 1945.

Bombing did not defeat Germany on its own, but it had a devastating impact on the German war effort. Some 400,000 were killed, and the major German cities reduced to rubble. Eight million people were evacuated and two million Germans were tied to air defence and the effort to clear the destruction. By 1944 over 54,000 guns were concentrated against the Allied bombing, and the defence effort used one third of the output of optical and electro-technical equipment and one fifth of all ammunition. This vast diversion of effort was compounded with the loss in 1944 of one third of the potential output of aircraft, tanks and trucks. Bombing forced an expensive and time-consuming dispersal of industry and resulted in high levels of absenteeism. For ordinary Germans bombing was the dominant issue on the home front from 1943 onwards. It was demoralizing for all who experienced it, producing not so much political protest as apathy and fear.

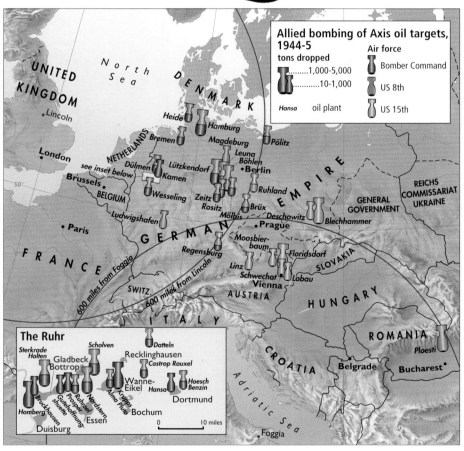

Allied bombing of Axis oil targets, 1944-5

tons dropped
- 1,000-5,000
- 10-1,000

Hansa oil plant

Air force
- Bomber Command
- US 8th
- US 15th

The Ruhr

The Allied bombing campaign against Germany, 1940-5

- ....major city bombed
- ....percentage of residential area of city destroyed

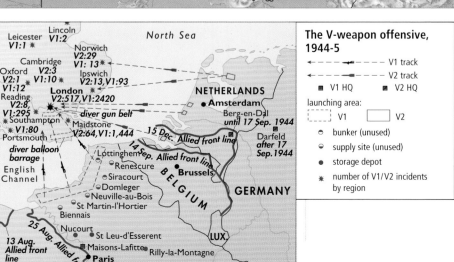

The V-weapon offensive, 1944-5

- ← — — — V1 track
- ← — — — V2 track
- ■ V1 HQ   ■ V2 HQ

launching area:
- ⌐ ¬ V1   ☐ V2

- ○ bunker (unused)
- ◡ supply site (unused)
- ● storage depot
- ✳ number of V1/V2 incidents by region

GENOCIDE IN EUROPE

The largest single body of civilian victims in the Second World War was Europe's Jewish population. They were systematically exterminated by the Hitler regime. By the end of the war only about 300,000 survived. Around six million perished.

Jewish genocide was the product of a state whose whole ideological outlook was rooted in racism. Hitler was a radical anti-semite, who had preached a final showdown with the Jewish people since the early 1920s. But there were those around him who wanted anti-semitism to be part of a general programme of ethnic cleansing and racial hygiene. Medical and scientific opinion was mobilized to support the sterilization – and eventually the extermination – of the physically and mentally handicapped, the hereditarily ill and those classified as "asocial". The search for a pure Germanic race was the obverse of the policies of discrimination and extermination directed at those regarded as un-German.

The formal persecution of the Jews began in September 1935 with the announcement of the Nuremberg Laws. These effectively denied Jews the same civil rights as other Germans. Over the next four years Jewish property was expropriated under a programme of "Aryanization", and during this time

A liberated prisoner from Buchenwald *(left)*, near Weimar. The inmates of this camp staged a revolt against their SS guards in April 1945 before the US army arrived.

A propaganda poster *(right)* showing a Soviet Jew murdering civilians. German authorities identified Bolshevism and Judaism as a common enemy, to be mercilessly rooted out.

Birkenau *(below)* was one of three concentration camps built around the town of Auschwitz (Oswiecim) in Poland. Founded in October 1941, it was used as an extermination centre using Zyklon-B gas. Some 1.2 to 1.5 million died at Auschwitz.

The German persecution of the Jews *(map below right)* moved from the discrimination against German Jews in the 1930s through the policy of isolation in ghettos in 1940 and 1941 to the final stage of physical extermination, aimed at the entire Jewish population of occupied and satellite Europe.

The persecution of the gypsies, 1939–45

- concentration camp with gypsy prisoners
- extermination camp with gypsy prisoners

1,000 estimated gypsy deaths 1939–45

GREAT BRITAIN
BELGIUM 500
FRANCE 15,000
SPAIN

THE FATE OF THE GYPSIES

Nazi race laws extended to cover the Sinti and Roma, popularly known as "gypsies". They had been subject to discrimination and harassment before 1933. The Nazi regime classified them as asocial and in 1936 a Reich Central Office for the Fight against the Gypsy Nuisance was set up. In September 1939 the decision was taken to round up Germany's 30,000 Sinti and Roma

half of Germany's Jewish population emigrated. A racist bureaucratic apparatus was developed under the head of the SS, Heinrich Himmler, which by 1939 began the official extermination of the disabled, euphemistically known as euthanasia.

The war changed the situation. Emigration, which was Hitler's preferred option for the "Jewish question", was cut off by the Allied blockade. Millions of Polish Jews were also brought under German control. The regime decided upon a policy of isolation. Jews and other "asocials" were rounded up and sent to designated ghettos, where they were denied adequate food and medical care. In the summer of 1940, after the defeat of France, Hitler toyed briefly with the idea of turning Madagascar into a vast tropical ghetto, but British control of the seas ruled it out.

In 1941 circumstances changed again. The war with the Soviet Union was defined by Hitler as a race war against Jewish-Bolshevism, and the armed forces and special SS death squads (the *Einsatzgruppen*) were instructed to murder Jews and communists indiscriminately. In the first two months of the campaign hundreds of thousands were killed with appalling brutality, some by native anti-semites in the Baltic States and the Ukraine, the large part by German soldiers, policemen and Himmler's agents. The precise date of Hitler's decision to exterminate all Jews – the so-called "Final Solution" – is unknown, but by the middle of July 1941, at the height of German victories in the east, a change was evident. Orders were sent out to end the programme of indiscriminate and public murders and to create a systematic killing programme, based on extermination centres using poison gas.

By the end of 1941 the system was in operation. At eight extermination centres millions of Jews from all over Europe were gathered. The young and fit were used for slave labour until they died from overwork or disease. The rest, including 1.5 million children, were sent straight to the gas chambers and then cremated. Their hair, glasses, gold fillings and shoes were collected and recycled for the war effort. Jewish property was seized by the state or stolen by the Nazi officials who ran the apparatus of death.

---

0   500 metres

woods

pits for burning bodies

woods

gas chamber and crematorium V

gas chamber and crematorium IV

'sauna' bathhouse

gas chamber and crematorium III

gas chamber and crematorium II

camp extension under construction

gypsy camp | men's camp | family camp

women's camp

SS guard dog kennels

SS barracks

Camp Commandant

quarantine camp

registration office

to Auschwitz

main gate

Birkenau death camp

perimeter fence

potato store

During the second half of the war Germany put pressure on its allies and satellites to release their Jewish populations. In Italy and Hungary this pressure was resisted until German occupation in 1943 and 1944. In the occupied zones the army and police were able to transport Jews to the eastern extermination centres with little resistance. In Warsaw in April 1943 the Jews in the ghetto rose in armed revolt, but the rising was brutally crushed. Those who survived the genocide did so by joining partisan groups, hiding with non-Jews or working as slave-labourers. Some of those who perpetrated the race programme were shot or hanged after the war, but many have evaded justice to this day.

and expel them to a ghetto in occupied Poland. In October Himmler ordered all soothsayers, many of whom were drawn from the Sinti and Roma population, to be rounded up in a concentration camp. Not until December 1942 was the decision taken to find a "final solution" to the fate of these two peoples. They were sent to a special camp at Auschwitz, where more than 20,000 men, women and

children were gassed. Dr Josef Mengele used Sinti and Roma children for experiments. In the areas occupied in the east, the Einsatzgruppen killed an estimated 250,000 more. Only a fraction of the European population of Sinti and Roma survived the war. Not until 1982 did the German authorities officially recognize that the Sinti and Roma had been the victims of genocide.

Prisoners using boots taken from dead captives *(above)* as fuel in Belsen concentration camp. The camp was used mainly for Jews. It was part of an empire of SS camps which in 1945 still housed over 700,000 prisoners.

In the Soviet Union an estimated seven million civilians lost their lives. These victims *(right)* on the Kerch Peninsula in July 1942 were murdered by German soldiers.

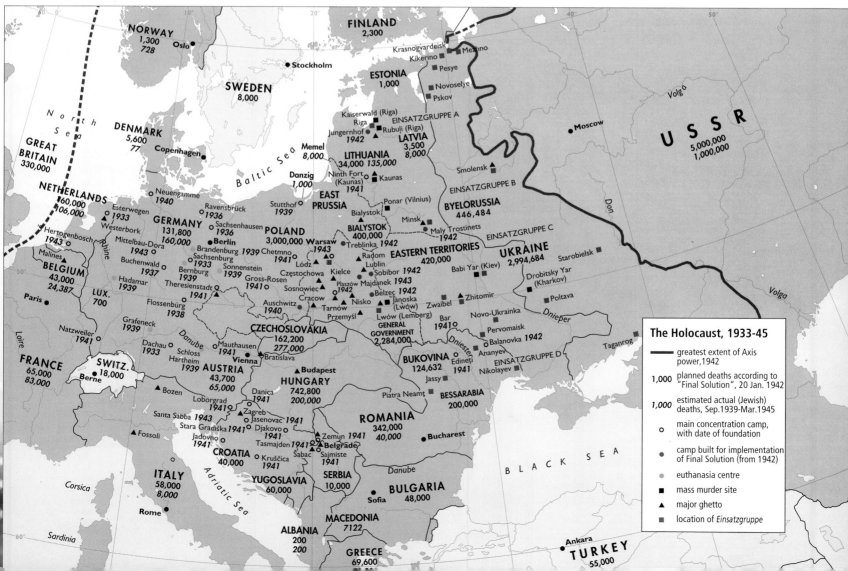

The Holocaust, 1933-45

▬▬▬	greatest extent of Axis power, 1942
1,000	planned deaths according to "Final Solution", 20 Jan. 1942
1,000	estimated actual (Jewish) deaths, Sep.1939-Mar.1945
○	main concentration camp, with date of foundation
●	camp built for implementation of Final Solution (from 1942)
●	euthanasia centre
■	mass murder site
▲	major ghetto
■	location of *Einsatzgruppe*

THE COSTS OF WAR

German genocide was the most conspicuous and terrible cost of the Second World War. The conflict was fought on an exceptional scale and with extraordinary ferocity. It fulfilled all the worst expectations of total war. Civilians no longer had any immunity from the conflict. In the east they were defined by German leaders as race enemies, to be exterminated or exploited. German and Japanese cities were blasted from the air on the grounds that the workers housed there sustained the fighting power of enemy forces at the front; Japanese soldiers were taught to regard the Chinese as racial inferiors. The laws of war, established over the previous half century by international agreement, were torn up. War became the instrument of a deeper ideological and racial conflict.

As well as the human cost, the war destroyed much of the physical and cultural environment in which it was fought. In the Soviet Union some 17,000 cities and 70,000 villages were destroyed. Half of Germany's housing stock in major cities was destroyed and much of the rest damaged. Italy lost one third of

Two elderly Germans *(below)* sit amidst the ruins of Berlin. During the war some 400,000 Germans were killed by the bombing and over four million dwellings destroyed. Almost nine million were evacuated to safer parts of Germany.

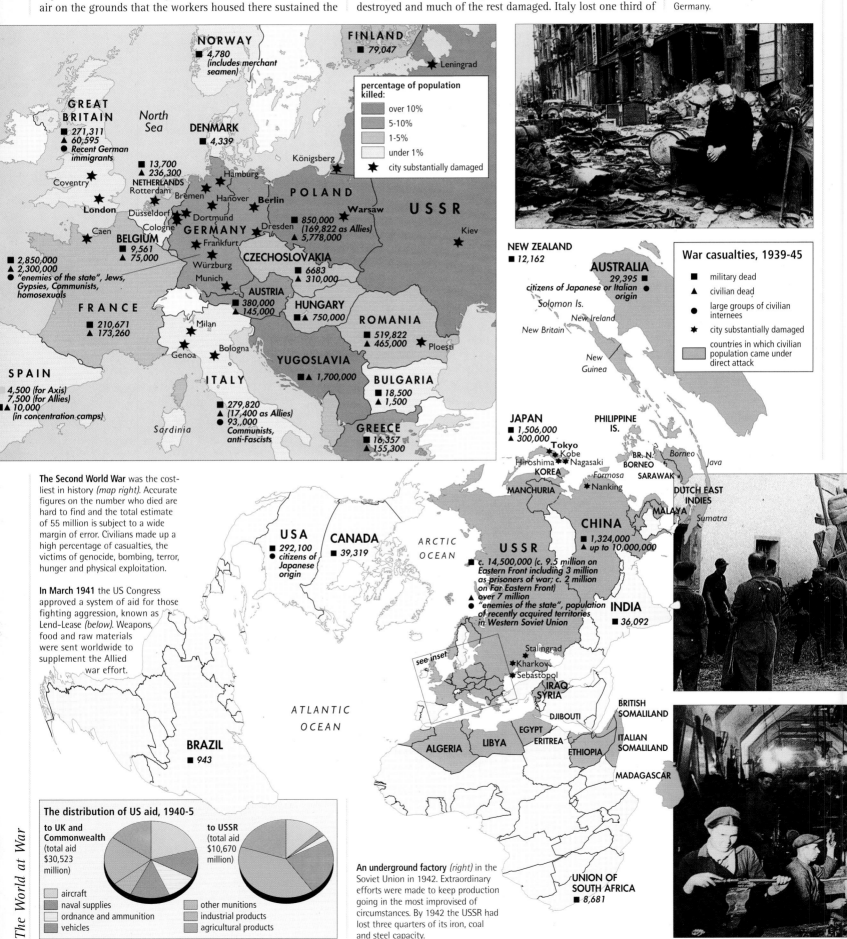

The Second World War was the costliest in history *(map right)*. Accurate figures on the number who died are hard to find and the total estimate of 55 million is subject to a wide margin of error. Civilians made up a high percentage of casualties, the victims of genocide, bombing, terror, hunger and physical exploitation.

In March 1941 the US Congress approved a system of aid for those fighting aggression, known as Lend-Lease *(below)*. Weapons, food and raw materials were sent worldwide to supplement the Allied war effort.

An underground factory *(right)* in the Soviet Union in 1942. Extraordinary efforts were made to keep production going in the most improvised of circumstances. By 1942 the USSR had lost three quarters of its iron, coal and steel capacity.

Map labels

percentage of population killed:
- over 10%
- 5-10%
- 1-5%
- under 1%
- ★ city substantially damaged

War casualties, 1939-45
- ■ military dead
- ▲ civilian dead
- ● large groups of civilian internees
- ★ city substantially damaged
- countries in which civilian population came under direct attack

NORWAY ■ 4,780 (includes merchant seamen)
FINLAND ■ 79,047
Leningrad

GREAT BRITAIN ■ 271,311 ▲ 60,595 ● Recent German immigrants
North Sea
DENMARK ■ 4,339
Königsberg
NETHERLANDS ■ 13,700 ▲ 236,300
Coventry
Hamburg
Rotterdam
Bremen
Hanover
Berlin
POLAND
London
Düsseldorf
Dortmund
Cologne
GERMANY
Dresden
Warsaw ■ 850,000 (169,822 as Allies) ▲ 5,778,000
USSR
Caen
BELGIUM ■ 9,561 ▲ 75,000
Frankfurt
Kiev
■ 2,850,000 ▲ 2,300,000 ● "enemies of the state", Jews, Gypsies, Communists, homosexuals
Würzburg
CZECHOSLOVAKIA ■ 6683 ▲ 310,000
Munich
FRANCE ■ 210,671 ▲ 173,260
AUSTRIA ■ 380,000 ▲ 145,000
HUNGARY ■▲ 750,000
Milan
ROMANIA ■ 519,822 ▲ 465,000
Ploesti
SPAIN 4,500 (for Axis) 7,500 (for Allies) ▲ 10,000 (in concentration camps)
Genoa
Bologna
YUGOSLAVIA ■▲ 1,700,000
ITALY ■ 279,820 ▲ (17,400 as Allies) 93,000 Communists, anti-Fascists
BULGARIA ■ 18,500 ▲ 1,500
Sardinia
GREECE ■ 16,357 ▲ 155,300

NEW ZEALAND ■ 12,162
AUSTRALIA ■ 29,395 ● citizens of Japanese or Italian origin
Solomon Is.
New Ireland
New Britain
New Guinea

JAPAN ■ 1,506,000 ▲ 300,000
Tokyo
Kobe
Hiroshima
Nagasaki
KOREA
PHILIPPINE IS.
BR. N. BORNEO
Borneo
Java
SARAWAK
Formosa
Nanking
MANCHURIA
DUTCH EAST INDIES
Sumatra
MALAYA

USA ■ 292,100 ● citizens of Japanese origin
CANADA ■ 39,319
ARCTIC OCEAN
CHINA ■ 1,324,000 ▲ up to 10,000,000
USSR ■ c. 14,500,000 (c. 9.5 million on Eastern Front including 3 million as prisoners of war; c. 2 million on Far Eastern Front) ▲ over 7 million ● "enemies of the state", population of recently acquired territories in Western Soviet Union
INDIA ■ 36,092
see inset
Stalingrad
Kharkov
Sebastopol
IRAQ
SYRIA
DJIBOUTI
EGYPT
ERITREA
ALGERIA
LIBYA
ETHIOPIA
BRITISH SOMALILAND
ITALIAN SOMALILAND
MADAGASCAR
ATLANTIC OCEAN
BRAZIL ■ 943
UNION OF SOUTH AFRICA ■ 8,681

The distribution of US aid, 1940-5

to UK and Commonwealth (total aid $30,523 million)
to USSR (total aid $10,670 million)

- aircraft
- naval supplies
- ordnance and ammunition
- vehicles
- other munitions
- industrial products
- agricultural products

its national wealth. European states under German rule paid out 150 billion marks – one quarter of Germany's war expenditure – as tribute to sustain the German war effort. The cultural heritage of Europe was the victim of war damage, such as that inflicted on the monastery of Monte Cassino, or of deliberate destruction, such as the burning down of the ancient university library in Naples by retreating German troops. At the end of the war the Western Allies set up vast collecting centres for lost and looted art. At Wiesbaden alone over 400,000 items were recovered, of which three quarters

A US airman *(above)* in 1943 trying on his British-made fire-fighting suit, one of the items supplied as return Lend-Lease by the British.

Across Europe cities in the path of the fighting, like this one *(above right)* in France, were destroyed. After the war French authorities calculated that France lost over one third of its national wealth

A German convoy *(below)* arrives at Passiria in northern Italy in 1944 loaded with looted art treasures. Thousands of priceless art works were lost or destroyed during the war.

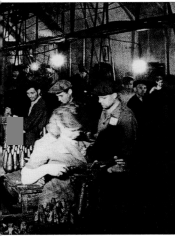

The new economic order in Europe, 1940-3

- ◾ (grey) Axis and Axis-occupied areas 1942
- ◖ coalfields and industrial regions
- ◖ other industrial regions
- ▲ crude oil plants
- ▼ synthetic oil plants
- 2,438 cost of German war effort borne by occupied or allied states (in RM million)
- ⚓ tonnage of merchant shipping seized by Axis Powers (in thousand tons)

mineral resources:
- ◼ bauxite
- ◇ chrome
- C copper
- L lead
- ◗ magnesite
- ◐ manganese
- ⊕ oil
- ☐ potash
- Z zinc

All the major warring states made exceptional demands on their industrial systems to turn out the weapons of war *(chart below)*. Up to two thirds of industrial output was devoted to war production.

During Germany's domination of Europe, plans were laid for a new economic order *(map above)* with Germany as the rich industrialized heartland surrounded by an outer circle of poorer agrarianized states.

Weapons production of the major powers 1939-45

	1939	1940	1941	1942	1943	1944	1945
Aircraft							
Britain	7,940	15,049	20,094	23,672	26,263	26,461	12,070
US	5,856	12,804	26,277	47,826	85,998	96,318	49,761
USSR	10,382	10,565	15,735	25,436	34,900	40,300	20,900
Germany	8,295	10,247	11,776	15,409	24,807	39,807	7,540
Japan	4,467	4,768	5,088	8,861	16,693	28,180	11,066
Major vessels							
Britain	57	148	236	239	224	188	64
US	–	–	544	1,854	2,654	2,247	1,513
USSR	–	33	62	19	13	23	11
Germany (U-boats only)	15	40	196	244	270	189	0
Japan	21	30	49	68	122	248	51
Tanks							
Britain	969	1,399	4,841	8,611	7,476	5,000	2,100
US	–	c.400	4,052	24,997	29,497	17,565	11,968
USSR	2,950	2,794	6,590	24,446	24,089	28,963	15,400
Germany	c.1,300	2,200	5,200	9,200	17,300	22,100	4,400
Japan	c.200	1,023	1,024	1,191	790	401	142

remained unclaimed five years after the end of the war.

As well as the loss of physical resources through war damage and expropriation, the war made great demands on the productive economies of every warring state. Up to two thirds of industrial output was devoted to war, and consumers everywhere, except in the United States, which had generous supplies of foodstuffs, were forced to eat rationed quantities of food and consumer goods. The greatest level of sacrifice was imposed in the Soviet Union, which lost half its meat and grain output with the occupation. The Soviet workforce consisted largely of women and young boys, forced to work long hours with poor food supplies, sometimes at the point of a gun. By 1944 half the native workforce in Germany consisted of women, many of them on German farms keeping up vital supplies of food. In Britain and the US over one third of the workforce was female by the end of the war.

The war encouraged the development of new technologies, particularly in aeronautics, rocketry and electronics. In Germany, rockets, jets and nuclear power were already in the process of development at the start of the war. They failed to make an impact on Germany's war effort only because of the confused nature of the wartime administration and Hitler's arbitrary intervention in the German research programme. Elsewhere the quality of weapons was often below that of Germany, but produced in much larger quantities using new mass-production techniques made possible by the flow of modern specialized machine tools from the US. In the Soviet Union spectacular improvements were achieved in the quality of weapons, which were produced in giant plants with relatively simple methods. The Soviet Union alone, despite wartime losses of industry and raw materials, outproduced Germany.

The war interrupted the normal development of the wider world economy. The United States was a net gainer. Here incomes rose by 75 per cent over the war and profits boomed. As European exports disappeared, the US picked up a large proportion of world trade. European economies, by contrast, were set back years. In the autumn of 1945 German industrial production was down to just 14 per cent of its level in 1936. Britain, though a victor power, had large debts and a declining trade position and by 1947 was close to bankruptcy. The economic balance of power moved decisively to the United States as a result of the war. In 1944 at a meeting at Bretton Woods, New Jersey, US officials got Allied powers to agree to a new economic order after the war, based on liberal trade and US-backed monetary co-operation.

THE GLOBAL IMPACT

The shift in the economic balance of power at the end of the Second World War in favour of the United States was matched by a corresponding shift in the political balance. In 1939 Europe was still thought to be the leading force in world affairs, and Hitler had ambitions to turn Germany into a global super-power. By 1945 the United States and the Soviet Union, both of which had stood aside from Europe's quarrels in the inter-war years, became major players in the world order as a result of their efforts to defeat the Axis powers.

In 1939 the war was not a global one but consisted of a number of different conflicts: a German-Polish war; a war between Germany and the French and British empires; and a war between Japan and China. Over the next two years the various conflicts merged. In 1940 Italy declared war on Britain and France, bringing the whole Mediterranean and Middle East into the war sphere. In 1941 the Balkans were drawn in. When Germany attacked the Soviet Union in June 1941 and Japan attacked the US in December, the war became genuinely global in extent, linked by the fact that the Axis faced a common enemy in the US and Britain, stretching across three oceans.

The three Axis states, Germany, Italy and Japan, had no formal military alliances. They collaborated poorly. Italy and Japan did not inform Germany of their attacks on Greece and Hawaii; Germany did not reveal the assault on the Soviet Union. When Italy surrendered in 1943 Germany treated the northern provinces as it treated the rest of occupied Europe. The Allies had no binding military alliance either, although Britain and the USSR signed a pact of mutual cooperation in May 1942. But the three major Allies were bound by an informal commitment to the unconditional surrender of the Axis, announced at the Casablanca Conference in January 1943 by President Roosevelt without prior consultation with his allies. They were also bound by the terms of the Atlantic Charter, first signed by Churchill and Roosevelt in August 1941, but signed subsequently by other states joining the Allied side.

The Atlantic Charter was a commitment to create a free world after the end of the war on the basis of the self-determination of peoples. It had a strong Wilsonian flavour, but after the failure of the First World War American leaders were committed to making the new post-war order work. Roosevelt referred to the Allied powers as the United Nations, a term first employed in January 1942. By the end of the war, 45 states, including the Soviet Union, had subscribed to the ideals of democratic freedom. In reality agreements were made between the three major Allies at a series of conferences – Teheran in November 1943, Yalta in February 1945, Potsdam in July of that year – which made a mockery of self-determination. Furthermore the Soviet Union was determined to safeguard its interests in Eastern Europe at the expense of popular nationalism, particularly in Poland.

Nonetheless the commitment to the principle of national independence had implications for the empires that Britain and France had gone to war in 1939 to defend. During the war the empires faced serious challenges. Nationalism, particularly in India, became a potent force in the face of British and French defeats. Japan and Germany encouraged independence movements or set up puppet regimes. Both the United States and the USSR were hostile to colonialism and keen to encourage decolonization. The moral authority of the remaining imperial states was blunted by the defeat of Italian, German and Japanese imperialism, and the empires were gradually relinquished over the 30 years following the war.

The defeat of the Axis exposed the transformation of the world order. The Soviet Union dominated Eastern Europe. The United States was now both willing and able to take the lead in the Western world. In April 1945 a conference was convened at San Francisco to establish formally a United Nations organization. Disagreements between the US and the USSR were resolved sufficiently for the new structure to be set up. The US, the USSR, China, Britain and France became permanent members of a security council. By the time the conference ended in June, Germany was defeated. The prospect now lay open for the establishment of a new world order based on principles of peaceful co-operation and national independence.

During the Second World War almost the entire globe was drawn into the conflict *(map right)* with the exception of four neutral states in Europe. The European empires drew on the resources and manpower of their overseas territories. Worldwide over 79 million men and women were mobilized in armed forces, and millions more drafted to war work.

ARCTIC OCEAN

ALASKA
Anchorage
Kodiak

Dutch Harbor

PACIFIC OCEAN

Pearl Harbor

San Diego

CANADA
▲ 10 Sep.1939
△ 8 Dec. 1941
↕ 0.78

USA
▲ 11 Dec. 1941
△ 8 Dec. 1941
↕ 11.49

MEXICO
▲ 22 May 1942

Key West

BAHAMAS

BR. HONDURAS

Guantanamo Bay

GUATEMALA
▲ 11 Dec.1941
△ 9 Dec.1941

Jamaica

HONDURAS
▲ 12 Dec. 1942

NICARAGUA
▲ 11 Dec. 1941

EL SALVADOR
▲ 12 Dec.1941

COLOMBIA
▲ 27 Nov. 1943

COSTA RICA
▲ 13 Jan. 1942 (Italy)
△ 8 Dec. 1941

PANAMA
▲ 13 Jan. 1942
△ 9 Dec. 1941

PERU
▲ △ 12 Feb. 1945

ECUADOR
▲ △ 2 Feb.1945

BOLIVIA
7 Apr. 1943 ▲ △

CHILE
▲ 14 Feb. 1945
△ 16 Feb 1945

Russian workers liberated in 1945 found a mixed reception on their return to the Soviet Union. Those who had served as auxiliaries in the German armed forces, like these pictured *(above)*, suffered imprisonment or execution on their return. Soviet POWs were regarded by Stalin as traitors for not fighting to the death.

The Allies had the advantage of being able to operate on a global scale after the US entry into the war. The USA sent goods to every theatre and established a worldwide network of bases and supply depots *(map right)*.

Throughout the British empire native labour was recruited, some of it, like the soldiers of the King's African Rifles pictured here *(right)*, to serve in military units, some as forced labour working on airfields and military bases.

THE BENGAL FAMINE

In 1943 over one and a half million Indians died of starvation in Bengal because of a famine largely induced by the disruptive effects of war. Bengal was a food-deficient area dependent on imports to meet the demands of its rapidly growing population. The Japanese occupation of Burma removed rice imports after April 1942 and pushed large numbers of refugees into the Bengali provinces. The demands of war led to transport shortages and rising food prices. Rather than export them, other regions kept back their surpluses to safeguard against wartime shortages. Poor harvests in January 1943 pushed prices up sixfold over the year. The authorities exacerbated the situation with a "Denial Policy", involving the purchase of rice surpluses to prevent them falling into Japanese hands in the case of invasion, and two thirds of local boats were requisitioned, making the transport of foodstuff to Bengal much more complicated. Between July and December 1943 thousands died. Estimates of deaths from famine ranged from 3.5 million to 1.5, the figure finally agreed by an official famine commission. Not until August 1943 were relief measures taken, too late to avert a disaster.

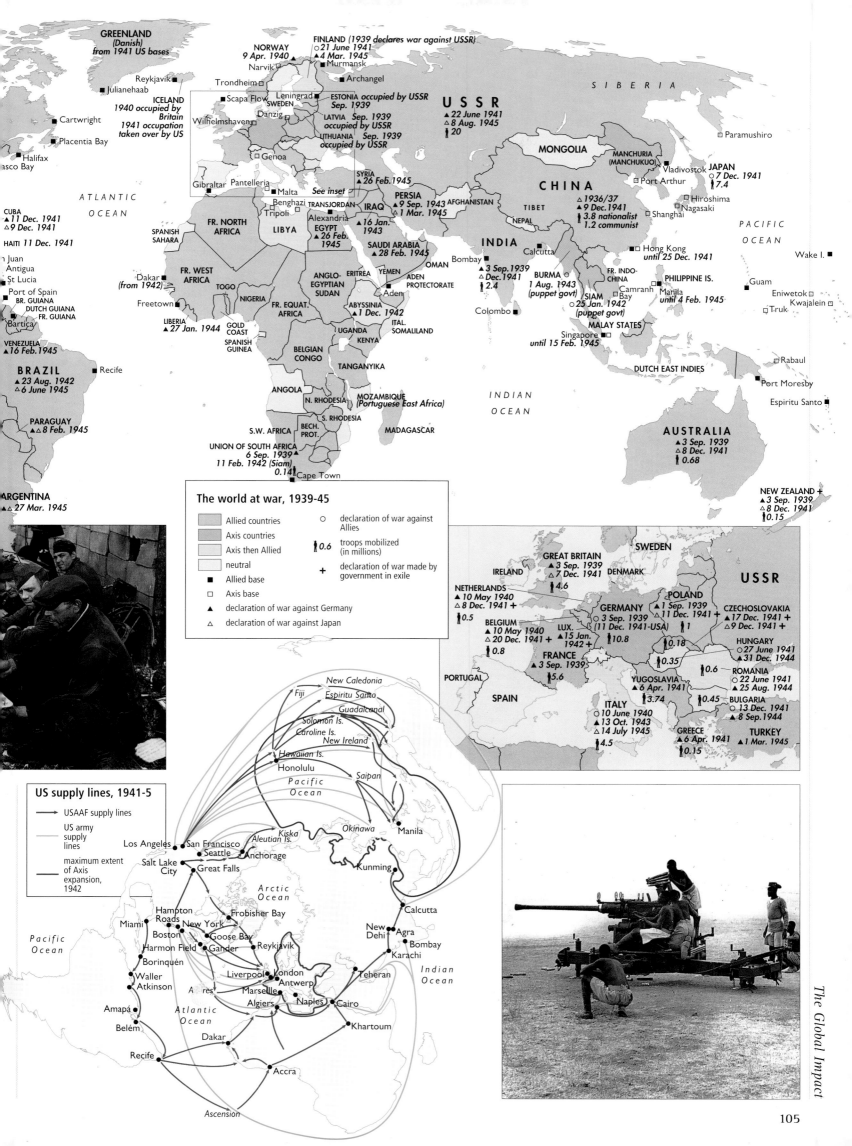

The world at war, 1939-45

US supply lines, 1941-5

THE END OF THE SECOND WORLD WAR opened up a new era in world affairs. The dominant position enjoyed by western Europe for more than a century disappeared. It was replaced by a system based around a growing hostility between the two major victors of the war: the United States and the Soviet Union. By the 1950s a state of Cold War existed between the American-led western bloc of democratic capitalist states and the Soviet-led communist world, which now stretched from central Europe to the Far East. The confrontation lasted for 40 years, during which fear of the use of nuclear weapons held both sides back from the brink of major war. In the shadow of the nuclear threat both sides enjoyed the most prolonged and expansionary economic boom in history, which transformed living-standards and life-styles in the developed world and hastened the pace of modernization, with all its costs, in every other region.

East German border guards, Berlin 14 August 1961

THE ROOTS OF THE COLD WAR

THE DEFEAT OF THE AXIS states in 1945 depended on the survival of an unlikely alliance between two democratic Western states and the communist Soviet Union. They were united by a common hostility to Hitler but by little else. The long history of tension and mistrust between the Soviet Union and the Western world began to resurface as victory over Germany drew closer.

The most serious issue was the political future of the reconquered lands. Though the three major Allies had agreed not to reach a separate peace with any of the enemy states, Britain and the United States accepted Italian surrender in 1943 without Soviet involvement. Stalin took this as his excuse to act on his own in the liberated states of Eastern Europe. As they were occupied one by one by the Red Army, the Soviet Union excluded Western intervention and set out to create a system of satellite states friendly to Soviet interests.

Though the Western Allies recognized that there was little they could do to prevent Soviet domination, there were two states in whose fate they had a real political interest: Poland and Germany. Poland, for which Britain had ostensibly gone to war in 1939, was the subject of tense negotiation at the Teheran Conference in November 1943. It was agreed that Poland must relinquish the areas seized by the Soviet Union in 1939 and should be compensated instead with territory carved out of eastern Germany. The settlement was confirmed at the Potsdam Conference in July 1945, by which time a pro-Soviet government had been installed in Warsaw. Although the Soviet Union paid lip service to the idea of establishing a popular democracy in Poland, in practice Polish communism was ruthlessly promoted. Elections in 1947 returned a communist majority and Poland came firmly within the Soviet sphere.

The German question was equally delicate. The Allies had agreed in 1943 to divide Germany into Allied zones of occupation. The demarcation lines were scrupulously observed by both sides, even though their respective forces had arrived at rather different points by the end of the war. It was agreed not to impose a peace settlement as had been done in 1919. Reparations were seized, mainly by the Soviet Union, and German political and military leaders put on trial at Nuremberg in November 1945 before an international military tribunal composed of the four victor states. Germany was completely disarmed. There was no agreement, however, on

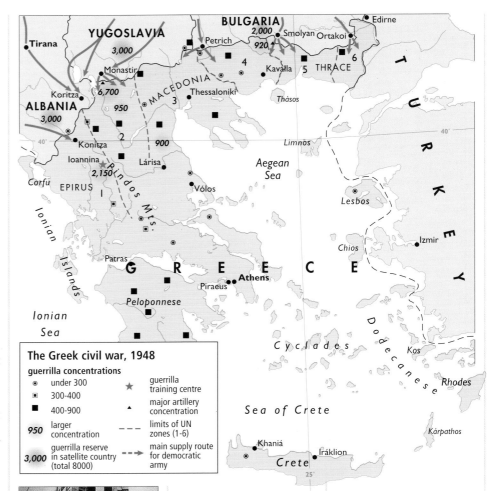

The Greek civil war, 1948

guerrilla concentrations

⊙	under 300	★	guerrilla training centre
▣	300-400	▲	major artillery concentration
■	400-900	- - -	limits of UN zones (1-6)
950	larger concentration		
3,000	guerrilla reserve in satellite country (total 8000)	⇢	main supply route for democratic army

The liberation of Greece in October 1944 provoked bitter conflict between the predominantly communist resistance, backed by the National People's Liberation Army (ELAS), and Western-backed monarchist forces *(map above)*. The war was waged for three further years until 1949. The eventual triumph of the right soured relations with the Soviet Union.

The Allies agreed to move Poland physically westwards after the war *(map right)*. Poland obtained East Prussia and large parts of eastern Germany but lost the eastern territories awarded at Versailles, which became part of the Soviet Union. Some three million Germans and three million Poles moved west with the frontiers.

The four Allied military chiefs, Field-Marshal Montgomery, Marshal Zhukov, General Eisenhower and the French General Koenig *(below)*, take the salute at a ceremony in Berlin in September 1945. Despite an outward show of solidarity, relations between the erstwhile Allies were already strained by disagreements over the political settlement in Eastern Europe and the future of Germany.

By 1945 communist anti-German resistance in Yugoslavia was strong enough to wage open warfare against the occupying power. The leader of the movement was Josip Broz, known as Tito, shown here *(above right)*. He formed a provisional communist government in March 1945. He maintained a policy of independence from the USSR after 1948.

the future political shape of Germany. Co-operation between the zones was limited, until in 1947 Britain and the United States created a joint area, known as Bizonia, which they began to see as the kernel of a new German state. With the Soviet Union vehemently opposed to a US-dominated capitalist state in Central Europe, the German issue became a symbol of the wider conflict forming between communist East and capitalist West.

This conflict was clear by 1946. On 5 March, in Fulton Missouri, Winston Churchill told an American audience that an "Iron Curtain" had descended across Europe, separating the democratic peoples of the West from the new communist bloc. The hopes which had been expressed by US statesmen for "one world" after the war based on self-determination and economic freedom were replaced by fears of a new polarization, christened by the American journalist Walter Lippmann the "Cold War".

Soviet leaders were equally anxious about the new world order and saw US ambitions as every bit as imperialist as those of European fascism. In 1946 Stalin ordered Western communist parties on to the political offensive. In Greece a civil war, which had begun during the Second World War, threatened a communist takeover. In 1946 the United States embarked on a strategy of "containment" aimed at restricting the further expansion of communism worldwide. The American president announced the Truman Doctrine, promising aid to any peoples resisting internal and external threats to democratic freedom. The first beneficiary was Greece, where US aid in 1947 helped to turn the tide against the communist guerrillas – a pattern to be repeated many times during the Cold War era.

Germany and Poland, 1945-9

Allied zones of occupation

- US
- British
- French
- Soviet
- ▬ boundaries between zones of occupation
- jointly occupied cities

territories lost by Germany, Poland and Czechoslovakia

- lost by Germany to Poland
- lost by Germany to USSR
- lost by Poland to USSR
- lost by Czechoslovakia to USSR

frontiers from 1947

- ▬ Poland, 1947
- ▬ Federal Republic of Germany, 1949
- - - - The Saar, 1949
- ▬ frontiers, 1947

1

DENMARK

• Hamburg

Königsberg •

EAST PRUSSIA to Poland

NETH.

• Cologne

G E R M A N Y

• **Berlin**

• Dresden

POLAND

RHINELAND independent state

• Frankfurt

SAAR to France

FRANCE

BAVARIA independent

• Stuttgart

SUDETENLAND to Czechoslovakia

CZECHOSLOVAKIA

• Munich

SWITZERLAND

AUSTRIA

HUNGARY

2

DENMARK

EAST PRUSSIA

NETH.

HANOVER-NORTHWEST-GERMANY

PRUSSIA

• Berlin

POLAND

G E R M A N Y

SAXONY

HESSE

SAAR

FRANCE

BADEN-WURTTEMBERG BAVARIA

CZECHOSLOVAKIA

SWITZERLAND

AUSTRIA

HUNGARY

Allied plans for the partition of Germany, 1941-43

1 Stalin's plan, 1941
2 Roosevelt's plan, Teheran Conference, 1943

▨ internationalized area
── Germany, 1919
── other frontiers, 1919

Berlin, 1945-90

American sector
British sector
French sector
Soviet sector

━━ city borders
━━ the Berlin Wall, 1961-89
── Autobahn
── international railway
⊕ airport
▪ headquarters
◻ allied HQ
→ air corridor

Bernau •

Havel

REINICKENDORF

Falkensee •

Tegel ⊕

PANKOW

SPANDAU

WEDDING

WEISSENSEE

TIERGARTEN

PRENZLAUER-BERG

CHARLOTTENBURG

MITTE

FRIEDRICHS-HAIN

Neuenhagen •

LICHTENBERG

Gatow

WILMERSDORF

KREUZBERG

SCHÖNE-BERG

Spree

Rüdersdorf •

ZEHLENDORF

Tempelhof ⊕

TEMPELHOF

STEGLITZ

NEUKÖLLN

KÖPENICK

Potsdam •

Schönefeld ⊕

Schmöckwitz •

During the war the Allies developed numerous plans for the post-war partition of Germany *(maps above)*. Their aim was to prevent a revival of German political and economic power. The American Morgenthau Plan, dreamt up by Roosevelt's treasury secretary, aimed at the ruralization of Germany. In 1946 the Allies agreed to a limitation of industry plan for Germany, but the extreme plans to agrarianize the economy were dropped. The onset of the Cold War hastened the reformation of a German state divided between East and West. Berlin was similarly partitioned into zones of occupation *(map left)* and was physically divided by a guarded wall built in 1961.

German losses from the peace

territory lost	population (1939 census)	% of Germans	area (sq. ml)
East Prussia	2,488,000	92.8	14,280
East Pomerania	836,000	99.0	6,800
West Pomerania	2,105,000	100.0	10,470
Liegnitz	2,720,000	100.0	8,110
Upper Silesia	1,527,000	57.0	3,750

DENMARK

LITHUANIA

Danzig Free State ceded to Poland, 1947

• Vilnius

Kaliningrad (Königsberg) •

UNION OF SOVIET SOCIALIST REPUBLICS

Bremerhaven •

• Hamburg

EAST POMERANIA

EAST PRUSSIA

Bremen •

NETHERLANDS

Elbe

Szczecin (Stettin) •

Gdańsk •

WEST POMERANIA

Oder

zones economically united in 1948

Potsdam •

Berlin •

Vistula

Poznan (Posen) •

Brest-Litovsk •

Essen •

• Dortmund

Düsseldorf •

(German Democratic Republic from 1949)

• Leipzig

Warsaw •

Cologne •

P O L A N D

Łodz •

Bonn •

G E R M A N Y

Dresden •

Liegnitz •

L'vov (Lwów) •

BELGIUM

Rhine

Wrocław (Breslau) •

LUXEMBOURG

Frankfurt am Main •

UPPER SILESIA

SAAR

Prague •

added to economically united zones in 1949

Cracow •

FRANCE

• Nuremberg

C Z E C H O S L O V A K I A

Brno •

Košice •

• Stuttgart

Danube

• Munich

Linz •

HUNGARY

Salzburg •

Vienna •

Bratislava •

ROMANIA

SWITZERLAND

LIECH.

Dniester

A U S T R I A

ITALY

• Graz

YUGOSLAVIA

THE COLD WAR CONFRONTATION: 1947-63

The Cold War was at its most dangerous in the years between 1947 and 1963. During this period both the major superpowers, the United States and the Soviet Union, struggled to gain the lead in an arms race based on nuclear weapons, while smaller powers – Britain, France and China – developed nuclear weapons of their own.

The arms race underpinned the political confrontation between the two blocs. Until August 1949 the United States had a monopoly on the new weapon. With the explosion of a Soviet atomic bomb and the development in the following decade of thermonuclear bombs with ever greater destructive capacity, the strategic balance altered. The more anxious each side became about its security, the more effort was put into stockpiling weapons capable of obliterating a great part of the globe. With the development of rockets in preference to long-range bombers as the means to carry nuclear weapons, the Soviet Union was able to bring the threat to bear on distant American cities. Equally, by 1956 it was calculated that the US nuclear arsenal could inflict 200 million casualties on the Soviet population.

In practice the Cold War was conducted at a lower level of confrontation, using conventional weapons, political pressure and propaganda. During 1947 and 1948 the states of Eastern Europe – Poland, Hungary, Czechoslovakia, Romania, Bulgaria and Yugoslavia – all became communist and pro-Soviet. In reaction to Western efforts to reform the German economy the Soviet Union attempted to blockade Berlin and force the West to abandon the city. Between June 1948 and May 1949, 275,000 flights were made to bring supplies to the Western zones of Berlin. The Soviet decision to end the blockade heralded a shift in the German policy of both sides. The Western states set up a German Federal Republic to replace their zones of occupation on September 1949, and in October the Soviet zone became the German Democratic Republic.

In response to the communist domination of one half of Europe, the Western states set up the North Atlantic Treaty Organization (NATO) in April 1949, which provided a framework for military co-operation in the face of a common enemy. US forces and equipment, including nuclear weapons, were stationed in NATO countries. In an effort to strengthen the alliance and to provide a clear military frontier in Europe, the Federal Republic of Germany was admitted to the NATO alliance in 1955.

The first serious test of the new anti-communist alliance came outside Europe. In 1950 war broke out in Korea between communist north and democratic south, following a partition agreed in 1948 between the United States and the USSR (see pages 130-1). US forces were dispatched to save the south and

large-scale re-armament began. Under pressure the United States' allies within Europe, as well as other United Nations states, gave assistance. The war dragged newly communist China in on the side of North Korea, and a long war of attrition set in until 1953. In this case "containment" was shown to work and the Korean partition was reimposed.

During the 1950s both sides avoided an open confrontation. Each used the promise of arms, money and political protection to win smaller states over. US influence was brought to bear worldwide as she came to assume the role of the world's policeman. In the late 1950s, following the successful launch of the first space satellite (in 1957), Krushchev, Soviet leader since 1956, embarked on a more aggressive strategy of expanding Soviet influence in the developing world. The strategy soon fell apart. The American fear of a "missile gap", prompted by the Soviet space programme, provoked a massive increase in US military procurement, which took the United States to a real lead in the arms race by the early 1960s.

The communist bloc was faced with its first serious crisis when China rejected Soviet collaboration in 1960. When Krushchev tried to recover the Soviet position by putting pressure again on Berlin and later by placing missiles in Cuba – thereby directly threatening the United States – firm US resistance forced him to back down. The Cuban crisis of 1962 marked a turning point in the Cold War. The following year the two sides agreed to a nuclear test-ban treaty, and the tension between the two blocs began to give way slowly to a mood of détente.

On 23 June 1948 the Soviet Union set up a blockade of the Western sectors of Berlin. The Western states mounted an airlift *(above)* until May 1949, which brought in 2.3 million tons of supplies.

Total ground forces in the Cold War, 1973	
NATO STATES	
Belgium	89,600
Britain	344,000
Canada in Europe	5,100
Denmark	39,800
W. Germany	485,000
Greece	160,000
Italy	427,500
Netherlands	112,200
Norway	35,400
Turkey	455,000
US in Europe	289,000
Total in Europe	**2,919,400**
Canada	77,900
US	1,963,900
Total Western bloc	**4,961,200**
WARSAW PACT BLOC	
Bulgaria	152,000
Czechoslovakia	190,000
East Germany	132,000
Hungary	103,000
Poland	280,000
Romania	170,000
USSR in Europe	520,000
Total in Europe	**1,547,000**
USSR	2,905,000
Total Eastern bloc	**4,452,000**

The US's nuclear monopoly ended in 1949 with the USSR's development of an atomic bomb. The US retained a lead in the number of intercontinental missiles until the 1970s *(chart below)* and even then early US development of multiple-warhead technology compensated for increasing numbers of Soviet missiles. For shorter-range missiles the nuclear balance tipped in the USSR's favour from the 1950s.

The high point of the Cold War was reached in 1962 when the United States forced the Soviet Union to abandon its programme to deploy missiles and nuclear warheads in Cuba *(map right)*, where a pro-communist revolution had occurred in 1959. In October 1962 the United States blockaded Cuba, and the Soviet Union, rather than risk all-out war, agreed on 26 October to withdraw its missiles.

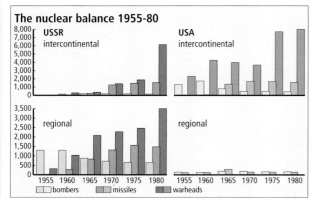

The nuclear balance 1955-80

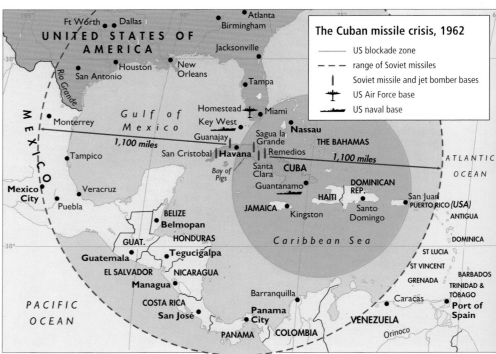

The Cuban missile crisis, 1962

- ——— US blockade zone
- - - - - range of Soviet missiles
- Soviet missile and jet bomber bases
- US Air Force base
- US naval base

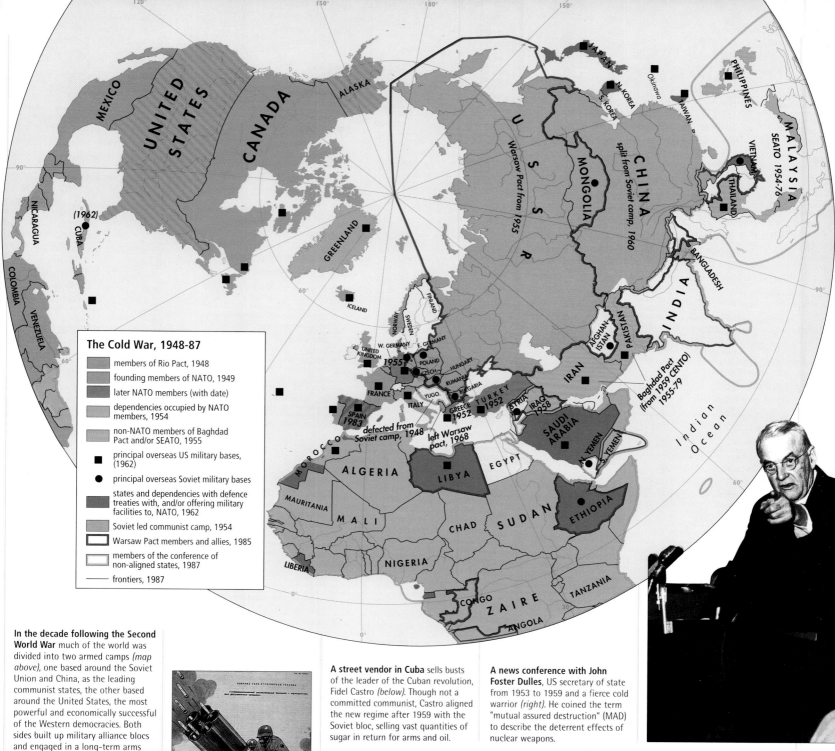

The Cold War, 1948-87

- members of Rio Pact, 1948
- founding members of NATO, 1949
- later NATO members (with date)
- dependencies occupied by NATO members, 1954
- non-NATO members of Baghdad Pact and/or SEATO, 1955
- ■ principal overseas US military bases, (1962)
- ● principal overseas Soviet military bases
- states and dependencies with defence treaties with, and/or offering military facilities to, NATO, 1962
- Soviet led communist camp, 1954
- Warsaw Pact members and allies, 1985
- members of the conference of non-aligned states, 1987
- — frontiers, 1987

Map labels: MEXICO, UNITED STATES, CANADA, ALASKA, NICARAGUA, CUBA (1962), COLOMBIA, VENEZUELA, GREENLAND, ICELAND, NORWAY, SWEDEN, FINLAND, UNITED KINGDOM, W. GERMANY, E. GERMANY, POLAND, CZECH., HUNGARY, FRANCE, ITALY, YUGO., RUMANIA, BULGARIA, GREECE 1952, TURKEY 1952, SPAIN 1983, MOROCCO, ALGERIA, LIBYA, EGYPT, MAURITANIA, MALI, CHAD, SUDAN, NIGERIA, LIBERIA, CONGO, ZAIRE, ANGOLA, TANZANIA, ETHIOPIA, N. YEMEN, S. YEMEN, SAUDI ARABIA, SYRIA, IRAQ 1958, IRAN, AFGHAN-ISTAN, PAKISTAN, INDIA, BANGLADESH, U.S.S.R, MONGOLIA, CHINA, N. KOREA, S. KOREA, JAPAN, Okinawa, TAIWAN, PHILIPPINES, MALAYSIA SEATO 1954-76, VIETNAM, (THAILAND), Indian Ocean

1955, defected from Soviet camp, 1948, left Warsaw pact, 1968, Warsaw Pact from 1955, split from Soviet camp, 1960, Baghdad Pact (from 1959 CENTO) 1955-79

In the decade following the Second World War much of the world was divided into two armed camps *(map above)*, one based around the Soviet Union and China, as the leading communist states, the other based around the United States, the most powerful and economically successful of the Western democracies. Both sides built up military alliance blocs and engaged in a long-term arms race. The respective forces of the Warsaw Pact and NATO are set out in the chart *(above left)*.

The Soviet poster *(right)* shows the US as a warmonger, trading in weapons. The world arms trade *(chart below)* was dominated by both the super-powers until the 1970s. Global arms sales totalled more than $21 billion in the 1960s, and more than $70 billion in the 1970s.

A street vendor in Cuba sells busts of the leader of the Cuban revolution, Fidel Castro *(below)*. Though not a committed communist, Castro aligned the new regime after 1959 with the Soviet bloc, selling vast quantities of sugar in return for arms and oil.

A news conference with John Foster Dulles, US secretary of state from 1953 to 1959 and a fierce cold warrior *(right)*. He coined the term "mutual assured destruction" (MAD) to describe the deterrent effects of nuclear weapons.

World arms exports, 1963-84

(Line chart, % share vertical axis 0–50, horizontal axis years 1963, 1968, 1973, 1978, 1984; lines labelled US, USSR, other NATO, other Warsaw Pact)

THE COLD WAR CONFRONTATION: 1963-85

In the 20 years following the Cuban missile crisis, the level of tension of the Cold War was reduced but by no means eradicated. The Sino-Soviet split and the economic revival of Europe created a more multi-polar system and also forced the Soviet Union to adjust to two rival major powers rather than one. The political and ideological conflict was shifted to Asia, Africa and Latin America. In 1963 the US committed its forces fully to the civil war in Vietnam (*see* page 130), and at its peak had 543,000 troops stationed in South East Asia. The US also actively supported anti-communist forces throughout Latin America, intervening in Nicaragua, Chile and, in 1983, the Caribbean island of Grenada. The USSR also increased its activities in the developing world, often by proxy. Soviet-backed Cuban forces participated in civil wars in the Horn of Africa and the former Portuguese colony of Angola.

There were, nonetheless, clear signs that the unstable super-power conflict of the 1950s was waning. In the United States and throughout the Western world there was strong hostility to US involvement in Vietnam (and more muted hostility to the Soviet intervention in Czechoslovakia in 1968).

During the 1960s and 1970s about 70 per cent of all satellites launched into space were for military purposes *(chart above)*. Satellites could provide photographic intelligence and electronic surveillance or act as communications centres for the military on the ground and in the air. Between 1945 and 1981 there were 1,321 nuclear tests, the great majority undertaken by the US and the USSR. Only France and China have persisted with tests above ground after the 1963 test-ban treaty *(chart above right)*.

The rise of an international peace movement challenged the ideological foundations of confrontation, which the post-war generation found difficult to understand. Links between the two opposing sides were also established. Arms limitation talks, begun in 1969, were successful in producing the first of a number of agreements on cutting back nuclear weapons development (SALT I). Also in 1969 the West German chancellor, Willy Brandt, launched closer links with the East German state, leading by 1971 to a Four-Power Agreement on Berlin and relaxing the harsh confrontation established there since the 1940s. The United States moved to mend its bridges with communist China. President Nixon signed an historic agreement with Mao's China during a visit to Peking in February 1972, which restored political and trade relations. Finally, in

ARMS LIMITATION NEGOTIATIONS

The great cost and destructive power of nuclear weapons encouraged the two super-powers to begin a cautious process of disarmament in the 1960s. In November 1969 talks began on a Strategic Arms Limitation Treaty, which finally came into force on 3 October 1972. When the Soviet leader, Leonid Brezhnev, visited Washington in June 1973, the two states signed an Agreement on the Prevention of Nuclear War. Talks continued in the 1970s on a second round of arms limitations (SALT II), but although the talks led to a final agreement signed in Vienna on

18 June 1979, limiting the number of long-range missiles, attitudes in the United States hardened against further disarmament and SALT II remained unratified. President Ronald Reagan publicly sustained a hostile stance towards the Soviet Union, but in 1982 initiated a fresh round of negotiations, the Strategic Arms Reduction Talks (START). Despite arms limitation, the modernization of both nuclear arsenals in the 1970s left both the US and USSR with greater destructive power in the 1980s than they had had in the 1960s. Serious force reductions only began with the onset of the Gorbachev administration in the Soviet Union in 1985.

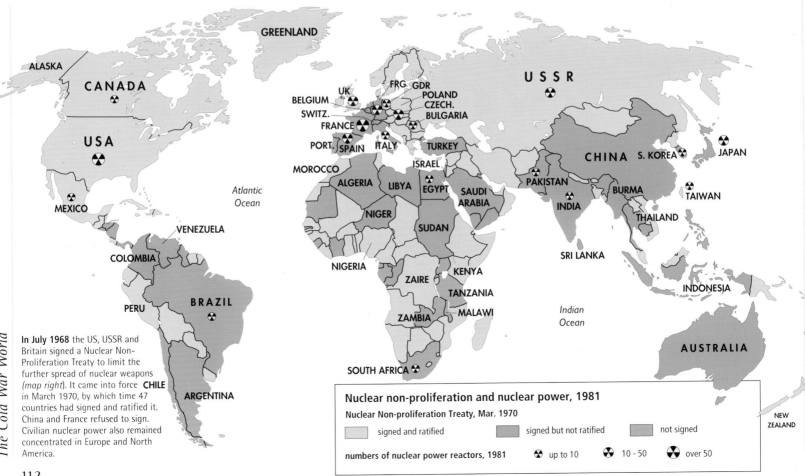

The Cold War World

In July 1968 the US, USSR and Britain signed a Nuclear Non-Proliferation Treaty to limit the further spread of nuclear weapons *(map right)*. It came into force in March 1970, by which time 47 countries had signed and ratified it. China and France refused to sign. Civilian nuclear power also remained concentrated in Europe and North America.

Nuclear non-proliferation and nuclear power, 1981

Nuclear Non-proliferation Treaty, Mar. 1970

    signed and ratified   signed but not ratified   not signed

numbers of nuclear power reactors, 1981   ☢ up to 10   ☢ 10 - 50   ☢ over 50

•	Nato bases	⊠	infantry division
•	Warsaw Pact bases	⊠	armoured infantry division
	US units	⊡	armoured division
	British units	⊠	armoured cavalry division
	French units	·	artillery brigade
	West German units	⊙	airborne division
	East German units	⬭	mountain division
	Russian units	⊞	corps HQ
		▭	army HQ

August 1975, after three years of negotiation on security and co-operation in Europe (CSCE), the Helsinki Accords were signed. Thirty-three European states, as well as Canada and the US, were party to the agreement, which recognized Europe's existing frontiers and committed all parties to observe human rights and improve communication between the power blocs.

Despite the sense that Helsinki had ushered in a new age of security, the Cold War had one last gasp. The administration of President Jimmy Carter (1976-80) produced a powerful mood of anti-communism, partly based on Democratic hostility to the Soviet human rights record, partly on renewed American fears that arms limitation had narrowed the gap dangerously between US and Soviet military strength. Once again that fear prompted a sharp increase in US military spending, and the development of a new range of weapons – the neutron bomb, multiple warhead systems, Cruise missiles – which destabilized the military balance and aroused Soviet anxieties. Carter's hostility was also based on the Soviet decision to invade Afghanistan in 1979 (see page 124).

There followed a five-year period of US posturing, first by Carter, then by his Republican successor, Ronald Reagan. Playing on US uncertainties about its position in the world following defeat in Vietnam in 1973, Reagan returned to the rhetoric of the 1950s. The Soviet Union became the "evil empire"; arms spending doubled between 1980 and 1985; and work was begun on the Strategic Defence Initiative (SDI), a space-based defence system. Behind the scenes Reagan's United States was less flamboyantly hostile. The basis was laid for what in 1985 became a period of true détente and the final end of the Cold War.

Nothing more clearly symbolized the Cold War era than the Berlin Wall *(below)*. Built in 1961 by the East German regime of Walter Ulbricht, it was designed to keep out "imperialist riff-raff and teddy boys", and to keep in the East German population which had been crossing to the West in large numbers in the 1950s. Between 1961 and the destruction of the wall in 1989 hundreds of East Germans tried to escape across the tangles of barbed wire. Many died in the attempt, victims of the wider Cold War in which divided Germany played a major part.

The United States's war in Vietnam provoked widespread popular hostility in the Western world in the late 1960s. Here *(left)* in Grosvenor Square, London, in March 1968, demonstrators clash with police in front of the US Embassy.

Throughout the Cold War the USSR deployed a larger number of short- and medium-range than intercontinental missiles *(map right)*. In 1979 the US stationed theatre nuclear weapons in Europe to face the Soviet challenge. This provoked strong resentment in Western Europe, which saw itself as the nuclear battlefield.

Divided Germany remained a critical area of confrontation, providing the "corridor" into Western Europe for any Soviet attack *(map above)*. NATO designated Germany the Central Front, and large forces were deployed on both sides.

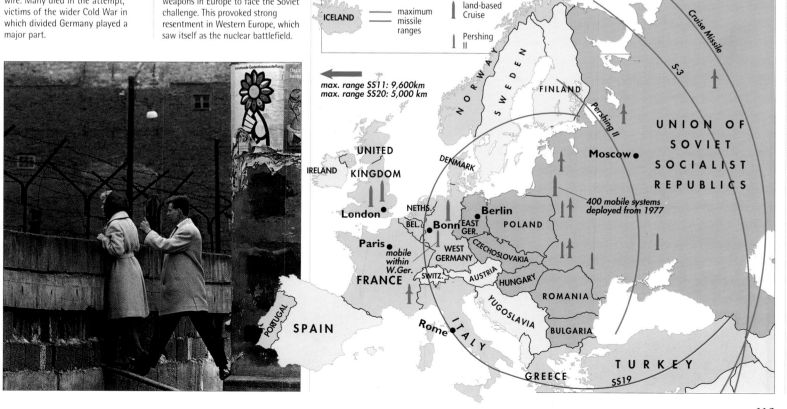

Intermediate-range land-based missiles in Europe after 1973

	NATO countries
	Warsaw Pact countries

NATO missiles
↑ S3
↑ land-based Cruise
↑ Pershing II

Warsaw Pact missiles
↑ SS11 ↑ SS20 ↑ SS19

——— maximum missile ranges

← max. range SS11: 9,600km
max. range SS20: 5,000 km

400 mobile systems deployed from 1977

mobile within W.Ger.

EUROPE: FROM RECONSTRUCTION TO UNION

The Europe which helped to generate the Cold War was a region of extraordinary desolation in 1945. More than 30 million people had been killed, and 16 million permanently resettled. Even a year after the war's end, industrial production was barely one third of the pre-war level and food production only half. The war also left a legacy of deep bitterness: collaborators with fascism were imprisoned, murdered or ostracized. In the ruins of war there were very real fears of social and economic collapse.

Revival depended on the US, the only state with the economic resources to invest in reconstruction. Through the United Nations Relief and Reconstruction Administration (UNRRA) and the International Bank for Reconstruction and Development (later the World Bank) the US pumped $17 billion into the revival of European economies. In spring 1947 a further programme was set up by the US secretary of state, George Marshall, in an effort to stabilize European politics and stimulate industrial revival. The European Recovery Programme provided $11.8 billion between 1948 and 1951 for the supply of raw materials, machinery and food and investment in recovery projects. In April 1948 the 16 nations receiving the assistance set up the Organization for European Economic Co-operation (OEEC) to co-ordinate the aid programme. The USSR refused US assistance and compelled the states of Eastern Europe under its control to do likewise. In 1947 and 1948 communist regimes were established throughout Eastern Europe and a programme of Soviet-style economic modernization imposed (see page 124). Eastern Europe was effectively closed off from the economic and political influence of the Western world, its development dependent on the interests of the Soviet Union.

In the rest of Europe economic reconstruction was the key to political stabilization. By 1950 output of goods was 35 per cent higher than in 1938; by 1964 it was 250 per cent greater. Europe embarked on the longest and largest economic boom in its history, made possible by the liberalization of trade and the greater degree of state economic management. After 1945 European governments adopted policies to stimulate economic growth along lines advocated by the British economist John Maynard Keynes. The state provided investment, gave subsidies for modernization, used tax policies to stimulate demand and redistributed income through national welfare systems. The provision of welfare was designed to avoid the desperate poverty of the 1930s and to link socialist labour movements more closely with the prevailing capitalist system. The growth of "mixed economies" combining private enterprise with state regulation and the establishment of welfare states helped to produce more effective parliamentary systems in Europe than had been possible before the war.

Democracy was restored in Italy in 1946, in West Germany in 1949, in Austria in 1955. There were widespread calls for political collaboration, even political unity. In 1949 a Council of

European economic trading blocs, 1957-85

- Benelux customs union, 1947

Council for Mutual Economic Assistance (COMECON)
- COMECON members 1949
- subsequent members

The European Economic Community (EEC)
- EEC members 1957
- joined January 1973
- joined January 1981

The European Free Trade Association (EFTA)
- EFTA members late 1972

In an effort to establish more liberal trade after 1945 European states established a number of trading blocs within which goods could circulate more freely *(map above)*. In 1949 the USSR set up Comecon to regulate and promote Soviet bloc trade. In 1957 the European Common Market (the EEC) was established, and two years later Britain set up a rival European Free Trade Association, which she finally abandoned in 1973.

By 1939 more than a quarter of Western Europe's population worked on the land, most of them on small peasant farms. After 1945 there was a remarkable revolution in agriculture *(map right)*. Yields rose with the application of modern technology and biology, and numbers working on the land fell by two thirds in a generation.

A Berlin housewife hunts for coal scraps *(below left)* to supplement the ration of ten kilos for the winter. Conditions throughout Germany in 1945-6 were exceptionally harsh.

Europe was set up under the leadership of the Belgian politician Paul-Henri Spaak, its ambition to create a politically united Europe. The only progress made in the 1950s was on economic integration. In 1952 a European Coal and Steel Community was established between France, Germany, Italy and the Benelux countries to rationalize and co-ordinate their heavy industry. In 1957 they moved towards full economic union when they signed the Treaty of Rome establishing the European Economic Community. Britain stood aside because of her links with the Commonwealth. For the states that did join, the 1960s saw a boom in trade between them, and a decline in the national tensions that had brought war twice since 1900.

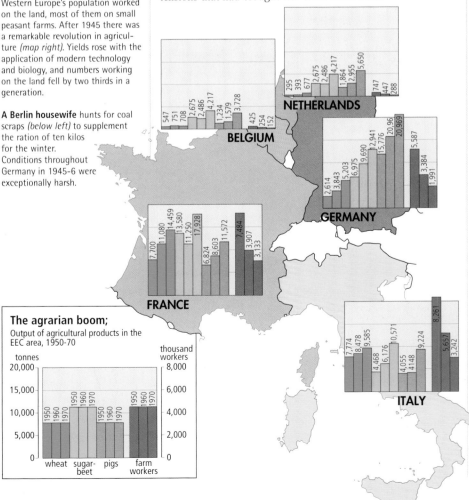

The agrarian boom;
Output of agricultural products in the EEC area, 1950-70

The French politician, Jean Monnet, *(below)*, who as a young man had helped to set up the League of Nations, became the foremost spokesman after 1945 for the rationalization and planned modernization of the European economy. He was a leading force behind the establishment of the European Coal and Steel Community in 1952 and a strong champion of the Common Market.

German workers restoring the Kurfurstendamm in Berlin in 1950 *(right)*, a project funded with Marshall Plan aid *(poster left)*. Marshall aid gave a much needed psychological boost to the faltering West European economies between 1947 and 1951. It was a positive statement of US intent to restore the economic health of Europe and thus contain the prospect of communist subversion. The economic aid came at a price. The British government was forced to cut back on social expenditure, which the US administrators of the Marshall aid programme distrusted.

In the spring of 1947 US Secretary of State George Marshall put his name to a European Recovery Programme backed by US money. The Marshall Plan *(map right)* was designed to stabilize democratic governments in Europe and revive the European trade area. It was granted to 16 Western countries, but it was also offered to the Soviet-dominated states of Eastern Europe, where Stalin insisted on rejection. In all, some US$11.8 billion was made available, in addition to funds offered under other recovery schemes. The scheme gave Europe access to foodstuffs, oil and machinery at a time when they lacked the dollars to buy from the United States directly.

The Marshall Plan, 1947-51

- applied for and received Marshall aid, with amounts (in US$)
- applied for Marshall aid but withdrew application
- did not apply

ICELAND $24 m

NORWAY $236 m

SWEDEN $118 m

FINLAND

DENMARK $257 m

UNITED KINGDOM $2.825 bn

IRELAND $116 m

NETH. $979 m

BELGIUM $546 m

LUX.

WEST GERMANY $1.297 bn

EAST GERMANY

POLAND

USSR

CZECHOSLOVAKIA

FRANCE $2.445 bn

SWITZ. Berne
applied for Marshall aid but received nothing

AUSTRIA $560 m

HUNGARY

ROMANIA

YUGOSLAVIA

PORTUGAL $50 m

SPAIN
refused Marshall aid because of Franco's pro-Axis sympathies

ANDORRA

ITALY $1.314 bn

BULGARIA

ALBANIA $515 m

GREECE $152 m

TURKEY

EUROPE: FROM ROME TO MAASTRICHT

The reduction in international tension in Western Europe owed much to the rapprochement between France and Germany which was achieved in the mid-1950s and sustained by the European commitment of successive German and French leaders, from Konrad Adenauer and Charles de Gaulle in the late 1950s to Helmut Kohl and François Mitterrand in the 1980s. Both states looked to build a strong European identity in the 1960s. De Gaulle was hostile to American influence in Europe and to the survival of Britain's wartime "special relationship" with the US. When Britain applied for membership of the EEC in 1962 de Gaulle vetoed her entry, and in 1966 he withdrew French forces from NATO.

It took another ten years before further progress was made on European integration, and by this time Britain, like the other colonial powers, had shed much of her global empire (see page 132). De Gaulle retired in 1969 and the new generation of politicians in EEC states were keen to extend the principle of integration. In 1973 Britain, Ireland and Denmark joined the EEC. After a period of growing economic crisis in the 1970s sparked by the increase in oil prices in 1973, and followed by rising unemployment and inflation, other states sought entry. Greece joined in 1981, Spain and Portugal joined five years later.

Despite the economic integration of Western Europe, little progress was made on political ties. Fears of losing sovereignty and arguments about currency reform and welfare policies divided the EEC members too sharply to fulfil the pledge made at the EEC Paris Summit in 1972 to produce political union by 1980. There also existed differences of opinion within the member states on the political future. While some groups favoured a supra-national Europe, there was an evident decline in the domestic consensus of the 1950s and 1960s that had sustained post-war welfare capitalism.

Although the post-1945 settlement left fewer issues of self-determination and minority nationalism than the Versailles Settlement after 1919, there remained areas of conflict which war and reconstruction left untouched (map below). The issue of Irish nationalism remained alive. In Northern Ireland the Irish Republican Army agitated for a union of Northern Ireland with the Irish Republic, and in the late 1930s, the late 1950s and the period from 1972 onwards conducted a campaign of violence on the British mainland to pressurize the British government. In Spain the Basque separatist movement, ETA, kept up a terrorist campaign against the Franco government. In Cyprus the Turkish and Greek communities resorted to civil war in 1974, which led to the island's partition. The growth of multi-national and supra-national organizations in Europe has not eroded the force of regionalism and irredentism.

Western Europe: separatism and nationalism, 1945-85

— territorial autonomy based on ethnic group, with date of autonomy

— separate administration/ autonomy for other reasons

✊ linguistic minorities or other communities whose members have used violence in pursuit of greater autonomy or other change of political status, with areas inhabited

✗ devolution rejected by referendum, with date

✓ devolution approved by referendum, with date

● ethnic or other communal based party delegated to national parliament in 1985

British soldiers in riot gear on the streets of Londonderry (left) in Northern Ireland in August 1985. It was here in 1972 that British troops opened fire on rioters on Bloody Sunday, killing 13 people. This was the signal for a state of near civil war in the province for the next three years, and resulted in direct rule from the British government in Westminster. The establishment of a Protestant-dominated assembly in 1982 provoked further violence between the Catholic and Protestant communities.

Map labels: Narvik, Trondhjem, Bergen, Oslo, Stockholm, Gothenburg, Åland Is. Swedes, NORWAY, SWEDEN, FAEROE ISLANDS Faeroe Islanders, 1948, Scots, SCOTLAND ✗1979, Glasgow, Edinburgh, Northern Irish Protestants, N. IRELAND, Northern Irish Catholics, Belfast, IRELAND, Dublin, Isle of Man, Newcastle, Cork, UNITED KINGDOM, Liverpool, Manchester, Welsh, WALES ✗1979, Birmingham, Bristol, London, North Sea, DENMARK, Århus, Copenhagen, Malmö, Baltic Sea, Amsterdam, NETH., Rotterdam, Hamburg, Bremen, Hanover, Berlin, EAST GERMANY, Leipzig, Dresden, POLAND, Flemings, mixed, Brussels, Walloons, BELGIUM, Belgian ethnic areas developing towards full autonomy, Flemings, Germans, Bonn, WEST GERMANY, Rhine, Elbe, Cherbourg, Channel Is., Seine, Paris, Rheims, Metz, Strasbourg, LUX., GERMANY, strongly autonomous regions, Orléans, Tours, Loire, Jurassiens (Swiss French), Jura Canton from 1979, Berne Canton to 1979, ✓1974, Munich, Stuttgart, Zurich, LIECHT., Vienna, CZECHOSLOVAKIA, strongly autonomous regions, AUSTRIA, HUN., Clermont-Ferrand, Lyon, Geneva, Berne, SWITZ., Germans, BOLZANO, FRANCE, Bordeaux, VAL D'AOSTA French, 1948, Turin, Milan, Venice, Trieste, FRIULI-VENEZIA GIULIA Friulians, Slovenes, 1963, TRENTINO-ALTO ADIGE Germans, Ladins, 1948, Nice, MONACO, Genoa, Bologna, Florence, SAN MARINO, Perugia, La Coruña, Gallegos 1981, ✓1980 GALICIA, Oviedo, Oporto, Basques, ✓1979 BASQUE COUNTRY Basques, 1979, Bilbao, NAVARRA, Toulouse, Basques, 1982, Saragossa, ANDORRA, ✓1979 CATALONIA Catalans 1979, Barcelona, Corsicans (Italians), Corsica, ITALY, Rome, PORTUGAL, Tagus, Lisbon, Salamanca, SPAIN, Madrid, other Spanish regions given some autonomy, 1979-83, Valencia, VALENCIA, Valencians 1982, Catalans, Balearic Is., Balearans (Catalans), 1983, Marseille, Sardinia, Sards 1948, Naples, Bari, Taranto, YUGOSLAVIA, ALBANIA, ✓1981 ANDALUSIA, Seville 1982, Cádiz, Malaga, Cartagena, Mediterranean Sea, Cagliari, Palermo, Sicily 1946, Messina, Catania

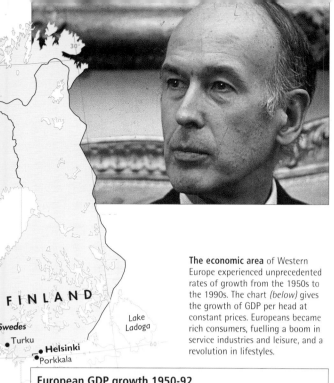

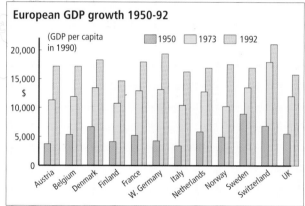

Valéry Giscard d'Estaing (*left*) became president of France in 1974. Candidate of the moderate right, he set about modernizing Fench society and encouraging European integration. His increasingly regal style of government, together with high inflation and unemployment, brought disillusionment, and in the 1981 presidential election he was succeeded by the socialist François Mitterand.

The success of the EEC in modernizing and rationalizing European agriculture produced regular food gluts in the 1970s and 1980s. Here (*below right*) Sicilian oranges have been turned, quite literally, into a food mountain. Under the Common Agricultural Policy farmers often found themselves the victims of bureaucratic decisions which challenged their livelihood. French peasants regularly resorted to physical protest. Here (*bottom right*) police attempt to remove a 20-ton lorry from a railway line in Brittany.

The crisis of national identity was in part a product of generational conflict. In the late 1960s Western Europe was hit by a wave of popular student-led protests against militarism and conservative values, much of which rejected the growth of consumerism and left-wing collaboration with capitalism. University reforms calmed down much of the student unrest, but in the 1970s terrorist groups emerged in Italy (the Red Brigades) and West Germany (the Baader-Meinhof gang), which waged a violent war against authority and the business community.

At the same time labour relations worsened as trade unions tried to retain the wage gains of the post-war period in an age of inflation. In a major strike-wave in 1972-5 over 150 million working days were lost in Britain, Italy, France and West Germany. States which had grown used to high levels of employment and government spending found Keynesianism no longer capable of sustaining growth. In the early 1980s Europe again faced high levels of unemployment while governments began to cut spending growth and to leave the economy to market forces. The result was further industrial unrest in France and Britain, where in 1984 a prolonged coal strike was used as an opportunity to undermine union influence. If by the late 1980s Western Europe was richer, more secure and more politically stable than it had been in the 1950s, there nevertheless remained underlying elements of crisis and uncertainty.

The economic area of Western Europe experienced unprecedented rates of growth from the 1950s to the 1990s. The chart (*below*) gives the growth of GDP per head at constant prices. Europeans became rich consumers, fuelling a boom in service industries and leisure, and a revolution in lifestyles.

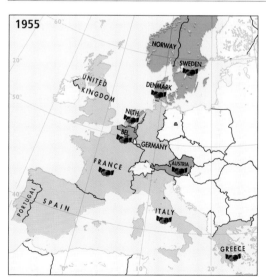

European GDP growth 1950-92

(GDP per capita in 1990)

legend: 1950 · 1973 · 1992

$: 20,000 / 15,000 / 10,000 / 5,000 / 0

Austria, Belgium, Denmark, Finland, France, W. Germany, Italy, Netherlands, Norway, Sweden, Switzerland, UK

Politics in Western Europe, 1955-85

political composition of governments

- right-wing dictatorship
- conservative/Christian Democrat
- socialist
- coalition government

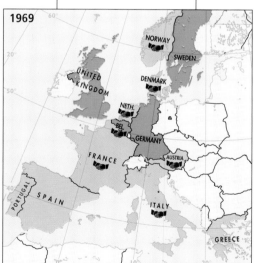

1955

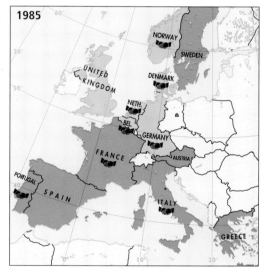

1985

1969

Turkish Federated State of Cyprus, estb. 1975; independent "Turkish Rep. of Northern Cyprus" from 1983

1975

ethnic distribution before 1974 · Turks · Greeks

In the 1950s West European politics were predominantly conservative – a reflection of the prevailing fear of the communist east (*maps above*). In the 1960s social democrat regimes appeared in Britain and Germany, and social democracy dominated the two states during the following decade. As socialist parties became more moderate they attracted support from the centre. Despite the swing to the right in Britain and Germany, led by the British conservative Margaret Thatcher as a reaction against growing state intervention and mounting budget expenditure, most of Western Europe in 1985 was ruled by moderate social-democrat governments.

THE SUPER-POWERS: US DOMESTIC POLICY TO 1985

Where Europe faced the devastation of war in 1945, the United States emerged from the conflict richer, more united and willing to play a full part in shaping the world economy and the international order. The wartime boom saw GNP grow by 50 per cent and average incomes increase by 75 per cent. For the next 25 years the US economy continued to boom, sustained by the application of science, a rapidly rising population and world demand for food and machinery from US producers. Americans were the world's richest consumers, enjoying a standard of living that was the envy of the rest of the world.

The United States' new role brought its problems. The spirit of pre-war isolationism had not entirely vanished, yet the wartime military-industrial complex was maintained after 1945, giving the military a degree of influence in American politics quite uncharacteristic of America's past. The Cold War, regarded by Americans as a contest between good and evil, fed back into domestic politics. In 1950 the Wisconsin senator Joseph McCarthy began an official campaign to root out communists in American politics and culture. McCarthyism had support from the nationalist right in the US, which resented the loss of China in 1949 and feared a vast fifth column of Soviet spies and agents. In 1954-5 anti-communism reached a frenzied finale of purges directed at anyone suspected of liberal sympathies. It ended only when McCarthy turned against the army. President Dwight Eisenhower, a wartime army hero, called a halt.

Many of those who hated communists hated blacks as well. For years the Southern black community had put up with segregation and a denial of civil rights. The war moved many to the north and west of the country, and eroded racism in the armed forces. In 1948 Truman made the forces fully open to American blacks. In 1954 the Supreme Court ruled that segregation in Southern schools should end. Across the South the non-black community prepared to fight. White citizens' councils were set up which engaged in acts of violence and intimidation, reaching a notorious climax in Little Rock, Arkansas in 1957, when black childen tried to enrol in the high school. Civil rights acts in 1957 and 1960 did little to dent this discrimination.

In 1960 Eisenhower was succeeded by a young Democrat,

John F. Kennedy, who promised a "New Frontier" in US politics. He symbolized the growing youth culture of the United States and the optimism of its prosperous classes. But he was also a strong Cold Warrior and did little for civil rights until his hand was forced by riots in Mississippi in 1963. That year Southern black groups launched a nationwide drive to end segregation, and 200,000 marched on Washington in August 1963 to meet the president. A new civil rights bill was drawn up, but three months later Kennedy was assassinated. Despite the new civil rights legislation in 1964 and 1965, which at last gave Southern blacks the vote, black communities in the west and north, numbering in 1960 some 7.5 million, reacted violently against impoverishment and ghettoization. From 1965 to 1968 rioting, arson and looting became familiar fare in America's cities.

In the 1970s the confident US of the post-war years was hit first by economic crisis following the sharp increase in oil prices in 1973, then by political crisis sparked off by the corruption of the Nixon administration, elected in 1968. The economic crisis was a profound psychological shock. The dollar was forced to float, in effect to be devalued, while the US faced a widening trade gap and intense competition from those very states she had helped to reconstruct in the 1940s. Income growth began to stagnate, and poverty and urban squalor became too large even for the world's richest state to solve.

The political crisis came partly from domestic hostility to the Vietnam War (see page 130), but it was largely sparked by the crisis surrounding the Nixon presidency. In 1972 Nixon aides began a campaign of "dirty tricks" against the Democrats in the presidential election. The subsequent cover-up organized by Nixon was exposed. The president resigned in 1974, a year after his vice-president, Spiro Agnew, had also resigned on corruption charges. The loss of the Vietnam War in 1973 left many Americans with a sense of uncertainty and a loss of direction, which was not assisted by the lacklustre presidencies of Gerald Ford (1974-6) and Jimmy Carter (1976-80). In 1980 Ronald Reagan won a Republican landslide on the promise of a nationalist revival, tax cuts and anti-Marxism. Though his presidency restored some of the battered morale of American white middle classes, the American consensus, like the European, had broken up. Blacks, Latins, women, gays were all organized in powerful and often radical lobby groups hostile to the mainstream white, male-dominated America of the old Cold War era.

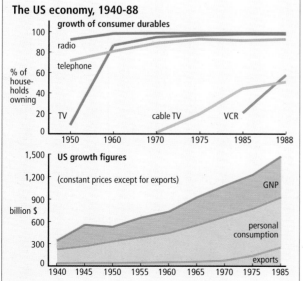

The US economy, 1940-88

growth of consumer durables

(% of households owning, 1950–1988; lines for radio, telephone, TV, cable TV, VCR)

US growth figures

(constant prices except for exports; billion $, 1940–1985; GNP, personal consumption, exports)

Growth of metropolitan areas, 1940-75

- 1940
- 1960
- 1975
} standard metropolitan statistical areas

— interstate highways

MARTIN LUTHER KING

Martin Luther King was the American Civil Rights movement's most prominent activist and victim. Born in Atlanta, Georgia, he took a doctorate in theology from Boston University and returned to the South, inspired by the example of Gandhi, to organize non-violent protest against the continued racism of the region. As leader of the Southern Christian Leadership Conference, he spread the practice of boycotts and sit-ins in shops, restaurants and schools. He was imprisoned 16 times during the campaign. By 1963, when he led the protest against segregation in Birmingham, Alabama, he had become the acknowledged leader of the movement. In 1964 he was awarded the Nobel Peace Prize. Over the next four years he took his campaign to the northern cities, where his oratory and organizational skills created a movement supported by black and white. By 1968 his Gandhian tactics were unpopular with the black urban radicals, who organized Black Power and rival bodies, and practised more violent confrontation. King himself was the victim of the hatreds he confronted. On 4 April 1968 he was murdered in Memphis, Tennessee, by a southern racist, James Earl Ray. There followed days of rioting in 125 cities, which needed 70,000 federal troops to suppress. King became the chief martyr of the cause of black freedom in the US.

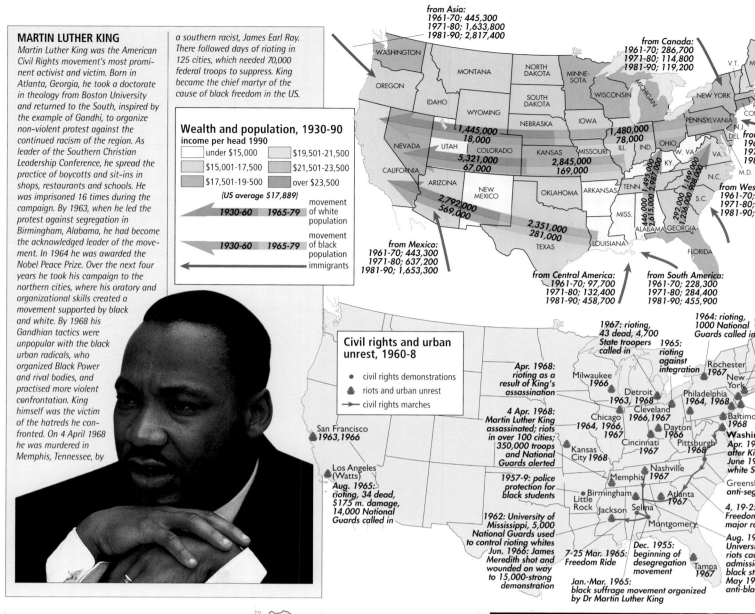

Wealth and population, 1930-90
income per head 1990

☐ under $15,000	☐ $19,501-21,500
☐ $15,001-17,500	☐ $21,501-23,500
☐ $17,501-19-500	☐ over $23,500

(US average $17,889)

→ 1930-60   1965-79 movement of white population

← 1930-60   1965-79 movement of black population

← immigrants

from Asia:
1961-70; 445,300
1971-80; 1,633,800
1981-90; 2,817,400

from Canada:
1961-70; 286,700
1971-80; 114,800
1981-90; 119,200

from Europe:
1961-70; 1,238,600
1971-80; 801,300
1981-90; 705,600

from West Indies:
1961-70; 519,500
1971-80; 759,800
1981-90; 892,700

from Mexico:
1961-70; 443,300
1971-80; 637,200
1981-90; 1,653,300

from Central America:
1961-70; 97,700
1971-80; 132,400
1981-90; 458,700

from South America:
1961-70; 228,300
1971-80; 284,400
1981-90; 455,900

Civil rights and urban unrest, 1960-8

● civil rights demonstrations
⚑ riots and urban unrest
→ civil rights marches

San Francisco
1963, 1966

Los Angeles (Watts)
Aug. 1965: rioting, 34 dead, $175 m. damage, 14,000 National Guards called in

Apr. 1968: rioting as a result of King's assassination

4 Apr. 1968: Martin Luther King assassinated; riots in over 100 cities; 350,000 troops and National Guards alerted

1957-9: police protection for black students

1962: University of Mississippi, 5,000 National Guards used to control rioting whites

Jun. 1966: James Meredith shot and wounded on way to 15,000-strong demonstration

7-25 Mar. 1965: Freedom Ride

Jan.-Mar. 1965: black suffrage movement organized by Dr Martin Luther King

Dec. 1955: beginning of desegregation movement

Milwaukee 1966

Detroit 1963, 1968

Chicago 1964, 1966, 1967

Kansas City 1968

Cincinnati 1967

Cleveland 1966, 1967

Dayton 1966

Pittsburgh 1968

Memphis 1967

Nashville 1967

Little Rock

Jackson   Selma

Birmingham

Atlanta 1967

Montgomery

Tampa 1967

1967: rioting, 43 dead, 4,700 State troopers called in

1965: rioting against integration

1964: rioting, 1000 National Guards called in

Rochester 1967

New York

Philadelphia 1964, 1968

Baltimore 1968

1967, 1968, 1964: racial rioting

Boston 1967

Newark *1967, 1968: rioting 26 dead*

New Jersey 1964

Washington DC
Apr. 1968: rioting after King's assassination; June 1968 black and white Solidarity March

Greensboro 1960: anti-segregation protests

4, 19-25 May 1961: Freedom Rides; first major racial confrontation

Aug. 1955: University of Alabama riots caused after admission of first black student; May 1963: anti-black riots

Starting in the 1950s there developed a widespread movement in the United States' Southern states for full civil rights for the black community *(map above)*. The movement began peacefully, but the savage reaction of Southern whites led to escalating violence. In northern and western cities black workers rioted in 1964 and 1965 against discrimination and poverty. When civil rights leader Martin Luther King was assassinated in 1968 a further wave of rioting swept the United States' major cities.

A young black American *(above right)* risks abuse and violence in one of the Freedom Rides organized in 1961 in defiance of Southern segregation laws. The first Freedom Ride in May 1961 ended with the burning of the bus by white opponents in Alabama.

During the post-1945 period there was large scale migration to US cities and, within the US, from the east coast to the west and from south to north *(maps left and top)*. The urban population grew from 69 million in 1950 to 158 million in 1990, when two thirds of all US citizens lived in metropolitan areas.

Veterans of the Vietnam war march in Washington in April 1971 *(right)* in protest against the war. By 1973 57,939 Americans had lost their lives in the conflict which cost the United States US$150 billion.

THE SUPER-POWERS: US FOREIGN POLICY TO 1985

At the dawn of the Cold War US statesmen were aware that they would have to shoulder responsibilities in the international arena that they had shunned before 1939. They hoped that the American new order would bring an age of international prosperity and of international collaboration. The first ambition proved easier to achieve than the second.

US leaders in 1945 were determined not to return to the bad old days of protectionism and economic nationalism of the 1930s. They persuaded the non-communist world to accept worldwide trade liberalization, which produced the General Agreement on Tariffs and Trade (GATT) in 1947. In Germany the US insisted on de-cartelization and anti-trust legislation, and the revived union organization was strongly influenced by the American desire to avoid a platform for labour radicalism. In Japan from 1948 the US occupiers began an economic reform programme, led by the president of the Bank of Detroit, Joseph Dodge, to revive Japan's economy along Western capitalist lines with, in the end, remarkable results.

US economic strength allowed the practice of "dollar diplomacy" worldwide. US firms took the lead in all areas of modern

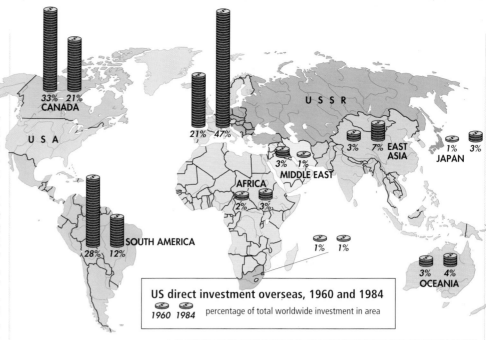

US direct investment overseas, 1960 and 1984
🪙 1960 🪙 1984   percentage of total worldwide investment in area

US troop dispositions 1945-80 (divisions)				
Year	Continental US	Europe	Asia/Pacific	overall army strength
1945	0	69	26	6,397
1950	7	1	4	666
1953	9	5	9	1,783
1960	8	5	4	1,044
1962	12	5	4	1,257
1964	10	5	4	1,162
1968	6	5	12	1,877
1972	9	4	3	1,009
1976	12	4	3	971
1980	12	4	3	963

The United States became the most important source of world investment funds between 1960 and 1984 *(map above)*. By 1984 US direct investment totalled US$223 billion, with more than 75 per cent placed in the other developed economies of the world.

In February 1972 President Richard Nixon *(right)* visited China and re-opened relations after a 22 year gap. He saw Mao Tse-tung in Peking, and in Shanghai signed an agreement between the two states on improving cultural and commercial relations.

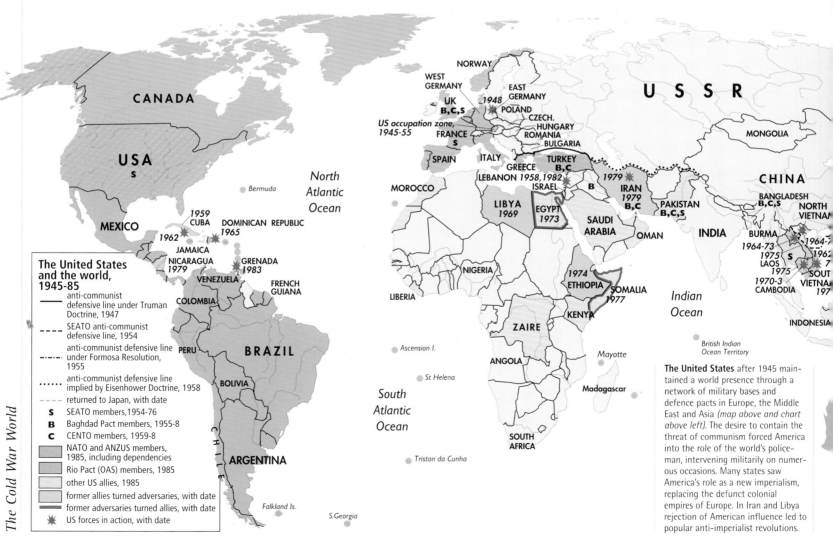

The United States and the world, 1945-85
— anti-communist defensive line under Truman Doctrine, 1947
- - - SEATO anti-communist defensive line, 1954
-·-·- anti-communist defensive line under Formosa Resolution, 1955
····· anti-communist defensive line implied by Eisenhower Doctrine, 1958
- - - returned to Japan, with date
S SEATO members, 1954-76
B Baghdad Pact members, 1955-8
C CENTO members, 1959-8
▨ NATO and ANZUS members, 1985, including dependencies
▨ Rio Pact (OAS) members, 1985
▨ other US allies, 1985
▨ former allies turned adversaries, with date
▨ former adversaries turned allies, with date
✳ US forces in action, with date

The United States after 1945 maintained a world presence through a network of military bases and defence pacts in Europe, the Middle East and Asia *(map above and chart above left)*. The desire to contain the threat of communism forced America into the role of the world's policeman, intervening militarily on numerous occasions. Many states saw America's role as a new imperialism, replacing the defunct colonial empires of Europe. In Iran and Libya rejection of American influence led to popular anti-imperialist revolutions.

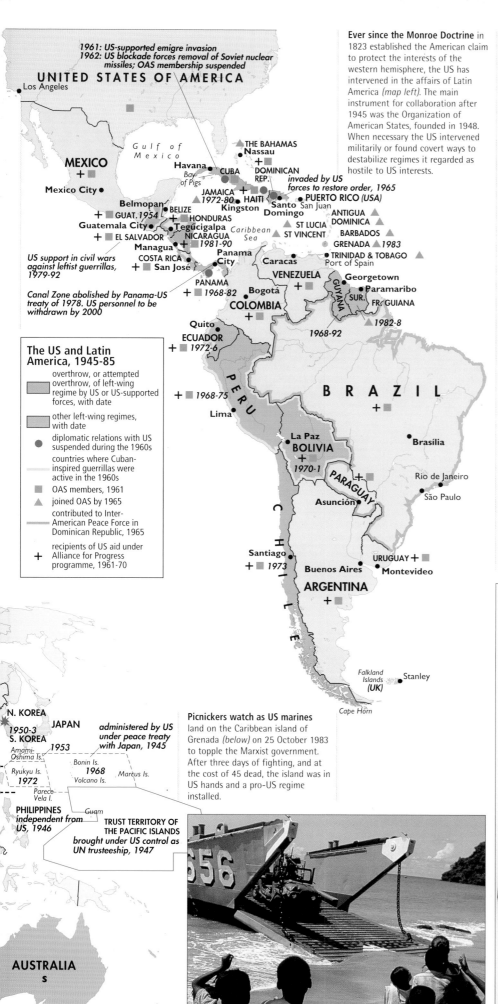

1961: US-supported emigre invasion
1962: US blockade forces removal of Soviet nuclear missiles; OAS membership suspended

UNITED STATES OF AMERICA

Los Angeles

Gulf of Mexico

MEXICO

Mexico City ●

THE BAHAMAS ▲
Nassau ●

Havana ● CUBA
Bay of Pigs

DOMINICAN REP.
invaded by US forces to restore order, 1965

Belmopan
GUAT. 1954 ■ BELIZE
Guatemala City ● ■ HONDURAS
EL SALVADOR ■ Tegucigalpa
Managua ● NICARAGUA ■ 1981-90
COSTA RICA ● San José ●

JAMAICA ▲ 1972-80
Kingston ●
HAITI
Santo Domingo ●
San Juan
PUERTO RICO (USA)

ANTIGUA ▲
DOMINICA ▲
ST LUCIA ▲
ST VINCENT ▲
BARBADOS ▲
GRENADA ▲ 1983
TRINIDAD & TOBAGO ▲
Port of Spain

Caribbean Sea

US support in civil wars against leftist guerrillas, 1979-92

Panama City
PANAMA + ■ 1968-82

Canal Zone abolished by Panama-US treaty of 1978. US personnel to be withdrawn by 2000

Caracas ●
VENEZUELA + ■
Bogotá ●
COLOMBIA +
1968-92

Georgetown ●
GUYANA
Paramaribo ●
SUR.
FR. GUIANA
▲ 1982-8

Quito ●
ECUADOR + ■ 1972-6

PERU + ■ 1968-75
Lima ●

B R A Z I L
+ ■
Brasília ●

La Paz ●
BOLIVIA + ■
1970-1

Rio de Janeiro ●
São Paulo ●

PARAGUAY +
Asunción ●

C H I L E

Santiago ●
+ ■ 1973

Buenos Aires ●

URUGUAY + ■
Montevideo ●

ARGENTINA + ■

Falkland Islands (UK)
Stanley ●

Cape Horn

The US and Latin America, 1945-85

- overthrow, or attempted overthrow, of left-wing regime by US or US-supported forces, with date
- other left-wing regimes, with date
- ● diplomatic relations with US suspended during the 1960s
- countries where Cuban-inspired guerrillas were active in the 1960s
- ■ OAS members, 1961
- ▲ joined OAS by 1965
- contributed to Inter-American Peace Force in Dominican Republic, 1965
- + recipients of US aid under Alliance for Progress programme, 1961-70

Ever since the Monroe Doctrine in 1823 established the American claim to protect the interests of the western hemisphere, the US has intervened in the affairs of Latin America *(map left)*. The main instrument for collaboration after 1945 was the Organization of American States, founded in 1948. When necessary the US intervened militarily or found covert ways to destabilize regimes it regarded as hostile to US interests.

N. KOREA
1950-3
S. KOREA

JAPAN
1953
administered by US under peace treaty with Japan, 1945

Amami-Oshima Is.
Ryukyu Is. 1972
Bonin Is. 1968
Volcano Is.
Marcus Is.
Parece Vela I.
Guam

PHILIPPINES
independent from US, 1946

TRUST TERRITORY OF THE PACIFIC ISLANDS
brought under US control as UN trusteeship, 1947

AUSTRALIA
S

NEW ZEALAND
S

Picnickers watch as US marines land on the Caribbean island of Grenada *(below)* on 25 October 1983 to topple the Marxist government. After three days of fighting, and at the cost of 45 dead, the island was in US hands and a pro-US regime installed.

technology, and in the 1950s and 1960s became household names: Boeing for aircraft; IBM for office equipment; General Electric for the household goods that transformed life styles throughout the Western world. Americanization was exported along with the dollar loans and American goods. American film studios provided much of the world's output. American popular culture, from Coca-Cola to rock and roll, provided icons of modernity.

There was a harsher side to the United States' world role. The Cold War forced the US to intervene militarily in Europe (Greece, 1947); Asia (Korea, 1950, Indo-China 1954-73); and the Middle East (Iran 1955, Lebanon 1958), where in January 1958 the Eisenhower Doctrine was proclaimed, committing the US to prevent the spread of communism in the region. To cement the American strategy of containing the Soviet and Chinese threat, the US made numerous defensive pacts. A peace treaty was signed with Japan in San Francisco in 1951, committing the US to defend Japan against her larger communist neighbours. Similar agreements were reached with Australia, New Zealand, the Philippines, South Korea and Taiwan. These broadened out into the SEATO alliance in 1954 – the Asian counterpart of NATO.

Containment in Latin America took a rather different form. In 1947 in Rio de Janeiro all 21 American republics signed the Inter-American Treaty of Reciprocal Assistance. The United States used collaboration as a means to limit political radicalism in the region, which it did by encouraging right-wing coups (Guatemala 1954, Chile 1973, Cuba – unsuccessfully – in 1961) or through direct intervention (Dominican Republic 1965, Grenada 1983). In 1960 Washington set up the Inter-American Development Bank to sweeten intervention with dollars.

During the 1970s détente altered the United States' role. Other areas of the Western world were able to provide more of their own security and no longer needed the same degree of financial assistance. As the threat of international communism receded, US statesmen began to play a more conciliatory role. In 1972 President Nixon, prompted by his national security adviser, Henry Kissinger, ended the diplomatic break with communist China. Following Kissinger's efforts for peace in the Middle East, President Carter succeeded in getting Egypt and Israel to reach a settlement in 1978.

ELVIS PRESLEY

In 1958 Elvis Aaron Presley (1935-77) arrived in West Germany as an ordinary GI, serving his time as a conscript with the US NATO forces facing the Soviet enemy. Elvis was the personification of the US abroad in the 1950s – a small part of the United States military presence overseas, he was the symbol worldwide of the new US youth culture. The son of a poor white sharecropper, he was born in Tupelo, Mississippi on 8 January 1935. In 1954 he cut his first record, and in 1956 the song "Heartbreak Hotel" launched him to global fame as the most successful of the new generation of rock and roll singers. He chose to do his military service from 1958 to 1960, an experience later recaptured in the film GI Blues. He had a record 18 number one hits in the United States, but was idolized everywhere. In Britain he holds the record for the greatest number of Top Ten hits. Popular music was the United States' most conspicuous export in the post-war years, from Frank Sinatra in the 1940s to Madonna in the 1980s. The invention of the long-playing record and the transistor radio made music from the United States more readily available. American performers were instantly recognized, and prompted an imitative youth culture wherever their music was heard, even in the communist East.

THE SUPER-POWERS: DOMESTIC POLICY IN THE USSR TO 1985

The Soviet Union also emerged from the Second World War as a super-power, but unlike the United States its new status was won at the cost of exceptional sacrifice. Almost half of Soviet territory was utterly devastated: 20 million were killed; seven million horses and 17 million cattle lost; 98,000 collective farms and 4.7 million houses destroyed. The Soviet Union had to start again on the economic revolution begun in the 1930s.

First Stalin had some scores to settle. Any individual or group classified as traitors was liquidated or sent to the network of labour camps and penal colonies run by Stalin's notorious head of police, Lavrenti Beria. The victims included the entire Party leadership in Leningrad, which had withstood the 900 days of siege during the war. In the last period of Stalin's rule 100,000 were purged each year from the Communist Party. Stalin also took revenge on the Soviet peasantry, who had succeeded during the war in freeing themselves from the grip of the collective farm organization. In September 1946 Stalin began a renewed campaign against peasant "profiteering". Their savings were eliminated, thousands brought back into the collectives and 14 million hectares of land returned to state-run farms.

While crushing peasant agriculture, Stalin launched a new era of state-directed industrial planning. The process of modernization continued its pre-war trajectory. By 1980 rural workers made up only one fifth of the labour force, where they had constituted more than half in 1945. The urban population was 69 million in 1950, but 186 million in 1988, by which time the rural population had fallen from 109 million in 1950 to 95 million. Priority went to heavy industry and the military at the expense of consumer goods.

The Soviet system underwent a minor revolution when, on 5 March 1953, Stalin died. There followed a period of collective leadership under Malenkov, Khrushchev and Molotov, but by the time of the 20th Party Congress in 1956, at which Stalin and Stalinism were denounced from the platform, Khrushchev emerged as the leading figure. The era of de-Stalinization had important if limited effects. Some eight to nine million political

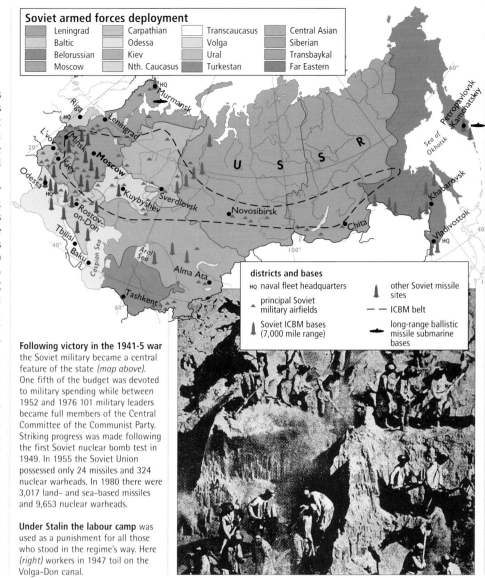

Soviet armed forces deployment

Leningrad	Carpathian	Transcaucasus	Central Asian
Baltic	Odessa	Volga	Siberian
Belorussian	Kiev	Ural	Transbaykal
Moscow	Nth. Caucasus	Turkestan	Far Eastern

districts and bases

- **HQ** naval fleet headquarters
- ▲ principal Soviet military airfields
- ◤ Soviet ICBM bases (7,000 mile range)
- ◣ other Soviet missile sites
- – – ICBM belt
- → long-range ballistic missile submarine bases

Following victory in the 1941-5 war the Soviet military became a central feature of the state *(map above)*. One fifth of the budget was devoted to military spending while between 1952 and 1976 101 military leaders became full members of the Central Committee of the Communist Party. Striking progress was made following the first Soviet nuclear bomb test in 1949. In 1955 the Soviet Union possessed only 24 missiles and 324 nuclear warheads. In 1980 there were 3,017 land- and sea-based missiles and 9,653 nuclear warheads.

Under Stalin the labour camp was used as a punishment for all those who stood in the regime's way. Here *(right)* workers in 1947 toil on the Volga-Don canal.

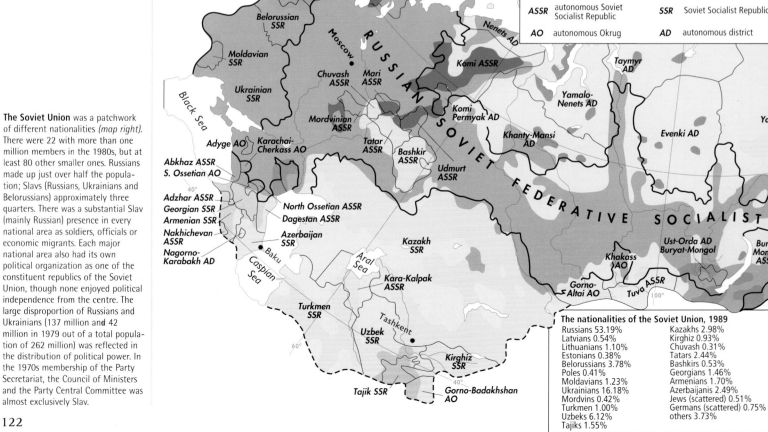

The Soviet Union was a patchwork of different nationalities *(map right)*. There were 22 with more than one million members in the 1980s, but at least 80 other smaller ones. Russians made up just over half the population; Slavs (Russians, Ukrainians and Belorussians) approximately three quarters. There was a substantial Slav (mainly Russian) presence in every national area as soldiers, officials or economic migrants. Each major national area also had its own political organization as one of the constituent republics of the Soviet Union, though none enjoyed political independence from the centre. The large disproportion of Russians and Ukrainians (137 million and 42 million in 1979 out of a total population of 262 million) was reflected in the distribution of political power. In the 1970s membership of the Party Secretariat, the Council of Ministers and the Party Central Committee was almost exclusively Slav.

Principal ethnic groups of the Soviet Union, 1989

Slav	Caucasian	Moldavian	Iranian
Turkic	Finno-Ugric	Baltic	others

- **ASSR** autonomous Soviet Socialist Republic
- **SSR** Soviet Socialist Republic
- **AO** autonomous Okrug
- **AD** autonomous district

The nationalities of the Soviet Union, 1989

Russians 53.19%	Kazakhs 2.98%
Latvians 0.54%	Kirghiz 0.93%
Lithuanians 1.10%	Chuvash 0.31%
Estonians 0.38%	Tatars 2.44%
Belorussians 3.78%	Bashkirs 0.53%
Poles 0.41%	Georgians 1.46%
Moldavians 1.23%	Armenians 1.70%
Ukrainians 16.18%	Azerbaijanis 2.49%
Mordvins 0.42%	Jews (scattered) 0.51%
Turkmen 1.00%	Germans (scattered) 0.75%
Uzbeks 6.12%	others 3.73%
Tajiks 1.55%	

prisoners were released from the labour camps. Censorship was relaxed and the ministry of the interior (home of the NKVD security police) closed down. Khrushchev attempted to reform agriculture and reverse peasant impoverishment, and to decentralize state economic planning in order to restore limited incentives to Soviet industry.

The results were disappointing. Economic decentralization produced chaos in both industry and agriculture. Wages stagnated, while expectations were raised. The relaxation of police terror opened up the issue of political dissidence and the extent to which the state could tolerate it and maintain communist ascendancy. In 1964 Khrushchev made one last attempt to secure change. At the 22nd Party Congress he announced a new Party Programme, the first since 1919. He promised genuine communism in 20 years: ten years building the material base for it and ten years redistributing the new product. His colleagues were unimpressed. Following the climb-down over Cuba (see page 110), Khrushchev was the victim of a palace coup. In October 1964 he gave way to another period of collective leadership, which by 1966 led to the emergence of

Agricultural land and land tenure 1940-79

year	state farms	collec-tives	sown area (mill. ha.)	farm workers (mill.)	pop-ulation (mill.)
1940	4,000	237,000	150.6	18.7	194
1950	5,000	124,000	146.3	20.5	179
1960	7,000	45,000	203.0	17.1	212
1970	15,000	34,000	206.7	14.4	242
1975	18,000	29,000	217.7	13.5	253
1979	21,000	26,000	217.3	12.8	262

The Virgin Lands campaign

distribution of virgin and long fallow lands

proposed protective tree belts

U S S R

Voronezh
Belgorod
Penza
Kharkov
Saratov
Donetsk
Kuybyshev
Orenburgh
Stepnoy
Uralsk
Cherkessk
Astrakhan
Guryev
Caspian Sea

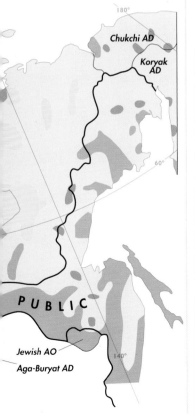

Chukchi AD
Koryak AD

P U B L I C

Jewish AO
Aga-Buryat AD

A Soviet "realist" painting from c. 1950 shows Stalin surrounded by the Soviet nationalities (top). Stalin, a Georgian himself, was deeply hostile to the smaller nations. During the war, eight small nationalities were forcibly transplanted for collaborating with the Germans, while in 1948 he authorized a fresh wave of official anti-semitism.

A May Day parade in Moscow, 1983 (right). After the war the Communist Party became the dominant presence in Soviet society. Under Stalin large numbers of workers and peasants entered the party, but by the 1980s it had become an organization which perpetuated a white-collar elite and gave fewer opportunities for social mobility.

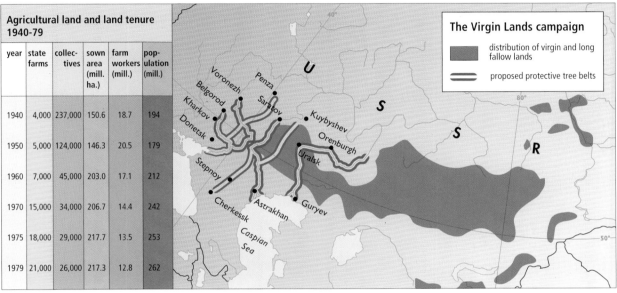

The Soviet economy 1950–85

year	steel (million tons)	coal (million tons)	exports (billion roubles)	cars (thousands)	televisions (thousands)
1950	27.3	261	1.6	65	12
1960	65.3	509	5.0	139	1726
1970	116.0	624	11.5	344	6682
1980	148.0	716	49.6	1327	7528
1985	155.0	726	72.7	1332	9371

Leonid Brezhnev as Party leader.

The Brezhnev years reversed much of the liberalization. The interior ministry and a new security service, the KGB, were restored to their central role in stamping out dissent. Censorship was reimposed. Central state planning was reintroduced to ensure some level of economic growth. In the 1970s Brezhnev balanced coercion with an increase in the output of consumer goods, but there persisted a gap between the popular political and economic aspirations of ordinary Soviet citizens and the ability of the regime to satisfy them. Neither of Brezhnev's successors in the 1980s, Yuri Andropov (1982-4) and Konstantin Chernenko (1984-5), was willing to risk comprehensive reform. By the 1980s Soviet society had reached stalemate.

THE SUPER-POWERS: THE USSR AND THE SOVIET BLOC TO 1985

When Hitler's armies reached the suburbs of Moscow in December 1941, communism looked spent as a force in world politics. Soviet victory four years later was a triumph for Stalin's brand of communism, and the USSR sought to export its politics to the areas liberated by the Red Army. This was a process carried out step by step, building on the existence of large and popular communist parties in much of Eastern Europe. A multi-party system was tolerated, but communist and socialist organizations were given special preference. In 1947 and 1948 political pluralism was eradicated and in Poland, Romania, Czechoslovakia, Albania, Bulgaria and Hungary people's republics were declared, ruled by a single pro-Soviet party bloc. In September 1947 a communist international organization, Cominform, was set up, dominated from Moscow. In June 1948 Yugoslavia under Tito was expelled from this for refusing to toe the Moscow line, although by 1957 better relations were restored between Yugoslavia and the post-Stalinist USSR.

The triumph of communism in China in 1949, in North Korea and, in 1954, in Vietnam, enhanced the international prestige and influence of the USSR. Khrushchev hailed communism as the "wave of the future" and confidently expected it to triumph throughout the developing non-capitalist world. In the 1950s the USSR and the Soviet European bloc sought to cement closer ties. In 1955 Egypt, under Colonel Nasser, obtained Soviet aid and technical assistance – a move that revolutionized the diplomacy of the region. There followed agreements with Afghanistan, Syria, Tunisia and Yemen, though none of these turned the recipient state into a Soviet bloc member. Developing countries distrusted Soviet motives almost as much as American.

There were significant limits by the 1960s to the further extension of Soviet influence. There were political problems in Soviet-dominated Europe. In 1953 there were strikes and riots in Poland and East Germany. In 1956 a more serious insurrection

In December 1979 the Soviet Union sent troops into Afghanistan to help the pro-Soviet government in Kabul (maps right), the first time that Soviet forces had been officially deployed abroad outside the European Soviet bloc. The Muslim Mujaheddin rebels (above) were armed by the West, and the Soviet forces found themselves fighting a fruitless anti-guerrilla war. Soviet forces were withdrawn by February 1989.

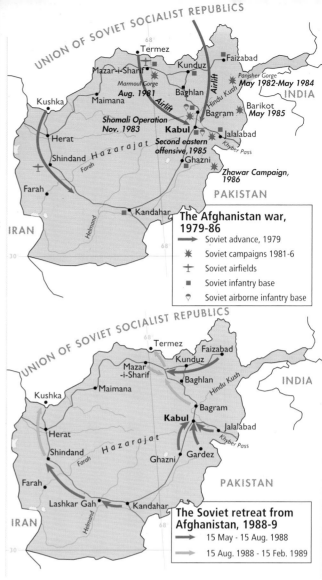

The Afghanistan war, 1979-86
- → Soviet advance, 1979
- ✳ Soviet campaigns 1981-6
- ✛ Soviet airfields
- ▪ Soviet infantry base
- ⬖ Soviet airborne infantry base

The Soviet retreat from Afghanistan, 1988-9
- → 15 May - 15 Aug. 1988
- → 15 Aug. 1988 - 15 Feb. 1989

Overseas aid to developing countries from the Soviet bloc, 1954-75

	1954-71 (mill. US$)	1972-5 (mill. US$)
Bulgaria	294	217
Czechoslovakia	1,241	679
GDR	834	346
Hungary	497	454
Poland	619	508
Romania	525	1,638
USSR	7,180	5,089

The USSR and the world, 1945-85
- USSR and client states, 1938
- annexed 1944-5
- territorial ambitions thwarted, 1945-6
- unsuccessful attempt to gain or maintain control by 1949
- control or influence secured by 1954
- Warsaw pact, 1955
- turned antagonistic to USSR from the early 1960s
- acknowledgement demanded from China that 19th-century Russian annexations were by unequal treaties
- ties from USSR loosened from mid 1960s
- brought into or kept in Soviet sphere in 1970s by military intervention
- other allies in 1985

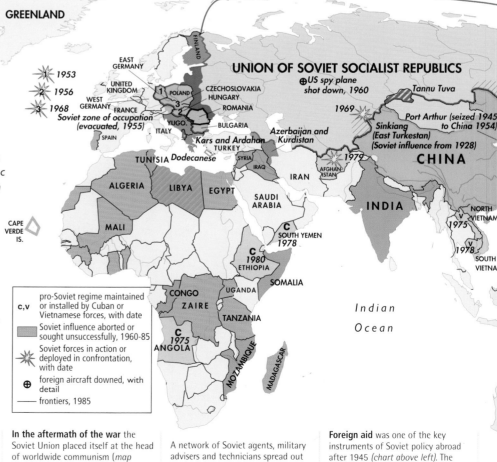

- c,v — pro-Soviet regime maintained or installed by Cuban or Vietnamese forces, with date
- Soviet influence aborted or sought unsuccessfully, 1960-85
- ✳ Soviet forces in action or deployed in confrontation, with date
- ⊕ foreign aircraft downed, with detail
- frontiers, 1985

In the aftermath of the war the Soviet Union placed itself at the head of worldwide communism (map above). It engineered the transition to communist regimes in Eastern Europe, and tried to dominate Chinese and East Asian communism in the 1950s.

A network of Soviet agents, military advisers and technicians spread out across the developing world, particularly in Africa and the Middle East, where the Soviet Union established military bases for a brief period.

Foreign aid was one of the key instruments of Soviet policy abroad after 1945 (chart above left). The European Soviet bloc sent more than US$20 billion worth of aid between 1954 and 1975 to a total of 36 countries in Africa, Asia and Latin America.

Between 1944 and 1948 a communist pro-Soviet bloc was created in Eastern Europe *(map right)*. Soviet forces were stationed throughout the bloc, which was bound together by economic agreements (COMECON, 1949) and a military alliance (Warsaw Pact, 1955). In June 1948 Tito's Yugoslavia broke ties with Moscow. All other attempts to challenge Soviet domination in the 1950s and 1960s – in East Germany, in Hungary and in Czechoslovakia – were crushed.

In January 1949 the Soviet bloc set up the Council for Mutual Economic Assistance (COMECON), which became the main regulator of trade between Soviet-bloc states *(chart far right)*. Until the 1970s over two thirds of all the trade of the USSR was with Eastern Europe.

Communist Eastern Europe to 1985

☐	Soviet zone of occupation in Austria, 1945-55
C	members of Cominform, 1947
——	Iron Curtain, 1948
·····	frontier incidents, 1950-2
——	frontier finalized by GDR-Polish treaty, 1950
——	Balkan Pact, 1954 (not functional from 1955)
——	Warsaw Pact from May 1955
➔	Soviet troop deployments in Hungary, 1956
▨	participated in invasion of Czechoslovakia, 1968
➔	Warsaw Pact troop deployments in Czechoslovakia, 1968
⇨	mass exodus of refugees
•	uprisings, 1953
•	uprisings, 1956
•	mass protests, 1968
•	mass protests and strikes, 1970-85
——	frontiers, 1950

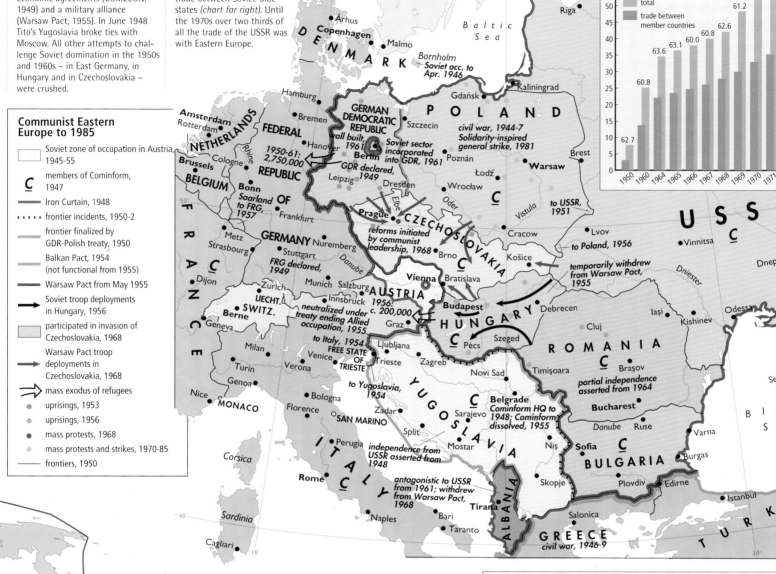

Foreign trade of Comecon members 1950-71

(chart, billion roubles)

■ total
■ trade between member countries

Values: 62.7, 60.8, 63.6, 63.1, 60.0, 60.8, 62.6, 61.2, 60.2, 60.9 (1950, 1960, 1964, 1965, 1966, 1967, 1968, 1969, 1970, 1971)

took place in Hungary under the leadership of the liberal communist, Imre Nagy. He promised multi-party elections and withdrew Hungary from the military alliance with the USSR. In November 250,000 Soviet troops and 5,000 tanks reimposed Soviet power. Two years later the first splits appeared in relations with China, and in 1960 Mao declared the USSR guilty of "bourgeois deviation" and placed China at the head of the world communist revolutionary movement. In Albania, the Far East and parts of Latin America communists followed China rather than the USSR. During the 1960s the European bloc states began to develop a more independent, nationalist form of development, which ended in 1968 with a further series of protests against Soviet hegemony and the Soviet invasion of Czechoslovakia.

Renewed efforts were made in the 1970s to curb Soviet influence. Cuba rallied to the side of the USSR after Castro's revolution in 1959. By the late 1970s there were an estimated 40,000 Cuban troops in Africa supporting communist regimes and guerrillas, and 8,000 Soviet military advisers. But elsewhere, in the Middle East, Latin America and Asia, the Soviet Union, like the US, found itself the victim of popular local nationalism and the Islamic revival, symbolized by the Iranian revolution of 1979 and Muslim revolt in Afghanistan, which led to Soviet intervention there in 1979. Communism in much of the developing world was persecuted violently; in the developed states it withered as a major electoral force. In France the Communist Party was the major party of the left from 1945 down to the early 1980s, when it was eclipsed by moderate social democracy. In Italy the left was dominated by moderate communism until the 1980s, when the social democrats increased their share of the vote. The collapse of the Soviet bloc in 1989 *(see pages 146-7)* proved fatal for Western communism.

THE PRAGUE SPRING

In the late 1960s Czechoslovakia became the focal point in the Soviet bloc for the conflict between Communist Party and popular demands for political participation. The Czech communist leadership under Antonín Novotný was the most Stalinist and pro-Soviet of all the communist blocs. When Novotný began a cautious liberalization in 1962, Czech intellectuals began limited criticism of the regime. Novotný's attempt in 1967 to clamp down was met with widespread hostility, some from within the Communist Party itself, led by the Slovak party boss, Alexander

Dubček. In January 1968 the reformers ousted Novotný and installed Dubček as party leader. In April he launched an action programme to end censorship and police repression and to introduce a more participatory political system. Hardliners in Moscow, Warsaw and Berlin, anxious about the effect of the Czech reform on their own regimes, pressed Dubček to reverse the liberalization. Popular feeling in Czechoslovakia forced his hand and he refused. On 21 August Warsaw Pact countries invaded. On 26 August Dubček was forced to sign the so-called Moscow Protocols, reversing his changes.

THE LONG BOOM, 1945-73

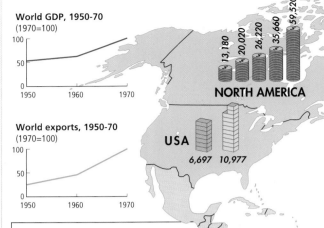

World GDP, 1950-70
(1970=100)

World exports, 1950-70
(1970=100)

NORTH AMERICA

13,180 20,020 26,220 35,660 59,520

USA
6,697 10,977

WESTERN EUROPE

18,981 32,962 48,287 74,166 129,547

2,941 7,462
FRANCE

The long boom, 1950-73

1950 1973

6,697 10,977

GDP per capita in US dollars, 1950 and 1973 (1980 prices)

7,540 9,240 9,950 12,670 17,330

1950 1955 1960 1965 1970

export growth by region, 1950-70, in US dollars (millions)

SOUTH AMERICA

7,540 9,240 9,950 12,670 17,330

Oil was a vital ingredient in the world boom, providing the material for the rapid growth of the petro-chemical industry and helping worldwide motorization. Oil production increased six-fold between 1950 and 1973 *(chart and map below left)*, with new fields opening in the Middle East, North Africa and Asia. Much of the increase came from the Persian Gulf states. In October 1973 they used their strong market position to reduce output and raise oil prices. Between 1970 and 1972 oil rose from $1.50 to $3.00, but at the end of 1973 Gulf oil fetched $18 a barrel. The price rise created havoc in the developed states and brought the long boom to an end. The temporary shortage of fuel *(picture below right)* was followed by sustainedworldwide inflation.

AUSTRALIA IN THE LONG BOOM

Australia was typical of the states which had been victims of the inter-war slump in demand for primary products. Yet the years of economic stagnation were followed by buoyant growth as the Australian economy was dragged along by the world boom. Between the 1940s and the 1970s

Australia reduced its long dependence on agricultural exports and shifted to manufacturing and services in the domestic economy. Primary production contributed 21 per cent to Australia's GNP in 1949, but only eight per cent in 1969. Wool made up 41 per cent of exports in 1959-60, but only 12 per cent in 1970-1. Exceptionally high levels of investment were pumped

into mining, communications and manufacturing. Labour supply was maintained by a steady stream of immigrants, who added over two per cent a year to the population between 1950 and 1970. Australia's changing economic base was reflected in a changing pattern of trade. Exports to Britain fell from 42 per cent of the total in 1949-50 to only 12 per cent

in the 1970s. Over the same period trade with Japan rose from 1.4 per cent of exports to 27 per cent. Australia became locked into the development of the Pacific Rim area at the expense of traditional links with Britain and Europe, and has continued to prosper from the transition.

	GDP (bn.$, 1980 prices)	Consumer expenditure (bn.$, 1980 prices)	Gross fixed capital formation (bn.$, 1980 prices)	Exports (bn.$, 1980 prices)	Coal (million tons)	Steel (million tons)	Vehicles (thousands)	Electricity (billion Kwh)
1950	40.44	26.77	10.12	4.74	24.25	1.28	58.00	9.51
1955	48.76	30.61	11.77	5.89	29.86	2.28	127.00	15.20
1960	59.43	35.83	15.02	7.62	37.11	3.62	205.00	23.20
1965	75.18	44.73	21.47	9.94	51.06	5.21	348.00	36.91
1970	101.54	58.16	27.68	15.68	73.15	6.87	478.00	56.15
1975	121.59	71.93	29.90	18.37	98.70	8.06	456.00	73.93

Annual average growth rates by region

- 1913-50
- 1950-73
- 1973-92

Western Europe: 1.4 / 4.7 / 2.2
US/Canada: 2.8 / 4.0 / 2.4
Southern Europe: 1.3 / 6.3 / 3.1
Eastern Europe: -0.4 / 1.6 / 4.7
Asia: 1.0 / 6.0 / 5.1
Africa: 3.0 / 2.8 / 4.4
World: 1.9 / 4.9 / 3.0

World crude oil production and price
(1970=100)

price

production

1950 1955 1960 1965 1970 1973 1974

THE NEW 'ENGLISH ELECTRIC' refrigerator

more storage space
snap-release ice
kitchen space
clear-view meat
frozen food locker
easy de-frosting
adjustable shelf
easy cleaning
standing room for quart
new vegetable containers
silent-running economy

CAPACITY 7·6 cu. ft.

Full of *Fresh* ideas!

BRINGING YOU ... BETTER LIVING

BARTLETT PEARS 22¢

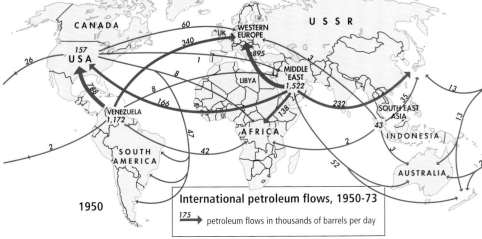

CANADA 60 UK WESTERN EUROPE USSR

340

157 USA 895 1

26 788 8 MIDDLE EAST 1,522 232 SOUTH EAST ASIA 35 13

VENEZUELA 1,172 166 LIBYA 138 43 INDONESIA 13

47 AFRICA 2

2 SOUTH AMERICA 42 52 AUSTRALIA 2

1950

International petroleum flows, 1950-73

175 → petroleum flows in thousands of barrels per day

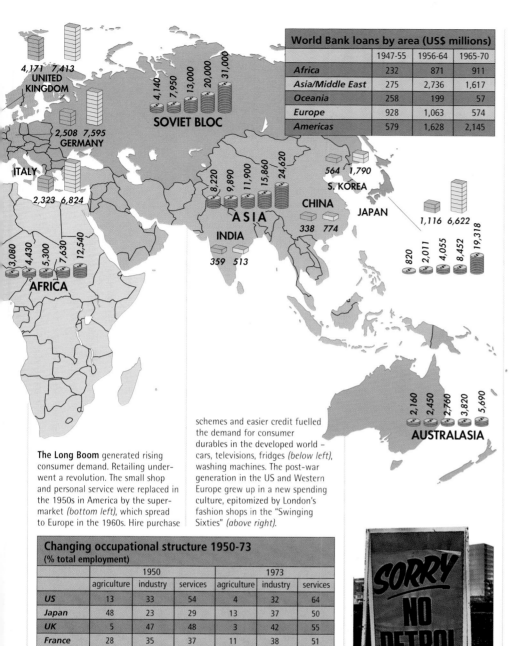

World Bank loans by area (US$ millions)			
	1947-55	1956-64	1965-70
Africa	232	871	911
Asia/Middle East	275	2,736	1,617
Oceania	258	199	57
Europe	928	1,063	574
Americas	579	1,628	2,145

UNITED KINGDOM 4,171 7,413

GERMANY 2,508 7,595

ITALY 2,323 6,824

SOVIET BLOC 4,140 7,950 13,000 20,000 31,000

AFRICA 3,080 4,430 5,300 7,630 12,540

ASIA
INDIA 359 513
8,220 9,890 11,900 15,860 24,620
CHINA 338 774
S. KOREA 564 1,790
JAPAN 1,116 6,622
820 2,011 4,055 8,452 19,318

AUSTRALASIA 2,160 2,450 2,760 3,820 5,690

The heart of the Long Boom was a sustained growth in manufacturing output which trebled between 1950 and 1970 *(chart below centre)*. The developed economies produced high-cost durables and capital goods, while the less developed world began to produce large quantities of consumer goods, such as textiles.

Increased prosperity was generally accompanied by a decline in the proportion of the workforce engaged in agriculture and an increase in the percentage employed in industry and, in particular, in services *(chart below left)*.

The Cold War struggle was carried out against a background of extraordinary economic revival from the devastation of the Second World War. From the 1940s to the early 1970s the world underwent the "Long Boom" – 25 years of almost uninterrupted growth at rates higher than any yet recorded. In Europe income per head grew faster in 20 years than in the previous 150. Unemployment rates throughout the developed world were low, rarely rising above three per cent. The business cycle was replaced by the economic miracle of almost continuous growth.

The boom had many causes. The war itself helped by creating boom conditions in the US, the world's largest economy. At the same time, European demand for food, materials and weapons stimulated output in developing countries. The war also forced the US to take the leading place in the world economy, a development which isolationist sentiment had precluded before 1939. The US became the world's major source of investment funds and aid. More significantly, US statesmen were committed to reversing the pre-war drift to protectionism and tariff wars by restoring a more open world market, with stable currencies.

The restoration of a healthy environment for trade was the single most important cause of the boom. In October 1947 23 countries signed the General Agreement on Tariffs and Trade, which launched a round of negotiations on reducing protection. The first round covered 45,000 tariff items, produced 123 agreements and covered about one half of world trade. By the mid-1950s the US alone had reduced tariff levels by 50 per cent, while a second major round in the early 1960s led to a further reduction of almost 50 per cent in tariffs between developed economies. World trade grew at six per cent a year between 1948 and 1960, and by nine per cent between 1960 and 1973. Europe's trade between 1950 and 1970 grew from $18 billion to $129 billion.

The second cause was technical. After 1945 the application of modern science to industry and agriculture produced a wave of new products and production methods. Mechanization and modern plant biology created a so-called Second Agricultural Revolution, which stimulated food production in the less developed regions. Modern mass-production methods and rational factory organization, both stimulated by the war, were applied across the developed world, raising productivity remarkably and producing a boom in profits and in wages.

The third factor was political. States had begun to intervene much more in regulating their economies before 1939 to cope with the recession. After the war there was a general desire to maintain regulation, or even to increase the level of planning and subsidy in order to avoid further slumps. State expenditure roughly doubled in the developed economies between the 1930s and the 1950s. State help was generally welcomed by firms keen to restore stability and by labour unions seeking to avoid high unemployment. By the early 1970s the non-communist economies were typically "mixed economies", combining private enterprise and state regulation to maximize growth.

It was just at this point that the post-war boom turned sour. The rate of innovation and of profit and productivity growth slowed down sharply. The developed economies began to ossify, with large state sectors, expensive welfare programmes and high-cost workforces. The sudden increase in oil prices in 1973 turned a modest rate of world inflation into a price boom, while increased competition and new technology created high levels of unemployment. Low growth and high inflation – "stagflation" – pricked the bubble of seemingly endless expansion.

The Long Boom generated rising consumer demand. Retailing underwent a revolution. The small shop and personal service were replaced in the 1950s in America by the supermarket *(bottom left)*, which spread to Europe in the 1960s. Hire purchase schemes and easier credit fuelled the demand for consumer durables in the developed world – cars, televisions, fridges *(below left)*, washing machines. The post-war generation in the US and Western Europe grew up in a new spending culture, epitomized by London's fashion shops in the "Swinging Sixties" *(above right)*.

Changing occupational structure 1950-73 (% total employment)						
	1950			1973		
	agriculture	industry	services	agriculture	industry	services
US	13	33	54	4	32	64
Japan	48	23	29	13	37	50
UK	5	47	48	3	42	55
France	28	35	37	11	38	51
Germany	22	43	35	7	47	46
Italy	45	29	26	18	38	44

From the 1940s to the early 1970s the world economy experienced an exceptional period of expansion *(chart far left)*. World output doubled and world exports quadrupled. The developing areas received increasing quantities of aid from the richer economies which helped to sustain their economic modernization *(map above and chart top right)*. The boom meant employment and high prices for raw material producers, like the Lyell open-cut lead mine *(far left top)* on the coast of Tasmania.

World manufacturing output 1950-1970 (1963=100)

1950 '51 '52 '53 '54 '55 '56 '57 '58 '59 '60 '61 '62 '63 '64 '65 '66 '67 '68 '69 '70

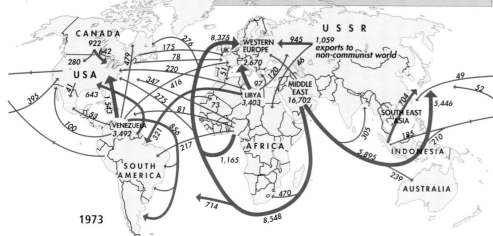

1973

CANADA 922
USA
VENEZUELA 3,492
SOUTH AMERICA
WESTERN EUROPE
UK 8,375
LIBYA 3,403
AFRICA
MIDDLE EAST 16,702
USSR 1,059 exports to non-communist world
SOUTH EAST ASIA
INDONESIA
AUSTRALIA

CHINA TO 1985

Communism in post-war China had different roots from that in the Soviet bloc and followed a very different course. When in 1945 the war with Japan ended, US negotiators tried to effect a reconciliation between the nationalist leader, Chiang Kai-shek, and the communist Mao Tse-tung. The following spring, when Soviet forces evacuated Manchuria, a full-scale civil war broke out between communists and nationalists. Early victories persuaded Chiang that he could smash communism. In April 1948 he was appointed president by a new national assembly. The communists worked to achieve the support of the peasantry, and their army swelled in number. In late 1948 the nationalists were defeated in Manchuria, and by September 1949 the nationalist cause had collapsed. Mao Tse-tung became head of a new communist republic in October 1949. Chiang fled to Taiwan.

In 1949 the Communist Party set up a "democratic dictatorship" under Mao. Although other parties were tolerated, the communists dominated. The 4.5 million party members grew to 17.5 million in 1961 (and 46 million in 1988). The first priority was agriculture. In September 1950 an Agrarian Reform Law was announced, under which 300 million peasants benefited from the redistribution of 700 million *mou* (equal to one sixth of an acre) of landlord estates. By 1957 the villages were collectivized, pooling resources but retaining private ownership. The slow pace of change encouraged Mao to gamble on what became known as the Great Leap Forward. Launched in February 1958, it was a programme to modernize China in three years, based around the Maoist concept of the people's commune. By November 1958 the countryside was organized into 26,000 communes, each responsible for abolishing private ownership and charged with delivering immense (and entirely unrealistic) increases in agricultural and industrial output. It was the age of the "backyard furnace", when 600,000 miniature blast furnaces were set up across China's villages and towns.

The Great Leap Forward was a grotesque failure: economic output declined sharply; a severe famine killed 20 million. Mao, who had stood down as chairman of the republic in 1959 to concentrate on developing his ideological input to the revolution, found himself isolated by moderate elements in the party, led by Liu Shao-ch'i, who wanted to replace Maoist utopias with socialist realism on the Soviet model. In 1961, despite Mao's fears that the Party was being taken over by a bureaucratic, revisionist elite, the Great Leap Forward was abandoned.

TIBET

After the foundation of the Chinese Republic in 1912 Tibet became virtually autonomous and remained in this condition until the communist victory in 1949. The new Chinese regime was determined to bring all the former parts of China under its control. In October 1950 the Chinese army, pictured (right) crossing a river during the invasion, seized control of Tibet, whose appeals to the West for help went unheeded. The Chinese clamped down on Tibetan separatism and provoked a popular rebellion against Chinese rule in March 1959. The revolt was brutally crushed and the Tibetan leader, the Dalai Lama, fled to India, where he was granted political asylum. China and India clashed in a number of serious border incidents along the Himalayan frontier, but Soviet refusal to back China calmed the crisis. The Chinese army continued to keep tight control in Tibet, and during the Cultural Revolution pursued a harsh policy of assimilation and Sinification, including the closing down of many Tibetan Buddhist monasteries, which had acted as the focus of anti-Chinese resistance. Since the 1950s a large influx of Han Chinese immigrants has had the effect of destroying many features of indigenous Tibetan culture. In 1989 the Dalai Lama was awarded the Nobel Peace Prize, but China remained impervious to Western criticism of its treatment of Tibet.

When the communists came to power in China the economy was badly damaged by 12 years of war and civil war. Mao initiated a forced state-led industrialization (map below), based on Five-Year Plans. The first from 1953-7 was overfulfilled; a second plan, even more ambitious, ran from 1958 to 1962. China's GNP doubled in the 1950s, and increased by one third again in the 1960s. Growth slowed down as a result of the social upheavals of the 1960s. Steel output rose from 1.5 million tons in 1952 to 18.6 million tons in 1960, but by 1976 the figure was only 21 million. Not until after Mao's death did China begin a sustained period of economic development.

Undeterred, in 1966, in alliance with the army and its leader, Lin Piao, Mao introduced a second revolutionary wave. Supported by enthusiastic young communists who shared Mao's fear that the revolutionary tide was ebbing, and who faithfully followed *The Thoughts of Chairman Mao*, (a collection of Mao's revolutionary beliefs, distributed by the million), Mao encouraged violent rooting out of deviationists. His wife, Chiang Ch'ing, led the Cultural Revolutionary Committee, which was responsible for imposing Maoist conformity and committed endless atrocities in the name of ideological purity.

In 1968 the violence was threatening the complete collapse of Chinese life. Lin Piao was called on by Mao to restore order. The following year a party congress unanimously elected Mao party chairman and *The Thoughts of Chairman Mao* were adopted as the official party line. Lin Piao was groomed as Mao's successor and the army came to play a prominent part in national politics. Once again Mao feared for his political

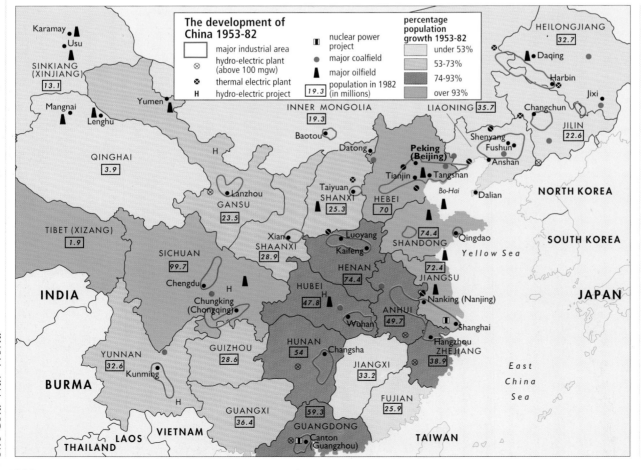

The development of China 1953-82

- major industrial area
- ⊗ hydro-electric plant (above 100 mgw)
- ◆ thermal electric plant
- H hydro-electric project
- ⬛ nuclear power project
- ● major coalfield
- ▲ major oilfield
- 19.3 population in 1982 (in millions)

percentage population growth 1953-82
- under 53%
- 53-73%
- 74-93%
- over 93%

After the defeat of Japan the Chinese nationalists and the Chinese communists competed for the administration of the liberated areas (map above right). The communists dominated Manchuria and the north. In 1947 full-scale civil war broke out between the two sides. Nationalist armies were defeated in Manchuria in 1948, and at Suchow from November 1948 to January 1949. In October 1949 a communist republic was proclaimed, and in May 1950 the nationalist remnants fled to Taiwan.

On 1 October 1949 the communist leader Mao Tse-tung (far right centre) announced the establishment of the People's Republic of China, with its capital at Peking. The new state was based on an Organic Law of the Government drawn up in September 1949. Under the law Mao became chairman of the republic.

The Chinese civil war, 1946-50

- occupied by communist armies at outbreak of civil war
- occupied July 1946-June 1948
- occupied July 1948-June 1949
- occupied by 1950
- communist guerrilla operations 1945-9
- → communist forces advance
- *Apr. 1946* date of capture by communists
- ★ battles, with date

U S S R

MONGOLIA

MANCHURIA

Harbin
Apr. 1946

Ch'ang-ch'un
(Changchun)
Oct. 1948 ★
Kirin (Jilin)
Jan. 1948

PLA forces

Mukden (Shenyang)
Nov. 1948

Chinchow (Jinzhou)
Oct. 1948

Kalgan (Zangjiakou)
Dec. 1948

Yellow River (Huang He)

Peking (Beijing)
Jan. 1949 ★

Tientsin (Tianjin)
Jan. 1949

KOREA

Shihkiachwang (Shijiazhuang))
Nov. 1947

Taiyuan
Apr. 1949

Sining (Xining)
Sep. 1949

Lanchow (Lanzhou)
Aug. 1949

Yenan (Yan'an)
Apr. 1948

Tsinan (Jinan)
Sep. 1948

Tsingtao (Qingdoa)
May 1949

Lùoyang
Apr. 1948

Kaifeng
June 1948

Suchow (Xuzhou)
Jan. 1949

Sian (Xi'an)
Aug. 1949

Nanking (Nanjing)
Apr. 1949

East China Sea

PLA advance repelled in Sichuan

C H I N A ★

Hankow (Hankou)
May 1949

Wuhu
May 1949

Shanghai
May 1949

Yangtze

Hangchow (Hangzhou)
May 1949

Chungking (Chongqing)
Nov. 1949

Nanchang
May 1949

Changsha
Aug. 1949

Foochow (Fuzhou)
Aug. 1949

Kweiyang (Guiyang)
Nov. 1949

Kweilin (Guilin)
Nov. 1949

Amoy (Xiamen)
Oct. 1949

Taiwan
held by nationalist forces, 1950

VIETNAM

Canton (Guangzhou)
Oct. 1949

Hong Kong (Br.)

Hoihow (Haikou)
Apr. 1950

Hainan

position. In 1970 he isolated Lin politically, and in 1971 foiled an attempted military coup, which ended with Lin Piao's death in September in a plane crash in Outer Mongolia. Mao, now a sick man, was unable to prevent his young and ambitious wife and her allies on the Cultural Revolutionary Committee from continuing a radical Maoist course. When Mao died on 9 September 1976, Chiang and her "Gang of Four" tried to seize control of the state.

By 1976 the army and much of the elite had had enough of ultra-leftism. Chiang and her co-conspirators were arrested, tried and expelled from the party. A more moderate leadership, keen to pursue economic modernization, emerged under Hua Guofeng. Thanks to détente with the US, which in October 1971 led to China's entry to the United Nations, China was able to base her modernization on closer links with the international community. In 1978 the Cultural Revolution was officially declared at an end. In the 1980s China began to unravel the legacy of Maoist oppression and economic mismanagement.

In January 1965, fearing that the revolutionary zeal of the movement was in decline, Mao decided on a new course in the Communist Party. In the summer of 1966 he launched the Great Cultural Revolution *(below)*. Supported by the army and by newly formed Red Guards *(hung-wei ping) (below left)*, Mao began a nationwide purge to rid the party of capitalists and revisionists. Victims were forced to wear dunce's caps and to recant in public *(above)*. The Cultural Revolution created chaos in China for over ten years.

THE COLD WAR IN ASIA

China was not the only area to come under communist control following the collapse of the Japanese empire. In Korea, which was divided in 1945 between Soviet and US occupation forces, a communist regime was installed in the Soviet north in September 1948. In South East Asia, under the influence of Maoist theory on mobilizing the peasant masses, communism was established in northern Vietnam by 1954 and extended to the whole of Indo-China by 1975.

In June 1950, confident that the United States would not intervene after it had withdrawn its forces in June 1949, North Korea invaded the south. After early victories communist forces were driven back to the Chinese frontier by a United Nations force supplied from 20 states but largely made up of US troops and aircraft. A Chinese counter-attack, which reached beyond Seoul by January 1951, threatened to turn the Korean conflict into a major war. When UN forces pushed the Chinese back to the original frontier between north and south, US President Truman accepted a Soviet suggestion for talks. After two years an armistice was finally signed at the village of Panmunjom, leaving the two states where they had been in 1950 except for 1,500 square miles granted to the south.

In South East Asia Vietnam became the focus of communist activity. In 1946 Vietnamese communists under Ho Chi Minh (the Vietminh) declared a Democratic Republic of Vietnam, but the French were determined to re-impose colonial control. In November 1946 French forces bombarded Ho's capital at Hanoi, killing 6,000 people. The Vietminh launched a guerrilla war against the French which ended in May 1954 in a spectacular French defeat at Dien Bien Phu. Under agreements reached at Geneva in July 1954, a communist state was established north of the 17th parallel, and a pro-American regime under Ngo Dinh Diem set up in the south with its capital at Saigon. Diem ignored the Geneva agreement and refused to hold elections in 1956. The north launched a guerrilla war in 1957, and in 1960, in reaction to the brutal Diem dictatorship, his South Vietnamese opponents set up the National Liberation Front, supported from Hanoi. The NLF, or Vietcong, controlled most of the countryside and were supplied with arms from the north. In 1963 Diem was assassinated by his own generals, and for the next 12 years South Vietnam suffered a harsh and destabilizing civil war.

In 1960 the United States stepped up its assistance to Vietnam, and its support of Laos and Cambodia, two royalist states which won independence from France in 1954. US intervention provoked Hanoi. In 1964 North Vietnamese gunboats fired at US destroyers in the Gulf of Tonking, and Congress passed a resolution which virtually amounted to a declaration of war. For the next nine years US military help kept alive the feeble pro-Western regimes in South Vietnam, Laos and Cambodia. When the US finally withdrew in 1973 the whole region fell to popular peasant-based communism. Vietnam was united in April 1975 following the fall of Saigon.

Cambodia (Kampuchea) came under the control of the communist Khmer Rouge, led by Pol Pot. Inspired by Mao's Cultural Revolution, he launched a campaign of extermination against enemies of the party. One fifth of the population, mainly urban, was slaughtered. In December 1978 Vietnam, which was closely aligned with the Soviet Union, invaded Cambodia and drove Pol Pot from power. China, which disliked Soviet support for Vietnam and Hanoi's policy of discrimination against its Chinese minority, invaded Vietnam in January 1979, but was forced to retreat. Vietnamese communism predominated throughout Indo-China, but political and racial conflict persisted into the 1990s.

Throughout South East Asia native communist movements fought against the restoration of colonial rule after 1945 *(map below right)*. The communist guerrilla war was defeated in Malaya in 1960, and communism suppressed in Indonesia and the Philippines in the 1960s. In Indochina communism achieved power in North Vietnam in 1954, but it took until 1975 before the communists controlled South Vietnam. The same year communist guerrillas in Cambodia, the Khmer Rouge, seized control with Chinese backing, and in Laos the communist Pathet Lao movement overthrew a coalition government. Vietnam, with Soviet support, extended its influence first into Laos in July 1977, then invaded Cambodia (Kampuchea) in December 1978. The United States tried to contain communism in Cambodia by heavy bombing of its capital, Phnomh Penh in 1973 *(top right)*. After US withdrawal government troops *(below left)* proved unable to resist the Khmer guerrillas, who were drawn largely from the Cambodian peasantry. The Khmer Rouge defeated the pro-American regime in April 1975 and imposed a savage rule.

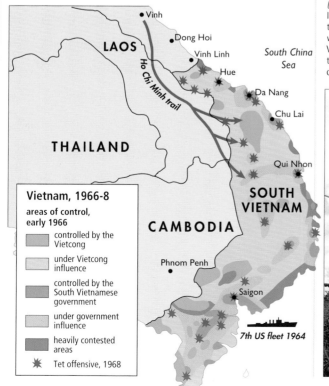

In 1954 agreement was reached at Geneva between the French government and Vietnamese insurgents on the independence of Vietnam. It was divided on the 17th parallel between a communist north and pro-Western south. From 1957 guerrilla war was waged in the south by communist forces, which led in 1960 to large-scale US assistance. The South Vietnamese government controlled the urban areas, the guerrillas most of the countryside *(map below left)*. In 1968 the guerrillas launched the "Tet" offensive against the cities which prompted gradual US withdrawal. In 1973 US troops left Vietnam altogether and two years later the whole country was united under communist rule.

American troops in Vietnam prepare to move out *(bottom)*. US forces reached a peak of 542,000 by February 1969, but by the end of 1972, shortly before the ceasefire negotiated by Kissinger in January 1973, their number fell to 25,000.

Vietnam, 1966-8

areas of control, early 1966

- controlled by the Vietcong
- under Vietcong influence
- controlled by the South Vietnamese government
- under government influence
- heavily contested areas
- ✳ Tet offensive, 1968

Vinh
Dong Hoi
Vinh Linh
LAOS
Ho Chi Minh trail
Hue
South China Sea
Da Nang
Chu Lai
THAILAND
Qui Nhon
SOUTH VIETNAM
CAMBODIA
Phnom Penh
Saigon
7th US fleet 1964

After the defeat of Japan in 1945 Korea was occupied by Soviet and US forces who divided the peninsula along the 38th parallel. In 1948 both sides established regimes sympathetic to their interests: a communist north under Kim Il Sung; and a pro-Western south under Syngman Rhee. In June 1950 the north launched a surprise attack on the south *(map top right)*. A United Nations force, composed chiefly of US troops and aircraft, and led by General MacArthur, drove the communist forces back to the Chinese border, but were then attacked by a force of 200,000 Chinese troops in October 1950. US forces, seen here *(above)* on the Taegu front, lost 33,000 men. In June 1951 negotiations began, and two years later the old frontier on the 38th parallel was restored and a demilitarized zone created between the two states.

The Cold War World

130

NORTH KOREA

In 1948 the new state of North Korea was set up by the Korean Workers' Party led by Kim Il Sung (right), who had spent time in Moscow, returning with the Red Army in 1945. During the 1950s Kim gradually established his personal dictatorship, which was finally secured in 1958. Kim imposed his own brand of communism on the people of North Korea. Political dissidents were sent to concentration camps. The rest of the population was forced to lead strictly regimented lives. Private ownership of cars or telephones was forbidden. North Korea's economy was closed off from the outside world and became reliant on Soviet and Chinese aid. Kim established a cult of personality which assumed bizarre proportions. His family was promoted to high office. Kim's son, Kim Jong Il, was groomed for the

dynastic succession. On Kim Il Sung's death in 1994, Kim Jong Il became the new dictator. The collapse of Soviet communism and the closer integration of Deng's China with the Pacific Rim countries has left North Korea isolated politically and economically.

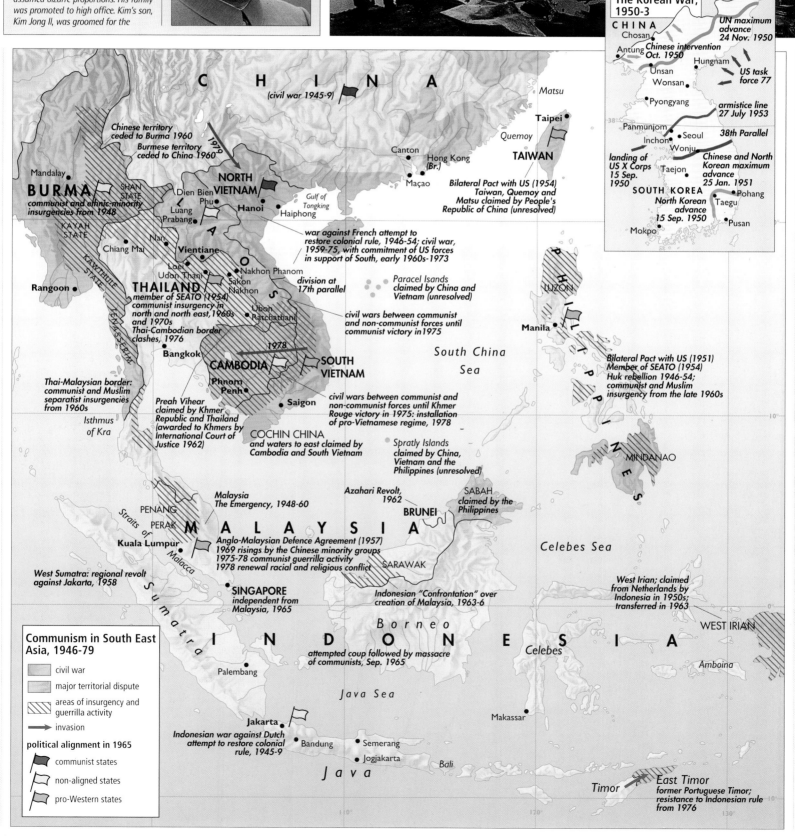

The Korean War, 1950-3

- CHINA
- Chosan
- Antung — Chinese intervention Oct. 1950
- UN maximum advance 24 Nov. 1950
- Unsan
- Hungnam
- Wonsan
- US task force 77
- Pyongyang
- armistice line 27 July 1953
- Panmunjom
- Seoul
- 38th Parallel
- Inchon — Wonju
- landing of US X Corps 15 Sep. 1950
- Taejon
- Chinese and North Korean maximum advance 25 Jan. 1951
- SOUTH KOREA
- North Korean advance 15 Sep. 1950
- Pohang
- Taegu
- Mokpo
- Pusan

CHINA (civil war 1945-9)

Matsu

Taipei
Quemoy

Canton
Hong Kong (Br.)
Maçao

TAIWAN

Bilateral Pact with US (1954) Taiwan, Quemoy and Matsu claimed by People's Republic of China (unresolved)

Chinese territory ceded to Burma 1960
Burmese territory ceded to China 1960

1979

NORTH VIETNAM

Dien Bien Phu
Luang Prabang
Hanoi
Haiphong
Gulf of Tongking

Mandalay

BURMA
communist and ethnic-minority insurgencies from 1948

SHAN STATE
KAYAH STATE

Chiang Mai
Nan

war against French attempt to restore colonial rule, 1946-54; civil war, 1959-75, with commitment of US forces in support of South, early 1960s-1973

Paracel Isands claimed by China and Vietnam (unresolved)

PHILIPPINES

LUZON

Vientiane

Rangoon

KAWTHULE STATE

TENASSERIM

Loei
Udon Thani
Nakhon Phanom
Sakon Nakhon

THAILAND
member of SEATO (1954) communist insurgency in north and north east, 1960s and 1970s
Thai-Cambodian border clashes, 1976

LAOS

division at 17th parallel

civil wars between communist and non-communist forces until communist victory in 1975

Ubon Ratchathani

Manila

Bilateral Pact with US (1951)
Member of SEATO (1954)
Huk rebellion 1946-54; communist and Muslim insurgency from the late 1960s

1978

Thai-Malaysian border: communist and Muslim separatist insurgencies from 1960s

Isthmus of Kra

Bangkok

CAMBODIA

Phnom Penh

SOUTH VIETNAM

Saigon

South China Sea

civil wars between communist and non-communist forces until Khmer Rouge victory in 1975: installation of pro-Vietnamese regime, 1978

Preah Vihear claimed by Khmer Republic and Thailand (awarded to Khmers by International Court of Justice 1962)

COCHIN CHINA
and waters to east claimed by Cambodia and South Vietnam

Spratly Islands claimed by China, Vietnam and the Philippines (unresolved)

MINDANAO

Malaysia The Emergency, 1948-60

Azahari Revolt, 1962

SABAH claimed by the Philippines

BRUNEI

Celebes Sea

PENANG
Straits of Malacca
PERAK

Kuala Lumpur

MALAYSIA

Anglo-Malaysian Defence Agreement (1957)
1969 risings by the Chinese minority groups
1975-78 communist guerrilla activity
1978 renewal racial and religious conflict

SARAWAK

West Irian; claimed from Netherlands by Indonesia in 1950s; transferred in 1963

West Sumatra: regional revolt against Jakarta, 1958

Sumatra

SINGAPORE independent from Malaysia, 1965

Indonesian "Confrontation" over creation of Malaysia, 1963-6

Borneo

INDONESIA

WEST IRIAN

Communism in South East Asia, 1946-79

- civil war
- major territorial dispute
- areas of insurgency and guerrilla activity
- invasion

political alignment in 1965
- communist states
- non-aligned states
- pro-Western states

Palembang

Celebes

Amboina

attempted coup followed by massacre of communists, Sep. 1965

Java Sea

Makassar

Jakarta

Indonesian war against Dutch attempt to restore colonial rule, 1945-9

Bandung
Semerang
Jogjakarta

Java

Bali

Timor

East Timor former Portuguese Timor; resistance to Indonesian rule from 1976

The Cold War in Asia

THE RETREAT FROM EMPIRE AFTER 1939

The Second World War created opportunities for communism, but it sounded the death knell for colonialism. By 1945 the Italian, Japanese and German empires were destroyed. The colonies of East Asia had been overrun by the Japanese and local nationalists were unwilling to return to colonial dependence. Neither the United States nor the Soviet Union was willing to tolerate the survival of an unreformed imperialism. Only the British empire, one of the major victors in 1945, survived relatively intact, and it was here, paradoxically, that the greatest concessions were extracted in the first post-war decade.

In India the war years ended the brief experiment in partial self-government begun in 1937. The nationalist Congress refused to participate in the war effort on the grounds that they had not been consulted about India's declaration of war on Germany. Members began a "Quit India" campaign which led to their arrest and imprisonment. By the end of the war it was clear to the British that they could not hold on to India on pre-war terms. A broad Muslim movement, backed by a 1940 League of Nations resolution approving a separate state for India's Islamic population, called for partition. Congress reluctantly agreed and two states – India and Pakistan – were granted independence on 15 August 1947.

In Britain's other Asian possessions there was a threat from nationalists and communists, inspired by Mao's example in China. In Malaya a long counter-insurgency war defeated the communists but brought the more moderate nationalist elements independence in 1957. In Burma, which had been occupied by the Japanese and granted a puppet government, British rule was violently rejected and independence was granted in 1948. The

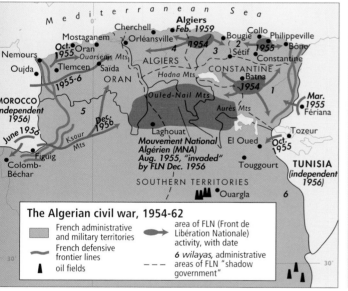

The Algerian civil war, 1954-62

- Shaded area: French administrative and military territories
- French defensive frontier lines
- ▲ oil fields
- ➤ area of FLN (Front de Libération Nationale) activity, with date
- - - - 6 wilayas, administrative areas of FLN "shadow government"

In 1954, disappointed at failing to win independence for Algeria after the war, Algerian nationalists set up the Front de Libération Nationale (FLN) and launched a guerrilla war which lasted until 1962 (map left). The French settler community fought back through their own organization, the OAS, which perpetrated atrocities against both Algerians and the French administration. High losses and high costs forced the French hand and independence came on 5 July 1962.

A demonstration in Algiers in December (below) greeted the visit of President Charles de Gaulle on a fact-finding mission during the civil war. De Gaulle was brought to power in 1958 to avert a military revolt over the Algerian issue, but he was a realist on the prospects of keeping Algeria French. He negotiated the ceasefire with the nationalists in March 1962, paving the way for independence.

In 1957 the British Gold Coast became the first of Britain's African colonies to achieve independence and with it a new name, Ghana. The independence ceremony (below) was attended by the Duchess of Kent and the retiring British governor-general in the presence of the leader of the Ghanaian nationalist movement, Dr Kwame Nkrumah.

In the 30 years after the end of the war in 1945 the colonial empires were relinquished by the European powers (map above). American and Soviet hostility to colonialism put pressure on Europe, but it was the impossibility of defending the empires against nationalist movements within the colonial areas that finally eroded Europe's imperialism.

The post-colonial world

- ░ territories independent since 1939, with dates
- **colonial possessions in 1996**
  - ● British
  - ● French
  - ● Dutch
  - ● Portuguese
  - ● Spanish
  - ● US
  - ● Australia
  - ● New Zealand
  - ● USA Australia New Zealand joint rule
- ⚑ states within British Commonwealth
- ⚑ states that broke away from Commonwealth
- ⚑ states within French Community
- ⚑ states that broke away from French Community
- ⚑ former colonial power
- ⚔ areas of colonial conflict
- ☐ stations and bases overseas
- ▬ members of abortive federations
- ●–●–● border conflict

Map labels:

IRAN · AFGHANISTAN · Kashmir · TIBET (absorbed by China 1965) · CHINA · Sikkim (annexed by India 1975) · S. KOREA · PACIFIC OCEAN

KUWAIT 1961 · BAHRAIN 1971 · to Oman · UAE 1971 · PAKISTAN 1947 left Commonwealth 1972; rejoined 1989 · NEPAL · BHUTAN · INDIA 1947

New Ireland New Britain 1975 (to Papua New Guinea) · KIRIBATI 1979 (formerly Gilbert Is.) · Fanning I. · NAURU · Ocean I. · Jarvis I. · TUVALU 1970 (formerly Ellice Is.) · Marquesas Is. · SOLOMONS IS. 1978 · Wallis and Futuna Is. · Manihiki I. · Rotuma · VANUATU 1980 (formerly New Hebrides) · WESTERN SAMOA 1962 · American Samoa · Tahiti Is. · New Caledonia · FIJI 1970 suspended from Commonwealth 1987 · Society Is. · TONGA 1970 · Cook Is. · Gambier I. · AUSTRALIA · Kermadec Is. · Pitcairn I. · Norfolk I. · Lord Howe I.

ARABIA · OMAN · 1963–67 · PEOPLES DEMOCRATIC REPUBLIC OF YEMEN 1967 · evacuated 1967 · Diu 1961 annexed by India · Daman 1961 to India · BANGLADESH 1971 (EAST PAKISTAN 1947) · BURMA 1948 · LAOS 1954 · NORTH VIETNAM 1954 · Hainan · 1946–54 · TAIWAN · Hong Kong · Macao

Goa 1961 annexed by India · Pondicherry 1954 to India · Karikal 1954 to India · Laccadive Is. · Andaman Is. to India · THAILAND · CAMBODIA 1953 · SOCIALIST REPUBLIC OF VIETNAM unified 1976 · SOUTH VIETNAM 1954 · PHILIPPINES 1946

1975 · SOMALIA 1960 · Mayotte administered by France since 1975, claimed by Comoros · CEYLON 1948 (SRI LANKA 1972) · MALDIVES 1965 · Gan evacuated 1967 · 1948–66 · MALAYA 1957 · MALAYSIA 1963 · Malaysian Federation 1963–5 · BRUNEI 1984 · Sabah · 1959–65 SINGAPORE · Sarawak · Borneo · Kalimantan · independent republic suppressed by Indonesia 1949–50 · PAPUA NEW GUINEA 1975

Sumatra · 1945–49 · Java · INDONESIA 1949 · South Moluccas · Dutch New Guinea to Indonesia 1963 · Portuguese Timor annexed by Indonesia 1976 · Timor 1976

MADAGASCAR 1960

AUSTRALIA · ATLANTIC OCEAN

BAHAMAS 1973 · Turks and Caicos Is. · MEXICO · Isla de Pinos Cayman Is. · CUBA · Sint-Maarten St. Martin · British Virgin Is. · BELIZE 1981 · Swan Is. (to Honduras) · HAITI · DOMINICAN REP. · Puerto Rico · Anguilla · ANTIGUA-BARBUDA 1981 · JAMAICA Federation of West Indies 1958–62 · American Virgin Is. · St. Barthélemy · Montserrat · HONDURAS · ST CHRISTOPHER-NEVIS 1983 · Guadeloupe · DOMINICA 1978 · Caribbean Sea · Martinique · NICARAGUA · ST LUCIA 1979 · Aruba · ST VINCENT 1979 · BARBADOS 1966 · Curaçao · GRENADA 1974 · Bonaire · TRINIDAD & TOBAGO 1962 · COSTA RICA · COLOMBIA · PANAMA · VENEZUELA · NEW ZEALAND

French and Dutch faced the same problems in the Far East. The Netherlands never regained control of the Dutch Indies after the Japanese left and independence was formally achieved in 1949. France attempted to pursue a strategy of assimilation with the metropolitan power, or the granting of associated status, to remove the stigma of colonial control. French rule was nonetheless rejected in Indo-China. The restored administration in Vietnam found itself in head-on confrontation with a mass communist and nationalist movement. Military defeat in 1954 at Dien Bien Phu (*see* page 130) persuaded French leaders to abandon the Far Eastern empire altogether.

The occupation of French North African colonies by British and American forces during the war also created problems when French rule was restored. Tunisia and Morocco were given independence in 1956. After eight years of brutal civil war between French settlers, Algerian nationalists and Islamic insurgents, and the French army, the French president, Charles de Gaulle, ended the conflict in 1962 by granting Algeria full independence rather than risk civil war at home. Italy's former possessions were placed under the United Nations: Libya became a new nation in 1951, Somalia in 1960. In the Middle East Britain abandoned Palestine, which formed the core of the new State of Israel set up in 1948 (*see* page 134). Civil wars in Cyprus and Aden precipitated British withdrawal in 1960 and 1967. In 1968 the British government announced the end of a British presence east of Suez.

In sub-Saharan Africa there were fewer challenges to European rule. Economic problems in Europe encouraged a vigorous exploitation of African resources, while in eastern and southern Africa there were large white settler communities anxious to obstruct the black independence movements. But here too violence forced the hand of the colonial powers. The Gold Coast was freed in 1957, Nigeria in 1960. The bloody Mau-Mau rebellion in Kenya was followed by independence for Kenya and Tanganyika by 1964 and, in central Africa, for Nyasaland (Malawi) and northern Rhodesia (Zambia). Belgium withdrew from the Congo in 1960, and France from its tropical African possessions between 1958 and 1960. Portugal was the last state to abandon empire. After a bitter guerrilla war Angola and Mozambique freed themselves from Portuguese rule by 1975. With the loss of the most significant colonial areas empire was at an end, leaving in its wake a legacy of political instability, religious and tribal conflict, impoverishment and oppression. Only a few of the new states were untouched by violence.

Independence for India brought a bitter religious war between Hindus and Muslims *(right)*, which left a million dead in 1947–8 and forced the partition of the country between Muslim Pakistan and (predominantly) Hindu India. Up to 15 million people made the arduous journey between the new states, the greatest mass exodus of the century. India was still left with a substantial Muslim minority and has been beset by constant communal tensions and intermittent violence ever since independence. Tensions between the two new states continued for many years.

The Retreat from Empire after 1939

THE MIDDLE EAST, 1945-67

The Middle East was the most unstable of the post-imperial regions after 1945. Ever since the collapse of Ottoman power in the early part of the century, the ambition of the Arab peoples had been to create new Arab nation-states. Arab nationalism produced an independent Iraq in 1932, and in 1936 an Egyptian state was established, although with a continued British military presence. The same year the French agreed to relinquish their mandate over Syria within three years. These gains were consolidated after 1945: Syria and Lebanon won full independence in 1945 and 1946; Transjordan was granted independence in March 1946, Libya in 1951. In 1951 Egypt abandoned the 1936 treaty with Britain and a schedule for the withdrawal of British troops was agreed. Despite efforts to keep a European military presence in an area of strategic concern to the West, by 1956 the Arab region was genuinely independent of European power.

The most intractable issue of all was the future of the Palestine mandate, granted to Britain by the League of Nations in 1920. Arab nationalists saw this as Arab land, and demanded its independence. But in November 1917 the British politician Arthur Balfour had published a declaration committing Britain to provide the Jewish people with a homeland in the Palestine region. The world Zionist organization, set up in the 1890s, wanted Britain to honour this pledge. As a result, in the 1930s the British administration found itself caught between a militant Arab nationalism, inspired by the idea of *jihad*, or holy war, and Jewish demands for a homeland. In 1937 Jewish activists set up Irgun Zvai Leumi (National Military Organization). In 1940 Abraham Stern founded the Fighters for the Freedom of Israel. Both groups undertook acts of terrorism against Arab and British targets. By 1945 there were 100,000 British troops in Palestine to resist the threat.

Reviled by both sides, the British struggled to maintain order. In July 1946 the King David Hotel in Jerusalem was blown up by Jewish terrorists and 91 killed. Public opinion in Britain turned against maintaining the mandate. In 1947 Britain asked the UN to resolve the issue, and on 29 November 1947 a UN resolution divided Palestine into a Jewish and an Arab state. Britain withdrew in haste in May 1948, as Jewish militia laid siege to Jaffa and Jerusalem. On 14 May David Ben-Gurion declared the foundation of the State of Israel and became its first prime minister, with Chaim Weizmann, leader of the world Zionist movement, as Israel's first president.

The following day the new state was invaded from Syria, Transjordan and Egypt, and by an Arab volunteer force. After fierce fighting a ceasefire was arranged early in 1949. Between 600,000 and 750,000 Palestinians became refugees, crowded into the Gaza Strip and the West Bank. Transjordan annexed the West Bank and the Arab state of Palestine disappeared. An uneasy truce followed. The Arab states were never reconciled to Israel's survival, which owed a good deal to American support.

Terrorist attacks on Israel continued until in October 1956, with British and French support, Israel invaded the Sinai Peninsula and defeated the Egyptian forces there. Under strong international pressure Israel withdrew, but the peninsula was placed under a UN force and demilitarized.

There followed a decade of uneasy peace. Israel's Arab neighbours embarked on programmes of economic modernization or political reform. Egypt and Syria established a United Arab Republic in 1958 as the core of a broader Arabist movement, but Arab unity proved skin-deep. The Republic broke up in 1961. A radical regime established in Iraq in 1958 under Abdel-karim Kassem threatened the independence of Kuwait in 1961 until restrained by the other Arab states. In Yemen a civil war between monarchists and republicans divided the Arab camp still more. Saudi Arabia backed the monarchists, Egypt the republicans. Nasser's attempt to place himself at the head of a broader Arab movement foundered on local nationalism and division between the royalist regimes (Saudi Arabia, Jordan, Libya) and more radical republican movements.

Nasser made one last attempt to revive the Pan-Arabist movement. Egypt and Syria cultivated close ties with the Soviet Union in the late 1960s, which furnished them with military advice and modern weaponry. Nasser hoped to use the relationship as a lever to complete the extinction of Israel, but instead he prompted strong American military assistance to Israel and the anti-Soviet Arab states, and yet further divisions in the Arab camp. In May 1967 Nasser demanded the removal of the UN peacekeeping force in Sinai, which soon followed. He then cut Israel off from sea routes to the Gulf of Aqaba. The absence of an Israeli response was interpreted as a sign of weakness. Nasser called for a final reckoning with Zionism.

Between the 1930s and the 1960s the Middle East and North Africa freed themselves from European rule and established independent Arab states *(map right)*, with the exception of the Jewish State of Israel, founded in 1948. The region has been constantly unstable, first because of anti-European and anti-Israeli conflicts, later because of the importance of Arab oil and the rise of militant Islamic movements throughout the region.

Gamel Abdul Nasser *(below left)* was the leader of the military coup which overthrew King Farouk of Egypt. In 1954 he became prime minister of the new Egyptian republic. Born in 1918, he became a career soldier and led the Egyptian battalion in the Arab-Israeli war of 1948. He negotiated the withdrawal of Britain from Egyptian bases and moved closer to the Soviet bloc in 1956 following US refusal to fund the Aswan Dam scheme. He saw himself as a revolutionary in the Arab cause, but was distrusted by more conservative Arab regimes. He led Egypt into war with Israel in 1967 when his forces were humiliated. His subsequent attempt to resign was reversed by popular demand. He died in 1970.

Following the declaration of an independent Jewish State of Israel on 14 May 1948, forces from the surrounding Arab states invaded former Palestine to re-establish Arab claims to the area *(map right)*. After seven months of war, a ceasefire was agreed with the Palestinians. Subsequent agreements left Israel with 21 per cent more land than she had controlled in 1948.

French paratroopers *(left)* in occupation of the Egyptian city of Port Said in early November 1956. British forces had left the Canal Zone five months before, but Nasser's decision to nationalize the Canal Company brought Anglo-French troops back to try to force his overthrow.

Israel: the War of Independence, 1948

- Jewish State as proposed by the United Nations, Nov. 1947
- territory conquered by Israel, 1948-9
- principal Arab attacks, May 1948
- Israel according to armistice agreements, 1949

Map labels: LEBANON, SYRIA, Safad, Acre, Haifa, Lake Tiberias, Nazareth, Plain of Sharon, Hadera, Netanya, Nablus, Jordan, WEST, Tel Aviv-Jaffa, JORDAN, Holon, Rehovot, Jerusalem, Ashod, BANK, Bethlehem, Hebron, Gaza, Dead Sea, Beersheba, Negev Desert, EGYPT, neutral zone, Eilat

The Middle East from 1945

- → invasion
- ✳ major conflicts
- ✊ guerrilla activity

TURKEY

1946 Withdrawal of French troops
1958-61 Union with Egypt (United Arab Republic)
1963 Ba'th Party seizes power
1967 Six Day War; Syria loses Golan Heights
1970 General Hafiz al-Assad seizes power
1973 October (Yom Kippur) War; Syria and Egypt attack Israel; Syrian forces expelled from Golan Heights and Israeli forces occupy Syrian territory
1976 Syrian forces intervene in Lebanese civil war
1980 Treaty of Friendship and Co-operation with USSR
1991 Peace made with Lebanon

1955 Anti-Soviet Baghdad Pact
1958 Hashemite dynasty overthrown in military coup; power seized by General Abdel-karim Kassem
1963 Kassem overthrown in military coup
1968 Ba'th Party seizes power
1972-5 Intermittent fighting between Kurds and government
1974-5 Iran-Iraq War; Iran withdraws support from Kurds; Kurdish rebellion collapses
1979 Saddam Hussein becomes president
1980-8 Iran-Iraq War
1990 Invasion of Kuwait by Iraq
1991 UN coalition expels Iraqi army from Kuwait; Shia and Kurdish rebels attempt overthrow of Saddam; massive reprisals ordered by Saddam; Western sanctions imposed; de facto independent Kurdish state established in north

USSR

CYPRUS

1963-74 Intermittent intercommunal clashes
1974 Turkish invasion and occupation of northern part of island

LEBANON

1969 Increase in Palestinian guerrilla activity
1975 Lebanese civil war breaks out
1976 Syrian invasion
1978 Israeli invasion
1982 Attack on Beirut by Israel
1985 Formal withdrawal of Israeli troops
1992 Christians boycott first elections for 20 years

Aleppo
Hama
SYRIA
Beirut
Damascus
IRAQ
Baghdad
• Teheran

Euphrates
Tigris

IRAN

1941 Abdication of Shah Reza Pahlavi following Anglo-Russian occupation of Iran; his son Mohammed Reza Pahlavi becomes Shah
1951 Nationalization of oil industry; deterioration in relations with UK
1953-4 Prime Minister Mossadeq becomes de facto ruler; the Shah flees but is later reinstated by royalist military forces with covert US support; oil dispute settled
1961 Shah declares "White Revolution"
1975 Algiers Agreement with Iraq acknowledges Iran's supremacy in Gulf
1978-9 Revolution; the Shah is exiled; Ayatollah Khomeini returns from exile; Iran becomes an Islamic Republic
1980-8 Iran-Iraq War
1989 Khomeini dies, Rafsanjani president
1995 US imposes economic sanctions

AFGHANISTAN

PAKISTAN

ISRAEL
Jerusalem
Amman
JORDAN

1952 Accession of King Hussein
1970 Attempted destruction of PLO by Jordanian Army (Black September)
1990 King Hussein refuses to join coalition against Iraq

Kuwait
KUWAIT

1990 Invaded by Iraq; Gulf Crisis
1991 Liberated by UN coalition forces

Persian Gulf

Cairo

EGYPT

1948 Leads Arab coalition against Israel
1952 Monarchy overthrown; military government led by Nasser after 1952
1956 Nationalization of Suez Canal Company; tripartite invasion by Britain, France, Israel
1958-61 Union with Syria (United Arab Republic)
1967,1973 Wars with Israel
1970 Nasser dies; Sadat becomes president
1979 Egyptian-Israeli peace treaty
1981 Sadat assassination; Mubarak becomes president
1990 Egypt sends troops to anti-Iraq coalition

Nile

SAUDI ARABIA

Medina •

BAHRAIN **QATAR**
Doha •
Riyadh • Abu Dhabi •
UAE
• Muscat

1951 Mutual Defence Assistance Agreement with US
1960 Organization of Petroleum Exporting Countries formed
1981 Gulf Co-operation Council formed (with Bahrain, Kuwait, Oman, Qatar and UAE)
1973 Saudi Arabia embargos oil exports to USA; oil price soars
1990 Base for UN Coalition attacks against Iraq
1992 Tentative steps towards political openess
1996 King Fahd temporarily steps down

• Mecca

Red Sea

1971 Created from former British-protected Trucial States after British evacuation

Arabian Sea

OMAN

1965-75 Marxist insurgency by People's Democratic Republic of Yemen defeated with British and Iranian help

SUDAN

1953 Anglo-Egyptian agreement on ending British condominium of 1899
1956 Sudan gains independence
1958 Coup by General Ibrahim Abboud
1963-72 Civil war between Arab Muslim rulers in north and Christian and animist Africans in south
1969 Abboud deposed; Colonel Gaafar Mohammed el-Nimeiri seizes power
1983 Civil war re-erupts; food shortages increase
1985 Military coup ousts Nimeiri
1989 Military coup; National Islamic Front in effective control
1990-1 Famine worsens; reports of military aid from Iran
1994-5 Ceasefire between feuding southern anti-government forces

Khartoum •

YEMEN ARAB REPUBLIC

1962-9 Civil war
1972-9 Intermittent war with Aden

San'a •

PEOPLE'S DEMOCRATIC REPUBLIC OF YEMEN

1967 Coup by National Liberation Front; civil war; Britain withdraws troops from Aden
1968 Ali Nasar Muhammad overthrown as president by Haidar al Attas

• Aden

YEMEN

1948 Assassination of Imam Yahya; his son takes power
1959 Creation of the Arab Emirates of the South (later the Federation of South Arabia)
1962 Civil war and revolution in San'a; Yemen Arab Republic (North Yemen) established
1967 Withdrawal of British forces; declaration of People's Democratic Republic of Yemen (South Yemen)
1990 YAR and PDRY united

Following the Egyptian government's decision to nationalize the Anglo-French Suez Canal in March 1956, British, French and Israeli forces attacked Egypt in October and November 1956 *(map right)*. The Israeli army quickly captured the Sinai Peninsula, and Anglo-French forces took Port Said and Port Fuad. After US and Soviet pressure they withdrew in December, and a UN peacekeeping force replaced them.

After 1945 Britain reluctantly allowed further Jewish immigration to Palestine from the decimated Jewish populations of Europe. Here *(left)* settlers are transferred by ship to the port of Haifa, where they faced an uncertain future.

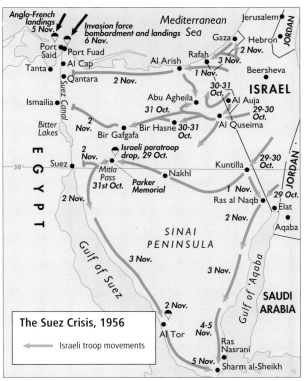

Anglo-French landings 5 Nov.
Invasion force bombardment and landings 6 Nov.
Mediterranean Sea
Jerusalem
JORDAN
Port Said Port Fuad
Gaza
Hebron
Tanta • Al Cap
Al Arish
Rafah 3 Nov.
2 Nov.
Qantara 2 Nov.
1 Nov.
Beersheva
Ismailia
Suez Canal
Abu Agheila
31 Oct.
30-31 Oct.
Al Auja
ISRAEL
Bitter Lakes
2 Nov.
Bir Gafgafa Bir Hasne 30-31 Oct.
Al Quseima
29-30 Oct.
E G Y P T
Suez
2 Nov.
Israeli paratroop drop, 29 Oct.
Kuntilla
29-30 Oct.
Mitla Pass 31st Oct.
Parker Memorial
Nakhl
1 Nov.
Ras al Naqb
29 Oct.
JORDAN
2 Nov.
Elat
2 Nov.
Aqaba
SINAI PENINSULA
Gulf of Suez
3 Nov.
3 Nov.
SAUDI ARABIA
2 Nov.
Al Tor
4-5 Nov.
Gulf of Aqaba
Ras Nasrani
5 Nov.
Sharm al-Sheikh

The Suez Crisis, 1956
→ Israeli troop movements

THE MIDDLE EAST, 1967-79

In June 1967 war broke out between Israel and her Arab neighbours but not the war Nasser had expected. Instead the Israeli government, nervous at Nasser's escalation, ordered a pre-emptive strike. On 5 June the Israeli air force attacked and destroyed the Egyptian air force on the ground. The army occupied the Gaza strip and the Sinai Peninsula in three days. Jordan, Iraq and Syria rallied to Egypt's support, but their forces were routed in a further three days of fighting. The West Bank and the Arab half of Jerusalem were seized by Israel from Jordan. The Golan Heights on the Israeli-Syrian border were occupied by Israel on 10 June, though not officially annexed until 1982.

The war precipitated a new and more serious Palestinian problem. Israel now had almost one million Arabs under her direct rule. Up to 250,000 fled from the West Bank into Jordan. 400,000 refugees lived in Lebanon. In the camps young Palestinians formed armed guerrilla movements dedicated to winning back the areas lost in 1967. The Palestine Liberation Organization (PLO), founded in 1964, assumed leadership of the struggle and used the camps as a base for a campaign of terrorism directed against Israel.

Within five years the Egyptian threat revived as well. Nasser's successor in 1970, Anwar Sadat, had ambitions to avenge Egypt's humiliation in 1967. In November 1972 he decided to attack Israel, again in collaboration with Syria. After 11 months of preparation they attacked on 6 October 1973, the Jewish holy day of Yom Kippur. After initial Arab

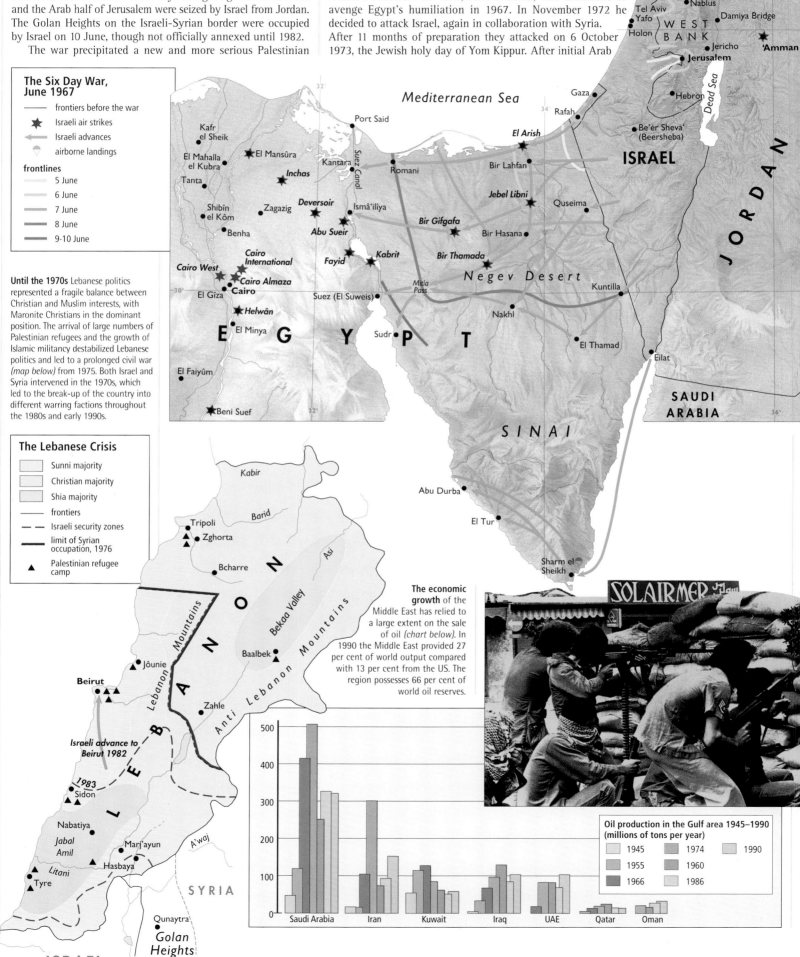

The Six Day War, June 1967

— frontiers before the war
★ Israeli air strikes
← Israeli advances
⚑ airborne landings

frontlines
— 5 June
— 6 June
— 7 June
— 8 June
— 9-10 June

Until the 1970s Lebanese politics represented a fragile balance between Christian and Muslim interests, with Maronite Christians in the dominant position. The arrival of large numbers of Palestinian refugees and the growth of Islamic militancy destabilized Lebanese politics and led to a prolonged civil war *(map below)* from 1975. Both Israel and Syria intervened in the 1970s, which led to the break-up of the country into different warring factions throughout the 1980s and early 1990s.

The Lebanese Crisis

▢ Sunni majority
▢ Christian majority
▢ Shia majority
— frontiers
--- Israeli security zones
▬ limit of Syrian occupation, 1976
▲ Palestinian refugee camp

The economic growth of the Middle East has relied to a large extent on the sale of oil *(chart below)*. In 1990 the Middle East provided 27 per cent of world output compared with 13 per cent from the US. The region possesses 66 per cent of world oil reserves.

Oil production in the Gulf area 1945–1990 (millions of tons per year)

▢ 1945 ▢ 1974 ▢ 1990
▢ 1955 ▢ 1960
▢ 1966 ▢ 1986

The Cold War World

In June 1967, following growing threats from her Arab neighbours, Israel sent out forces – seen here (below) in the Sinai desert – to launch a pre-emptive strike against Egypt, whose forces were defeated in three days. Attempts by Syria, Jordan and Iraq to help Egypt led to the rapid defeat of their forces (map left), and a five-fold increase in the territory under Israeli control.

The Yom Kippur war hit Israeli forces with almost complete surprise (below right). The Israeli air force restored the position in the Golan Heights and when Egyptian forces moved into Sinai away from their missiles defence system, Israeli air-ground co-operation broke the Egyptian line and brought Israel's army to within 50 miles of Cairo (map below). On 24 October a cease-fire came into operation after the US put its forces on nuclear alert.

gains in Sinai and on the Golan Heights, Israel successfully counter-attacked. An armistice was agreed on 24 October under American and Soviet pressure. Egypt bowed to reality. Agreement was reached with Israel in 1974 and 1975 on the disengagement of forces in Sinai and in 1977, despite widespread hostility from the rest of the Arab world, Sadat finally sought a peace settlement. On 19 November he flew to Israel, whose parliament was dominated by the hardline Likud bloc led by Menachem Begin, where he offered to recognize Israel and sign a peace treaty. In September 1978 the two sides met in the US at Camp David, where, in the presence of President Jimmy Carter, Sadat and Begin agreed to peace between their states.

Sadat's move provoked outrage in the Arab world. Egypt was thrown out of the Arab League and exposed to a political and economic boycott. In practice this division simply added one more conflict to an Arab world which was anything but united. Civil war broke out in Jordan in 1970 between Palestinians and King Hussein's army. Despite Syrian assistance for the refugees, hundreds were forcibly expelled. During the 1970s violence briefly flared between Egypt and Libya, between Iran and Iraq, and between Iraq and Syria. In Lebanon divisions between

Christians and Muslims, between Lebanese and Palestinians and between Sunni and Shia forms of Islam created a microcosm of the wider tensions in the Arab world. In April 1975 full-scale civil war broke out in Lebanon, and in 1976, anxious about Israeli reaction, Syria occupied Lebanon and introduced an Arab peace-keeping force. In 1978 Israel invaded the south in retaliation against PLO incursions and Lebanon was effectively partitioned.

The Lebanese crisis highlighted a new development in the Middle East: the rise of revolutionary Islam. By the 1970s the Cold War tension between conservative and radical Arab states gave way to a new tension between secular Arab nationalism and Islamic fundamentalism. The Islamic movement dated back to the foundation of the Muslim Brotherhood in 1928, but it was given new life by the failures against Israel in 1967 and 1973. Radical movements spread throughout the Arab world: in 1974 in Lebanon, Imam Musa al-Sadr founded Amal, committed to the violent defence of Shia Islam; in November 1979 a group led by a self-styled Mahdi (messiah) seized the Grand Mosque in Mecca and held it until expelled by Saudi police. Militant Muslims declared a jihad (holy war) against Arab secularists and Westernizers and Western "cultural imperialism".

In October 1973 Egypt and Syria launched a surprise attack on Israeli-held territory. After initial success, Egyptian forces were surrounded by an Israeli counter-attack (map below left). In the Golan Heights (map below right) Israel extended the territory under its military control.

The Yom Kippur War, 1973: the Suez Canal

☐ occupied by Israel at the end of the Six Day War, 1967
— de facto frontier before the war
➜ Egyptian advances
━ furthest Egyptian advance into Israeli-held territory
➜ Israeli counter-offensives
▨ Israeli territory held by Egypt at ceasefire, 24 Oct.
▨ Egyptian territory held by Israel at ceasefire, 24 Oct.

The Yom Kippur War, 1973: the Golan Heights

— de facto frontiers before the war
━ occupied by Israel at the end of the Six Day War, 1967
◀ Arab advances
━ furthest Arab advance into Israeli territory
◀ Israeli counter-offensives
━ Syrian territory held by Israel at ceasefire, 24 Oct.

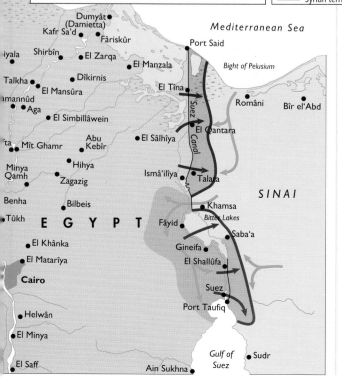

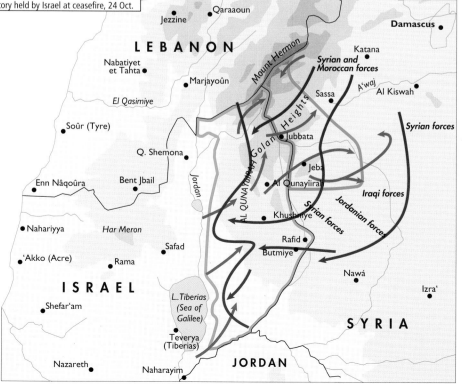

AFRICA SINCE INDEPENDENCE

The residue of colonialism lingered on in Africa far longer than in the Middle East. Most of Africa was independent by 1965, but a handful of colonies survived, and in southern Africa white settlers still monopolized both the political system and economic wealth. Portugal still clung to its colonies in Angola, Mozambique and Guinea-Bissau despite bitter civil wars with nationalist forces. In 1974 Portugal's one-party system was finally overthrown and the new democratic regime granted independence the following year.

White domination of the south took longer to alter. In 1965 the white population of southern Rhodesia, led by Ian Smith, issued a Unilateral Declaration of Independence in the

RWANDA

In the scramble for Africa the two central African kingdoms of Rwanda and Burundi were conquered by Germany. After the Great War they came under Belgian control and were administered in collaboration with the local aristocracy, the Tutsi. In Rwanda the majority of the population were Hutus and they expected independence in 1962 to end Tutsi domination. Civil war broke out before the Belgians left, and resulted in the overthrow of the Tutsis when the colonial link was severed. An attempted coup the following year by Tutsi exiles led to a massacre of their people in Rwanda. In neighbouring Burundi a Hutu rebellion in 1972-3 led to the deaths of 150,000 Hutus and 10,000 Tutsis. In the early 1990s the ethnic conflict flared up again in an acute form. The Rwandan president, Juvenal Habyarimana, reached an agreement with the Tutsi-led Rwandan Patriotic Front, but when his plane was shot down by rocket fire in April 1994 a blood-bath ensued. The Hutu army and armed militia massacred up to 500,000 Tutsi; an estimated two million – mostly Hutus – from a population of eight million fled to neighbouring states between April and July 1994. The UN intervened in May to cope with the humanitarian disaster. The RPF declared a pyrrhic victory in July.

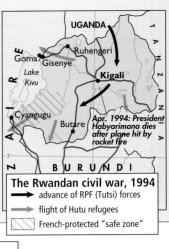

The Rwandan civil war, 1994

Apr. 1994: President Habyarimana dies after plane hit by rocket fire

→ advance of RPF (Tutsi) forces
→ flight of Hutu refugees
▨ French-protected "safe zone"

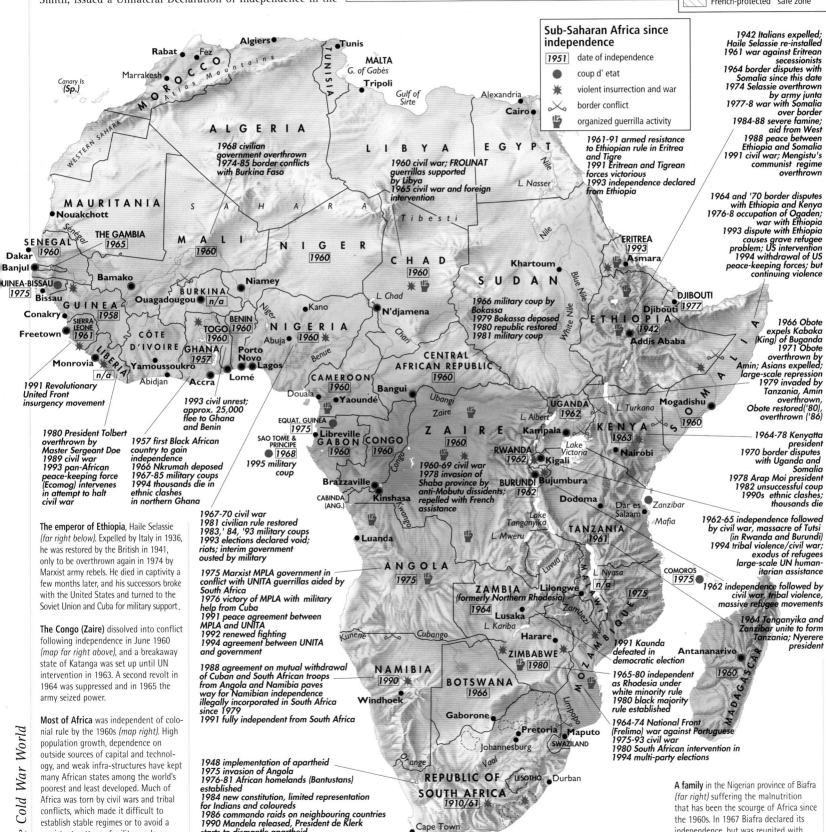

Sub-Saharan Africa since independence

1951	date of independence
●	coup d' etat
✳	violent insurrection and war
✕	border conflict
✊	organized guerrilla activity

ALGERIA
1968 civilian government overthrown
1974-85 border conflicts with Burkina Faso

LIBYA
1960 civil war; FROLINAT guerrillas supported by Libya
1965 civil war and foreign intervention

1942 Italians expelled; Haile Selassie re-installed
1961 war against Eritrean secessionists
1964 border disputes with Somalia since this date
1974 Selassie overthrown by army junta
1977-8 war with Somalia over border
1984-88 severe famine; aid from West
1988 peace between Ethiopia and Somalia
1991 civil war; Mengistu's communist regime overthrown

1961-91 armed resistance to Ethiopian rule in Eritrea and Tigre
1991 Eritrean and Tigrean forces victorious
1993 independence declared from Ethiopia

1964 and '70 border disputes with Ethiopia and Kenya
1976-8 occupation of Ogaden; war with Ethiopia
1993 dispute with Ethiopia causes grave refugee problem; US intervention
1994 withdrawal of US peace-keeping forces; but continuing violence

1966 military coup by Bokassa
1979 Bokassa deposed
1980 republic restored
1981 military coup

1966 Obote expels Kabaka (King) of Buganda
1971 Obote overthrown by Amin; Asians expelled; large-scale repression
1979 invaded by Tanzania, Amin overthrown, Obote restored ('80), overthrown ('86)

1964-78 Kenyatta president
1970 border disputes with Uganda and Somalia
1978 Arap Moi president
1982 unsuccessful coup
1990s ethnic clashes; thousands die

1962-65 independence followed by civil war, massacre of Tutsi (in Rwanda and Burundi)
1994 tribal violence/civil war; exodus of refugees large-scale UN human-itarian assistance

1962 independence followed by civil war, tribal violence, massive refugee movements

1964 Tanganyika and Zanzibar unite to form Tanzania; Nyerere president

1991 Revolutionary United Front insurgency movement

1980 President Tolbert overthrown by Master Sergeant Doe
1989 civil war
1993 pan-African peace-keeping force (Ecomog) intervenes in attempt to halt civil war

1957 first Black African country to gain independence
1966 Nkrumah deposed
1967-85 military coups
1994 thousands die in ethnic clashes in northern Ghana

1993 civil unrest; approx. 25,000 flee to Ghana and Benin

1995 military coup

1967-70 civil war
1981 civilian rule restored
1983,' 84,' 93 military coups
1993 elections declared void; riots; interim government ousted by military

The emperor of Ethiopia, Haile Selassie (far right below). Expelled by Italy in 1936, he was restored by the British in 1941, only to be overthrown again in 1974 by Marxist army rebels. He died in captivity a few months later, and his successors broke with the United States and turned to the Soviet Union and Cuba for military support.

The Congo (Zaire) dissolved into conflict following independence in June 1960 (map far right). A breakaway state of Katanga was set up until UN intervention in 1963. A second revolt in 1964 was suppressed and in 1965 the army seized power.

1960-69 civil war
1978 invasion of Shaba province by anti-Mobutu dissidents; repelled with French assistance

1975 Marxist MPLA government in conflict with UNITA guerrillas aided by South Africa
1976 victory of MPLA with military help from Cuba
1991 peace agreement between MPLA and UNITA
1992 renewed fighting
1994 agreement between UNITA and government

1988 agreement on mutual withdrawal of Cuban and South African troops from Angola and Namibia paves way for Namibian independence illegally incorporated in South Africa since 1979
1991 fully independent from South Africa

1991 Kaunda defeated in democratic election

1965-80 independent as Rhodesia under white minority rule
1980 black majority rule established

1964-74 National Front (Frelimo) war against Portuguese
1975-93 civil war
1980 South African intervention in
1994 multi-party elections

Most of Africa was independent of colonial rule by the 1960s (map right). High population growth, dependence on outside sources of capital and technology, and weak infra-structures have kept many African states among the world's poorest and least developed. Much of Africa was torn by civil wars and tribal conflicts, which made it difficult to establish stable regimes or to avoid a persistent pattern of military rule or one-man dictatorships.

1948 implementation of apartheid
1975 invasion of Angola
1976-81 African homelands (Bantustans) established
1984 new constitution, limited representation for Indians and coloureds
1986 commando raids on neighbouring countries
1990 Mandela released, President de Klerk starts to dismantle apartheid
1993 multi-racial council, free elections
1994 election victory for ANC; Mandela president

A family in the Nigerian province of Biafra (far right) suffering the malnutrition that has been the scourge of Africa since the 1960s. In 1967 Biafra declared its independence, but was reunited with Nigeria after a costly three-year civil war.

Country labels on map with independence dates:
MOROCCO; ALGERIA; TUNISIA; MALTA; LIBYA; EGYPT; MAURITANIA; MALI 1960; NIGER 1960; CHAD 1960; SUDAN; ERITREA 1993; DJIBOUTI 1977; ETHIOPIA 1942; SENEGAL 1960; THE GAMBIA 1965; GUINEA-BISSAU 1975; GUINEA 1958; SIERRA LEONE 1961; LIBERIA n/a; CÔTE D'IVOIRE 1960; GHANA 1957; BURKINA n/a; TOGO 1960; BENIN 1960; NIGERIA 1960; CAMEROON 1960; CENTRAL AFRICAN REPUBLIC 1960; EQUAT. GUINEA 1975; SAO TOME & PRINCIPE 1975; GABON 1960; CONGO 1960; ZAIRE 1960; UGANDA 1962; KENYA 1963; SOMALIA 1960; RWANDA 1962; BURUNDI 1962; TANZANIA 1961; ANGOLA 1975; ZAMBIA (formerly Northern Rhodesia) 1964; MALAWI n/a; MOZAMBIQUE 1975; COMOROS 1975; ZIMBABWE 1980; NAMIBIA 1990; BOTSWANA 1966; SWAZILAND; LESOTHO; MADAGASCAR 1960; REPUBLIC OF SOUTH AFRICA 1910/61

Cities: Rabat, Fez, Algiers, Tunis, Tripoli, Alexandria, Cairo, Marrakesh, Nouakchott, Dakar, Banjul, Bissau, Conakry, Freetown, Monrovia, Bamako, Ouagadougou, Niamey, Kano, N'djamena, Khartoum, Asmara, Djibouti, Addis Ababa, Mogadishu, Abuja, Porto Novo, Lagos, Lomé, Accra, Abidjan, Yamoussoukro, Douala, Yaoundé, Bangui, Libreville, Brazzaville, Kinshasa, Kampala, Nairobi, Kigali, Bujumbura, Dodoma, Dar es Salaam, Zanzibar, Luanda, Lilongwe, Lusaka, Harare, Antananarivo, Windhoek, Gaborone, Pretoria, Maputo, Johannesburg, Durban, Cape Town

Geographic features: Canary Is (Sp.), Atlas Mountains, Western Sahara, SAHARA, Tibesti, L. Nasser, L. Chad, Nile, Blue Nile, White Nile, Senegal, Niger, Benue, Chari, Ubangi, Zaire, Congo, L. Albert, L. Turkana, Lake Victoria, Kwango, Kwilu, Lake Tanganyika, L. Mweru, Luvua, Lake Nyasa, Kunene, Cubango, Zambezi, L. Kariba, Limpopo, Vaal, Orange, Mafia, Comoros, Cape of Good Hope, Gulf of Sirte, G. of Gabès

The Rwandan civil war, 1994 inset (top right): UGANDA, Ruhengeri, Gisenyi, Goma, Lake Kivu, Kigali, Cyangugu, Butare, BURUNDI, TANZANIA

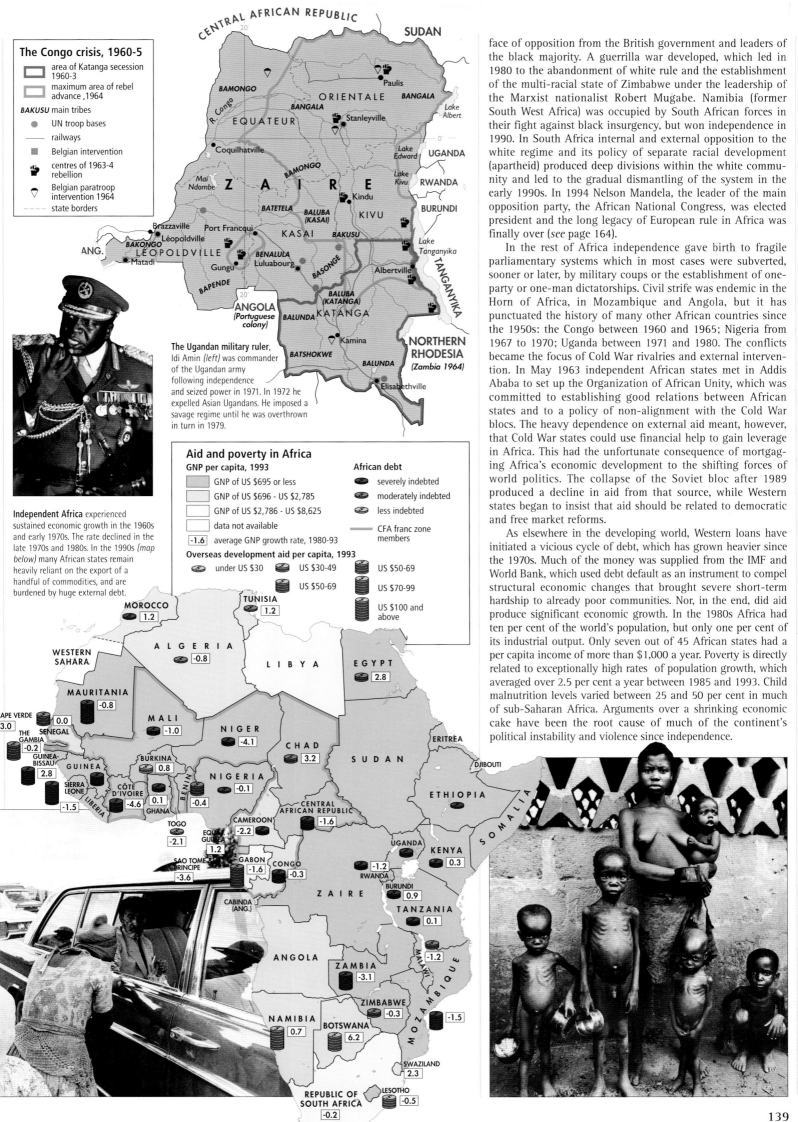

The Congo crisis, 1960-5

- area of Katanga secession 1960-3
- maximum area of rebel advance, 1964
- *BAKUSU* main tribes
- ● UN troop bases
- — railways
- ■ Belgian intervention
- ✊ centres of 1963-4 rebellion
- ▽ Belgian paratroop intervention 1964
- ---- state borders

The Ugandan military ruler, Idi Amin *(left)* was commander of the Ugandan army following independence and seized power in 1971. In 1972 he expelled Asian Ugandans. He imposed a savage regime until he was overthrown in turn in 1979.

Independent Africa experienced sustained economic growth in the 1960s and early 1970s. The rate declined in the late 1970s and 1980s. In the 1990s *(map below)* many African states remain heavily reliant on the export of a handful of commodities, and are burdened by huge external debt.

Aid and poverty in Africa

GNP per capita, 1993
- GNP of US $695 or less
- GNP of US $696 - US $2,785
- GNP of US $2,786 - US $8,625
- data not available
- -1.6 average GNP growth rate, 1980-93

Overseas development aid per capita, 1993
- under US $30
- US $30-49
- US $50-69

African debt
- severely indebted
- moderately indebted
- less indebted
- CFA franc zone members
- US $50-69
- US $70-99
- US $100 and above

Map labels (GNP growth rates):
MOROCCO 1.2 · TUNISIA 1.2 · ALGERIA -0.8 · EGYPT 2.8 · MAURITANIA -0.8 · CAPE VERDE 0.0 · SENEGAL 3.0 · THE GAMBIA -0.2 · GUINEA-BISSAU 2.8 · MALI -1.0 · NIGER -4.1 · CHAD 3.2 · BURKINA 0.8 · GUINEA · SIERRA LEONE -1.5 · CÔTE D'IVOIRE -4.6 · NIGERIA -0.1 · GHANA 0.1 · BENIN -0.4 · CAMEROON -2.2 · CENTRAL AFRICAN REPUBLIC -1.6 · UGANDA · KENYA 0.3 · TOGO -2.1 · EQUAT. GUINEA 1.2 · SAO TOME PRINCIPE -3.6 · GABON -1.6 · CONGO -0.3 · RWANDA -1.2 · BURUNDI 0.9 · TANZANIA 0.1 · ANGOLA · ZAMBIA -3.1 · MALAWI -1.2 · ZIMBABWE -0.3 · MOZAMBIQUE -1.5 · NAMIBIA 0.7 · BOTSWANA 6.2 · SWAZILAND 2.3 · LESOTHO -0.5 · REPUBLIC OF SOUTH AFRICA -0.2

---

face of opposition from the British government and leaders of the black majority. A guerrilla war developed, which led in 1980 to the abandonment of white rule and the establishment of the multi-racial state of Zimbabwe under the leadership of the Marxist nationalist Robert Mugabe. Namibia (former South West Africa) was occupied by South African forces in their fight against black insurgency, but won independence in 1990. In South Africa internal and external opposition to the white regime and its policy of separate racial development (apartheid) produced deep divisions within the white community and led to the gradual dismantling of the system in the early 1990s. In 1994 Nelson Mandela, the leader of the main opposition party, the African National Congress, was elected president and the long legacy of European rule in Africa was finally over (*see* page 164).

In the rest of Africa independence gave birth to fragile parliamentary systems which in most cases were subverted, sooner or later, by military coups or the establishment of one-party or one-man dictatorships. Civil strife was endemic in the Horn of Africa, in Mozambique and Angola, but it has punctuated the history of many other African countries since the 1950s: the Congo between 1960 and 1965; Nigeria from 1967 to 1970; Uganda between 1971 and 1980. The conflicts became the focus of Cold War rivalries and external intervention. In May 1963 independent African states met in Addis Ababa to set up the Organization of African Unity, which was committed to establishing good relations between African states and to a policy of non-alignment with the Cold War blocs. The heavy dependence on external aid meant, however, that Cold War states could use financial help to gain leverage in Africa. This had the unfortunate consequence of mortgaging Africa's economic development to the shifting forces of world politics. The collapse of the Soviet bloc after 1989 produced a decline in aid from that source, while Western states began to insist that aid should be related to democratic and free market reforms.

As elsewhere in the developing world, Western loans have initiated a vicious cycle of debt, which has grown heavier since the 1970s. Much of the money was supplied from the IMF and World Bank, which used debt default as an instrument to compel structural economic changes that brought severe short-term hardship to already poor communities. Nor, in the end, did aid produce significant economic growth. In the 1980s Africa had ten per cent of the world's population, but only one per cent of its industrial output. Only seven out of 45 African states had a per capita income of more than $1,000 a year. Poverty is directly related to exceptionally high rates of population growth, which averaged over 2.5 per cent a year between 1985 and 1993. Child malnutrition levels varied between 25 and 50 per cent in much of sub-Saharan Africa. Arguments over a shrinking economic cake have been the root cause of much of the continent's political instability and violence since independence.

SOUTH ASIA TO 1985

South Asia, like Africa, was faced with the same problems of constructing new modern nation-states when the imperial presence ended in the 1940s. When India and a separate Muslim Pakistan were established as independent states in August 1947, the national frontiers had not even been clearly defined. The subsequent settlement led to the transfer of as many as 15 million people from one state to the other, and religious conflict between Muslims, Hindus and Sikhs that may have cost the lives of as many as 500,000 people. The resettlement of the refugees was a major burden on the infant states, and tensions between resident and immigrant communities still persist. In India the new state also had to define the role of the 600 small princely states that survived on the sub-continent. Most joined either India or Pakistan, but the Muslim Nizam of Hyderabad had to be forcibly absorbed into India in 1948. In Kashmir, which was predominantly Muslim but was ruled by Hindus, the two new states fought a war from October 1947 to December 1948, when the UN imposed a ceasefire and divided the area between India and Pakistan. The solution pleased neither side, and a second war in 1965 was also fought on the issue of the frontier, following efforts by Pakistan to infiltrate troops into the Indian-held parts of Kashmir. A peace settlement restoring the status quo was reached at Tashkent, but the Kashmiri region remained an area of friction.

Even within the new states ethnic, religious and linguistic disputes persisted. In India around ten per cent of the post-partition population was Muslim. India also contained 23 major linguistic groups, with Hindi-speakers making up 40 per cent of the whole. The Nehru government's attempt to make Hindi the national language was resisted in the Dravidian south. In 1956 the states of India were reorganized along linguistic lines. In the north east the tribal peoples, led by the Nagas and Mizos, pressed for separate statehood. In the Punjab the rise of militant Sikhism led to demands in the 1980s for an independent Sikh state of Khalistan.

Tensions between the western and eastern halves of Pakistan, brought about by Bengali resentment at their economic and political subordination and their dislike of the military dictatorship imposed in 1958 by General Ayub Khan, led to their partition into two separate states in 1971. In Ceylon, the growing domination of the Buddhist majority provoked hostility from the Hindu Tamils of the north of the island. Though the island was a democracy, 800,000 Tamil labourers were nonetheless disenfranchised as descendants of Indian migrants. During the 1980s Tamil militants sought a separate Tamil state, Eelam, and fought a terrorist war against the Sinhalese authorities which prompted Indian intervention from 1987 to 1990. In Burma democracy gave way to a socialist military revolution in 1962, and military rule became the norm.

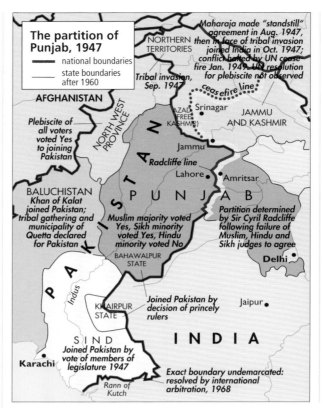

The partition of Punjab, 1947
— national boundaries
— state boundaries after 1960

AFGHANISTAN

NORTHERN TERRITORIES

Maharaja made "standstill" agreement in Aug. 1947, then in face of tribal invasion joined India in Oct. 1947; conflict halted by UN cease fire Jan. 1949. UN resolution for plebiscite not observed

Tribal invasion, Sep. 1947

ceasefire line

AZAD (FREE) KASHMIR

Srinagar

JAMMU AND KASHMIR

Plebiscite of all voters voted Yes to joining Pakistan

NORTH WEST PROVINCE

Jammu

Radcliffe line

Lahore

Amritsar

BALUCHISTAN
Khan of Kalat joined Pakistan; tribal gathering and municipality of Quetta declared for Pakistan

P U N J A B

Muslim majority voted Yes, Sikh minority voted Yes, Hindu minority voted No

Partition determined by Sir Cyril Radcliffe following failure of Muslim, Hindu and Sikh judges to agree

BAHAWALPUR STATE

Delhi

P A K I S T A N

Indus

KHAIRPUR STATE

Joined Pakistan by decision of princely rulers

Jaipur

S I N D
Joined Pakistan by vote of members of legislature 1947

I N D I A

Karachi

Exact boundary undemarcated: resolved by international arbitration, 1968

Rann of Kutch

Two days after Indian and Pakistani independence on 15 August 1947, a Boundary Commission reported on the partition of the Punjab *(map left)*, recommending a division that forced six million Muslims to cross into Pakistan and 4.5 million Sikhs and Hindus to seek refuge in India.

General Zia-ul-Haq seized power in July 1977 *(below)*. A devout Sunni Muslim, he reinstated basic Islamic law, including the *zakat* tax, which Shia Muslims refused to pay.

The Sikh community numbered around ten million in the 1980s, concentrated in the main in Indian Punjab, around the holy city of Amritsar. In 1984 the demand for a Sikh homeland led to violent clashes. Sikhs gather here *(below)* around the dead body of their leader.

The modern state of Bangladesh, formed in 1972 from the eastern half of Pakistan, was the product of the division of Bengal in 1947 *(map below left)*. The largely Muslim province was the home of the Muslim League, set up in 1906 to fight for a separate Muslim identity. Division led to the transfer of 1.6 million Hindus to India, and approximately 1.2 million Muslims from Calcutta and eastern India into East Bengal.

The economic development of the region has been hampered by high population growth, low incomes and an initial dependence on external sources of capital, technology and financial aid. Under Nehru India embarked on a series of Five-Year Plans, which helped to create a small industrial core – India produced ten million tons of steel in 1980 and 100 million tons of coal – but most Indians remained in peasant agriculture or traditional crafts and trades.

Despite economic underdevelopment and sharp social divisions, south Asian states did not develop broad communist movements, nor did they become locked into the Cold War struggles. India under Nehru was closer to the USSR than to the United States, a by-product of Indian-Chinese tension, while Pakistan was more closely aligned with the West. Nehru was, however, also a pioneer of the Non-Aligned Movement of countries, which chose not to take sides in the East-West split, and his relatively poor relations with the United States in the 1960s and 1970s stemmed in part from his desire to distance India from crude Westernization, a view echoed by Burma and Pakistan in the 1980s under the Islamic military dictator General Zia-ul-Haq. The issues of South Asian politics were never internationalized to the same extent as those of the Middle East or East Asia.

East Pakistan, 1947-71
— national boundaries
— state boundaries after 1960

NEPAL

BHUTAN

COOCH BEHAR

BIHAR

Rangpur

Brahmaputra

(Maharaja joined India. Separate state of Indian Union)

ASSAM
To India subject to Sylhet plebiscite

Radcliffe line August 1947

Mymensingh

SYLHET
Voted to join Pakistan by plebiscite

E A S T
P A K I S T A N

Rajashahi

East Bengal to Pakistan, Aug. 1947. Renamed East Pakistan, 1953. Independent as Bangladesh, 1971.

TRIPURA
(Maharaja joined India)

INDIA

WEST BENGAL
To India, Aug. 1947

Dacca

Comilla

B E N G A L

EAST BENGAL

Ganges

Calcutta

Barisal

Radcliffe line August 1947

ORISSA

Chittagong

Mouths of the Ganges

BURMA

Bay of Bengal

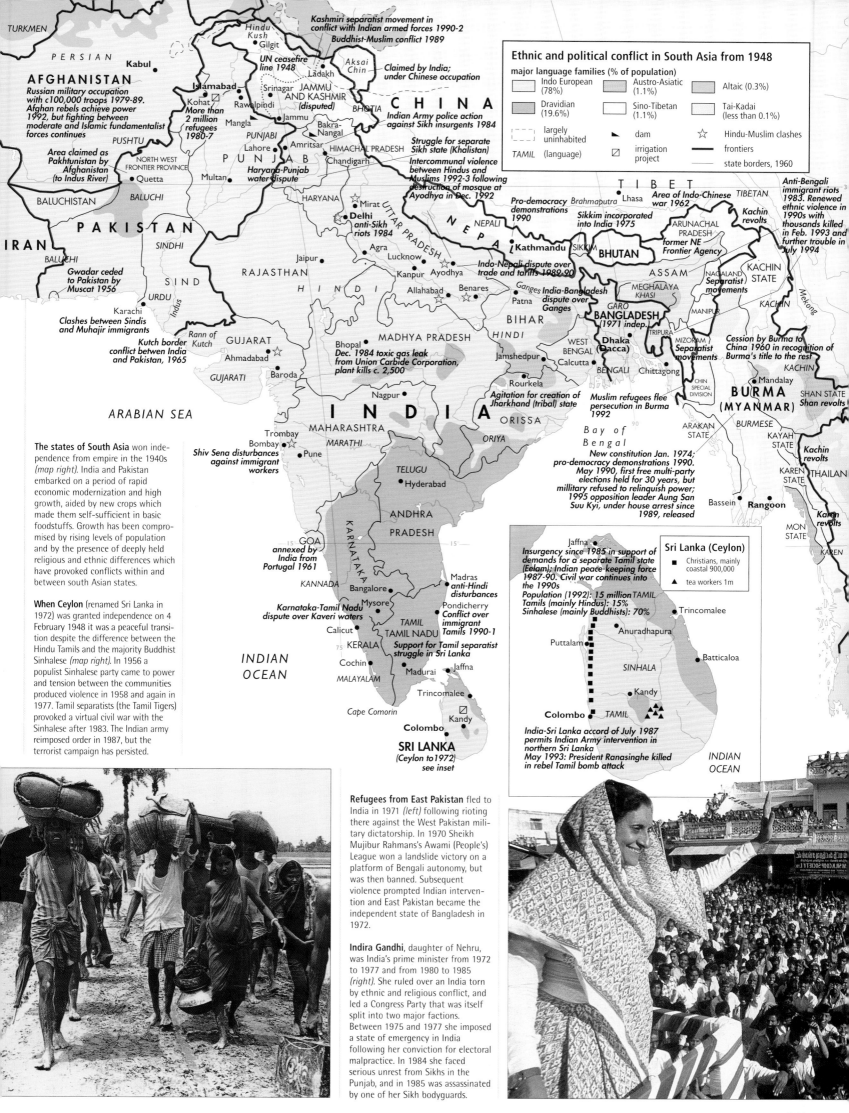

TURKMEN

AFGHANISTAN
Russian military occupation with c100,000 troops 1979-89. Afghan rebels achieve power 1992, but fighting between moderate and Islamic fundamentalist forces continues

Hindu Kush
• Gilgit
• Kabul

Area claimed as Pakhtunistan by Afghanistan (to Indus River)

PUSHTU

NORTH WEST FRONTIER PROVINCE

IRAN

BALUCHISTAN
BALUCHI

Gwadar ceded to Pakistan by Muscat 1956

• Quetta

PAKISTAN

SINDHI

Clashes between Sindis and Muhajir immigrants

• Karachi

SIND

Islamabad
Kohat □
More than 2 million refugees 1980-7
• Rawalpindi
Mangla
• Jammu
Bakra-Nangal
• Lahore
• Amritsar
Chandigarh
Multan •

PUNJABI
PUNJAB
Haryana-Punjab water dispute

Kashmiri separatist movement in conflict with Indian armed forces 1990-2
Buddhist-Muslim conflict 1989
UN ceasefire line 1948
Ladakh
Srinagar
JAMMU AND KASHMIR (disputed)

Aksai Chin
Claimed by India; under Chinese occupation

CHINA

BHOTIA

Indian Army police action against Sikh insurgents 1984

Struggle for separate Sikh state (Khalistan)

HIMACHAL PRADESH

URDU

Indus

Kutch border conflict betwen India and Pakistan, 1965

Rann of Kutch

GUJARAT
• Ahmadabad
• Baroda
GUJARATI

RAJASTHAN
• Jaipur

HARYANA
☆ • Mirat
• Delhi
anti-Sikh riots 1984

Intercommunal violence between Hindus and Muslims 1992-3 following destruction of mosque at Ayodhya in Dec. 1992

HINDI

• Agra
• Lucknow
UTTAR PRADESH
• Kanpur
• Ayodhya
Allahabad •
• Benares ☆
☆

Indo-Nepali dispute over trade and tariffs 1989-90

NEPAL
Pro-democracy demonstrations 1990
Lhasa •
Brahmaputra

TIBET
TIBETAN
Area of Indo-Chinese war 1962
Sikkim incorporated into India 1975
• Kathmandu
SIKKIM
BHUTAN

Anti-Bengali immigrant riots 1983. Renewed ethnic violence in 1990s with thousands killed in Feb. 1993 and further trouble in July 1994

Kachin revolts

ARUNACHAL PRADESH
former NE Frontier Agency

KACHIN STATE
KACHIN

Ganges
India-Bangladesh dispute over Ganges

ASSAM
MEGHALAYA
GARO KHASI

NAGALAND *Separatist movements*

MANIPUR

Mekong

BHOPAL • Bhopal
Dec. 1984 toxic gas leak from Union Carbide Corporation, plant kills c. 2,500

MADHYA PRADESH

BIHAR
• Patna
HINDI

Agitation for creation of Jharkhand (tribal) state

• Jamshedpur
• Rourkela

WEST BENGAL
• Calcutta

BANGLADESH (1971 indep.)
Dhaka (Dacca)
BENGALI
• Chittagong

TRIPURA
MIZORAM *Separatist movements*

CHIN SPECIAL DIVISION

Muslim refugees flee persecution in Burma 1992

Cession by Burma to China 1960 in recognition of Burma's title to the rest

KACHIN

• Mandalay

BURMA (MYANMAR)
BURMESE

SHAN STATE
Shan revolts

MAHARASHTRA
• Nagpur
MARATHI

Bombay • ☆
Trombay
Shiv Sena disturbances against immigrant workers
• Pune

ARABIAN SEA

INDIA

ORISSA
ORIYA

TELUGU
ANDHRA PRADESH
• Hyderabad

Bay of Bengal
90

ARAKAN STATE

New constitution Jan. 1974; pro-democracy demonstrations 1990. May 1990, first free multi-party elections held for 30 years, but millitary refused to relinquish power; 1995 opposition leader Aung San Suu Kyi, under house arrest since 1989, released

KAYAH STATE
Kachin revolts
KAREN STATE
THAILAN

• Bassein • Rangoon

MON STATE

Karen revolts
KAREN

GOA
annexed by India from Portugal 1961

KARNATAKA
KANNADA
• Bangalore
Karnataka-Tamil Nadu dispute over Kaveri waters
• Mysore

Madras •
anti-Hindi disturbances
Pondicherry
Conflict over immigrant Tamils 1990-1

INDIAN OCEAN

• Calicut
KERALA
• Cochin
MALAYALAM

TAMIL
TAMIL NADU
Support for Tamil separatist struggle in Sri Lanka
• Madurai
• Jaffna
• Trincomalee

Cape Comorin

• Kandy
Colombo
SRI LANKA
(Ceylon to1972)
see inset

Ethnic and political conflict in South Asia from 1948

major language families (% of population)

Indo European (78%)	Austro-Asiatic (1.1%)	Altaic (0.3%)
Dravidian (19.6%)	Sino-Tibetan (1.1%)	Tai-Kadai (less than 0.1%)

- - - - largely uninhabited
TAMIL (language)
◣ dam
□ irrigation project
☆ Hindu-Muslim clashes
━━ frontiers
──── state borders, 1960

Sri Lanka (Ceylon)

Insurgency since 1985 in support of demands for a separate Tamil state (Eelam); Indian peace-keeping force 1987-90. Civil war continues into the 1990s
Population (1992): 15 million TAMIL
Tamils (mainly Hindus): 15%
Sinhalese (mainly Buddhists): 70%

■ Christians, mainly coastal 900,000
▲ tea workers 1m

• Jaffna
Puttalam •
• Anuradhapura
• Trincomalee
SINHALA
• Batticaloa
• Kandy
Colombo *TAMIL*
▲▲▲

India-Sri Lanka accord of July 1987 permits Indian Army intervention in northern Sri Lanka
May 1993: President Ranasinghe killed in rebel Tamil bomb attack

INDIAN OCEAN

The states of South Asia won independence from empire in the 1940s *(map right)*. India and Pakistan embarked on a period of rapid economic modernization and high growth, aided by new crops which made them self-sufficient in basic foodstuffs. Growth has been compromised by rising levels of population and by the presence of deeply held religious and ethnic differences which have provoked conflicts within and between south Asian states.

When Ceylon (renamed Sri Lanka in 1972) was granted independence on 4 February 1948 it was a peaceful transition despite the difference between the Hindu Tamils and the majority Buddhist Sinhalese *(map right)*. In 1956 a populist Sinhalese party came to power and tension between the communities produced violence in 1958 and again in 1977. Tamil separatists (the Tamil Tigers) provoked a virtual civil war with the Sinhalese after 1983. The Indian army reimposed order in 1987, but the terrorist campaign has persisted.

Refugees from East Pakistan fled to India in 1971 *(left)* following rioting there against the West Pakistan military dictatorship. In 1970 Sheikh Mujibur Rahmans's Awami (People's) League won a landslide victory on a platform of Bengali autonomy, but was then banned. Subsequent violence prompted Indian intervention and East Pakistan became the independent state of Bangladesh in 1972.

Indira Gandhi, daughter of Nehru, was India's prime minister from 1972 to 1977 and from 1980 to 1985 *(right)*. She ruled over an India torn by ethnic and religious conflict, and led a Congress Party that was itself split into two major factions. Between 1975 and 1977 she imposed a state of emergency in India following her conviction for electoral malpractice. In 1984 she faced serious unrest from Sikhs in the Punjab, and in 1985 was assassinated by one of her Sikh bodyguards.

LATIN AMERICA FROM 1939

Latin America after 1945, unaffected by decolonization, pursued its own course. The major issue for all Latin American countries was the drive to industrialize. But where other developing areas became integrated with the long boom in the industrialized world (*see page 126*), Latin American states continued the policies of economic nationalism – tariffs, import substitution, state economic management – inherited from the 1930s.

The desire to industrialize presented both economic and political difficulties. Latin American economies had long depended on the export of food and raw material. By shifting to domestic industrialization the regimes tried to limit that traditional export dependence, alienating the rich agrarian elites who controlled the commodity trade. In Colombia a virtual civil war – La Violencia – which left over 200,000 dead was fought from the 1940s to the 1960s between industrial modernizers and agrarian conservatives. The build-up of industry also needed a strong domestic market, but over two thirds of the Latin American population were poor rural workers. Efforts were made from the 1940s to redistribute land or to reform agricultural practices, but this generated conflict with the rural elites and in Chile and Peru led to a fall in agricultural output. Millions of peasants moved into the cities, where poor living conditions and soaring unemployment created yet another set of problems.

In the 1960s the high rates of industrial expansion suddenly fell away. State preference for industry led to a decline in export earnings from the traditional commodity trade. Between 1948 and 1960 Latin America's share of world trade fell from 10.3 to 4.8 per cent. In 1961 the Kennedy administration launched an Alliance for Progress with Latin America to encourage economic restructuring and social reform, but with escalating inflation (over 46 per cent a year in Brazil between 1960 and 1970), high unemployment and growing class conflict, little was achieved. Rapid urban growth and village decay created the conditions for the spread of communism, which had been a tiny force until the 1960s. Communist reformers, inspired by Castro's revolution in Cuba in 1959, offered an alternative model of state-directed growth and social transformation. Under the twin pressures of economic crisis and revolutionary threat many Latin American states faced political collapse. The limited constitutionalism practised since 1945 gave way to military rule in Brazil (1964-85), Argentina (1966-84), Peru (1968-80), Chile (1973-89) and Uruguay (1973-85). Where the military did not rule directly, they conducted campaigns against communist guerrillas – including the Argentine "Che" Guevara, killed in Bolivia in 1967 – or helped prop up authoritarian party systems. By 1975 only Colombia, Venezuela and Costa Rica had elected governments.

Harsh rule did little to alleviate the problems. During the 1970s Latin American regimes survived on accumulating massive debts with the developed world. Like the states of Eastern Europe, Latin America built inefficient state-dominated

Latin America: economic development, 1940-91

direct US investment in Latin America (US$ m)

1929 1943 1960 1979 1991

Chief exports of Latin America, 1955-90

coffee chief exports 1955
coffee chief exports 1990
▲ represents over 50 per cent of total exports
● represents over 25 per cent of total exports
50% % of population engaged in agriculture (1963)

Venezuela: oil production 1940-91 (million barrels)

186	325	500	700
1940	1945	1950	1955
1,041	1,040	791	780	865
1960	1979	1980	1990	1991

From 1945 Latin America underwent its own industrial revolution. Population shifted to the cities, where it was housed in vast sprawling impoverished suburbs. By 1988 more than two thirds of Latin America's population lived in urban areas. Industrial growth *(map above)* made most strides in Mexico, Brazil and Argentina, but was spread throughout the region, backed by large US investments or by aid from the more developed world. By 1984 Latin America's total debt was $360 billion.

The Argentine seizure of the Falkland Islands in April 1982 was a classic ploy on the part of a beleaguered military regime to whip up popular support. Early enthusiasm *(picture above right)* evaporated rapidly in the face of Britain's refusal to be cowed, however, and, with economic and diplomatic crises looming, Galtieri's government fell soon afterwards, paving the way for a return to democracy.

PERU: THE SHINING PATH

The most radical expression of nationalist Marxism in Latin America came from the Shining Path (Sendero Luminoso) guerrilla movement in Peru. Founded in 1980 by a philosophy professor at the University of Ayacucho, Abimael Guzmán, Shining Path adopted a mixture of Maoist political and military tactics and the indigenismo of the 1920s, which sought to root political struggle in the culture of the Indian peoples of the region. Shining Path flourished in the central Andean highlands, where its brand of savage violence against officials, soldiers and ordinary peasants brought it control of large areas. The movement blew up trains and power stations, cut off urban electricity supply, and signalled its presence with flaming crosses set on the hillside. In 1992 Guzmán – known by his guerrilla name of Chairman Gonzalo – was captured and placed in a cage in Lima, from which he appealed to his followers for a truce with the government of Alberto Fujimori. In October he was

sentenced to life imprisonment, and the movement fragmented. An estimated 500 Shining Path guerrillas, followers of a breakaway group led by Oscar Ramírez, survive in the

Andean highlands. The popularity of Maoism has been eaten away by 13 years of savagery from guerrillas and army alike.

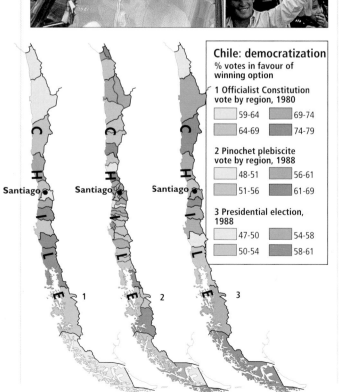

Chile: democratization

% votes in favour of winning option

1 Officialist Constitution vote by region, 1980

| | 59-64 | | 69-74 |
| | 64-69 | | 74-79 |

2 Pinochet plebiscite vote by region, 1988

| | 48-51 | | 56-61 |
| | 51-56 | | 61-69 |

3 Presidential election, 1988

| | 47-50 | | 54-58 |
| | 50-54 | | 58-61 |

Santiago ●   Santiago ●   Santiago ●

1   2   3

Communist aid to Latin America (US$m)

Country	total 1954-73	USSR	East Europe	China
Argentina	1176	44	595	133
Bolivia	61	30	31	-
Brazil	312	85	227	-
Chile	423	238	120	65
Colombia	7	2	5	-
Ecuador	15	-	15	-
Guyana	26	-	-	26
Peru	223	28	153	42
Uruguay	45	20	25	-
Venezuela	10		10	

Map labels:

UNITED STATES OF AMERICA
US intervention
Dominican Republic Grenada 1983
Cuba 1921-23, 1933, 1961
Haiti 1915-34, 1944, 1965-6
Panama 1903-18, 1989
Nicaragua 1912-33
Guatemala 1954
Mexico 1914

MEXICO 5,000
Mexico City
Puebla   Veracruz
BELIZE
GUAT.
HONDURAS 300
Guatemala City
Tegucigalpa
EL SALVADOR   NICARAGUA 100 M
San José   Managua 125 M
COSTA RICA 1,000 M
PANAMA 500 M
Panama City

Havana
CUBA 125,000 M
HAITI n/a M
DOMINICAN REP. 1,400
PUERTO RICO (USA)

Cuban-inspired guerrilla movements 1959-68

Caribbean Sea

GRENADA
Port of Spain
Cartagena
Caracas 8,000 M
VENEZUELA
Georgetown 100 M
Bogotá 11,000 M/P
COLOMBIA
GUYANA   Paramaribo
SUR.   Cayenne
FR. GUIANA
Macapá
Amazon
Belém
Manaus
Fortaleza

Quito
ECUADOR 1,200 M/P
Guayaquil
Piura
3,500 M/P
PERU
Trujillo
Huánuco
Callao
Lima   Cuzco
Arequipa
La Paz 3,200 M/P
BOLIVIA
Sucre
PARAGUAY
Asunción
Antofagasta
Copiapó   Tucumán
Córdoba   Santa Fé
Valparaíso
Santiago   Mendoza
ARGENTINA 70,000 M
Buenos Aires
Montevideo
URUGUAY 22,000 M
Concepción
Valdivia   Bahía Blanca
Osorno
Rawson
Comodoro Rivadavia
Santa Cruz
Stanley
Tierra del Fuego

BRAZIL 7,000 M/P
Brasília
Salvador (Bahía)
Recife
Belo Horizonte
Rio de Janeiro
São Paulo
Santos
Florianópolis
Porto Alegre
Rio Grande
Fray Bentos
Uruguay

120,000 M

Right-hand margin annotations:

Cuban revolution 1959

Rómulo Betancourt 1945-48, ⚪
1959-64; Carlos Andrés Pérez 1974-9

Rafael Caldera 1969-74; ◇
Luis Herrera Campins 1979;
Jaime Lusinchi 1984

Getulio Vargas 1930-45; 1950-4 ☆
João Goulart 1961-4

Modernizing militarism 1964 ▽

Civilian rule 1985; ⚪
democratization 1986

Military dictatorship ▽
Stroessner 1954
Rodriguez 1989

Batllismo ⚪
1903-33

Military 1973 ▽

Tupamaros ⬡

Civilian rule 1985; ⚪
democratization 1986

Montoneros ⬡

Juan Domingo Perón 1943-55; 1973-4 ☆

Military 1976-83 ▽

Democratization and civilian rule 1984 ⚪

Falkland Islands (Islas Malvinas)
occupied by Argentina 1982:
occupation ended by UK
Task Force June 1982

Left-hand margin annotations:

⬛ Mexican revolution 1910-40
◇ Zapatista revolt, 1994

⬛ Guatemalan revolution 1944-54

▽ ◇ Military Junta 1979

⬛ ⚪ Sandinista revolution 1979-90; democratization 1990

⚪ Figueres 1948

⚪ Liberal-Conservative Pact, 1957

▽ Intermittent militarism to 1978

⚪ Election of reformist government 1978

☆ Radical militarism 1968
⬛ Sendero Luminoso from 1980
⚪ Return to civil rule 1980 President Fujimori suspends constitution 1992

⬛ Bolivian revolution 1952-64
⬡ Che Guevara (killed 1967)
▽ Military 1980; democratization 1982

◇ Eduardo Frei 1964-70
⬛ Salvador Allende (Popular Unity) 1970-3
▽ Pinochet 1973-88
⚪ Democratization 1989

Lower text columns:

For much of the period from 1945 Latin American states were ruled by military dictatorships or forms of single-party rule. During the 1960s communism emerged as the major political opposition (map above), backed by the Soviet bloc or by China, and assisted by the one communist state in the region, Fidel Castro's Cuba. As popular movements confronted repressive elites throughout the continent, aid programmes (chart above right) were seen as a way of increasing communist influence and trade. From the 1970s, however, substantial strides have been made towards greater democracy, notably in Chile, Brazil and Argentina.

In 1970 the socialist Salvador Allende won the Chilean presidential election and embarked on three years of reform, until he was overthrown by General Augusto Pinochet in 1973. In the 1980s Chile slowly worked back to democracy. In 1980 Pinochet won backing for a new constitution (maps left), but a plebiscite in 1988 on extending Pinochet's rule was defeated by 58 per cent to 42 per cent. In December 1989 the Christian Democrat, Patricio Aylwin, won the presidential election with 55 per cent of the votes. The picture (left) shows General Pinochet in September 1985.

economies and controlled the consequences with large bureaucracies and police oppression. In the 1980s the decaying system collapsed. In 1982 Mexico defaulted on its international debt, which totalled over $85 billion. All the other debtor states declared insolvency. Wealthy Latin Americans, fearing financial collapse, sent their money abroad, making a bad situation worse. The creditor states insisted on rescheduling payments and on financial stringency. Governments cut spending programmes: living standards fell sharply. By 1987 Latin America had paid back $121 billion at the cost of impoverishing the continent.

The result was political crisis. Without economic growth to support them the dictatorships had nothing to offer. They toppled one by one as popular reformist movements, backed by new urban classes demanded, and won, free elections. The new generation of politicians rejected the strategy of state-backed industrialization and autarky practised since the 1930s, and looked to market reforms and liberalization to bring Latin America back into the wider world system. Between 1988 and 1990 new governments committed to democracy and economic reform appeared in Mexico, Argentina, Chile, Brazil, Uruguay and Peru, at almost exactly the same time as the statist regimes of Eastern Europe started down the same path.

Latin America from 1939

THE 1980s AND EARLY 1990s witnessed fundamental changes in world politics and in the balance of economic power. The Cold War confrontation disappeared with the collapse of the Soviet bloc from 1989 and the eclipse of Soviet communism. At least partly as a result, popular democracy made strides in many parts of the world: in the former Soviet republics themselves; in Latin America; and in southern Africa. As important, the economic balance of power began to shift. In China and around the Pacific Rim, new economic powers emerged to challenge the long-held monopoly of the developed industrial world beginning the reversal of one of the central features of the century: Western economic imperialism. At the same time, two contrary pressures in world affairs developed. On the one hand, there was a move towards greater globalization: in communications, in finance, in manufacturing, and, through the activities of the UN and other international organizations, in politics, too. On the other hand, political fragmentation and conflict accelerated. The revival of nationalism, the growth of religious fundamentalism and the spread of terrorism and corruption have all contributed to a more violent, less stable world.

TOWARDS THE NEW
WORLD ORDER

THE COLLAPSE OF COMMUNISM IN THE EASTERN BLOC

IN MARCH 1985 MIKHAIL GORBACHEV became leader of the Soviet Union. A young and popular member of the communist Politburo, he saw clearly that the Soviet bloc had reached a critical turning point. Over the following five years he tried to modernize socialism through a package of economic and political reforms. The result was the collapse of the USSR and the disappearance of the Soviet bloc in Europe.

The Soviet system in the 1980s faced critical choices. The escalating cost of modern defence systems made it difficult to keep up in the arms race without reducing domestic living standards, which had stagnated in much of the Soviet bloc. Gorbachev seized the initiative in 1985 in the face of hard-line opposition. He sought to establish serious disarmament talks so that the Soviet Union could run down its massive military commitment without risking its security. The resources this freed were intended to satisfy the population's demands for economic reform and improved living standards. Disarmament was not an immediate success. In October 1986 Gorbachev met President Reagan at Reykjavik, but final agreement on arms control was only achieved in Washington on 8 December 1987. The treaty removed one fifth of existing nuclear weapons, including most intermediate range nuclear weapons. Further cuts in the long-range nuclear arsenal were announced by both sides in 1988. The programme of economic reform could not be achieved without a measure of political reform. In 1988 the Soviet system became a limited democracy (see page 163).

Gorbachev's plans profoundly affected the rest of the Eastern bloc. Gorbachev regarded the other communist states as a drain on the Soviet economy. He encouraged them to think about economic and political reform in order to reduce their dependence on the Soviet Union. The change in Soviet attitudes came

GERMAN REUNIFICATION

The East German communist regime of Erich Honecker sought to avoid making reforms in 1989, but as the Hungarian regime liberalized, thousands of East Germans, posing as tourists, sought asylum in Hungary, Poland and Czechoslovakia and a possible escape route to the West. Early in October, in spectacular celebrations for the 40th anniversary of the East German state, the Soviet leader Mikhail Gorbachev told the hard-liners to make concessions. Demonstrators risked police violence to support the Soviet leader's line. On 9 October Honecker issued the police with a shoot-to-kill order. They refused and massive demonstrations in the major cities forced Honecker's resignation on 18 October. Popular hostility reached fever pitch. On 4 November half a million marched in East Berlin. Five days later the notorious Berlin Wall was opened amidst scenes of wild celebration (right) and the government collapsed. By the time popular elections were held in March 1990 there were widespread calls for union with West Germany. The two states, a direct product of the Cold War conflict 40 years before, were formally reunited on 3 October 1990.

Mikhail Gorbachev, seen here *(left)* receiving the Ronald Reagan Freedom Award from the former US president in 1992, played the central part in ending the Cold War tension with the USA and in precipitating political reform throughout the Soviet bloc. He resigned as president in December 1991, swept aside by the very forces of popular politics he had set in motion.

The Polish union leader, Lech Walesa *(bottom right)*, addressing a crowd. As leader of the free trade union Solidarity, he led the resistance to communism from 1980 and became Polish president in 1990. He lost office to a former communist minister in elections in 1995.

Between 1989 and 1991 the Soviet bloc was transformed from a monolithic communist empire into a patchwork of independent states, most of which became multi-party democracies *(map below)*. The process began in Poland and Hungary in January 1989, when talks began with non-communist opposition parties, but accelerated in September with the flight of thousands of East Germans to Hungary, Poland and Czechoslovakia. Between October and December communist regimes were replaced in East Germany, Czechoslovakia, Bulgaria and Romania. The Soviet Union broke up into its constituent parts during 1991 and was officially dissolved on 31 December 1991.

Mar. 1990: Congress of Estonia formed, declares Soviet rule illegal
Mar. 1991: referendum endorses independence
Aug. 1991: independence declared
Sep. 1991: independence recognized by USSR

1989: mass anti-communist demonstrations
Mar. 1991: referendum endorses independence
Aug. 1991: independence declared
Sep. 1991: independence recognized by USSR

1989: mass anti-communist demonstrations
Mar 1991: independence declared
Apr.-June 1990: economic embargo imposed by USSR
Sep. 1991: independence recognized by USSR

Mar. 1985: Mikhail Gorbachev becomes leader of Communist Party; initiates perestroika and glasnost, loosens Soviet control of satellite states
June 1991: Boris Yeltsin elected president of Russian Federation
Aug. 1991: hard-line communist coup against Gorbachev fails
Nov. 1991: Communist Party declared illegal
Dec. 1991: USSR dissolved

from 1985: Solidarity leads opposition to communism
June 1989: partially free elections
Sep. 1989: Solidarity-led government takes office
Jan. 1990: Communist Party dissolved
Oct. 1991: free elections

June 1989: Popular Front founded.
Aug. 1991: independence declared
Dec. 1991: founder member of CIS

Sep. 1989: mass exodus of political refugees reach the West via Hungary; communist leadership in crisis
Oct.-Nov. 1989: widespread demonstrations against leadership
9 Nov. 1989: Berlin Wall breached
Mar. 1990: free elections
July 1990: currency union with West Germany
Oct. 1990: reunified with West Germany

from 1988: anti-government demonstrations
Nov. 1989: mass demonstrations end communist rule
Apr. 1990: new constitution adopted; becomes a federation
June 1990: free elections

1989: opposition mass-movements emerge
Aug. 1991: independence declared
Dec. 1991: referendum endorses independence; founder member of CIS

Dec. 1989: economic war between Belgrade government and Slovenia
Apr. 1990: free elections
June 1991: independence declared; Yugoslav army attempts to regain control of Slovenia
July 1991: Brioni Agreement ends fighting in Slovenia; Yugoslav army withdraws

from 1987: communist regime relaxes control
Sep. 1989: allows East Germans to travel to the West
Oct. 1990: communist rule ends peacefully
Mar.-Apr. 1990: free elections

June 1989: Popular Front wins 75% of votes in election
Aug. 1991: independence declared

Apr.-May 1990: free elections
Dec. 1990: Serbian-inhabited areas declare independence
June 1991: independence declared; fighting in Slovenia spreads to Croatia as Serbs attempt to extend territory in Croatia and Bosnia

Dec. 1989: mass demonstrations lead to armed uprisings and overthrow of Ceaucescu regime
June 1991: free elections
Nov. 1991: new constitution adopted

Nov. 1991: independence declared

1987: mass strikes against wage freeze and falling living standards; growing Serb militancy against minorities
July 1990: provincial autonomies abolished
1990-1: increasing tension between Belgrade government and Slovenia and Croatia

Nov. 1989: President Zhivkov removed from office
June 1990: free elections
July 1991: fresh elections following adoption of new constitution

Nov. 1988: mass demonstrations against Russification
Mar. 1991: referendum endorses independence
Apr. 1991: independence declared

Sep. 1989: economic embargo imposed by Azerbaijan
Sep. 1991: referendum endorses independence; independence declared

Jan.-May 1990: democratic reforms initiated by leadership
Mar. 1991: free elections

UNITED KINGDOM
North Sea
SWEDEN
DENMARK
NETHERLANDS
BELGIUM
GERMANY
Berlin
Bonn
Elbe
Seine
FRANCE
SWITZ.
ITALY
Sardinia
Mediterranean Sea
Sicily
AUSTRIA
CZECHOSLOVAKIA
Prague
Bratislava
SLOVENIA
Zagreb
CROATIA
HUNGARY
Budapest
YUGOSLAVIA
Sarajevo
Belgrade
ALBANIA
Tirana
Skopje
Sofia
BULGARIA
Bucharest
ROMANIA
MOLDOVA
TRANSNISTRIA
GAGAUZIA
Dniester
UKRAINE
Kiev
Don
POLAND
Warsaw
Baltic Sea
RUSSIAN FED.
Vilnius
LITHUANIA
Riga
LATVIA
Tallinn
ESTONIA
FINLAND
Minsk
BELARUS
Moscow
RUSSIAN FEDERATION
Black Sea
TURKEY
GEORGIA

at a difficult time for the other Eastern bloc states, whose economic development had been adversely affected by recession in the West and by reductions in trade and aid resulting from renewed Cold War pressures. Economic modernization slowed in the 1980s, and provoked growing popular unrest, particularly in Poland. There a military dictatorship was set up in 1981 to suppress the democracy movement and to forestall possible Soviet military intervention. In Romania the isolated and impoverished regime of Nicolae Ceauçescu became yet more extravagantly repressive. In East Germany the *Stasi* (security police) clamped down on any signs of dissent.

There was little popular opposition in the 1980s, but there was limited enthusiasm for the regimes even among elements in the communist movements. In Czechoslovakia Václav Havel's Charter 77 kept alive the struggle for civil rights. In Poland the outlawed Solidarity Union maintained a network of Catholic and working class opposition. When in 1989 Gorbachev put pressure on his communist partners to grasp the nettle of reform, popular protest grew rapidly. Without Soviet backing and generally unwilling to provoke civil war, the communist regimes crumbled one by one: opposition parties were legalized in Hungary in January 1989; in August 1989 Poland established the first non-communist government since 1948. Demonstrations ended communist rule in Czechoslovakia in November 1989 and in East Germany in October. Ceauçescu fought to the end using his *Securitate* agents to stamp out resistance until he was shot by an army firing squad on Christmas Day 1989 and replaced by a National Salvation Front government, composed largely of former communists.

By 1990 multi-party elections had brought coalition governments to power throughout the Eastern bloc, committed to democratic reform and economic liberalization. Exposed to market pressures the area plunged into economic decline, regarded by the West as a developing, rather than a developed, region. High unemployment and rural poverty contributed to the revival of ethnic and religious conflicts with deep roots, conflicts that the communist regimes had only papered over for 40 years.

In March 1989 the two Cold War blocs began negotiations on reducing the number of conventional weapons in Europe. The Conventional Forces in Europe Treaty was signed on 9 November 1990, committing both sides to reductions in weapons throughout the region *(map above right)*. By this time the Soviet Union had already promised its former allies in Eastern Europe a phased withdrawal of Soviet forces. The last left in August 1994.

Communism rapidly crumbled as a political force in Eastern Europe after 1989 *(map right)*. The revival of nationalism led to the demise of East Germany which united with the western part in October 1990, and of Czechoslovakia, whose Slovak minority won full independence in January 1993. Ethnic conflict in Yugoslavia led to the fragmentation of the federation in 1991. The collapse of the Soviet bloc and harsh economic measures taken by new governments led to steep economic decline and a renaissance in the fortunes of the socialist parties which arose from the ruins of communism.

CFE arms reduction agreements, 1990

Limits under CFE (NATO, Warsaw Pact)

	Tanks	ACVs	Artillery
	7,500	11,250	5,000
	10,300	19,260	9,100
	11,800	21,400	11,000
	4,700	5,900	6,000
to be placed in storage	3,500	2,700	3,000
Totals	20,000	30,000	20,000

— division between Eastern and Western blocs

Central and Eastern Europe, 1992-6

- independent, 1992
- independent, 1993
- ruled by former communist parties under new names, May 1996
- other formerly communist countries or regions of the USSR
- ▲ applied for membership of the EU
- ethnically based territorial autonomy granted, with details
- ethnically based parties represented in parliament (with name of minority)

aspiration for ethnically based autonomy thwarted (with name of minority)
aspiration for historically/regionally based autonomy thwarted (with name of area)
— territorial autonomy abolished, with dates
○ Soviet troop withdrawals (with dates)
■ members of Visegrad group
✷ states in civil war during part of the period

3,420	per capita GDP, 1995 in US$
2.2%	1992 percentage change in GDP
5.9%	1995
26.5	inflation, 1995
— frontiers, 1996

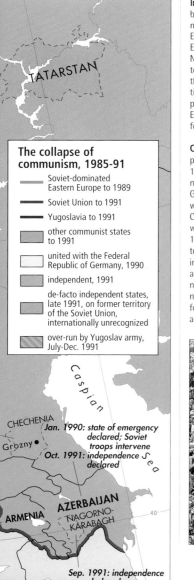

The collapse of communism, 1985-91

- Soviet-dominated Eastern Europe to 1989
- Soviet Union to 1991
- Yugoslavia to 1991
- other communist states to 1991
- united with the Federal Republic of Germany, 1990
- independent, 1991
- de-facto independent states, late 1991, on former territory of the Soviet Union, internationally unrecognized
- over-run by Yugoslav army, July-Dec. 1991

TATARSTAN

CHECHENIA — *Jan. 1990: state of emergency declared; Soviet troops intervene*
Grozny — *Oct. 1991: independence declared*

Caspian Sea

ARMENIA
AZERBAIJAN
NAGORNO-KARABAGH
Sep. 1991: independence declared

THE COLLAPSE OF THE SOVIET UNION

The tidal wave that swept away the communist regimes of Eastern Europe was the direct consequence of the forces of change inspired by Mikhail Gorbachev's leadership of the Soviet Union. Within two years the wave engulfed both Gorbachev and the system he had tried to reform.

Gorbachev did not set out to destroy the system but to make it work better. He pinned his faith on *glasnost* (openness) and *perestroika* (restructuring). The apparatus which ran the Soviet Union proved resistant to both, but more particularly to Gorbachev's plans to de-centralize the economy and to encourage more economic individualism. This resistance led Gorbachev in 1988 to rally reformist elements in the party. At the Party Congress in June the decision was taken to replace the Supreme Soviet with a Congress of People's Deputies, two thirds of whom would be popularly elected. The Congress then elected a 450-strong parliament in May 1989 and Gorbachev became president. The system was far from fully democratic, but it excited expectations of more fundamental change.

Arguments about the pace of reform and growing economic uncertainty led to a downturn in economic growth in 1990. There was growing nationalist unrest as the Russian-dominated republics of the USSR sensed the opportunity to emulate the Eastern European states' assertion of independence. In 1991 Gorbachev dithered between more radical reform and a return to old-fashioned authority. Russian troops were sent into the Baltic States and the Caucasus to hold the crumbling structure together.

The crisis could not be reversed. In April 1990 the military element of the Warsaw Pact had been scrapped, and on 1 July 1991 the Warsaw Pact as a whole was wound up. The COMECON trade bloc was ended in June 1991. Meanwhile, the president of the Russian Republic, Boris Yeltsin, elected in May 1990, urged Gorbachev to give the Soviet republics more independence. A reactionary backlash was not long coming. On 19 August 1991 a group of hard-line communists put Gorbachev under house arrest in the Crimea and attempted a coup in Moscow. Yeltsin suppressed the revolt, but the collapse of the coup signalled the end of the existing order. The non-Russian republics declared their independence. When the Ukraine staged a referendum in December 1991 that resulted in an overwhelming vote for independence, Gorbachev bowed to reality. The USSR was dissolved on 31 December, to be replaced by a Commonwealth of Independent States (the CIS), co-operating on military and economic issues but no longer controlled from Moscow.

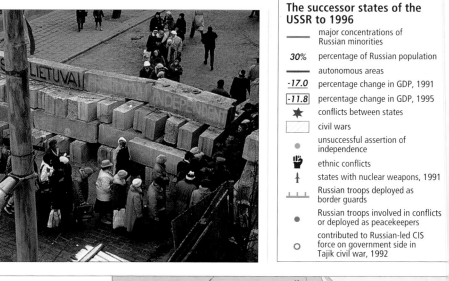

The successor states of the USSR to 1996

- major concentrations of Russian minorities
- *30%* percentage of Russian population
- autonomous areas
- *-17.0* percentage change in GDP, 1991
- *-11.8* percentage change in GDP, 1995
- ★ conflicts between states
- civil wars
- unsuccessful assertion of independence
- ✊ ethnic conflicts
- states with nuclear weapons, 1991
- Russian troops deployed as border guards
- Russian troops involved in conflicts or deployed as peacekeepers
- ○ contributed to Russian-led CIS force on government side in Tajik civil war, 1992

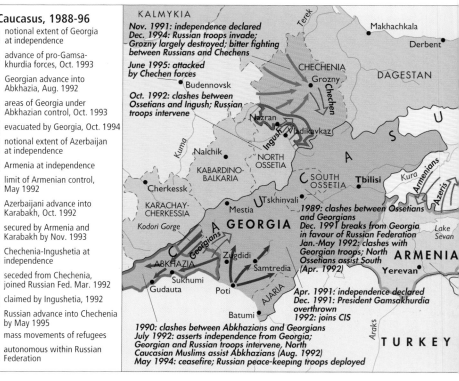

The Caucasus, 1988-96

- notional extent of Georgia at independence
- → advance of pro-Gamsakhurdia forces, Oct. 1993
- → Georgian advance into Abkhazia, Aug. 1992
- areas of Georgia under Abkhazian control, Oct. 1993
- evacuated by Georgia, Oct. 1994
- notional extent of Azerbaijan at independence
- Armenia at independence
- limit of Armenian control, May 1992
- → Azerbaijani advance into Karabakh, Oct. 1992
- secured by Armenia and Karabakh by Nov. 1993
- Chechenia-Ingushetia at independence
- seceded from Chechenia, joined Russian Fed. Mar. 1992
- claimed by Ingushetia, 1992
- → Russian advance into Chechenia by May 1995
- → mass movements of refugees
- autonomous within Russian Federation

Nov. 1991: independence declared Dec. 1994: Russian troops invade; Grozny largely destroyed; bitter fighting between Russians and Chechens

June 1995: attacked by Chechen forces

Oct. 1992: clashes between Ossetians and Ingush; Russian troops intervene

1989: clashes between Ossetians and Georgians Dec. 1991 breaks from Georgia in favour of Russian Federation Jan.-May 1992: clashes with Georgian troops; North Ossetians assist South (Apr. 1992)

Apr. 1991: independence declared Dec. 1991: President Gamsakhurdia overthrown 1992: joins CIS

1990: clashes between Abkhazians and Georgians July 1992: asserts independence from Georgia; Georgian and Russian troops intervene; North Caucasian Muslims assist Abkhazians (Aug. 1992) May 1994: ceasefire; Russian peace-keeping troops deployed

a KARACHAY-CHERKESSIA 42%
b KABARDINO-BALKARIA 39%
c NORTH-OSSETIA 30%
d CHECHEN-INGUSHETIA 23%

During 1991 the former USSR began to disintegrate under the impact of Gorbachev's economic and political reforms. The Soviet republics which had made up the Moscow-dominated USSR took greater responsibility for local affairs. The Baltic States and the Ukraine began to demand genuine independence, which was achieved by the end of 1991. In December 1991 the republics met at Alma Alta in Kazakhstan and set up the Commonwealth of Independent States. Within the Commonwealth the Russian authorities set up a Russian Federation composed of 21 republican and 69 other defined areas *(map left)*. Since then, Tatarstan won effective independence in 1994, and Chechenia fought a war against Russian forces from 1994 in pursuit of independence.

The Russian Federation, 1991-6

- the Russian Federation
- constituent republics within the Russian Federation, Mar. 1992 (in national languages)
- united with Russia by treaty, 15 Feb. 1994
- independence declared, Nov. 1991; at war with Russia from Dec. 1994
- entered into close political and economic union with Russia, 2 Apr. 1996 (Commonwealth of Sovereign States, CIS)
- entered into close economic union with Russia, 30 Mar. 1996
- other members of the CIS
- *60%* percentage of Russians in other members of the Russian Federation

referendum decides for independence, Mar. 1992

30% ESTONIA
34%
-19.3 3.8 • Moscow
9%
13%
LITHUANIA LATVIA
-35.0 2.5 BELARUS
POLAND -9.6 -10.0
-33.8 1.0

RUSSIAN FEDERATION
-18.7 -2.6

22% UKRAINE
-17.0 -11.8

MOLDOVA
-29.1 -3.0 TRANSNISTRIA
ROMANIA GAGAUZIA
CRIMEA
Tatars and Russians,
1991 and '92
tensions between Russia
and Ukraine over status
of Black Sea Fleet

BULGARIA Black Sea

-40.3 -15.0 ABKHAZIA
GEORGIA CHECHENIA

6% S. OSSETIA
AJARIA -22.6 -17.2
TURKEY ARMENIA AZERBAIJAN
-52.3 5.0 -1.5% 5.5% Armenians and Azerbaijanis,
NAKHICHEVAN KARABAGH 1988 and 1990

war between Armenia and
Azerbaijan over Karabagh

IRAQ IRAN

Akmola
• to be capital
from 2000

KAZAKHSTAN
-13.0 -8.9
38%

Kazakhs and
Lezgians, 1989

KARAKALPAKSTAN
UZBEKISTAN
8%

Uzbeks and
Meshketians, KYRGYZSTAN
1989-90 22%
Kyrgyz and
Kyrgyz and Uzbeks,1990
Tajiks, 1989
TAJIKISTAN
TURKMENISTAN GORNO-
BADAKHSHAN

IRAN 10%

independent 1991, founder member
of CIS (8 Dec. 1991)
independent 1991, joined CIS late
Dec. 1991
independent 1991, joined CIS 1993
independent 1991,
de facto independent 1991

Caspian Sea

Sumgait
Kuba • Baku
Jan. 1990: rioting in Baku:
Soviet troops intervene
Oct. 1991: independence declared
Sep. 1992: unsuccessful attempts
to subjugate Nagorno-Karabakh
and Armenia (to May 1994)

AZERBAIJAN

Lenkoran •

NAGORNO-
KARABAKH
Stepanakert
Shusha
Lachin Corridor
Kelbadzhar
Corridor
IRAN

NAKHICHEVAN (Azer.)
• Nakhichevan
Jan. 1990:
independence declared

Sep. 1991: independence declared;
Azerbaijani forces intervene;
Armenian-supported Karabagh
forces clash with Azeris; Armenian
invasion of Azerbaijan follows
May 1994: ceasefire

Lithuanians man a barricade in the
capital, Vilnius, in 1991 *(above left)*.
The Baltic state became a democratic
republic in March 1991, and then
seceded from the USSR. In early 1991
Gorbachev sent in Russian forces, but
in August 1991, during a political
crisis in Moscow, Lithuania and
Estonia declared their independence.
Latvia became independent a month
later. None of the Baltic States joined
the Confederation of Independent
States which replaced the USSR.

The break-up of the Soviet Union
saw widespread ethnic and political
conflict in the Caucasus region *(map
left)*. Armenia and Azerbaijan fought
over the Christian Armenian enclave
in Nagorno-Karabakh; Georgia fought
to keep South Ossetia within her
boundaries and to prevent the
independence of Abkhazia, but in
both cases was faced with Russian
intervention to keep the peace.
Chechenia was invaded by Russian
forces in December 1994 to prevent
its complete independence.

With the collapse of the USSR,
around 25 million ethnic Russians
were left outside the Russian
Federation, formed in March 1992
(map above). They found themselves
the victims of discrimination once
they no longer had the protection that
Russia had given them under the old
Soviet system. Ethnic conflict in
Moldova, the Baltic States and
Kazakhstan contributed to the
instability of the new states.

Alexander Solzhenitsyn (1918-),
novelist and perhaps the Soviet
Union's most famous dissident *(above)*
became an international symbol of
opposition to the communist regime.
His novel *One Day in the Life of Ivan
Denisovich* (1962) and his study of
the Soviet prison camp system *The
Gulag Archipelago* (1973-6) received
wide circulation abroad and his
criticism of Soviet repression eventually caused the regime to revoke his
citizenship and deport him. He lived in
exile in Switzerland and the US,
returning only after the collapse of
the Soviet Union. He was no ardent
supporter of Russia's new rulers,
however; his almost mystical belief in
Russia's fundamental strength and
destiny as the leader of the Slavs sits
ill with the fragmentation of Russia
and the economic hardship accompa-
nying the reforms.

With no country left to rule, Gorbachev slipped into obscurity.
Russia embraced a form of presidential democracy, with Yeltsin
as its first president. Some republics – Belarus, Uzbekistan,
Kazakhstan, Turkmenistan – maintained reformed communist
governments; the others adopted some form of democracy. In
Russia democracy gave rise to a fragmented collection of small
political parties across the political spectrum. In the parliamentary elections of 1993, Yeltsin's supporters controlled only a fraction of the new house. The largest share of the vote, over 20 per
cent, went to the extreme right-winger Vladimir Zhirinovsky. By
1996 he had faded, and the recently discredited Communist Party,
led by Gennady Zyuganov, was once again a major political force.

The revival of communism in Russia was a response to the
years of crisis since 1991. Market economics, introduced throughout the former Soviet bloc, brought with them unemployment,
low wages, a decline in welfare and a wave of economic crime
and corruption. The fragmentation of the USSR gave rise to
numerous crises in the "Near Abroad", the circle of states once
part of the Union. Arguments between Russia and the Ukraine
over the Soviet Black Sea Fleet and the nuclear arsenal on non-
Russian soil were resolved without military conflict. But in the
Caucasus conflict has been endemic since 1988. These pressures
have brought Russia into a position of predominance in the CIS,
despite its own economic and political fragility. Boris Yeltsin
emerged from 1992 as the de facto spokesman for the CIS in the
international arena. Fear of Russian imperialism did not disap-
pear with the advent of the reform movement.

CONFLICT IN THE CAUCASUS
*The dissolution of the Soviet Union
brought into the open violent racial
and religious tensions throughout the
Caucasus region. Old rivalry between
Christian Armenia and Muslim
Azerbaijan flared up over the fate of
Armenians living in the enclave of
Nagorno-Karabakh in Azeri territory.
Armenian forces, with Soviet backing,
took control of the area and created a
land link with Armenia by taking over
Muslim areas. In Georgia independence
brought three separate conflicts. The
first, with South Ossetia, whose inhabitants wanted to join with North Ossetia
in the Russian Federation, was ended in
1992 by Russian military intervention.*

*The second involved war with Abkhazia,
an autonomous Georgian province
which wished for independence. Here,
too, Russian forces imposed a ceasefire. The third conflict was an internal
civil war between rival Georgian forces,
which forced the president, Zviad
Gamsakhurdia, to flee to Chechenia in
1992. Chechenia itself, a republic within
the Russian Federation, demanded
independence in 1994, and provoked a
full-scale war with Russia. The battles
for the capital, Grozny (above), echoed
the terrible battles of the Second World
War. In May 1996 a ceasefire was
signed between the two sides following
the assassination of the Chechen
nationalist leader, Dzhokhar Dudayev.*

Russia established a fragile
democracy in 1992. In October 1993
hard-line parliamentary delegates
hostile to further reform tried to
depose President Boris Yeltsin. He
ordered in tanks and special forces
and bombarded the White House, the
Russian parliament. On 4 October the
rebels surrendered *(right)*. There were
140 deaths in the fighting.

THE YUGOSLAV CIVIL WAR, 1990-6

The greatest casualty of the collapse of the communist bloc was Yugoslavia, which for years had successfully evaded domination by Moscow. Yugoslavia was a federation of republics (Serbia, Croatia, Macedonia, Bosnia-Herzegovina, Slovenia and Montenegro) held together by an over-arching communist apparatus. As long as Tito, the founder of communist Yugoslavia, was alive, the federation worked well. After his death in 1980, divisions began to appear along historic lines of ethnic diversity.

Yugoslavia faced desperate problems in the 1980s. The economy was in decline, burdened by $18 billion of international debt and rising inflation and unemployment. The trigger for conflict was the emergence of an aggressive Serb nationalism in 1987, following the choice of Slobodan Milošević as leader of Serbia. He suppressed the Albanian minority in Kosovo, and then set out to expand Serbia's influence in the federation as a whole, where there were large Serb minorities. He quickly revived old hatreds: Slovenia and Croatia moved towards separatism, and communist influence evaporated. In 1990 multi-party elections were held in all the republics, which brought nationalists to the fore in Slovenia and Croatia and paved the way for their simultaneous declaration of independence on 25 June 1991. By then ethnic conflict had already broken out between Croats and the Serb minority in Krajina, who were anxious for their future under the Croat leader Franjo Tudjman. There followed a brief war between Slovenia and the rump Yugoslav state in early July 1991, and then a prolonged and bitter conflict between Croats and Serbs which was brought to a halt in January 1992 following American diplomatic intervention. By then UN peacekeepers were in Croatia, and its independent existence was internationally recognized. On 27 April 1992 Serbia and Montenegro formed a new Federal Republic of Yugoslavia and prepared to fight to retain a grip on the key republic of Bosnia-Herzegovina.

Bosnia was ruled by a multi-ethnic government, led by Alija

Muslim families arrive at the UN camp at Tuzla in July 1995 following the fall of Srebrenica *(below right)*. About 12,000 made the 60-mile trek. They were a small part of a vast refugee problem. By August 1995 2.3 million people were registered as refugees in the former Yugoslav area, 1.37 million in Bosnia.

The former Yugoslav federal republic of Bosnia and Herzegovina declared its independence in December 1991. There followed four years of civil war between the Muslim, Serb and Croat populations *(map below)*, with interventions from Croatia and Serbia. In 1992 the United Nations sent peacekeeping forces, and in November 1995 NATO intervened to keep the warring peoples apart and impose a peace settlement.

The Contact Group plan
- - - - frontlines, 1 Jan. 1995

division of territory under Contact Group plan
- Serb territory
- Croat-Muslim Federation territory
- Sarajevo, under UN administration

The Yugoslav civil war, 1991-5
- → Croatian advances, Jan. 1993
- → Federation of Bosnia and Herzegovina advances, Oct.-Nov. 1994
- - -→ Croatian and Federation of Bosnia and Herzegovina advances, spring 1995
- → Bosnian Serb advances, summer 1995
- → Croatian and Federation of Bosnia and Herzegovina advances, Aug.-Oct. 1995

- Croatia, June 1991
- overrun by the Yugoslav army and Croatian Serb forces by Dec. 1991
- Bosnia-Herzegovina, Mar. 1992
- secured by Yugoslav army and Bosnian Serb forces by Dec. 1992
- controlled by Bosnian Croat forces, Dec. 1992
- under Bosnian government control, Dec. 1992
- overwhelmingly or largely Muslim, 1991. No significant Muslim presence by 1996
- Autonomous Province of Western Bosnia, Sep. 1993-Aug. 1994
- remained under control of breakaway Serbian forces, Oct. 1995
- to be returned to Croatian control by Nov. 1996 under Zagreb agreement
- UN UN designated "safe areas"

Map labels (ethnic divisions map):

AUSTRIA

SLOVENIA
Maribor
Ljubljana
Varaždin

HUNGARY

Zagreb
Karlovac
Rijeka
Sisak

C R O A T I A

Subotica
Zenta
Sombor

ROMANIA

Osijek
VOJVODINA
Vukovar
Novi Sad

Bihać

Belgrade
Pančevo

Banja Luka
Brčko

Y U G O S L A V I A
Tuzla

Zadar
Knin

Vitez
Srebrenica
Kragujevac

B O S N I A -

S E R B I A

Sarajevo
Pale

Split

HERZEGOVINA
Goražde
Kruševac

Niš

Mostar

Novi Pazar

MONTENEGRO

Priština

BULGARIA

Dubrovnik
Podgorica
KOSOVO

ALBANIA

Kumanovo
Skopje

MACEDONIA

Bitola

GREECE

The ruined bridge at Mostar *(left)* was one of the enduring images of the Bosnian civil war. Mostar was at the centre of the conflict between Croats and Muslims in Bosnia.

The collapse of the Eastern bloc in 1989 opened up the deep ethnic divisions in Yugoslavia as the individual Yugoslav republics began to argue for independence *(map right)*. In June 1991 Croatia and Slovenia seceded, and in December 1991 Bosnia and Macedonia followed suit. Serbia and Montenegro formed a new Federal Republic of Yugoslavia on 27 April 1992.

In April 1994 new efforts were made to resolve the Bosnian war by a small circle of states known as the Contact Group. The US, Russia, France, Britain and Germany proposed a division of Bosnia *(map left)* with a UN area around the capital of Sarajevo. Like the plans of the European Union and the UN, the Contact Group plan foundered on the absence of any real ability to coerce the warring factions.

In October 1995 the sides in the Bosnian civil war agreed to a ceasefire. President Clinton invited the parties to a preliminary peace conference at the Wright-Patterson air force base in Dayton, Ohio. The settlement arrived at *(map below)* created a miniature Yugoslavia in the region, with separate Croat-Bosnian and Serb-Bosnian republics within the state of Bosnia.

Yugoslavia: ethnic divisions, 1991
areas populated mainly by

- Serbs and Montenegrins
- Croats
- Muslims
- Slovenes
- Macedonians
- Albanians
- Hungarians
- Bulgarians
- Romanians, Slovaks
- — — Yugoslav republican boundaries
- - - - Yugoslav provincial boundaries

Map (Dayton Peace Agreement):

controlled by separatist Serb administration

CROATIA
Bihać
Banja Luka
Sava
Orašje
(US)
Bosna
Vrbas
BOSNIA-
Tuzla
Drina
Mrkonjic Grad
Travnik
Zenica
SERBIA
(Br.)
Vitez
Srebrenica
Dinaric Alps
Gornji Vakuf
HERZEGOVINA
Sarajevo
Goražde
(Fr.)
Mostar
MONTENEGRO

The Dayton Peace Agreement, 1995
- Serb Republic
- Croat-Muslim Federation
- transferred to Serb control under Dayton Agreement
- transferred to Croat/ Muslim control under Dayton Agreement

NATO peace implementation forces
- ✱ NATO Allied Rapid Reaction Corps HQ
- ■ NATO divisions
- boundaries between divisions

The ancient Croat city of Dubrovnik, seen here in flames *(bottom left)* in October 1991, was the scene of bitter fighting between Croat and Serb forces following Croatia's declaration of independence in June 1991. By the end of the year Serbia had accepted Croat secession and fighting between the two sides petered out.

The president of the Bosnian Serbs, Radovan Karadzić, in conference with his military chief, Ratko Mladić *(below)*, at the height of the civil war in January 1993. For their part in atrocities against Muslims during the war, both were indicted as war criminals.

Izetbegović, keen to stay in the federation. Serb agitation destabilized the republic and pushed the Muslim and Croat populations towards independence, which was recognized internationally in March 1992. Full-scale civil war broke out in April 1992. It was fought with a ferocious savagery as each ethnic element sought to "cleanse" the areas under its control of the opposing ethnic group's population. The Muslims were caught between the Croats, who had ambitions to create a state of Herceg-Bosna in the Croat areas of Bosnia, and the Serbs, who set up a Serb Republika Srpska, intending to partition Bosnia altogether. By 1993 the Serbs contolled around 70 per cent of Bosnia and the Muslims around 10 per cent, mainly in the cities. From 1992 to November 1995 the Bosnian Serbs besieged the capital, Sarajevo, held by the Muslim-led government, but the Muslim enclave resisted, short of food and weapons and suffering high losses.

Against the odds, Bosnia survived. In 1992 international intervention from the European Community and the UN kept open a life-line to Bosnia. During much of 1993 and 1994 talks were held intermittently in an attempt to find a solution. In Serbia itself arguments developed between the political leaders over the future of Bosnia, and the expense of the war, while the weak state of the Serb economy made it difficult to complete the task of dividing Bosnia. In March 1994 President Clinton succeeded in getting Bosnian Muslims and Croats to form a federation. The same month the Russians put pressure on Serbia and Croatia to terminate hostilities. The Bosnian Serbs refused all compromise. In the summer of 1995 their military commander, Ratko Mladić, began one last assault on the Muslim enclaves. His spectacular success provoked first Croat armed intervention, then, in August 1995, the armed intervention of NATO. In November, facing military defeat, and no longer fully suported by Milošević, the Bosnian Serbs bowed to American-backed pressure to accept a settlement, leaving a fragile Bosnian state divided between the three races and utterly devastated by four years of war.

EUROPEAN UNION IN THE AGE OF MAASTRICHT

While the Soviet bloc disintegrated after 1989, Western Europe moved towards greater integration. In the early 1980s the expanded European Economic Community was stagnating. In the attempt to reform it the ideal of full economic and political union, first championed in the 1940s, came closer to realization. The impulse to reform came from the economic crises of the 1970s and early 1980s: growth rates in the EEC were only half the levels of the 1957-73 period. To cope with the crisis member states had introduced new restrictions on trade and capital movements which challenged the very nature of the market. Growing doubts about the effectiveness of Community institutions, whose procedures were cumbersome and long-winded, contributed to the sense that a new departure was needed.

In July 1981 the European Parliament set up an Institutional Committee, headed by Altiero Spinelli, which recommended a new treaty for the community to supplement the founding Rome Treaty. The proposals in the draft treaty were discussed at an Intergovernmental Conference in Luxembourg at the end of 1985 and formed the basis of the Single European Act, which was ratified by the member states in 1987. The Act paved the way for a full European Union in which remaining economic barriers would be removed, steps taken towards political union, and foreign and defence policies merged. Two new commissions were established, one under the energetic Community president, Jacques Delors, to work out the basis for European monetary union, the second to establish a framework for political union.

The proposals provoked strong argument, particularly on the prospect of creating a genuine monetary union, and on the Social Chapter, a proposal to merge the welfare provisions of the member states into a single format. In 1989 Margaret Thatcher won exemption for Britain from the social policy clauses of the Single European Act. By the end of 1991 the work of the two commissions was finished, and on 9 December 1991 the heads of government met in the Dutch city of Maastricht to draw up a Treaty on European Union. Over 300 individual pieces of legislation were necessary to complete economic union. The date for the end of economic frontiers was set for 31 December 1992, but Britain continued to stall on monetary union and a final summit in Edinburgh was needed to win over the waverers. Britain ratified the Treaty in June 1993 by a parliamentary margin of three votes, and a fully integrated market was created.

The prospects for political union were less fruitful. A great many states, including those recently freed from communism, were keen to associate economically with the new union, but even within the 12 member states – 15 from 1995 – there were strong reservations about political merger. The problems of creating common foreign and defence policies were exposed by the arguments over the Yugoslav crisis, when Germany broke ranks and unilaterally recognized an independent Croatia and Slovenia. Though the European Union set up a peace-keeping initiative under the British politician David Owen, its fruits were meagre. The issue of national sovereignty, which broke up the USSR and Yugoslavia, was still strong enough in Western Europe to make Delors's vision of a federal political union a distant ambition.

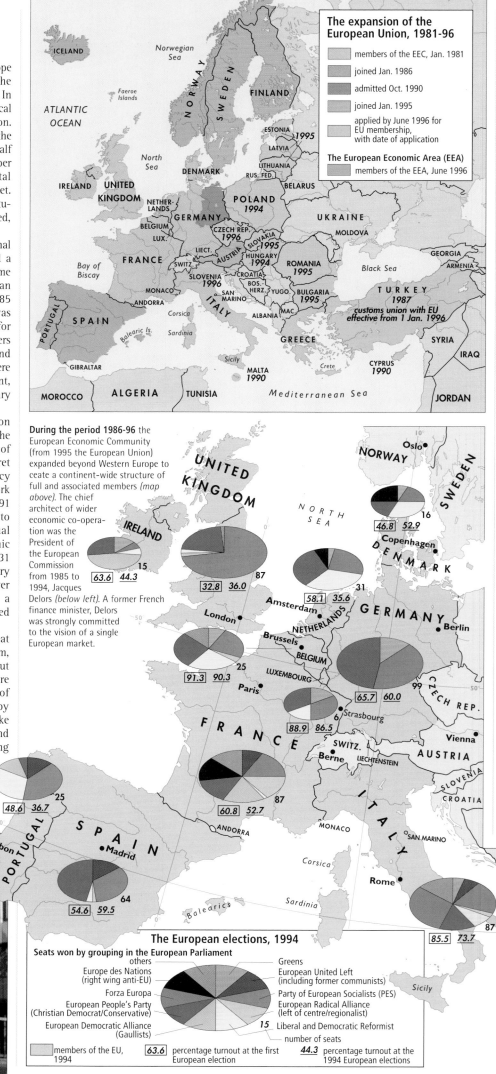

The expansion of the European Union, 1981-96

- members of the EEC, Jan. 1981
- joined Jan. 1986
- admitted Oct. 1990
- joined Jan. 1995
- applied by June 1996 for EU membership, with date of application

The European Economic Area (EEA)
- members of the EEA, June 1996

During the period 1986-96 the European Economic Community (from 1995 the European Union) expanded beyond Western Europe to ceate a continent-wide structure of full and associated members (map above). The chief architect of wider economic co-operation was the President of the European Commission from 1985 to 1994, Jacques Delors (below left). A former French finance minister, Delors was strongly committed to the vision of a single European market.

The European elections, 1994

Seats won by grouping in the European Parliament

- others
- Europe des Nations (right wing anti-EU)
- Forza Europa
- European People's Party (Christian Democrat/Conservative)
- European Democratic Alliance (Gaullists)
- Greens
- European United Left (including former communists)
- Party of European Socialists (PES)
- European Radical Alliance (left of centre/regionalist)
- Liberal and Democratic Reformist

15 number of seats

- members of the EU, 1994
- 63.6 percentage turnout at the first European election
- 44.3 percentage turnout at the 1994 European elections

None the less, a great deal was achieved between 1985 and 1996, when the Maastricht Treaty was due for review. The free movement of people, goods, capital and services throughout the Union was established in 1993. The collapse of communism strengthened the appeal of the Western European model of economic collaboration. Despite differences over the issue of a European security force or a common foreign policy, the European Union began to talk with a more united voice on political issues than ever before.

The decision whether to join the European Economic Community, or to accept its transformation, aroused strong political feelings. Since 1971 referenda have commonly been used to resolve the issues *(chart left)*. Norway rejected membership in 1971, but re-applied in 1994 only to have its voters reject entry again. There is also opposition from nationalist groups hostile to the loss of sovereignty and rule by the Brussels bureaucracy, a view widely evident in Britain, where the government of John Major refused a referendum on Europe over acceptance of the Maastricht Treaty. In Denmark the Treaty was rejected in 1992 by the tiny margin of 40,000 votes, but a second referendum in May 1993 produced 58 per cent in favour.

Right from the foundation of the EEC, a Franco-German axis has underpinned the organization. Chancellor Kohl and President Mitterand, here seen at the 1994 Franco-German summit *(bottom right)*, continued the tradition.

Referenda on European membership, 1971-94

	Country	Issue	Decision
1994			
	Norway	whether to join	no
	Finland	whether to join	yes
	Sweden	whether to join	yes
	Austria	whether to join	yes
1993			
	Denmark	Maastricht Treaty	yes
	France	Maastricht Treaty	yes
	Ireland	Maastricht Treaty	yes
	Switzerland	whether to take part in European Economic Area	no
1987			
	Ireland	Single European Act	yes
1986			
	Denmark	Single European Act	yes
1982			
	Greenland	(1979 gained internal autonomy from Denmark) voted to withdraw, implemented 1985	
1975			
	UK	whether to remain in EEC	yes
1972			
	Denmark	whether to join	yes
	Ireland	whether to join	yes
	France	whether EEC should be enlarged	yes
1971			
	Norway	whether to join	no

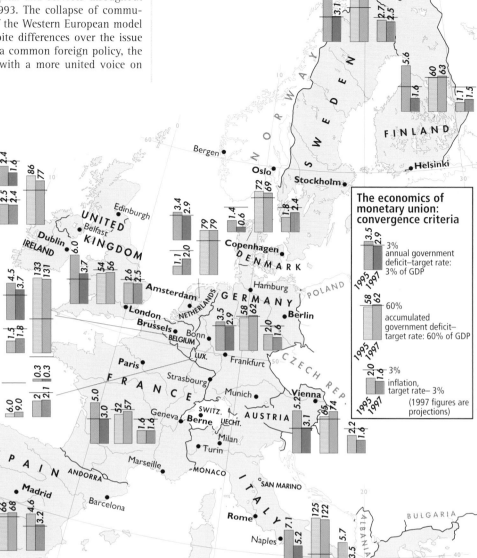

The economics of monetary union: convergence criteria

- 3% annual government deficit–target rate: 3% of GDP (1995, 1997)
- 60% accumulated government deficit–target rate: 60% of GDP (1995, 1997)
- 3% inflation, target rate– 3% (1995, 1997)

(1997 figures are projections)

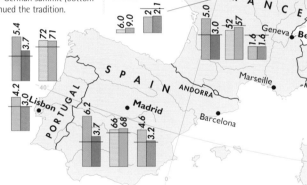

European Economic Community trade flows, 1980-92

% of total exports or imports

- Export trade to other EEC countries
- Import trade from other EEC countries
- Imports to EEC countries from external world
- Exports from EEC countries to external world

(axis: 30, 40, 50, 60 / years 1950, 1989, 1992)

One of the central aims of the Maastricht Treaty was to create the framework for full monetary union in the European Union by 1999. Wide differences in wealth and economic outlook between the 15 members *(map above)* make it unlikely that the criteria for the single currency will by met by more than a handful of countries. The trade ambitions of the Treaty have been more successful. Trade between member states has grown at the expense of trade with external markets *(chart left)*. The European Union now embraces the largest and richest market in the Western world.

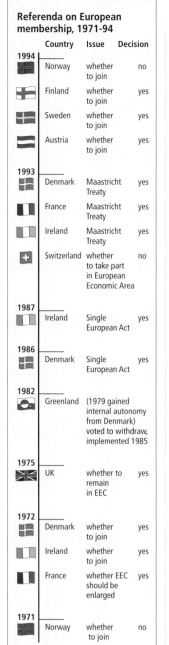

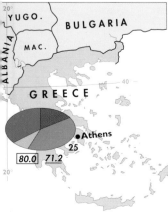

Direct elections to the European Parliament in Strasbourg began in 1979. By 1994 voter participation in European elections, even in states with compulsory voting, steadily declined *(map left)*. The parliament has only limited powers within the European Union, where major legislative decisions are taken by the Council of Ministers, directly responsible to their national parliaments. Although not strictly partisan, the European Parliament has divided into political groupings on the conventional right-left spectrum.

EUROPEAN MONETARY UNION

In October 1977 the EEC president, Roy Jenkins, proposed the establishment of a European monetary system to cope with the unstable currency markets ushered in by the oil crisis of 1973. After two years of negotiation the European Monetary System was set up in March 1979, based on the European Currency Unit (ECU) and the Exchange Rate Mechanism (ERM), through which national currencies were linked to a central rate with only minor fluctuations permitted. The object was to increase European monetary solidarity by getting the strong currencies in the Community to support the weaker ones through intervention by central banks working with a European Monetary Fund. Britain did not join the ERM until October 1990, but following a wave of speculative activity in the summer of 1992 both Britain and Italy were forced to devalue and leave the ERM. Under the Maastricht Treaty monetary union and a single currency were to be established by 1999 at the latest, as long as the member states had, by 1997, achieved a number of "convergence criteria": price stability, interest rate stability, exchange rate stability and low government debt. In 1995 the new common unit of currency was christened the Euro. British opposition, the diversity of economic performance within the 15 member states and the problem of monetary relations between those qualified to join and those failing to meet the convergence criteria made it unlikely that full monetary union would be achieved before 2000.

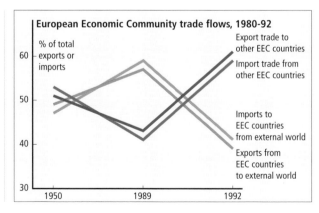

European Union in the Age of Maastricht

THE NEW ECONOMIC SUPER-POWERS: GERMANY

Few people in 1945 would have predicted how rapidly Germany was to recover from the disastrous economic effects of the Second World War. In the year of defeat, industrial production sank to one third of the pre-war level. The German currency once again experienced high inflation. Germany was occupied and divided by states determined to extract reparation and to limit the German industrial revival. In 1946 the victor states agreed a "Level of Industry Plan" designed to reduce German steel production to eight million tons (a quarter of its wartime level), to restrict trade and to prohibit the output of a range of modern products.

Germany's economic future depended on political developments outside German control. The onset of the Cold War produced a territorial division of Germany into two states: a Federal Republic (West Germany) set up in co-operation with the Western Allies; and a much smaller rump German state set up under Soviet domination, the German Democratic Republic (East Germany). The Western state, set up in 1949, was soon integrated with the wider capitalist economy. In 1950 the Federal Republic (FRG) was freed from most of the post-war restrictions on economic development, and embarked on the "Economic Miracle". Eastern Germany (the GDR) adopted the Stalinist model. Development there was in general above the standard for the rest of the Soviet bloc, but far below the economic achievements of the West.

In the 1950s and 1960s the Federal economy was dominated by the theory, usually associated with Economics Minister Ludwig Erhard, of the social market economy. This was an economy committed to avoiding too much state direction while building up an effective welfare state. Economic revival became the central ambition of German society. Freed from 30 years of war and peacetime restrictions, the German peoples enthusiastically embraced the rush for growth. The achievement was remarkable. The national product grew almost four-fold in 20 years. The Federal economy grew faster on average between 1950 and 1980 than any other European economy. There were brief downturns in 1965-6, in the mid-1970s and again in the early 1980s, but the trend was continually upward. By the 1980s the Federal Republic was a major force in the world economy, exporting large quantities of capital, supporting the IMF, and acting as the central industrial economy in the EC. In 1986 Germany became the world's largest exporter.

Foreign trade was key to German success. In the 1950s reconstruction set up a vigorous demand for just the goods in which Germany specialized – high quality engineering products, chemicals, electro-technical goods, vehicles. The state gave the

When the Wall came down in Berlin, East Germans flooded into the prosperous shopping areas in the Western half of the city. Most of them came to window-shop (below). The gap between incomes in the East and West widened steadily from the 1950s (chart top right). By the time of unification, Eastern workers earned about one third the wages of their Western counterparts. The immediate effect of unity was to cut industrial output in the Eastern areas by half and to push unemployment to over a million. State funds for welfare and modernization programmes in the East raised the German budget to 59 per cent of the national product in 1993.

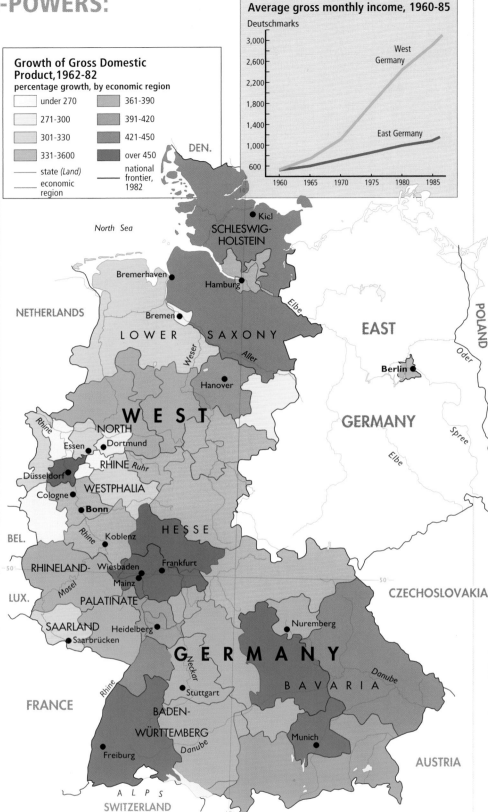

Growth of Gross Domestic Product, 1962-82
percentage growth, by economic region

- under 270
- 271-300
- 301-330
- 331-3600
- 361-390
- 391-420
- 421-450
- over 450
- state (Land)
- economic region
- national frontier, 1982

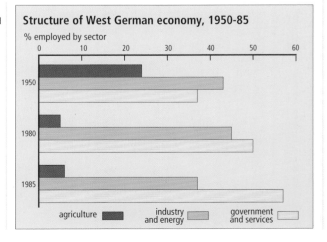

Average gross monthly income, 1960-85
Deutschmarks

West Germany

East Germany

In the 1950s, the West German economy was catching up lost ground after defeat and occupations. From the 1960s to the 1980s, West Germany became a wealthy country, overtaking other European states. The most successful regions more than quadrupled output in 20 years (map above). The high growth areas were not the traditional industrial heartlands, but the more rural north and south of the country. New industries moved into the less-developed regions. New service sectors grew up there, no longer reliant on the coal and iron ore fields. The structural changes in the economy (chart right) meant that by 1985 more than half of all employment was in services and government.

Structure of West German economy, 1950-85
% employed by sector

agriculture | industry and energy | government and services

1950
1980
1985

Towards the New World Order

export industries tax and investment concessions, while German industry concentrated on effective marketing and after-sales service to compete with well-established rivals. From 1951 the balance of trade remained in permanent surplus. Thanks to steady productivity growth, German prices were kept at competitive levels despite regular revaluations of the Mark since 1961.

Low inflation was also a central plank in Federal economic policy. Wage growth was modest in the 1950s and 1960s as workers accepted the need for growth and employment as a greater priority than high levels of consumption. Memories of the hyper-inflation of 1923 ran deep in German society. Financial stability was accepted as an essential element in the state's economic strategy. Inflation remained low, and German goods enjoyed a permanent advantage on world markets as a result.

In 1990 the Federal economy faced a new challenge. The GDR collapsed and was brought into the Western state. The Eastern provinces were poor by Federal standards, their industries uncompetitive. The costs of the transition brought a brief reverse in German growth, as the Eastern provinces experienced the pressures of the free market. The East has now begun to embark painfully on its own "economic miracle".

High levels of exports in the 1950s and 1960s laid the foundation for the German economic miracle. Most German exports went to Western Europe and the US. The developing world has been much less important in German trade growth *(map right)*: Asia and the Far East have been insignificant markets. From the 1970s, Germany also became a major exporter of capital. German firms set up branches abroad. German investors bought up foreign shares. German businesses had to become more multi-national to compete with giant American or Japanese firms. Labour shortages in the 1960s produced a flood of migrant workers, reaching more than three million by the early 1970s. With rising unemployment in the 1980s, many returned *(map below)*, but the *Gastarbeiter* remained a permanent feature of German society, prompting race attacks and nationalist hostility with echoes of the 1930s.

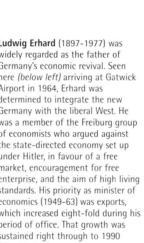

German trade and investment, 1988-89

352 → German exports, 1989 (bn DM)

🪙 *Spain 9.1*   direct foreign investment, 1988 (bn DM)

- EEC
- other Europe
- Communist Europe
- Communist Asia
- other Asia
- Oceania
- U.S.A./Canada
- Africa
- Latin America

Foreign workers in Germany, 1968-84

net inward migration (in 1000s)

net outward migration

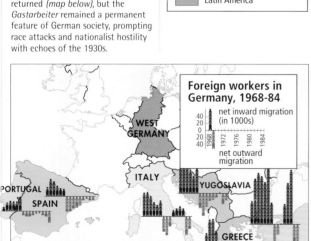

Ludwig Erhard (1897-1977) was widely regarded as the father of Germany's economic revival. Seen here *(below left)* arriving at Gatwick Airport in 1964, Erhard was determined to integrate the new Germany with the liberal West. He was a member of the Freiburg group of economists who argued against the state-directed economy set up under Hitler, in favour of a free market, encouragement for free enterprise, and the aim of high living standards. His priority as minister of economics (1949-63) was exports, which increased eight-fold during his period of office. That growth was sustained right through to 1990 *(chart below)*.

German exports 1950-90

Deutschmarks (billions)

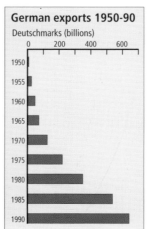

VOLKSWAGEN

The story of the Volkswagen symbolizes the remarkable recovery of German economic fortunes after the disaster of 1945. But the car was the offspring of the Third Reich, chosen by Hitler as the design for the "People's Car" to bring cheap motoring to the masses. The war interrupted production, but it was started up again in 1946 even though British engineers declared it commercially unviable. The car became an immediate hit. It was cheap and simple to maintain, and marketed with great ingenuity. In 1960 the one-millionth car rolled off the assembly line at Wolfsburg (above). Thousands were exported.

So successful was the car abroad (where it was affectionately nicknamed the "Beetle") that Volkswagen set up overseas production lines. In the 1960s over 400,000 were sold every year in the US, the only overseas car sold in volume on the American market. Hollywood turned Hitler's dream car into the lovable "Herbie". Demand for the Beetle declined in the 1980s and Volkswagen diversified into a range of higher quality and more expensive vehicles. After 1989 the company took the lead in helping to modernize the motor industry of Eastern Europe by linking up with the Czech Skoda works. The Beetle itself is still produced in Brazil.

Norway 0.9, Sweden 1.0, Britain 10.5, Denmark 1.0, Ireland 0.9, Netherlands 10.3, Lux. 7.1, Belgium 8.2, Austria 6.7, France 14.5, Switzerland 8.8, Italy 8.7, Portugal 0.7, Spain 9.1, Greece 0.6, Turkey 0.3

Australia and New Zealand 2.6, Japan 4.1, Hong Kong 0.9, Singapore 1.1, Mexico 2.5, Canada 4.9, U.S.A. 49.7, India 0.4, Iran 0.1, Venezuela 0.2, Argentina 1.9, Brazil 8.9, Algeria 0.2, Nigeria 0.2, Egypt 0.5, Libya 0.6, South Africa 2.2

THE NEW ECONOMIC SUPER-POWERS: JAPAN

Japan found herself in much the same position as Germany at the end of the Second World War. Not only had her major cities been burnt down during the bombing offensive, her home islands were occupied by American forces, her economy was in ruins, and the Allies intended Japan to pay reparations to the countries she had occupied during the war. The United States wanted to prevent the revival of a strong Japanese industrial economy, and to keep Japan disarmed. Forty years later Japan was one of the world's new economic super-powers and had the fourth largest armed forces in the world.

The American occupiers contributed significantly to this revival. On 6 March 1946 Japan was forced to accept a new constitution, which permitted free labour unions, created a genuine parliamentary system and excluded war as an option in settling disputes. The enforced modernization of Japanese society and politics was accompanied by generous aid to revive the economy. With the Korean War, the United States changed tack and began to encourage the rebuilding of Japanese industry to supply the war effort. Reparations were shelved, a peace treaty was signed at San Francisco in 1951 and Japan was hailed as the Asian capitalist bulwark against the march of Asian communism.

The Japanese authorities seized the opportunity. During the 1950s they pursued a conscious policy of "catching up" with the West. In 1955 Prime Minister Ichiro Hatoyama launched a "Five-Year Plan for Economic Self-Reliance", which was masterminded, as was all subsequent expansion, by the Ministry of International Trade and Industry (MITI). The industrial strategy was based on priority for selected growth areas in heavy industry and high-technology sectors. Development aid, subsidies, export bounties and tariffs were all used to secure high domestic growth and a vigorous export performance. Free labour unions, initially run by Japanese communists, gave way

During the 1970s and 1980s Japan's economic presence abroad expanded dramatically *(map right and chart above right)*. By 1987 overseas investments were worth $139 billion, three quarters of it placed in the 1980s. Foreign trade totalled only seven billion Yen in 1970, but was 40 billion by 1993. The US trade deficit with Japan reached a yawning $60 billion by 1987, over a third of the total trade deficit. The great bulk of Japanese exports, over 75 per cent in the mid-1980s, was made up of machinery and equipment. The remarkable boom transformed the face of Japan's cities, as skyscrapers housed the mushrooming new corporations *(right)*.

Much of Japan's economic miracle was based on the help given to the industrializing countries of the Pacific Rim. Japanese planners designated development corridors *(map left)*, where investment was concentrated and communications developed.

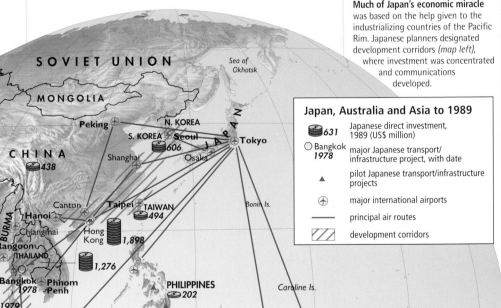

Japan, Australia and Asia to 1989

631	Japanese direct investment, 1989 (US$ million)
○ Bangkok 1978	major Japanese transport/infrastructure project, with date
▲	pilot Japanese transport/infrastructure projects
✈	major international airports
——	principal air routes
▨	development corridors

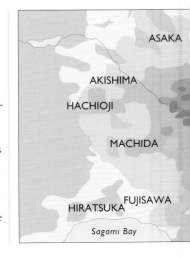

In 1995 the Japanese city of Kobe was struck by a major earthquake *(above)*, which left many dead. One of the world's centres for the production of microchips, its destruction had worldwide implications for the computer industry.

Between 1979 and 1986 land prices in Tokyo rocketed as businesses sought prestigious office sites in the capital. In 1986 prices in much of central Tokyo increased 100 per cent in 12 months *(map right)*, but the slow-down of the economy burst the bubble and land prices began to fall thereafter.

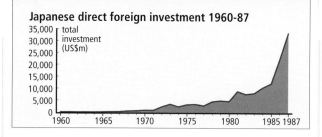

Japanese direct foreign investment 1960-87

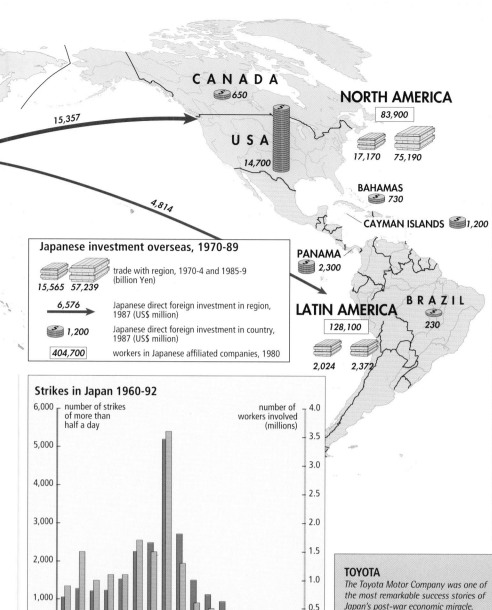

C A N A D A
650

NORTH AMERICA
83,900

15,357

U S A
14,700

17,170   75,190

4,814

BAHAMAS
730

CAYMAN ISLANDS 1,200

PANAMA
2,300

LATIN AMERICA   B R A Z I L
128,100   230

2,024   2,372

Japanese investment overseas, 1970-89

15,565 57,239   trade with region, 1970-4 and 1985-9 (billion Yen)

6,576   Japanese direct foreign investment in region, 1987 (US$ million)

1,200   Japanese direct foreign investment in country, 1987 (US$ million)

404,700   workers in Japanese affiliated companies, 1980

to "enterprise unions", which created close bonds between management and workforce and helped to achieve remarkable growth in productivity. Japan's output grew at 9.5 per cent a year in the 1950s and at 10.5 per cent a year in the 1960s.

Japanese society exhibited a number of features favourable to high post-war growth. Great emphasis was placed on technical education. In 1974 Japan had 330,000 engineering students at university, one fifth of all students. In Britain in the same year there were just 24,000. The ethos of the large enterprise discouraged individualism and encouraged loyalty to the firm and effective collaboration to achieve growth targets. The level of strike activity was tiny. Workers accepted low wages and rigid work discipline. Economic achievement became one of the defining features of post-war Japanese politics.

When Japan was hit by the oil crisis of 1973 these strengths enabled her to adapt quickly. In 1971 MITI produced "Visions for the 1970s", which formed the basis for the reorientation of the economy away from heavy industry and mass consumer products towards the sunrise sectors. In 1969 Japanese engineers produced the first factory robots; in the 1970s they fought tenaciously for a leading share in the new computer market. By the late 1970s Japanese output and exports had been restructured. Steel, ships and chemicals gave way to high-quality technology – machinery and electronic equipment – which brought a second "economic miracle" in the 1980s. In the 1990s Japan's problems are those of success: high expectations from her workforce; large balance-of-trade surpluses; and growing competition from those very economies around the Pacific Rim that Japanese economic strength had stimulated into imitation.

Strikes in Japan 1960-92

number of strikes of more than half a day

number of workers involved (millions)

Japanese exports, 1947-93

(billion Yen)

TOYOTA

The Toyota Motor Company was one of the most remarkable success stories of Japan's post-war economic miracle. Producing a mere 11,000 vehicles in 1950, the corporation produced 5.6 million at its peak in 1985, 1.9 million of these being produced in Toyota's 33 overseas production plants. The business was founded in 1937 on a forest site at Koromo-cho, near Nagoya, by Risaburo and Kiichiro Toyoda. The name of the area was later changed to Toyota City, and it became the centre of the Toyodas' industrial empire. In 1959 they built the Motomachi plant, one of the largest and most modern in the world, where they pioneered an extreme version of the rationalized, time-and-motion production long associated with car manufacture. In ten years they increased output 1,000 per cent and became Japan's largest car maker.

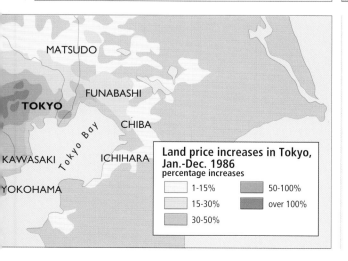

MATSUDO

FUNABASHI

TOKYO

CHIBA

KAWASAKI

ICHIHARA

Tokyo Bay

YOKOHAMA

Land price increases in Tokyo, Jan.-Dec. 1986
percentage increases

1-15%   50-100%
15-30%   over 100%
30-50%

The great bulk of Japan's strikes lasted for less than half a day and were largely token in character. Longer strikes involved few workers and had little effect on growth *(chart top left)*. By 1993 strikes involved only 60,000 out of a workforce of 49 million.

During the 1980s Japanese business began to expand operations overseas in addition to pushing export sales to new heights *(chart above left)*. By 1987 two thirds of Japanese TV sets were made outside Japan, and 50 per cent of its audio equipment, almost all of it in Asia *(chart right)*.

Japanese overseas production, 1979-87

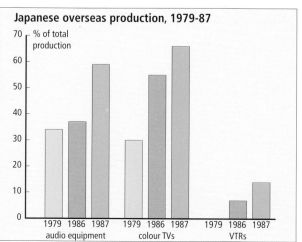

% of total production

1979 1986 1987   1979 1986 1987   1979 1986 1987
audio equipment   colour TVs   VTRs

CHINA AFTER MAO

One economy which Japan helped to stimulate was the Chinese, as it struggled to cope with the legacy of Mao's communism. The decision taken by Mao's successors, Hua Guofeng and Deng Xiaoping, to launch China on the path of economic liberalization in 1978 was taken in the hope of attracting financial and technical aid from the capitalist West. The government created a number of Special Economic Zones (SEZ) in areas close to Hong Kong, Macao and Taiwan, and later in the Yangtze valley. The SEZs were administrative zones in which a package of inducements was used to encourage foreign investment and technology transfer. In the early 1980s investment was modest, but by 1992 $36 billion had been invested, and China's annual GNP growth had averaged ten per cent for ten years.

The drive for prosperity produced dramatic results in only five years. The peasant communes were reformed in 1979 with the introduction of the Responsibility System, which gave farmers individual plots of land on 15-year leases and the right to retain any surpluses. By 1984, 98 per cent of the peasantry farmed their land under the system and the communes faded away. Agricultural output rose 49 per cent in five years, and China became a net food exporter. Private businesses were also permitted. In 1978 there were 100,000; by 1985 there were 17 million. The object was not yet to create a full market economy, but to replace a centralized command economy with one based on "planning through guidance". By 1984 about one third of industrial output was still governed by state planning, 20 per cent came from the market sector and approximately half from "guided production". At the same time foreign trade doubled. The boom was not, however, without its difficulties. It led to high inflation, large external deficits, high rural migration to the cities and public corruption. In 1986 a period of retrenchment was called for by hardline elements in the Communist Party.

These problems emerged alongside calls for greater political freedom. In 1978-9 there were demands for a "Fifth

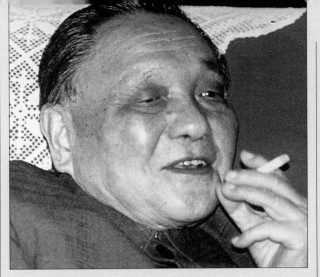

DENG XIAOPING

The son of a peasant from Sichuan province, Deng Xiaoping is the great survivor of Chinese politics, and has been the major figure in the Chinese Communist Party since 1978. His road to power was a difficult one. He joined the party in 1925 and took part in the Long March with Mao in 1935 (see pages 74-5). During the civil war he was a successful military commander. He became a Politburo member in 1955. Following the failure of the Great Leap Forward, Deng favoured a more pragmatic economic policy and formed an alliance with state Chairman Liu Shaoqi. Both became victims of Mao's Cultural Revolution. Deng was forced into internal exile. Briefly rehabilitated by Zhou Enlai in 1973, the Gang of Four forced him again into exile. After Mao's death he was fully readmitted to power in July 1977 by Mao's chosen successor, Hua Guofeng. He put himself at the head of the economic modernization drive, and within three years had become the dominant figure in Chinese politics. He moved China away from Maoist state planning towards greater economic liberalization under the slogan "economics in command". He made few political concessions, however, for fear of alienating more conservative elements in the Party. In the early 1990s he inspired a further wave of liberalization, opening up all of China except Tibet to outside economic influence. Despite increasing frailty, Deng continues to play the role of elder statesman to a new generation of "Dengist" reformers.

The Shanghai steelworks *(above right)*, part of the industrialization drive of the post-Mao era. Steel output doubled between 1978 and 1987 after years of stagnation. Shanghai itself was one of 14 coastal cities targeted for development, and the value of its industrial output trebled in the 1980s.

Production of key commodities

commodities	pre-1949 peak	1986
steel (tons)	923,000	52,050,000
coal (tons)	61,875,000	870,000,000
electricity (kilowatt-hours)	5,955,000	445,500,000
crude oil (tons)	nil	131,000,000

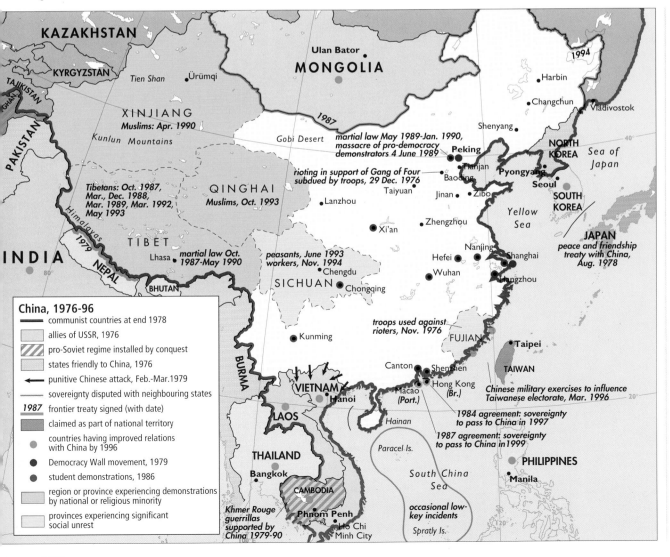

China, 1976-96
- ——— communist countries at end 1978
- allies of USSR, 1976
- pro-Soviet regime installed by conquest
- states friendly to China, 1976
- → punitive Chinese attack, Feb.-Mar.1979
- ——— sovereignty disputed with neighbouring states
- *1987* frontier treaty signed (with date)
- claimed as part of national territory
- ● countries having improved relations with China by 1996
- ● Democracy Wall movement, 1979
- ● student demonstrations, 1986
- region or province experiencing demonstrations by national or religious minority
- provinces experiencing significant social unrest

From the 1970s communist China succeeded in normalizing relations with the outside world after years of isolation. Japan recognized China in 1972 following the visit of Japanese premier Kakuei Tanaka, and the United States finally switched allegiance from Taiwan to China in the spring of 1979, when formal diplomatic relations were renewed. Frontier treaties were signed with neighbouring states, and agreement reached over the transfer of Hong Kong and Macao to Chinese sovereignty. In December 1978 an "Open Door" policy was launched, to give China access to science, technology and capital from the West to boost her modernization drive in the 1980s. The Open Door also admitted the ideas and culture of the West, and led to growing unrest which peaked in 1989 *(map left)*.

China experienced an economic revolution from the late 1970s, following the decision to pursue the "Four Modernizations", which were written into a new state constitution in March 1978 (agriculture, industry, national defence, science and technology). The object was to turn China into a leading modern state by the year 2000. A massive programme of investment produced significant gains in output, but led to high inflation and balance of trade deficits. In the mid-1980s a policy of retrenchment stabilized the economy, before the regime launched more thorough market reforms and a new expansionary wave from 1990 *(map right)*. Domestic production and foreign trade have boomed *(charts above and above right)*.

Emigration from Hong Kong, 1980-91

Year	Number
1980	22,400
1981	18,300
1982	20,300
1983	19,800
1984	22,400
1985	22,300
1986	19,000
1987	30,000
1988	45,800
1989	42,000
1990	62,000
1991	58,000

thousands (10 20 30 40 50 60 70)

Chinese foreign trade, 1976-91 (US$ billion)

■ exports
□ imports

('77 '79 '81 '83 '85 '87 '89 '91)

A student demonstrator stands in front of tanks in Peking in June 1989 *(below)*. He was one of thousands who occupied Tiananmen Square for six weeks in May and June to demonstrate for greater democracy. On 3-4 June the army violently ejected the protestors. Initial reports suggested a death toll of 3,000, later revised to 400-800. The Chinese authorities admitted 23 deaths.

In October 1984, China and Britain reached agreement over the future of Hong Kong, ruled by the British since 1842. Sovereignty was to be restored to China in 1997 in return for a guarantee that for 50 years China would respect Hong Kong's existing economic and social system. These promises did not allay worries about the transition to communist rule and failed to prevent a sharp rise in emigration from Hong Kong from the late 1980s *(chart above)*.

Modernization" – democracy – to accompany the Four Modernizations of economic expansion. Dissidents ran considerable risks posting up protests on a wall in Peking. This so-called "Democracy Wall" movement was swiftly shut down and its leader, Wei Jingsheng, sentenced to 15 years in jail. In December 1986 there was renewed violence when students demonstrated for democracy in 15 major Chinese cities, encouraged by criticisms of the Party made by astrophysicist Fang Lizhi. Deng's response, prompted by bitter memories of his treatment during the Cultural Revolution *(see page 128)* was to clamp down hard. The party secretary general, Hu Yaobang, was sacked for his openly liberal views. The Party launched an "Anti-Spiritual Pollution" movement to counter the seeping effects of "bourgeois liberalism" and its alleged Western allies. This did nothing to satisfy the students and when Hu Yaobang died in April 1989 they seized the opportunity to occupy Tiananmen Square in Peking to honour his memory and press for reforms. In June the army was sent in to crush the movement to prevent any emulation of the political crisis sweeping communist Eastern Europe.

In the wake of Tiananmen, China acted to reassure its own populace and the outside world that it was a responsible government which offered stability and economic security. A wave of anti-corruption measures followed. Many of China's trading partners, with an eye on a vast and rapidly expanding market, set aside human rights concerns and continued to trade and invest with enthusiasm. In the early 1990s the regime of Jiang Zemin launched a new wave of economic reforms, opening up most of China and offering a model of a "socialist market economy" to replace the now outmoded state-run models of the past. Growth in 1993 reached 13 per cent. By the mid-1990s China was an economic super-power in the making, with the world's largest armed forces and a communist-dominated political system. Communism survived the process of economic change more successfully than in the Soviet bloc, but the issue of democratization remained a live one in a society experiencing growing wealth and personal freedom.

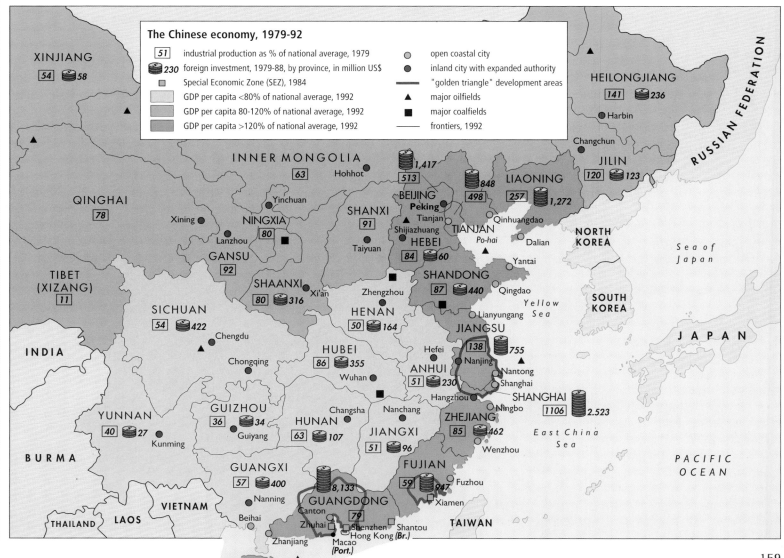

The Chinese economy, 1979-92

- *51* industrial production as % of national average, 1979
- *230* foreign investment, 1979-88, by province, in million US$
- ■ Special Economic Zone (SEZ), 1984
- GDP per capita <80% of national average, 1992
- GDP per capita 80-120% of national average, 1992
- GDP per capita >120% of national average, 1992
- ● open coastal city
- ● inland city with expanded authority
- — "golden triangle" development areas
- ▲ major oilfields
- ■ major coalfields
- — frontiers, 1992

China after Mao

159

THE ECONOMIC REVOLUTION IN ASIA: THE PACIFIC RIM

Japan and China have been the two largest players in the transformation of the Asian Pacific Rim from an economic backwater in the 1950s to the hub of the world trading economy in the 1990s. The east Asian region quadrupled its per capita income in 25 years, a record unparalleled in economic history. In 1995 Hong Kong and Singapore ranked among the ten richest states in the world per capita, ahead of France and Britain.

The economy of the Pacific Rim developed in a number of waves. Japan launched Asian prosperity in the 1950s. Spurred by her example, the so-called "Four Dragons" of the region – South Korea, Hong Kong, Singapore and Taiwan – began a second wave of expansion. All four states lacked the advantages necessary for the normal path of industrialization. They were short of capital, they were overpopulated and they possessed little arable land and few raw material reserves. They opted instead to concentrate on export-led growth, using cheap labour and borrowed capital to undercut established textile and light consumer goods producers. Their success was phenomenal. By 1976 they produced 60 per cent of the world's manufacturing exports.

When the growth of the developed world slowed down in the 1970s, Japan and the Four Dragons began to invest heavily in the developing states of South East Asia – Malaysia, Thailand, Indonesia and the Philippines – which were rich in raw materials and food supplies. These states, organized in the Association of South East Asian Nations (ASEAN) from 1967, then embarked on their own version of the Asian economic miracle, emulating their richer neighbours to the north by producing low-price exports in huge volumes. In 20 years Malaysia became the world's largest exporter of manufactured goods. The region developed an increasingly integrated economy: Japan and the Four Dragons produced high-quality, high-cost manufactures; ASEAN produced more of the low-price consumer goods aimed predominantly at Asian markets. The states of the Pacific Rim became each others' best markets. When communist China and Vietnam began to expand and modernize their economies in the 1980s, further huge new markets opened up. The rest of the Pacific Rim – Australia, New Zealand, western Canada and the United States – were drawn in as consumers and suppliers for the world's fastest growing economic arena.

The success of the new industrial giants in Asia owed something to favourable economic circumstances. The emergence of rich overseas markets in the developed world and the globalization of trade and finance created a healthy framework for rapid export growth. Modern electronic technology was easily transferred between states, and products based around the microchip were particularly suitable for economies that lacked a heavy industrial base. There were also important advantages enjoyed by many Asian societies. They began with a cheap labour force, willing to work long hours for low pay and flexible in the face of new technologies. The Confucian ethic, with its emphasis on frugality, group loyalty, respect for hierarchy and for educational achievement, has been a stimulus for high savings and low labour unrest. Governments spent less on welfare and infrastructure and more on education and export subsidy. In the 1990s, 80 per cent of 18-year olds in Taiwan and 85 per cent in South Korea were still in full-time education. Literacy rates throughout the Pacific Rim were considerably higher than those in South Asia, Africa or Latin America.

The economic revolution in Asia has altered the balance of the world economy, which for much of the century was dominated by the United States and Europe. The World Bank has estimated that by the year 2002 the "Chinese Economic Area" – China, Hong Kong, Taiwan, Singapore – will have a GNP larger than that of the United States. Pacific Rim states in Asia began in the 1980s to invest their surpluses in the states of Europe and America on which they had relied for economic aid in the 1960s. The world economy has gone full circle over the course of the century.

The Pacific Rim

- - - - - fibre-optic links
- major air connections
- major container ports
- major container-shipping routes
- development corridors
- 5.2 4.9 3.6 average growth rates 1960-65 1965-85 1985-92
- 52 proportion of imports from other Pacific Rim producers, 1992

The Pacific Rim area is the fastest growing economic region in the world *(map above)*. Based on a complex and modern communications network and high-growth urban centres, the region is one of high investment, large-scale manufacturing output and aggregate growth rates well above those of the rest of the developed world.

A view of Hong Kong at night *(left)* showing the Plaza Tower and the Bank of China. Over the last 20 years the urban areas of Hong Kong have grown rapidly, spreading from Hong Kong Island and Kowloon into the New Territories on mainland China, where one third of the population now lives.

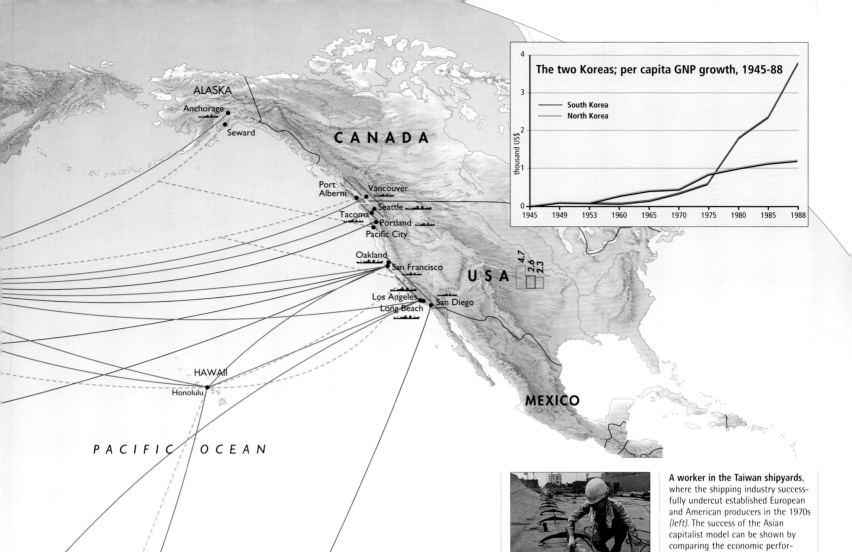

The two Koreas; per capita GNP growth, 1945-88

— South Korea
— North Korea

(chart axis: thousand US$ 0–4; years 1945 1949 1953 1960 1965 1970 1975 1980 1985 1988)

A worker in the Taiwan shipyards, where the shipping industry successfully undercut established European and American producers in the 1970s *(left)*. The success of the Asian capitalist model can be shown by comparing the economic performance of North and South Korea *(chart above)*. South Korea's growth rate in the 1980s was three times that of her communist neighbour.

Percentage productivity growth in the Pacific Rim 1960-80

Country	1960s	1970s	1980s
Japan	9.3	4.0	2.8
Taiwan	6.7	6.0	4.7
South Korea	6.5	5.5	6.1
Hong Kong	5.6	5.8	4.8
Singapore	7.0	3.9	4.7
Philippines	2.3	2.0	-2.2
Thailand	5.2	3.8	2.5
Malaysia	3.8	5.4	2.0
Indonesia	2.3	3.8	0.5
USA	1.4	-0.7	1.1

A traffic jam in Bangkok *(above)*. Rapid modernization produced the same problems of environmental pollution and urban sprawl that first affected the developed world earlier in the century. High population growth has exacerbated the problems of urban overcrowding.

One of the reasons for the rapid growth of the Pacific Rim area has been the very high ratio of savings to Gross Domestic Product *(chart below left)*. Savings were diverted to investment or private welfare schemes, relieving the government of high spending.

A trader on the Hong Kong Stock Exchange *(left)*. By 1990 Hong Kong was the second richest state on the Asian Pacific Rim, with a per capita GNP of over $11,000. Its long-term future remains uncertain, following re-unification with communist China in 1997.

The productivity record of Pacific Rim states showed remarkable rates of growth from the 1960s to the 1980s *(chart above right)*, as modern technology and production methods were introduced to relatively underdeveloped societies. Rates slowed in the 1980s as the first wave of industrial modernization was completed.

Pacific Rim: savings ratios

Country	1960s	1980s
Japan	23.2	30.8
Taiwan	21.2	37.0
Korea	17.5	27.2
Singapore	14.1	42.4
Hong Kong	11.3	30.2
Malaysia	19.8	26.4
Philippines	19.4	18.1
Thailand	20.6	16.3
Indonesia	2.6	22.1
USA	18.6	17.7

SINGAPORE

The island of Singapore has a long history as a trading centre dating back to the early 1820s, when it was first developed by the British. Since independence in 1965 it has generated a boom in trade and output quite out of proportion to its size. With a population of only 2.9 million and an area of 625 square kilometres, Singapore rose by 1996 to being the seventh most prosperous state in the world. Its success has been boosted by long-term political and social stability. The Peoples' Action Party (PAP) has been in power since 1959, with Lee Kuan Yew (pictured right) as prime minister from 1959 to 1993. Singapore's economic success was based on attracting multi-nationals to use the island as a subsidiary for financial and trading services and for the manufacture of products for the Asian market. The economy has been closely regulated by the state, although state expenditure is low – less than 20 per cent of GDP.

Singapore's citizens contribute 20 per cent of their earnings under a compulsory scheme to provide each with a personal welfare fund. The resulting high savings ratio has been used to power Singapore's growth, while keeping state debt and expenditure low. Now European states are exploring the Singapore model for their own development.

The Economic Revolution in Asia: the Pacific Rim

161

THE SPREAD OF DEMOCRACY

Rapid economic development has not always been accompanied by political reform. The post-war American ideal of democratic capitalism which emerged in Japan, Italy and Germany under the influence of an American occupying power made little headway elsewhere until the late 1980s.

Democracy took time to develop fully even in Western Europe. In Spain and Portugal army-backed dictatorships survived for 20 years after 1945. When General Franco died in 1975, his chosen successor, King Juan Carlos, transformed Spain into a parliamentary democracy, with the first free elections since 1936 held in 1977. In Portugal the death of Salazar in 1970 paved the way for the gradual evolution of a democratic system, though not until 1982 was full civilian government established. In Greece a military coup in 1967 ushered in a period of seven years of dictatorship, which ended in 1974 with a National Salvation Government under Constantine Karamanlis. Paradoxically, the coming of democracy to the former Soviet bloc has coincided with the emergence in Western Europe of new movements of the extreme right, highly critical of the existing parliamentary system. In Italy the Movimento Sociale Italiano, an avowedly ultra-right party with fascist roots, won 12 per cent of the vote in the elections of 1994. In France the Front National led by Jean-Marie Le Pen has become a mainstream party despite accusations of neo-fascism. Overtly neo-fascist groups, small and violent, wait in the wings.

Where democracy has a long history, there has been a conspicuous decline in political enthusiasm. In the United States, presidents are elected with little more than half the country voting, an almost continuous decline since the post-war peak of 64 per cent in the Kennedy campaign year of 1960. In Europe turnout in elections for local government or for the European parliament can be as low as a third. Membership of the main political parties has steadily declined since 1945. Democracy was pursued as an ideal outside Europe and America by relatively few, which explains its limited success in the years since 1945.

The situation changed with dramatic speed in the 1980s and 1990s. In Latin America the military and the wealthy

Democracy has had a chequered career worldwide since 1945, but from the mid-1980s the pace of democratization has speeded up significantly *(maps below and below right)*. The most conspicuous break-throughs occurred throughout the Soviet bloc between 1989 and 1991, in Southern Africa between 1990 and 1994, and in Latin America, where in December 1989 the US toppled the Panamanian dictator, General Manuel Noriega, and in April 1990 Nicaragua experienced its first peaceful and democratic transfer of power. In Asia democracy returned to the Philippines in 1986 and to Pakistan in 1988. In Cambodia, which was occupied by Vietnamese forces in 1978, a fragile democracy was restored in May 1993, despite the boycott of the elections by the communist Khmer Rouge. The Cambodian elections were protected and monitored by a UN force *(below right)*, including Japanese military engineers, serving overseas for the first time since the end of the Second World War. This was one of a number of elections in which the UN has played the role of policeman, acknowledging the determination of the world community since the 1980s to enforce the UN's founding commitment to democratic freedom.

DEMOCRACY IN PAKISTAN

The fate of democracy in Pakistan since the partition of 1947 has been closely bound up with that of the Bhutto family. In the 1960s Zulfikar Ali Bhutto was a prominent pro-army politician with ambitions to profit from the failure of the Pakistani dicta-tor Ayub Khan. In 1967 he founded the Pakistan People's Party, and when Ayub failed to keep Bangladesh from seceding in 1971 Bhutto ousted him as president with the backing of the army. A fanatical anti-Hindu and admirer of Napoleon, Bhutto ruled Pakistan for six years, ending martial law, introducing a democratic

constitution in 1973 and seeking popular backing for an "Islamic Socialism". Distrusted by the army and disliked by Islamic fundamentalists, he was overthrown in July 1977 and executed two years later. His daughter, Benazir, fled into exile, but returned to Pakistan in 1987, where she rallied the PPP to fight elections following the death of the new military strong-man, General Zia. She was elected prime minister in 1988, but overthrown by an army plot in 1990. In October 1993 new elections brought her back to power, but democracy remains hostage to the ambitions of the army, the bureaucracy and a wealthy elite.

Towards the New World Order

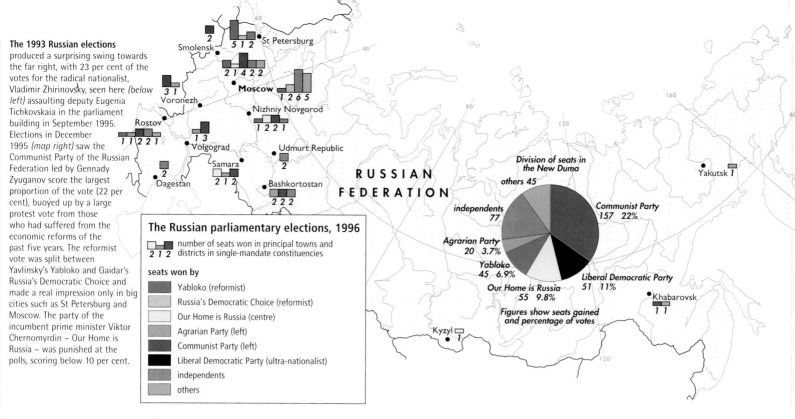

see inset left

The 1993 Russian elections produced a surprising swing towards the far right, with 23 per cent of the votes for the radical nationalist, Vladimir Zhirinovsky, seen here *(below left)* assaulting deputy Eugenia Tichkovskaia in the parliament building in September 1995. Elections in December 1995 *(map right)* saw the Communist Party of the Russian Federation led by Gennady Zyuganov score the largest proportion of the vote (22 per cent), buoyed up by a large protest vote from those who had suffered from the economic reforms of the past five years. The reformist vote was split between Yavlinsky's Yabloko and Gaidar's Russia's Democratic Choice and made a real impression only in big cities such as St Petersburg and Moscow. The party of the incumbent prime minister Viktor Chernomyrdin – Our Home is Russia – was punished at the polls, scoring below 10 per cent.

The Russian parliamentary elections, 1996

	number of seats won in principal towns and
2 1 2	districts in single-mandate constituencies

seats won by

- Yabloko (reformist)
- Russia's Democratic Choice (reformist)
- Our Home is Russia (centre)
- Agrarian Party (left)
- Communist Party (left)
- Liberal Democratic Party (ultra-nationalist)
- independents
- others

Division of seats in the New Duma

- others 45
- independents 77
- Agrarian Party 20 3.7%
- Yabloko 45 6.9%
- Our Home is Russia 55 9.8%
- Communist Party 157 22%
- Liberal Democratic Party 51 11%

Figures show seats gained and percentage of votes

elites were forced to make concessions to popular demands for political freedom. In Africa a fragile democracy has appeared in Kenya, Zambia, Malawi and Namibia. In Asia, Pakistan, the Philippines, Cambodia and Taiwan have achieved, or moved close to, genuine democratic rule, watched closely by the army and the established political elites.

The changing fortunes of democracy have a number of causes. The economic downturn in the developed world in the 1970s and the slow growth that followed have had their effects in the developing world, where dictatorial regimes had survived on the back of the post-war boom. Economic crisis or decline has fuelled political protest. There has also been a great rise in literacy and remarkable improvements in communications, which have helped to raise political expectations and

spread ideas of emancipation or revolution. The slow decline of Soviet communism and its disappearance in 1989-91 ended fears by local elites that democratization meant revolutionary Marxism. Finally, world opinion, expressed through the UN or through pressure from the United States and Europe, highlighted oppressive regimes and acts of tyranny. Sanctions, economic boycotts and the threat of international isolation, have helped to force the hand of reform, while UN observers and the world's television cameras have helped to sustain it thereafter. Yet in the end a flourishing democracy depends on a powerful consensus about the nature of society and its purposes, and it is that prior consensus which is lacking in many of the areas where democracy has taken its first faltering steps.

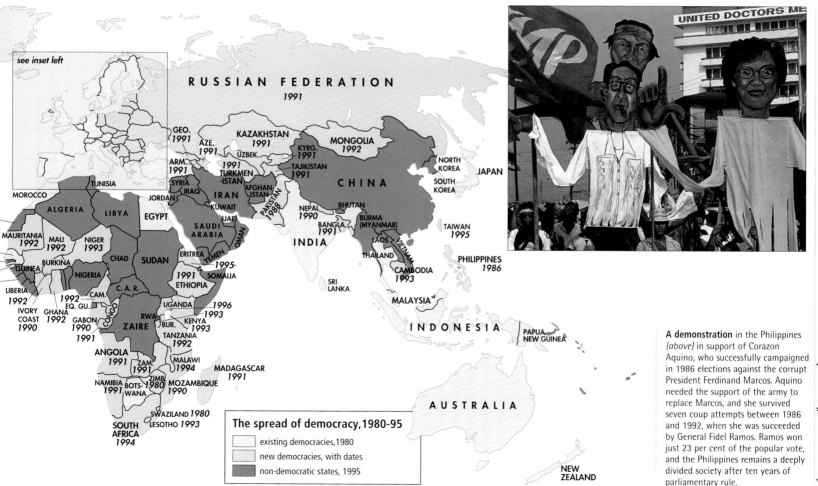

The spread of democracy, 1980-95

- existing democracies, 1980
- new democracies, with dates
- non-democratic states, 1995

A demonstration in the Philippines *(above)* in support of Corazon Aquino, who successfully campaigned in 1986 elections against the corrupt President Ferdinand Marcos. Aquino needed the support of the army to replace Marcos, and she survived seven coup attempts between 1986 and 1992, when she was succeeded by General Fidel Ramos. Ramos won just 23 per cent of the popular vote, and the Philippines remains a deeply divided society after ten years of parliamentary rule.

The Spread of Democracy

DEMOCRACY AND CIVIL WAR IN AFRICA

Nowhere has democracy been harder to establish than in sub-Saharan Africa. A concept imported by the colonial powers, the European model of parliamentary rule often meant little in the context of linguistic, tribal and religious conflict which the independent states of Africa inherited from their European masters.

The one exception, South Africa, was an anomaly, for it contained a large white settler community, British and Dutch (Afrikaner), in which parliamentary government for whites was fully established in 1926. It was in this unlikely setting that democracy made its most significant gain in Africa when in 1994 its 27 million black population was admitted to the parliamentary system. The struggle for emancipation of the non-white population went back to the 1920s, but it gathered momentum with the election of the Afrikaner Nationalist Party government in 1948. The nationalists adopted the idea of separate development – apartheid. Economic, political and military power was concentrated in white hands. Black "Bantustans" were set up where the black population was concentrated in poverty-stricken townships.

From the 1970s, when the policy of separate development reached its climax, the white regime faced growing pressures. Popular protest grew in the townships, backed by a campaign of violence waged by the ANC and the Pan-African Congress. The independence of Angola and Mozambique, and the transfer to black rule in Zimbabwe, presented South Africa with a hostile frontier across which guerrillas could operate, and drew the regime into a brutal counter-insurgency war. In the 1980s the government of P.W. Botha became increasingly militaristic and dictatorial in the fight against internal and external enemies. The white population found its own democracy under threat, while foreign opinion hardened against the regime's abuse of human rights. In 1986 economic sanctions were imposed by the USA and the EC. The resulting economic isolation damaged the South African economy at just the time when the tide of violent protest was rising to a level the security forces could barely contain.

The result was a slow move towards reform and stabilization. The Coloured and Asian communities were given a share of power in 1984. In 1988 Botha agreed to negotiate a settlement of the anti-guerrilla war in Angola and Namibia. In September 1989 Botha was succeeded by F.W. de Klerk, who, with the backing of important sections of the white community, opened the door to reform. In 1990 the banned opposition parties were legalized and the ANC leader Nelson Mandela freed from prison. Sanctions were lifted abroad and the ANC abandoned violence. A referendum in 1991 among the white population gave de Klerk a two-to-one majority in favour of a new democratic constitution. The following year a Convention for a Democratic South Africa drafted a transitional constitution and power passed to a multi-racial Executive Council. The whole process was rejected by the extreme wing of the Afrikaner movement, and by the large Zulu minority led by Chief Mangosuthu Buthelezi, whose people feared for their ethnic identity in a state dominated by non-Zulus. Buthelezi was persuaded to rejoin the democratic process, but not before an estimated 15,000 people had died in township clashes.

In April 1994 multi-racial and multi-party elections gave Mandela's ANC 63 per cent of the vote, the Nationalist Party 20 per cent, and Chief Buthelezi's Inkatha Freedom Party ten per cent. Mandela became president and de Klerk vice-president, a post he resigned in May 1996. Democracy in South Africa had the advantage of a more literate and prosperous black community than elsewhere in sub-Saharan Africa and a large industrial economy. The rest of Africa, subject to periods of famine, a rising tide of disease, growing impoverishment and poor links with the rich developed world, had a sorry foundation on which to build democratic institutions. Nonetheless progress was made. In Zambia, Malawi, Kenya, Uganda and Zimbabwe popular elections were held and several long-serving dictators dismissed by their electorates. Now that the South African issue has been resolved, the prospects for democracy in the region have been significantly strengthened.

The former British dominion of South Africa has been transformed during the 1990s from rule by a white minority to a multi-racial democracy *(maps below)*. From 1948 the ruling Afrikaner National Party, representing the Dutch settler communities living predominantly in the Transvaal and the Orange Free State, established the policy of apartheid, or separate development for the different races. In 1959 black "homelands" or Bantustans were created, four of which were granted independence from the late 1970s, but not recognized internationally. Widespread economic and political protest in the 1970s, particularly in the big black townships around Johannesburg, were violently suppressed, but in the 1980s, in the face of international sanctions and domestic criticism, some of the apartheid system was relaxed. In 1984 the Asian and Coloured (mixed-race) communities were given separate parliamentary assemblies and responsibility for their affairs. In 1990 the ban on parties was lifted and the African National Congress, led by Nelson Mandela, and Chief Buthelezi's Inkatha Freedom Movement (representing the Zulus of Natal), collaborated with the white government on a new multi-racial constitution. In 1994 the ANC won an overwhelming electoral victory, marred by persistent violence between ANC and IFP activists *(bottom left)*.

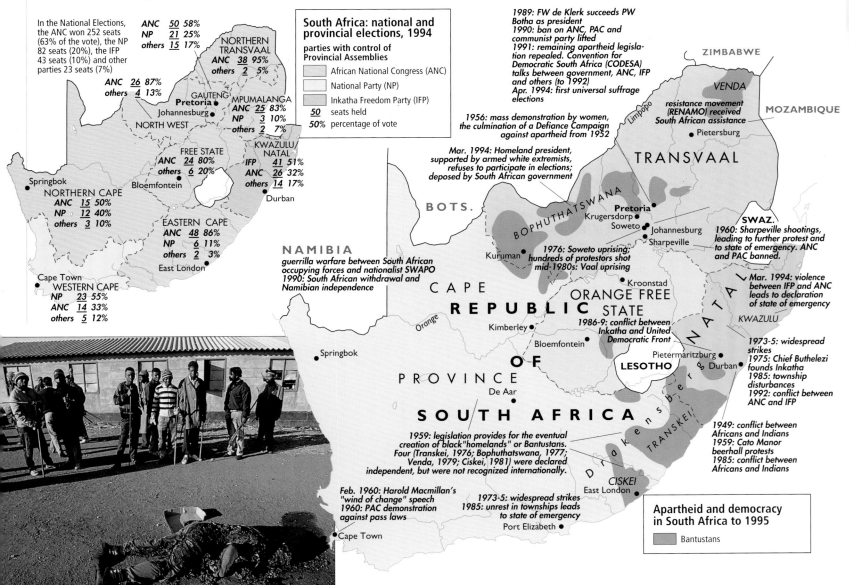

In the National Elections, the ANC won 252 seats (63% of the vote), the NP 82 seats (20%), the IFP 43 seats (10%) and other parties 23 seats (7%)

ANC	50	58%
NP	21	25%
others	15	17%

NORTHERN TRANSVAAL
ANC	38	95%
others	2	5%

ANC	26	87%
others	4	13%

GAUTENG
Pretoria
Johannesburg

MPUMALANGA
ANC	25	83%
NP	3	10%
others	2	7%

NORTH WEST

FREE STATE
ANC	24	80%
others	6	20%

Bloemfontein

KWAZULU/NATAL
IFP	41	51%
ANC	26	32%
others	14	17%

Durban

Springbok

NORTHERN CAPE
ANC	15	50%
NP	12	40%
others	3	10%

EASTERN CAPE
ANC	48	86%
NP	6	11%
others	2	3%

East London

Cape Town

WESTERN CAPE
NP	23	55%
ANC	14	33%
others	5	12%

South Africa: national and provincial elections, 1994

parties with control of Provincial Assemblies

- African National Congress (ANC)
- National Party (NP)
- Inkatha Freedom Party (IFP)

50 seats held
50% percentage of vote

1989: FW de Klerk succeeds PW Botha as president
1990: ban on ANC, PAC and communist party lifted
1991: remaining apartheid legislation repealed. Convention for Democratic South Africa (CODESA) talks between government, ANC, IFP and others (to 1992)
Apr. 1994: first universal suffrage elections

1956: mass demonstration by women, the culmination of a Defiance Campaign against apartheid from 1952

Mar. 1994: Homeland president, supported by armed white extremists, refuses to participate in elections; deposed by South African government

ZIMBABWE

VENDA

resistance movement (RENAMO) received South African assistance

MOZAMBIQUE

Pietersburg

TRANSVAAL

BOTS.

Pretoria
Krugersdorp
Soweto
Johannesburg
Sharpeville

SWAZ.

1960: Sharpeville shootings, leading to further protest and to state of emergency. ANC and PAC banned.

Kuruman

1976: Soweto uprising; hundreds of protestors shot
mid-1980s: Vaal uprising

NAMIBIA
guerrilla warfare between South African occupying forces and nationalist SWAPO
1990: South African withdrawal and Namibian independence

CAPE

Orange

Kimberley

Bloemfontein

REPUBLIC

ORANGE FREE STATE
1986-9: conflict between Inkatha and United Democratic Front

Kroonstad

LESOTHO

Pietermaritzburg

Durban

NATAL

Mar. 1994: violence between IFP and ANC leads to declaration of state of emergency

KWAZULU

1973-5: widespread strikes
1975: Chief Buthelezi founds Inkatha
1985: township disturbances
1992: conflict between ANC and IFP

Springbok

OF

PROVINCE

De Aar

SOUTH AFRICA

1959: legislation provides for the eventual creation of black "homelands" or Bantustans. Four (Transkei, 1976; Bophuthatswana, 1977; Venda, 1979; Ciskei, 1981) were declared independent, but were not recognized internationally.

Feb. 1960: Harold Macmillan's "wind of change" speech
1960: PAC demonstration against pass laws

1973-5: widespread strikes
1985: unrest in townships leads to state of emergency

Port Elizabeth

Cape Town

CISKEI

East London

Drakensberg

TRANSKEI

1949: conflict between Africans and Indians
1959: Cato Manor beerhall protests
1985: conflict between Africans and Indians

Apartheid and democracy in South Africa to 1995

Bantustans

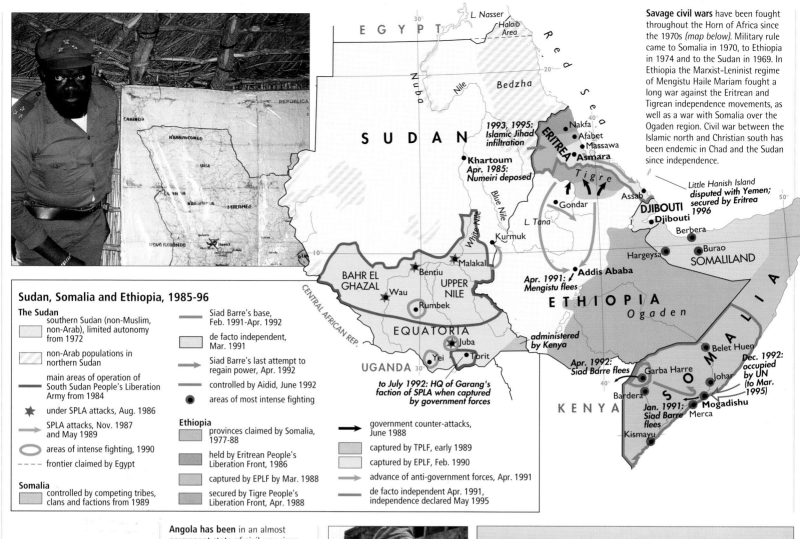

Savage civil wars have been fought throughout the Horn of Africa since the 1970s *(map below)*. Military rule came to Somalia in 1970, to Ethiopia in 1974 and to the Sudan in 1969. In Ethiopia the Marxist-Leninist regime of Mengistu Haile Mariam fought a long war against the Eritrean and Tigrean independence movements, as well as a war with Somalia over the Ogaden region. Civil war between the Islamic north and Christian south has been endemic in Chad and the Sudan since independence.

Sudan, Somalia and Ethiopia, 1985-96

The Sudan

- southern Sudan (non-Muslim, non-Arab), limited autonomy from 1972
- non-Arab populations in northern Sudan
- main areas of operation of South Sudan People's Liberation Army from 1984
- ★ under SPLA attacks, Aug. 1986
- SPLA attacks, Nov. 1987 and May 1989
- areas of intense fighting, 1990
- - - - frontier claimed by Egypt

Somalia

- controlled by competing tribes, clans and factions from 1989

- Siad Barre's base, Feb. 1991-Apr. 1992
- de facto independent, Mar. 1991
- → Siad Barre's last attempt to regain power, Apr. 1992
- controlled by Aidid, June 1992
- ● areas of most intense fighting

Ethiopia

- provinces claimed by Somalia, 1977-88
- held by Eritrean People's Liberation Front, 1986
- captured by EPLF by Mar. 1988
- secured by Tigre People's Liberation Front, Apr. 1988
- → government counter-attacks, June 1988
- captured by TPLF, early 1989
- captured by EPLF, Feb. 1990
- → advance of anti-government forces, Apr. 1991
- de facto independent Apr. 1991, independence declared May 1995

Angola has been in an almost permanent state of civil war since the struggle for independence from Portugal began in the early 1960s *(map below)*. Following independence in 1975 the Marxist MPLA took over power, while the South African-backed UNITA movement of southern Angola, led by Jonas Savimbi *(top)*, waged a guerrilla campaign against them. A temporary truce in 1991 and a general election in 1992 did not prevent the resumption of civil war. In November 1993 UNITA forces ended a nine-month siege of Cuito, in which 35,000 people died *(right)*.

The Angolan civil war, 1975-95

- Bakongo ethnic area
- Bambundu ethnic area
- Ovimbundu ethnic area

held in February 1975 by

- FNLA (Bakongo based, pro-Western)
- MPLA (Bambundu based, pro-Soviet)
- UNITA (Ovimbundu based, pro-Chinese then pro-Western)
- → advance of FNLA-UNITA forces, 1975
- area secured by MPLA, mid-1976
- → MPLA-supported incursions
- north-western limit of UNITA activity, 1976 to May 1991 ceasefire
- ⇒ MPLA attack on UNITA, 1990
- zones of intense fighting from 1992 elections to Lusaka peace accord, 1994
- - - - northern limit of South African incursions, 1976-88

NELSON MANDELA

Nelson Rolihlahla Mandela was born into the ruling Tembu family in Transkei in South Africa in 1918. He studied law in Johannesburg, where he met the nationalist politician, Walter Sisulu, by whom he was strongly influenced. He joined the African National Congress in 1944 and helped to found its Youth League. He preached non-violent civil disobedience against the imposition of apartheid, and in 1952 launched the Defiance Campaign which led to his imprisonment along with 8,500 others. Banned from any political activity, he developed the so-called M-plan for an underground network of ANC cells to keep the fight for political freedom alive in a police state. In 1961, frustrated at the failure of civil disobedience, he adopted the idea of armed struggle and was instrumental in founding the Spear of the Nation movement, which mounted a campaign of bombing. In 1962 he was captured and given life imprisonment on Robben Island. In 1990 he was released, aged 72, and became the leader of the ANC, and, in 1994, South Africa's first president elected by universal franchise – a tribute to his exceptional strength of character, his tolerance and his position as a symbol worldwide of the struggle for political emancipation.

Democracy and Civil War in Africa

THE MIDDLE EAST TO 1991

In the Middle East war, civil conflict and repression persisted just as they did in Africa. The major cause was no longer the Arab-Israeli conflict, which had produced four wars in a generation, but the threat posed to the whole region by revolution in Iran. In the 1970s Iran was ruled by Shah Reza Pahlavi, whose politically corrupt and vicious regime was kept in power by Western support and the SAVAK secret police. He was opposed by communists and by Islamic Shia fundamentalists. One of their leaders, Grand Ayatollah Ruhollah Khomeini *(see panel)*, called in public from his exile in Iraq for the faithful to rise up in revolution. Hostility to the shah reached boiling point in 1978, and on 16 January 1979 he left Iran. In February Khomeini declared Iran an Islamic Republic and imposed a militant Islamic regime. The revolutionary wave produced

On 2 August 1990 Iraq invaded the oil state of Kuwait and on 8 August proclaimed the union of the two states. Following condemnation of Iraq's action by the UN, a coalition of states undertook Operation Desert Storm in January and February 1991 to drive Iraqi armies out of Kuwait *(maps below and below right)*. The bulk of the military forces taking part in the operation were provided by the US. On 28 February Iraqi leaders agreed to a ceasefire and Iraq withdrew unconditionally.

widespread killings and imprisonment of political opponents and Westernizers, and the imposition of a harsh Koranic law.

According to Khomeini's theology, the revolution had to be exported. Rather than a world of nation-states, he sought a broader Islamic religious community, or *umma*: Iran had a sacred duty to lead the worldwide struggle to disseminate the message of Islam. During the 1980s, Iran destabilized the whole of the Middle East in pursuit of this goal. Terrorism and subversion were aimed at Bahrain and Kuwait in the early 1980s. In Syria support for militant Islamic opponents of the regime of Hafez al-Assad led to the massacre of 15-20,000 fundamentalists in the city of Hama in 1982. In Lebanon Iran backed the Party of God (Hezbollah), founded in 1983 to wage terrorist war on Israel.

The greatest efforts were reserved for Iran's immediate neighbour Iraq, where there existed a sizeable Shia community

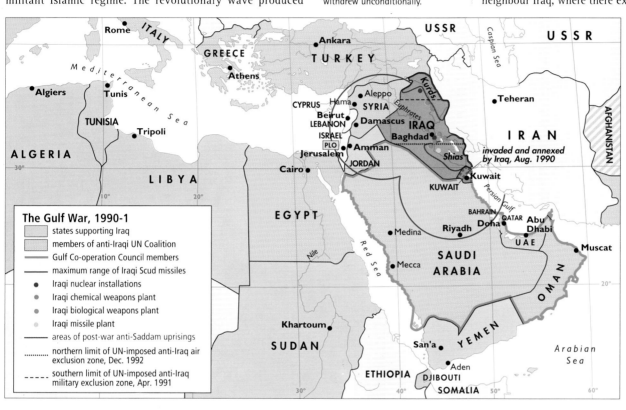

The Gulf War, 1990-1

- states supporting Iraq
- members of anti-Iraqi UN Coalition
- Gulf Co-operation Council members
- maximum range of Iraqi Scud missiles
- Iraqi nuclear installations
- Iraqi chemical weapons plant
- Iraqi biological weapons plant
- Iraqi missile plant
- areas of post-war anti-Saddam uprisings
- northern limit of UN-imposed anti-Iraq air exclusion zone, Dec. 1992
- southern limit of UN-imposed anti-Iraq military exclusion zone, Apr. 1991

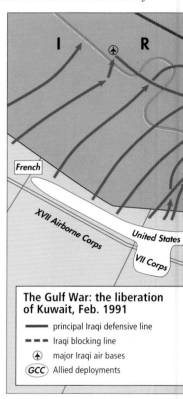

The Gulf War: the liberation of Kuwait, Feb. 1991

- principal Iraqi defensive line
- Iraqi blocking line
- major Iraqi air bases
- *GCC* Allied deployments

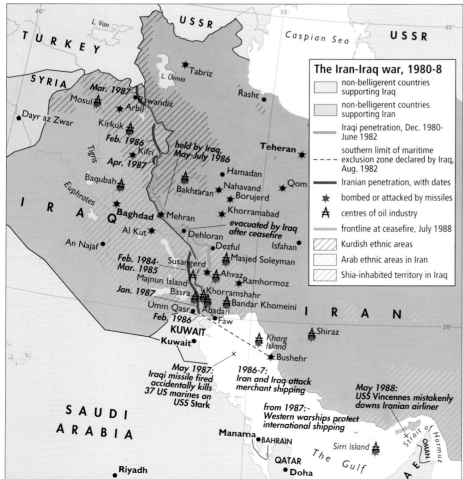

The Iran-Iraq war, 1980-8

- non-belligerent countries supporting Iraq
- non-belligerent countries supporting Iran
- Iraqi penetration, Dec. 1980-June 1982
- southern limit of maritime exclusion zone declared by Iraq, Aug. 1982
- Iranian penetration, with dates
- bombed or attacked by missiles
- centres of oil industry
- frontline at ceasefire, July 1988
- Kurdish ethnic areas
- Arab ethnic areas in Iran
- Shia-inhabited territory in Iraq

In September 1980 Iraq invaded Iran and sparked an eight-year war of attrition *(map left)*, which ended in July 1988 when Iran sought an armistice. The war cost an estimated 400,000 dead and 750,000 wounded, and burdened both states with massive debts. The contested area was small and the outcome inconclusive. It has been estimated that the war cost Iran $644 billion and Iraq $452 billion in damage and loss of oil revenues.

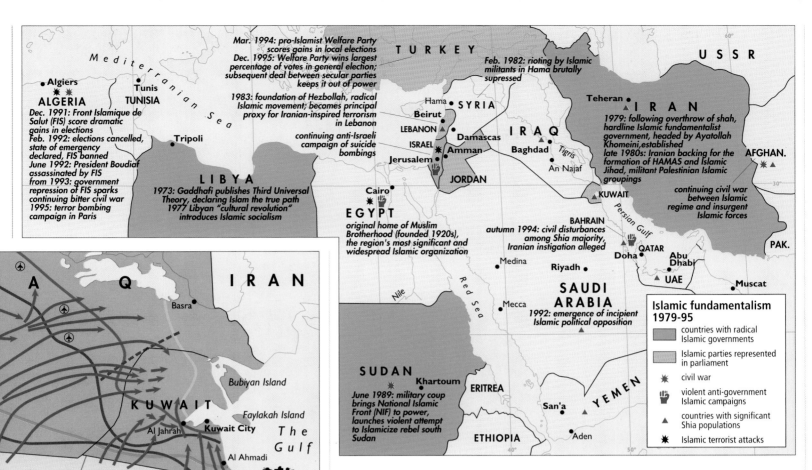

Mar. 1994: pro-Islamist Welfare Party scores gains in local elections
Dec. 1995: Welfare Party wins largest percentage of votes in general election; subsequent deal between secular parties keeps it out of power

1983: foundation of Hezbollah, radical Islamic movement; becomes principal proxy for Iranian-inspired terrorism in Lebanon

continuing anti-Israeli campaign of suicide bombings

Feb. 1982: rioting by Islamic militants in Hama brutally supressed

T U R K E Y

U S S R

M e d i t e r r a n e a n   S e a

• Algiers
ALGERIA
Tunis •
TUNISIA

Dec. 1991: Front Islamique de Salut (FIS) score dramatic gains in elections
Feb. 1992: elections cancelled, state of emergency declared, FIS banned
June 1992: President Boudiaf assassinated by FIS
from **1993:** government repression of FIS sparks continuing bitter civil war
1995: terror bombing campaign in Paris

Tripoli •

L I B Y A
1973: Gaddhafi publishes Third Universal Theory, declaring Islam the true path
1977 Libyan "cultural revolution" introduces Islamic socialism

Hama •
Beirut •
LEBANON
ISRAEL
Jerusalem •

S Y R I A
Damascus •
Amman •
JORDAN

I R A Q
Baghdad •

Tigris

An Najaf •

Teheran •

I R A N
1979: following overthrow of shah, hardline Islamic fundamentalist government, headed by Ayatollah Khomeini, established
late **1980s:** Iranian backing for the formation of HAMAS and Islamic Jihad, militant Palestinian Islamic groupings

AFGHAN.

continuing civil war between Islamic regime and insurgent Islamic forces

KUWAIT

PAK.

Cairo •
E G Y P T
original home of Muslim Brotherhood (founded 1920s), the region's most significant and widespread Islamic organization

BAHRAIN
autumn 1994: civil disturbances among Shia majority, Iranian instigation alleged

Doha •
QATAR
Abu Dhabi •
UAE

Muscat •

Persian Gulf

Nile

Red Sea

Medina •

Mecca •

Riyadh •

S A U D I   A R A B I A
1992: emergence of incipient Islamic political opposition

Islamic fundamentalism 1979-95

◾ countries with radical Islamic governments
◽ Islamic parties represented in parliament
✳ civil war
✊ violent anti-government Islamic campaigns
▲ countries with significant Shia populations
✳ Islamic terrorist attacks

S U D A N
Khartoum •
ERITREA
June 1989: military coup brings National Islamic Front (NIF) to power, launches violent attempt to Islamicize rebel south Sudan

ETHIOPIA

San'a •
Y E M E N

Aden •

A Q

I R A N

Basra •

Bubiyan Island

K U W A I T

Al Jahrah •
Kuwait City •

Faylakah Island

The Gulf

Al Ahmadi •

Allied Amphibious Taskforce

Egyptian
Saudi

British

Syrian   Kuwaiti

United States   Saudi

GCC

Allied frontlines
— day 1
— day 2
— day 3
— day 4
→ Allied advances

Tapline Road
Joint Command-North

Marines Central Command

S A U D I   A R A B I A

Joint Command-East

The Middle East witnessed a revival of religious fundamentalism from the 1970s *(map above)*. The Islamic revival was prompted by popular hostility to Westernization and secularization, but was triggered by the battle-lines over the Israel/Palestine issue and then by the Iranian revolution in 1979. Fundamentalists argue for a return to Islamic law and values and use the Islamic *jihad* (holy war) as an instrument to promote the Islamic cause.

Demonstrators proclaim the Islamic message in the streets of Algiers, 20 April 1990 *(below)*. Islam itself remained divided between its Sunni and Shia branches, a conflict that led in August 1987 to the massacre of 275 Iranian pilgrims in Mecca by Saudis. Shia agitation in Iraq in 1980 contributed to the decision by the Iraqi leader, Saddam Hussein *(left)*, to invade Iran, the centre of Shia Islam. In the 1990s both Sunni and Shia communities focused their efforts on promoting the Islamic reform movement throughout North Africa and the Middle East.

AYATOLLAH KHOMEINI

Ruhollah Khomeini (1900–89) was the inspiration behind the Iranian revolution of 1979 and the first head of the new Islamic Republic. He became an Islamic theologian and taught from 1919 in the Iranian holy city of Qom. He championed the view that the clergy should have the chief voice in politics and became an enemy of the shah. In 1962 he became Grand Ayatollah, one of the six spiritual heads of the Shia Muslims, and the following year launched an active campaign against the shah. He was imprisoned in Teheran and then released following violent popular demonstrations. The following year he was exiled to Turkey, and in 1965 settled in the holy city of Najaf in Iraq. He gathered around him Islamic radicals who approved his arguments for clerical government. In 1978 he issued an edict deposing the shah, and was exiled to Paris following Iranian pressure on the Iraqi government to restrict his activities. In February 1979, he returned in triumph to Iran, from where, after a year of rioting and

violence, the shah had fled. He became head of the new Islamic state and embarked on a wave of political repression. An estimated 10,000 opponents were executed and 40,000 imprisoned, including other senior Islamic theologians accused of plotting to kill Khomeini. In 1983 the traditional Islamic penal code was restored. In 1989 he issued an edict exhorting Muslims to execute the British writer Salman Rushdie for publishing The Satanic Verses, *a book considered blasphemous by devout followers of Islam. The same year, after ten years of clerical despotism, Khomeini died.*

---

ruled by Saddam Hussein and the pan-Arabist Ba'ath movement. Rather than risk an internal Islamic revolution, Hussein invaded Iran on 23 September 1980. Khomeini called the faithful to battle and thousands of poorly armed and trained Islamic militia – *Basij* – swarmed to the call. A long war of attrition, in which neither side made any substantial gains, ended in the summer of 1988 from mutual exhaustion. Hussein saw himself as the defender of the Arab world against Iranian extremism, playing the role Nasser had occupied in the 1950s as leader of the Arab cause.

The war with Iran had, however, bankrupted Iraq. In 1988 there was an $80 billion foreign debt and vast reconstruction costs. Hussein turned to his tiny oil-rich neighbour, Kuwait. An ultimatum was sent, asking Kuwait to give Iraq a gift of $30 billion and an annual $10 billion subsidy, to stop using Iraqi-claimed oilfields and to launch an Arab "Marshall Plan" for Iraq. When Kuwait refused, on 2 August 1990 Iraq invaded and annexed the emirate and its vast wealth. Although the West had supported Iraq in its war against Iran, the invasion of Kuwait posed a serious crisis in a region where the West had large oil interests. On the day of the invasion the UN passed Resolution 660 calling on Iraq to withdraw immediately from Kuwait or face military force.

Saddam Hussein refused to abandon Kuwait, partly because he did not believe that the UN could unite sufficiently to wage all-out war, partly because he could not risk the loss of face and consequent domestic political crisis. On 16 January 1991 a coalition of forces, including a number of Arab states, launched the operation to remove Iraqi forces from Kuwait. After a month of air strikes, the coalition began a ground offensive. Between 24 and 27 February Kuwait City was recaptured with the loss of 150 coalition soldiers. Iraqi losses were estimated at more than 200,000. Iraq was compelled to accept humiliating armistice terms. An estimated $170 billion of damage was done to Iraqi targets destroyed by coalition aircraft. The war split the Arab world, with Egypt, Syria, Saudi Arabia and Morocco sending troops to help the UN Coalition and Jordan, Yemen, Libya, Sudan, Algeria, Tunisia and the PLO giving moral support, but no military help, to Iraq. Saddam Hussein survived the post-war crisis at home and remained in power over an impoverished and isolated state. His attempts to halt Islamic fundamentalism and then to pose as the new pan-Arab leader failed. After Khomeini's death in 1989, Iran's effort to export the revolution subsided, but radical Islam became a significant and violent political force in the 1990s in Algeria, Egypt, Sudan and among the Palestinian diaspora.

THE SEARCH FOR PEACE IN THE MIDDLE EAST

As the Iranian crisis grew in the 1980s, Israel's position in the politics of the Middle East began to alter. The decision by Egypt to recognize Israel in the peace settlement reached between them in 1979, although it initially provoked an Arab boycott of Egypt, created a framework which permitted the gradual easing of tension between Israel and the Arab states around her. Jordan restored relations with Egypt in 1983, while in Amman in November 1987 most of the other Arab states were persuaded to do likewise. Formal recognition of Israel took longer, but in July 1994 King Hussein of Jordan ended the long-running state of war and reached a comprehensive peace settlement that September. Syria, which had led the campaign against Egyptian recognition of Israel throughout the 1980s, finally restored relations with Cairo in December 1989 and agreed in 1995 to talks with Israel.

The most significant breakthrough came with the recognition by the PLO in November 1988 of two states in Palestine, one Jewish, one Arab. In early 1989 exploratory talks began in Tunis on a Palestinian settlement. The reasons for the revolution in the PLO position lay in its declining fortunes. In 1982 Israel occupied southern Lebanon in an effort to end Palestinian attacks across her borders. The military defeat led to the transfer of the PLO to a base in Tunisia, remote from the seat of the conflict. The PLO failure also alienated many younger Palestinians, who began to turn to more radical groups – Hezbollah, Hamas and Islamic Jihad – all of which remained focused on the physical destruction of Israel. The weakness of the PLO also encouraged the Palestinians in Gaza and the West Bank to take the political conflict into their own hands. In December 1987 violent clashes between Arabs and Israeli forces in the Gaza Strip launched the Intifada, a popular and often violent rebellion in the occupied areas against the Israeli occupation. The PLO claim that a Palestinian state and a

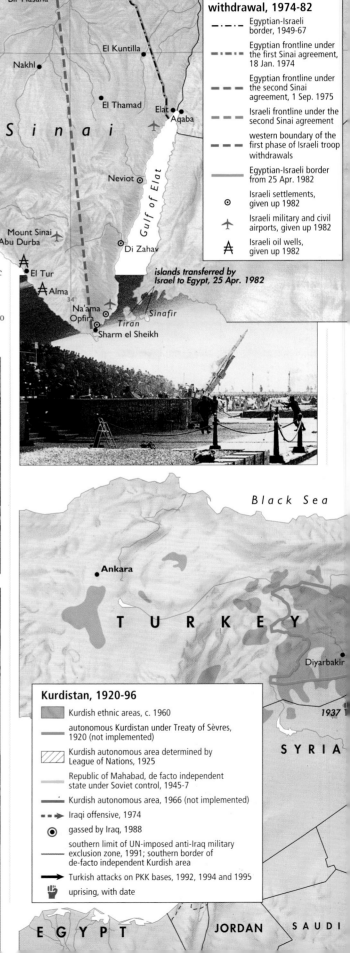

Rafah salient: 11 Israeli settlements established by 1978, with Jewish population of 2,000

Sinai: the Israeli withdrawal, 1974-82

–··–··– Egyptian-Israeli border, 1949-67

–·–·–·– Egyptian frontline under the first Sinai agreement, 18 Jan. 1974

– – – Egyptian frontline under the second Sinai agreement, 1 Sep. 1975

– – – Israeli frontline under the second Sinai agreement

– – western boundary of the first phase of Israeli troop withdrawals

–– Egyptian-Israeli border from 25 Apr. 1982

⊙ Israeli settlements, given up 1982

✈ Israeli military and civil airports, given up 1982

Å Israeli oil wells, given up 1982

islands transferred by Israel to Egypt, 25 Apr. 1982

On 6 October 1981, President Sadat of Egypt was shot down by six Egyptian soldiers from a radical Islamic group while he took the guard of honour *(right)* at celebrations for the eighth anniversary of the Yom Kippur war. The revolutionaries had planned to overthrow the regime and install an Islamic Republic.

Between 1974 and 1982 Israel withdrew her forces from the Sinai Peninsula *(map above right)* in phases agreed with Egypt. The end of tension with Egypt did nothing to alter the conflict between Palestinians and Israelis. In the West Bank city of Nablus in February 1988 Palestinians clash with the Israeli army *(below)*. Jewish settlers from the Israeli occupied areas demonstrate their anger against their own government in Hebron in 1994 *(right)*.

Kurdistan, 1920-96

▨ Kurdish ethnic areas, c. 1960

–– autonomous Kurdistan under Treaty of Sèvres, 1920 (not implemented)

▨ Kurdish autonomous area determined by League of Nations, 1925

–– Republic of Mahabad, de facto independent state under Soviet control, 1945-7

–– Kurdish autonomous area, 1966 (not implemented)

▪▪▶ Iraqi offensive, 1974

◉ gassed by Iraq, 1988

–– southern limit of UN-imposed anti-Iraq military exclusion zone, 1991; southern border of de-facto independent Kurdish area

➤ Turkish attacks on PKK bases, 1992, 1994 and 1995

✊ uprising, with date

Jewish state could live side by side was part of a concerted effort by Yasser Arafat and the PLO leadership not to lose touch with the grass roots of the liberation movement.

Little progress was made between the PLO and Israel, not because there were not circles in Israel willing to tackle the Palestine issue but because the Israeli government of Yitzhak Shamir was opposed to any idea of losing control over the occupied areas. In October 1991 the American president, George Bush, succeeded in setting up a summit at Madrid that brought the parties in the conflict together, including the PLO, which was represented by the Gaza politician Haydar al-Shafi. Progress accelerated when Shamir was replaced by the Labour prime minister, Yitzhak Rabin, in July 1992. In the spring of 1993 the PLO and the Israeli government met secretly to agree a basis for a settlement. In September both sides signed a Declaration of Principles on interim self-government for the Gaza Strip and the West Bank. In May 1994 Israel withdrew from Jericho and the Gaza Strip, and the Palestinian National Authority, with Arafat at its head, took over the running of these areas. Repeated terrorist attacks aimed at disrupting the peace process derailed the second stage of the so-called Oslo Accords, which were not agreed until September 1995, but which led to the Israeli withdrawal from some areas of the West Bank and the extension of limited Palestinian self-rule until 1999.

The agreements met with bitter opposition from several quarters. Hamas, Islamic Jihad and Hezbollah kept up a campaign of terror. Even moderate Palestinians were divided over Arafat's change of heart. Jewish fundamentalists were also profoundly hostile. The Gush Emunim (Bloc of the Faithful), set up after the 1973 war, remained implacably opposed to any loss of territory in the sacred land of Israel. An ultra-nationalist gunman was responsible for the murder of 29 Muslims at prayer in Hebron in February 1994. Rabin himself was the victim of a Jewish extremist in November 1995. The issue of Jewish settlement on the West Bank and the future of Jerusalem were not settled at Oslo, and became major stumbling blocks following the election in May 1996 of a right-wing Likud government under Binyamin Netanyahu.

The stabilization of the Middle East in the mid-1990s pushed violence to the fringes of the Islamic world, or underground among the numerous terror groups working to destroy the peace. In Algeria, in the Sudan and in Turkey, nationalist or religious violence persists. Even at the core, the differences between the religious and ethnic elements involved have produced a fragile peace at best.

The Israeli-Palestinian agreement, 1993-6

under full Palestinian control from May 1994
under full Palestinian control from 1995
under Palestinian administrative control from 1995
■ Jewish settlements in occupied territories
— patrolled by the Israeli military
— patrolled by joint Israeli-Palestinian forces
☆ Israeli police posts
▪ co-ordination offices
— East Jerusalem

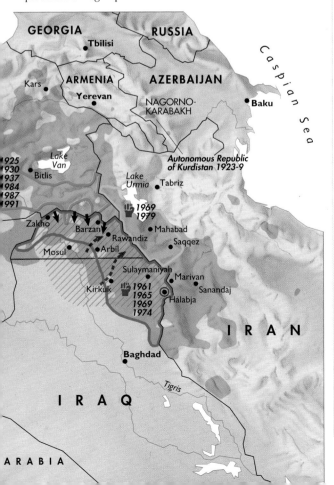

In 1993 the Israeli government and the PLO met in secret to draw up a political agreement. The Oslo Accords of September 1993 formed the foundation of the settlement negotiated in 1994 on the phased withdrawal of Israeli troops and limited Palestinian self-rule (map above). The PLO was installed in the Gaza Strip and Jericho in the summer of 1994 and an interim agreement on the West Bank was made in September 1995, granting the PLO local authority over approximately one third of the area.

The issue of independence for the Kurds goes back to the unredeemed promise of an independent homeland made by the Treaty of Sèvres in 1920. Since 1961 the Kurds, supported by Kurdish minorities in Iran and Turkey, have fought against the Iraqi regime for national independence (map left). After the Gulf War millions became refugees from Saddam Hussein, from whom they were eventually protected by troops of the anti-Hussein coalition.

YITZHAK RABIN
On 13 September 1993 the Israeli prime minister, Yitzhak Rabin, met President Clinton and the PLO leader Yasser Arafat at the White House in Washington to sign an agreement on the Palestinian question (picture above). Rabin was the architect of a compromise to which many Israelis – and Palestinians – remained determinedly opposed. He was born in Jerusalem in 1922, and became a career soldier after fighting in the first Arab-Israeli war in 1948. He was army chief-of-staff from 1964-8, and then ambassador to Washington for five years. On his return he succeeded Golda Meir as prime minister. He returned as Labour prime minister in July 1992, while relations between Israel and the Palestinians were deadlocked. He negotiated the Oslo Accords and forced through the first stages of the agreement in 1994 and 1995. On 4 November 1995 he was assassinated as he left a rally for peace in Tel Aviv: his assassin, a Jewish fundamentalist, claimed to the police to have been obeying the will of God. Rabin's successor, the veteran Shimon Peres, pledged to continue Rabin's work in bringing about peace. Major issues remained unresolved, particularly the future of East Jerusalem, taken over by Israel in 1967, and the fate of numerous Jewish settlers in the West Bank area. When Peres was defeated in the general election of May 1996 by the right-wing Likud leader, Binyamin Netanyahu, prospects for resolving outstanding differences declined sharply.

THE UNITED NATIONS

Throughout the years of crisis in the Middle East the United Nations has played a prominent part, from the Arab-Israeli wars of 1948 through to the problems of Lebanon and Palestine in the 1990s. The UN has not succeeded in preventing wars in the region, but it has succeeded in the Middle East and elsewhere in containing violence and monitoring its aftermath.

The roots of the United Nations Organization lie in the Second World War. Roosevelt's secretary of state, Cordell Hull, who was later awarded the Nobel Peace Prize for his efforts, worked behind the scenes to turn the wartime anti-Axis coalition into a permanent world organization. At Dumbarton Oaks in Washington in August 1944 the major powers drew up a preliminary charter for an organization that Roosevelt insisted should be called the United Nations.

The founding conference of the UN was held in San Francisco in April 1945. It was agreed that the organization should have a general assembly, a smaller security council with permanent Great Power membership, and a permanent secretariat. It differed little in structure from the League of Nations, which co-operated in the development of its successor. After two months the charter was agreed. The American millionaire John D. Rockefeller offered a free site in New York for a UN building, and the permanent headquarters was established there as a sign of American commitment to world peace. Membership was to be open to all, but exceptions soon emerged. Switzerland remained neutral and did not join; Japan and Italy were admitted after US pressure on the USSR in 1955; the two German states did not join until 1975. The two Korean states refused to join, since each claimed the other's territory. Taiwan kept its membership as the Republic of China following the Chinese revolution, but in 1971 communist China took its place and Taiwan was formally expelled.

The UN's primary purpose was to keep the peace. It was almost immediately involved in the Greek civil war (1947) and in 1948 in the war between Israel and its Arab neighbours. The monitoring organization set up – UNTSO – is still there, 50 years later. The greatest test came in 1950 when North Korea invaded the South. The Security Council immediately voted to take military action, but without the presence of the Soviet delegate. The USSR declared the UN intervention illegal, but rather than risk an open breach remained in the UN system. Since Korea the UN has been actively involved in most conflicts and has played some part in resolving them, its success dependent on the goodwill of those involved rather than on military strength.

The key element in the activity of the UN was support from the USA, which contributed disproportionately to the UN budget. American governments worked closely with the UN in the 1950s, but during the secretaryship of U Thant (1961-71) and the Austrian Kurt Waldheim (1971-81) the USA distanced itself from the UN, ignoring its resolutions and tending to act unilaterally. This change in attitude stemmed to some extent from strong UN

Officials prepare the name plates in April 1949 for the United Nations General Assembly meeting *(below)*. The assembly had 55 members, most of whom had signed the original United Nations declaration during the war. Unlike the League of Nations, whose activities the UN took over, the new organization could send a multi-national peace-keeping force to impose its decisions. Danish soldiers *(bottom)* patrol the Sinai Desert in 1957 following the Suez crisis.

An airport funeral ceremony for a Bangladeshi soldier killed in the fighting in former Yugoslavia, December 1994 *(above right)*. By the end of 1994 129 UN personnel had been killed in the conflict and there were over 38,000 UN troops committed. The United Nations Protection Force (UNPROFOR) was sent first to Croatia in February 1992, and then to Bosnia in March 1992, and Macedonia in November 1992. Sanctions were imposed on Serbia/Montenegro for violation of UN Security Council resolutions.

criticism of American involvement in Vietnam, coupled with the shifting balance in the General Assembly towards the developing world and its problems. But it also stemmed from the declining success rate of UN intervention. In the mid-1980s Congress moved to cut the US budget contribution, creating a serious crisis for the new secretary-general, the Peruvian Javier Pérez de Cuellar. In 1988 President Reagan finally gave a new endorsement to the UN, following its successes in Afghanistan, in terminating the Iran-Iraq war, and in winning independence for Namibia. In the 1990s under Secretary-General Boutros Boutros Ghali, the UN has sought to become a major political force, as during the Gulf conflict, or less successfully, in the former Yugoslavia.

The UN, like the League of Nations, has also played a key part in economic, social and cultural questions. Under the aegis of its Economic and Social Council, the UN has helped in providing development aid, educational aid and cultural collaboration, and has taken initiatives on human rights, refugee problems and issues of drugs, health and the environment. It is pre-eminently in these areas of global concern, on which eight per cent of its budget is spent, that the UN has succeeded in becoming an indispensable inhabitant of the global village.

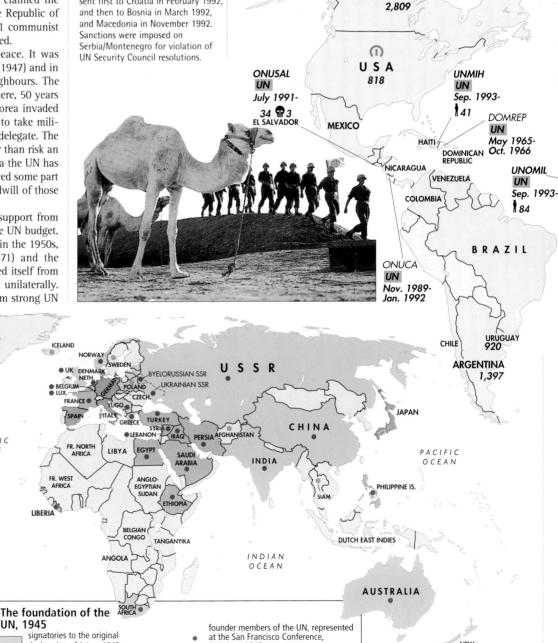

The United Nations Educational, Scientific and Cultural Organization (UNESCO) was set up in Paris in November 1945. It has played a growing part in encouraging educational development worldwide *(chart far right)* together with the World Bank. The UN organization itself was set up at San Francisco in April 1945 *(map right)* when 50 states sent a total of 282 delegates.

The foundation of the UN, 1945

- signatories to the original declaration of 1 Jan. 1942
- states signing the declaration between June 1942 and Mar. 1945
- founder members of the UN, represented at the San Francisco Conference, Apr.-June 1945
- states deliberately excluded from joining
- additional members, 1946-7

UN member states' contributions arrears (at 30 April 1996)

country	accumulated arrears (US$ millions)
US	1,591.3
Russian Federation	401.4
Ukraine	245.1
Japan	137.5
Germany	66.1
Belarus	62.9
Iran	25.2
Brazil	25.1
Venezuela	16.8
Yugoslavia	15.8
Poland	15.7
Kazakhstan	13.5
Georgia	12.0
France	11.7
Italy	10.1

The secretary general of the UN, Dag Hammarskjöld, in Stockholm shortly after his election in January 1953 *(below)*. He was instrumental in organizing UN intervention in the Suez crisis and in the Congo civil war. It was while on a visit to the 20,000 UN troops in the Congo that he was killed in an air crash. He was succeeded by the Burmese diplomat, U Thant.

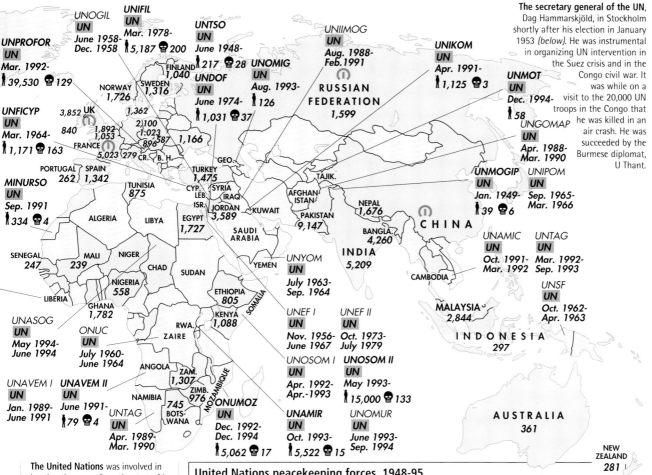

Map labels (peacekeeping operations):

UNPROFOR UN Mar. 1992- 39,530 ☠129
UNFICYP UN Mar. 1964- 1,171 ☠163
MINURSO UN Sep. 1991 334 ☠4
UNASOG UN May 1994-June 1994
UNAVEM I UN Jan. 1989-June 1991
UNAVEM II UN June 1991-
UNTAG UN Apr. 1989-Mar. 1990
UNOGIL UN June 1958-Dec. 1958
UNIFIL UN Mar. 1978- 5,187 ☠200
UNTSO UN June 1948- 217 ☠28
UNDOF UN June 1974- 1,031 ☠37
UNOMIG UN Aug. 1993- 126
UNIIMOG UN Aug. 1988-Feb.1991
UNIKOM UN Apr. 1991- 1,125 ☠3
UNMOT UN Dec. 1994- 58
UNGOMAP UN Apr. 1988-Mar. 1990
UNMOGIP UN Jan. 1949- 39 ☠6
UNIPOM UN Sep. 1965-Mar. 1966
UNYOM UN July 1963-Sep. 1964
UNEF I UN Nov. 1956-June 1967
UNEF II UN Oct. 1973-July 1979
UNOSOM I UN Apr. 1992-Apr.-1993
UNOSOM II UN May 1993- 15,000 ☠133
UNOMUR UN June 1993-Sep. 1994
UNAMIC UN Oct. 1991-Mar. 1992
UNTAG UN Mar. 1992-Sep. 1993
UNSF UN Oct. 1962-Apr. 1963
ONUC UN July 1960-June 1964
ONUMOZ UN Dec. 1992-Dec. 1994 5,062 ☠17
UNAMIR UN Oct. 1993- 5,522 ☠15

Countries providing troops (where over 200):
FINLAND 1,040; NORWAY 1,726; SWEDEN 1,316; 1,362; UK 840; 1,892; 1,053; 2,100; 1,023; 896; 587; 1,166; 1,171; FRANCE 5,023; 279; CR. B. H.; PORTUGAL 262; SPAIN 1,342; TUNISIA 875; TURKEY 1,475; CYP.; LEB.; ISR.; SYRIA; IRAQ; JORDAN 3,589; EGYPT 1,727; SAUDI ARABIA; PAKISTAN 9,147; NEPAL 1,676; BANGLA. 4,260; INDIA 5,209; RUSSIAN FEDERATION 1,599; MALAYSIA 2,844; INDONESIA 297; AUSTRALIA 361; NEW ZEALAND 281; SENEGAL 247; MALI 239; NIGERIA 558; GHANA 1,782; KENYA 1,088; ETHIOPIA 805; ZAM. 1,307; ZIMB. 976; BOTSWANA 745; 3,852; 3,852; GEO.; TAJIK.; AFGHANISTAN; CHINA

United Nations peacekeeping forces, 1948-95

forces still operational, Dec. 1994

UN UN peacekeeping force with dates of operation	**UNOSOM II** operational force	countries providing UN peacekeeping troops, late 1994 with number of troops provided (where over 200)
	👤 size of force	
	☠ fatalities suffered to 1994	Ⓤ permanent member of the Security Council

The United Nations was involved in keeping the peace from the start of its formal life. In January 1946 at the first Security Council meeting Iran asked the UN to compel the USSR to remove its forces stationed there during the war, which it successfully did. The UN has since been involved in keeping the peace across the world *(map above)*. Its efforts have been mixed, since its powers to deploy force have been used sparingly. Though having stronger powers on paper, it has depended, like the League of Nations before it, on the goodwill of the major players in the international system to be effective. Peacekeeping has been expensive. The UN action in Yugoslavia in 1994 cost $1.6 billion, but by the end of the year almost half the promised contributions were still outstanding. By 1996 member states' arrears in payment, both to the peacekeeping and regular budgets had reached crisis proportions. A total of $2.8 billion in dues was unpaid, over half of that by the United States *(chart top)*.

UN and World Bank expenditure on educational development, 1970-90 (US$ million)

Organization	1970	1975	1980	1985	1990
UNDP	4	29	31	16	18
UNPF	–	1	3	4	8
UNICEF	10	25	34	33	57
UNWFP	–	77	103	121	157
UNESCO	22	57	78	88	73
UN Total	36	189	249	262	313
World Bank	80	224	440	928	1,487

UNDP United Nations Development Programme   **UNPF** United Nations Population Fund
UNICEF United Nations Children's Fund   **UNWFP** United Nations World Food Programme
UNESCO United Nations Educational, Scientific & Cultural Organization

THE WORLD TOWARDS THE MILLENIUM

The increasing role of the United Nations in world politics reflected the emergence of issues that were genuinely global in extent, to which individual nation states could only offer partial solutions. The UN highlighted these issues in a series of multi-national conferences from the 1970s: the World Population Conference in 1974; the World Food Conference the same year; a World Conference on the Status of Women in 1975, followed up with further congresses in 1980, 1985 and in 1995 in Peking. Disarmament, drugs and racism were tackled in the 1980s. In 1992 Rio de Janeiro hosted an "Earth Summit", when delegates from 185 nations assembled to agree a programme to tackle global pollution.

A central strand linking all these themes was the issue of development. In the 1950s programmes of social and economic modernization were widely regarded as the simple answer to the problem of progress in the non-industrialized world. By the 1990s development strategies had had mixed results, and led in some cases to a decline in economic performance and severe social dislocation. In 1980 the Organization of African Unity held an economic summit at Lagos, where a plan of action was adopted to create a common market in Africa by the year 2000, and to achieve African self-sufficiency in food, clothing and energy by 1990. The plan was immediately compromised by economic decline in much of sub-Saharan Africa, where the per capita annual income in the 27 "Least Developed Countries" fell over the 1980s from $220 to $200. By 1989 the developing world had accumulated $1,146 billion of debt with the developed states. Even in the rich industrial regions the 1980s saw the onset of severe poverty for socially disadvantaged groups.

In the developed world economic power was wielded increasingly by multi-national enterprises. The UN has estimated that there are now 37,000 multi-nationals with

The success of the European Economic Community from 1957 in stimulating trade between its members encouraged the development of regional economic groups (map right). The US, which took the lead after 1945 in liberalizing world trade, finally established its own trade bloc in North America in 1993. The developing world has also received help through the UN Conference on Trade and Development (UNCTAD) established in 1964, which meets every four years for a summit on trade issues.

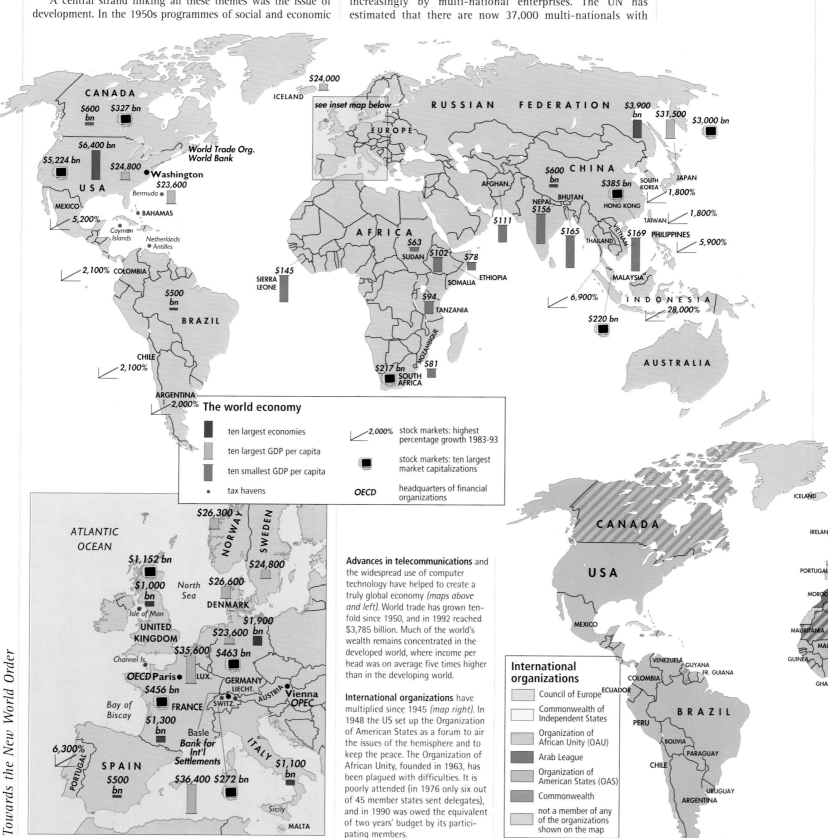

The world economy

- ten largest economies
- ten largest GDP per capita
- ten smallest GDP per capita
- tax havens
- 2,000% stock markets: highest percentage growth 1983-93
- stock markets: ten largest market capitalizations
- OECD headquarters of financial organizations

Advances in telecommunications and the widespread use of computer technology have helped to create a truly global economy (maps above and left). World trade has grown ten-fold since 1950, and in 1992 reached $3,785 billion. Much of the world's wealth remains concentrated in the developed world, where income per head was on average five times higher than in the developing world.

International organizations have multiplied since 1945 (map right). In 1948 the US set up the Organization of American States as a forum to air the issues of the hemisphere and to keep the peace. The Organization of African Unity, founded in 1963, has been plagued with difficulties. It is poorly attended (in 1976 only six out of 45 member states sent delegates), and in 1990 was owed the equivalent of two years' budget by its participating members.

International organizations

- Council of Europe
- Commonwealth of Independent States
- Organization of African Unity (OAU)
- Arab League
- Organization of American States (OAS)
- Commonwealth
- not a member of any of the organizations shown on the map

see inset map below

Economic groups

North American Free Trade Area (NAFTA)

Common Market of the Southern Cone (Mercosur)

Caribbean Community (CARICOM)

Central American Common Market (CACM)

Economic Community of West African States (ECOWAS)

Organization for Economic Co-operation and Development (OECD)

Colombo Plan

Organization for Petroleum Exporting Countries (OPEC)

Economic Community of Central African States (CEEAC)

South African Development Co-ordination Conference (SADCC)

Association of South East Asian Nations (ASEAN)

not a member of any of the organizations shown on the map

European Union (EU)

European Economic Area (EEA)

European Free Trade Association (EFTA)

The floor of the London International Futures Exchange *(left)*. Despite the relative decline of the British economy since the war, London has remained one of the world's leading financial centres, and Britain remains one of the inner group of G7 states, which meet regularly to discuss general issues facing the world economy. International collaboration through summit conference has become the hallmark of late 20th century politics. Here East meets West *(above right)* as British prime minister, John Major, shakes the hand of China's premier, Li Peng, at preliminary talks before the start of a Europe-Asia Summit in Bangkok, March 1996.

200,000 foreign affiliates worldwide: they account for over 73 million jobs. The 100 largest businesses control one third of all foreign investment: 38 are in the European Union, 29 in the US and 16 in Japan. They have an overriding influence on the pattern of the global economy. China now has 50,000 foreign affiliates which play a central part in the regime's commitment to quadruple China's GNP by the year 2000.

The rapid spread of multi-national enterprise since the 1970s has been the result of a number of factors. The liberalization of trade made the global market a reality; large trade and oil surpluses built up in the 1970s provided ready capital; above all the globalization of finance and communications removed physical disadvantages to trans-national operation. By 2000 "low earth orbit" satellites will provide voice and data telecommunication anywhere in the world. Modern technology, including the Internet and e-mail systems, have transformed the possibilities of global production and exchange.

The trend towards "globalization" in the last quarter century has challenged the position of the nation state, which has remained the basic unit of political organization throughout the century. Many issues can only be tackled now by multi-national agencies: sovereignty has been compromised through economic integration; the internationalization of the productive economy has taken power away from national governments. These developments have helped erode the legitimacy and power of the modern nation state. Crime and corruption have also become globalized: terrorism has its own international networks; while in much of the world there are sizeable minorities, religious, ethnic, social, who do not acknowledge allegiance to the state whose territory they inhabit. The rise in violent warlordism indicates a weakening of the state after a century of growing state responsibility. Whether the forces of economic integration and global collaboration can cope with the forces of fragmentation remains perhaps the most serious issue of the new millennium.

The World Towards the Millenium

THE 20TH CENTURY has witnessed a revolution in the lives of ordinary
people. Improvements in health, in land and air communications, in
telecommunications and in education have produced a larger, more literate and
more informed population. In 1900 most people lived on the land. By the year
2000 a majority of people will live in cities, with all the problems of amenity,
overcrowding and pollution that cities generate. The costs of social and
industrial transformation have to be weighed against gains in wealth and
opportunity, which have been unevenly spread between the developed and the
developing areas of the world. The United States in 1995 boasted one million
millionaires. In Africa, South Asia and Latin America millions still live at a
bare level of subsistence, their traditional ways of life disrupted or torn up by
relentless modernization. Rapid change, and the management of that change,
are the hallmarks of 20th-century life.

Astronaut John Young of *Apollo 16* salutes
the American flag on the lunar surface

THE POPULATION EXPLOSION

THE 20TH CENTURY is the century of population increase. At its start the global population was 1.6 billion. Now it is close to six billion. This increase is rooted partly in growing agricultural efficiency (seen, for example, in communist China and in post-independence India) and partly in the opening up of new areas of cultivation (such as the American West). Such developments have combined with changes in medical provision to increase life expectancy and, particularly, to lower infant mortality. Life expectancy increased first in the prosperous countries of North America and western Europe, then in eastern and southern Europe, and finally in the less developed countries. In the 1930s, a Frenchman could expect to live almost 60 years while an Indian had a life expectancy of less than 24 years. Population growth of this kind placed increasing pressure on the land and encouraged peasants to go to the cities in search of work. Such migrations then further stimulated economic growth and set the cycle of population increase to work again.

However, demographic growth has not been universal or uniform. Population increase is usually linked to particular social circumstances. Societies in which people expect to be dependent on their children in old age encourage large families. So do rapidly industrializing countries, in which there is a strong demand for labour (especially when that demand is for child labour). Land inheritance systems may play a large part in influencing choices about the size of families. In France, legislation ensured that land was divided equally among all heirs. Peasants had to limit the size of families to avoid their land being divided into plots too small to support them. Partly as a result, the population of France hovered at just under 40 million throughout the period from 1870 to 1940. By contrast labourers on the great estates of southern Italy, Spain or Hungary had no land and consequently no incentive to limit population. Peasants in Germany practised primogeniture and could afford to have large families without dividing the family land. Most striking of all, peasants in Russia had a positive incentive to have large numbers of children because communal land was distributed among families according to the number of children that they had to work it: the Russian population increased by over 50 per cent between 1880 and 1910.

Control over the size of families was exercised by various means. In Brittany one third of men married women older than themselves (and consequently nearer to the end of their child-bearing lives). In many countries unmarried women kept house for their brothers or were packed off into convents. Deliberate birth control for the sexually active also became easier in the 20th century with the use of condoms and, from the 1960s, the contraceptive pill. A variety of agencies interested themselves in the control of population. The Catholic Church opposed birth control and succeeded in having its views enshrined in legislation in countries such as Ireland and Italy. States that were preoccupied with the military uses of a large population sometimes sought to prevent birth control. France became particularly obsessed with

The chart (below) shows the growth of world population since 8000 BC. The most rapid population growth has been concentrated in the period since 1750, when improved agricultural productivity and medical knowledge began to make themselves felt. Population growth has been particularly rapid in the latter part of the 20th century, with dramatic growth in the last 50 years concentrated in the Third World. Improvements in healthcare and decreases in infant mortality in developing countries (chart bottom right) mean that the children of large families (below) are much more likely to survive and have families of their own.

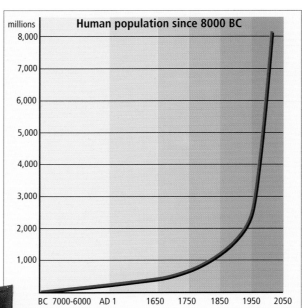

longest doubling time: Austria (6,930 years)

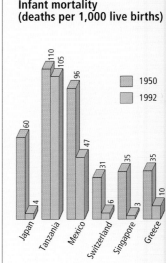

**Infant mortality
(deaths per 1,000 live births)**

	1950	1992
Japan	60	4
Tanzania	110	105
Mexico	96	47
Switzerland	31	6
Singapore	35	3
Greece	35	10

जिन्दगी में रंग दो बच्चों के संग

TO ENJOY LIFE ONLY ONE OR TWO CHILDREN

This poster (left) encourages citizens of the Andaman Islands to practise birth control. The poster was part of a wider campaign by the Indian government to reduce the birth rate (a campaign that was mirrored in many other parts of the Third World). Such campaigns met with mixed results: the population of the Andaman Islands increased by almost 50 per cent in the 1980s.

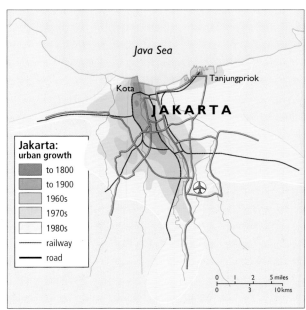

World population doubling time

shortest doubling time: Kenya (17 years)

under 20 years		45-99 years	
20-24 years		100-999 years	
25-29 years		over 1,000 years	
30-44 years		population static or decreasing	

The map *(below)* shows the rate of growth of Jakarta in Indonesia. As in many parts of the Third World the original city has become surrounded by shanty towns that house people who have migrated to the city in search of work. Such migrants settle on the outskirts of towns where there is space and some possiblity of access to work or trade. However, areas populated in this way rarely possess the amenities that make urban life tolerable.

Even in a comparatively prosperous country like the United Kingdom, degrees of population density vary. The wealthy often seek housing in rural or suburban areas of low population density. Other areas, such as the Highlands of Scotland, have lost population due to absence of economic opportunity or migration. These processes leave high population density in working class quarters of inner cities *(above)*, where people are housed in high-rise flats.

The map *(above)* shows the length of time that it will take for population to double in various parts of the world. Population is increasing fastest in the least wealthy countries. The German population is actually expected to drop while Kenya's population will double in just 17 years. These figures can mislead. Projections that the population of Rwanda would double in 18 years have been undermined by a genocidal civil war.

Urbanization is taking place at very different paces *(map left)*. Developed countries, which experienced rapid urbanization in the first half of the century, are now comparatively stable. Urban populations have doubled between 1976 and 1991 in many parts of South America, Asia and Africa. By the end of the century more than 50 per cent of the world's population will live in cities.

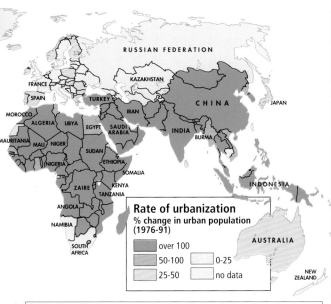

Rate of urbanization
% change in urban population (1976-91)

over 100			
50-100		0-25	
25-50		no data	

Jakarta: urban growth

	to 1800
	to 1900
	1960s
	1970s
	1980s
	railway
	road

The charts *(left)* show age profiles of the populations of the United Kingdom, Japan and Malaysia. The last of these countries shows a classic Third World pattern with a population that is largely made up of those under 20 years of age. The populations of the United Kingdom and Japan, by contrast, contain a much larger proportion of old and middle aged people. The number of women among the very old tends to exceed the number of men. A youthful country like Malaysia is expected, over time, to develop an age profile similar to that of more developed countries. The ageing of a population can bring with it severe problems of financial support and healthcare for the elderly.

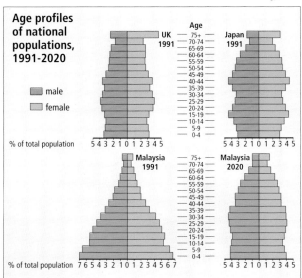

Age profiles of national populations, 1991-2020

- male
- female

UK 1991

Japan 1991

Malaysia 1991

Malaysia 2020

% of total population

the issue after the casualties and low birth rate of the First World War: in 1920 all forms of birth control were made illegal in France, and this legislation was not revoked until the late 1960s. Stalin's Russia and Mussolini's Italy were driven by similar preoccupations when they launched their "battles for births".

Sometimes bids were made to raise birth rates among certain parts of the population while lowering them among others. Eugenicists from the early part of the century onwards believed that the strength of the "race" was being undermined by the fact that the poorest, and therefore "least fit", were having the most children. The government of Nazi Germany institutionalized such thinking, encouraging high birth rates among the "racially fit" and abortion and sterilization among those that it regarded as least desirable. After 1945, as concerns focused more on economics, states began to seek low birth rates. The Chinese government tried to dissuade its citizens from having more than a single child. Efforts to influence population size are now linked to the problems of securing adequate economic and social development, which became a principal theme at the 1995 UN Cairo summit on population.

The Population Explosion

177

RICH AND POOR

Extremes of wealth and poverty have attracted increasing concern across the course of the 20th century. In 1900 nations and individuals still measured their worth in non-monetary terms. Countries were concerned with power, particularly with the power to conduct war. Such power might be linked to economic success but it was not a direct reflection of such success – Britain's hegemony in trade and finance was of less obvious use than Germany's growing industrial strength. By the end of the century nations measured their success in almost entirely economic terms. A military super-power like Russia was obliged to humble itself before the might of the International Monetary Fund while a state with almost no military ambitions – Japan – was widely seen as successful.

Similar changes occurred in the way people thought about wealth. At the beginning of the century, status, rank and caste (concepts that were often linked to the military power described above) were more important than wealth. European aristocrats regarded breeding and the possession of land as the key sources of their prestige. If they squandered their inherited wealth at gaming tables they could rely on protection through entail laws (which prevented land from being sold) or through marriage to wealthy heiresses: 28 daughters of American millionaires married into the French aristocracy between 1870 and 1914. By the end of the century this had changed. America had ceased to export heiresses and started to export, via films and television, the values of a society without aristocracy.

These changes accompanied a growing enthusiasm for the

In the developed Western countries, poverty was felt most severely by those who had lost contact with ordinary social structures, and particuarly by the homeless. The homeless (left) were usually unemployed and often unable to gain access to state benefits. They were also often the victims of family break-up and the closing down of residential mental hospitals.

GDP per capita, 1992

over $15,000	$500-1,999
$5,000-15,000	$0-499
$2,000-4,999	no data

-5.3 annual average growth rate, 1980-92

CANADA

UNITED STATES 1.7

NICARAGUA -5.3

PERU -2.8

BRAZIL 0.4

ARGENTINA -0.9

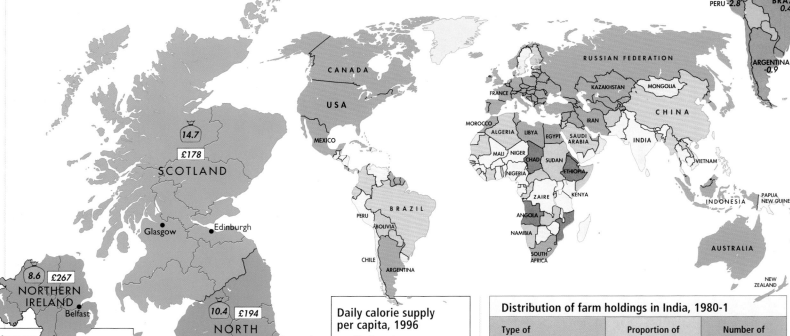

Income distribution in the United Kingdom

% of households with gross weekly income of £65 or less, by region

6% or less	
6-8%	
8-10%	
over 10%	

% of households with gross weekly income of £550 or more

£176 average annual expenditure per head on income support

SCOTLAND
14.7
£178
Glasgow • Edinburgh

NORTHERN IRELAND
8.6 £267
Belfast

NORTH
10.4 £194
Newcastle

NORTH WEST
£198
Liverpool
Manchester
15.0

YORKSHIRE HUMBERSIDE
£160
Hull
Sheffield

WALES
11.5
£176
Cardiff

WEST MIDLANDS
£177
Birmingham

EAST MIDLANDS
£139

EAST ANGLIA
£100
Norwich

SOUTH WEST
£128
Bristol

SOUTH EAST
25.0 £146
London
• Southampton

Daily calorie supply per capita, 1996

as a % of needs (1989 or latest figure)

above 120	80-89
110-120	below 80
100-109	no data
90-99	

CANADA
USA
MEXICO
PERU
BOLIVIA
BRAZIL
CHILE
ARGENTINA

RUSSIAN FEDERATION
FRANCE
KAZAKHSTAN
MONGOLIA
MOROCCO
ALGERIA LIBYA EGYPT SAUDI ARABIA
IRAN
CHINA
MALI NIGER CHAD SUDAN
INDIA
NIGERIA
ETHIOPIA
VIETNAM
ZAIRE KENYA
ANGOLA
INDONESIA
PAPUA NEW GUINEA
NAMIBIA
SOUTH AFRICA
AUSTRALIA
NEW ZEALAND

Distribution of farm holdings in India, 1980-1

Type of holding	Proportion of area operated	Number of holdings
marginal: below 1ha	12.2%	50.5 million
small: 1-2ha	14.1%	16.1 million
semi-medium: 2-4ha	21.2%	12.5 milion
medium: 4-10ha	29.7%	8.1 million
large: 10ha and over	22.8%	2.1 million

The map (left) shows the distribution of income inequality in Britain. Incomes are highest in the south east, a centre for company headquarters and for lucrative financial jobs. They are lowest in areas that have been damaged by the decline of traditional industries. It is worth noting that the distribution of income within each area varies sharply: income inequalities in the south east are often much sharper than those in the less prosperous parts of the country.

The map (above) shows daily calorie intake per person as a percentage of total needs in various parts of the world. Such intake is highest in the most developed regions – though it is notable that in areas such as Portugal that are comparatively underdeveloped in terms of per capita production people may still eat more food than necessary. Indeed the health conscious San Francisco yuppie will generally consume fewer calories than a Siberian steel worker.

The map *(below)* shows the distribution of Gross Domestic Product (GDP) per head across the world. The average inhabitant of north Europe, North America and Japan can expect a yearly income of over $15,000, while the average inhabitant of sub-Saharan Africa will have an income of less than $500. Such statistics do not take into account dramatic differences in the cost of living between, say, Japan and Peru, nor variations in the strength of a country's currency against the US dollar. Nor do they allow for differences in income distribution within individual countries.

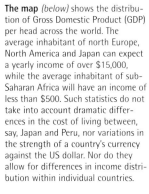

measurement of wealth and poverty. The debate about reparations after the First World War encouraged many nations to calculate their wealth with a new zeal. Discussion of the international distribution of wealth was changed by the decolonizations of the 1960s. Newly created states in Africa were usually very poor, and membership of institutions such as the United Nations gave them some capacity to get their plight discussed, if not remedied. Interest was also focused on gulfs between rich and poor within nations. Mobilization of resources during the two world wars obliged the wealthy states to take an interest in the diet and accommodation of their poorest citizens if only to ensure that they had effective soldiers. The

The map *(above)* attempts to illustrate global variations in the quality of life. It does so by charting levels of literacy, life expectancy and infant mortality. The translation of such figures into a "quality of life index" is a highly subjective process. It is arguable that quality of life is dependent on matters such as personal freedom and family structure as much as on more obviously measurable variables. It may be that the hard work and social discipline that allows states such as Singapore and Japan to produce healthy and well-educated populations would be seen as compromising "quality of life" in other parts of the world.

As income inequalities in countries like the UK and the US widened during the 1980s, some of the rich began to indulge in the kind of conspicuous consumption that had not been seen since before the First World War. Such consumption revolved around holidays, fashion and entertaining *(below)*.

Quality of life
physical quality of life index, 1992

90 high	calculated from figures of life expectancy, infant mortality and literacy
70 moderate	
30 low	
poor	

Poverty took various forms around the world. In Third World countries the poor often gathered in shanty towns, such as the one near São Paulo in Brazil *(left)*, on the outskirts of the cities in which they hoped to gain work. If such shanty towns provided Western photographers with spectacular examples of poverty, many of their inhabitants saw them as providing opportunities for advancement unavailable in the countryside.

The table *(left)* shows the changing levels of inequality in various countries. The pursuit of free market economics often increases income inequalities. The greatest increase in inequality was registered in countries such as New Zealand and Great Britain that implemented radical free market economics which ran against their previously entrenched welfare state traditions.

interest in wealth and poverty that has marked recent history has not produced any consensus about how such conditions are defined. Infant mortality and calorie intake may provide some kind of indication of living standards in the Third World. However, infant mortality may also be high in an area such as the South Bronx in New York, which is wealthy by Third World standards. Whole areas of human activity which take place outside the market economy may escape quantification: it was estimated in the early 1950s that over 25 per cent of agricultural production in France was consumed by the farmers themselves.

Most significantly assessment of wealth and poverty is almost always a matter of relativity. Outside sub-Saharan Africa few have experienced an absolute decline in their fortunes. Rapid economic growth after the Second World War masked issues of relative deprivation because almost everyone derived benefits from increasing prosperity. Slower economic growth since the oil crisis of 1973 has meant that awareness of, and conflicts over, distribution of wealth have become more intense. Debate has raged over whether large and prosperous trading blocs, such as the European Union and the United States, should open their frontiers to imports (and thus sacrifice manufacturing jobs at home) or close their frontiers (and thus impede economic growth in the Third World). Conflicts in domestic politics have been marked by a resurgence of free market economics and a widening gap between the richest and poorest parts of the population. The globalization of the economy has created empires of wealth which stretch across frontiers. The rich in the US, Japan and Germany have much in common; the poor of Ethiopia and of Europe almost nothing.

Rich and Poor

MIGRATION

At the beginning of the 20th century, millions of Europeans were on the move. Poles from the Russian and Austro-Hungarian empires went to work on the East Prussian great estates or sought industrial employment in the Ruhr. Over two million Russians went east to Siberia. Italians went to Switzerland and France. Millions of Europeans headed for Australia, Argentina and, most importantly, the United States. Migrants usually came from comparatively poor agricultural areas of Europe (especially southern Italy and the Russian empire). They sought an escape from poverty that had been exacerbated by population growth and the declining price of grain (itself the product of transatlantic trade). Their passage was made easier by the new technologies of travel, especially cheap steam ships.

The reception that awaited immigrants varied. Germany treated such people as "labour imports": they had few rights and their stay was clearly limited. France, whose low birth rate meant that she had a chronic need for labour, was more tolerant and indeed absorbed immigration so successfully that France in the 1980s showed few signs that one in three of its population was descended from immigrants. America was the country where immigrants aroused most interest and concern. Officials on New York's Ellis Island sought to test intelligence, literacy and health. An initial anxiety about the potentially damaging impact of immigrants from Poland and southern Italy gradually gave way to the concept of the "melting pot" through which the United States would be enriched by waves of new culture. Immigrants themselves often continued to live in relatively closed communities. Many of them did not speak the language of their adopted country and most of them had moved from agricultural work in the countryside to industrial work in a city.

Emigration had an important effect on Europe. In Calabria, the departure of young men left a population in which there were three women for every two men. Underemployment that had plagued certain regions was alleviated and in some areas – such as eastern Germany – new immigrants were imported to replace workers who had left. Many immigrants intended to return to their native lands (about half of the Italians who went to the United States returned) and used money saved abroad to buy land. Sometimes the links between the Old World and the New produced bizarre political projects: in the 1930s the Polish government discussed the possibility of establishing a colony in South America and in 1945 some Sicilians seriously proposed that their island be made part of the United States. Emigration often acted as an alternative to revolution. The young dynamic men who left were those who might normally have been most politically radical; the old and female population left behind was usually seen as conservative. Furthermore, remittances from abroad and the purchase of land sometimes created a more stable social structure. In the civil wars in Italy, Finland and Germany after the First World War, areas that had provided many emigrants usually fought on the Right.

After the First World War, migration (as opposed to flight from political persecution) became less common. The United States imposed sharp limits on entrants and work was harder to find almost everywhere as the impact of economic depression began to be felt after 1929. In 1930, the number of people returning to Germany was greater than the number who left and in the middle part of that decade France encouraged some immigrant workers to return to their country of origin.

The next great wave of migration came with the economic growth of the 30 years after the Second World War. Workers from the southern periphery of Europe (Portugal, Greece,

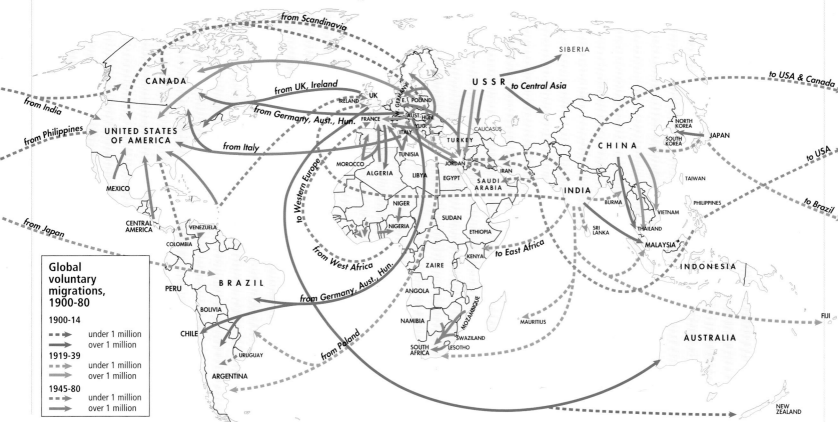

Global voluntary migrations, 1900-80

1900-14
- ---▶ under 1 million
- ──▶ over 1 million

1919-39
- ---▶ under 1 million
- ──▶ over 1 million

1945-80
- ---▶ under 1 million
- ──▶ over 1 million

southern Italy) sought work in the factories of Turin, St Etienne and Stuttgart. France and Great Britain took in substantial numbers from the colonies (or former colonies) of Africa, Asia and the Caribbean. Less well documented migrations took place in communist countries. Substantial numbers of people, such as ethnic German inhabitants of the Soviet Union, were moved for political reasons, while citizens of the most impoverished communist countries, such as Vietnam, were allowed to work in the richer ones, such as the German Democratic Republic. After the economic downturn that began with the oil crisis of the 1970s migration into prosperous countries became harder. Furthermore, the lives of individual immigrants became more difficult as they faced increasing problems in finding work, as well as the growing racism of the societies in which they lived.

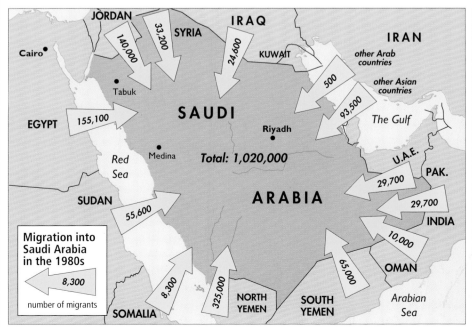

Migration to Britain from the Caribbean, 1955-74

number of migrants (net)

70,000
60,000
50,000
40,000
30,000
20,000
10,000
0
-10,000

1955   1960   1965   1970   1974

UNITED STATES OF AMERICA

Ft Worth • Dallas   • Birmingham

Jacksonville

Houston   New Orleans

San Antonio

Rio Grande

Gulf of Mexico

Tampa

Miami   THE BAHAMAS

Nassau

MEXICO

Monterrey

Tampico

Mexico City   Veracruz

Puebla

BELIZE

Belmopan

GUATEMALA   HONDURAS

Guatemala City   Tegucigalpa

San Salvador

EL SALVADOR   NICARAGUA

Managua

COSTA RICA

San José

Havana   CUBA

Guantanamo   HAITI   DOMINICAN REP.

Kingston   Port-au-Prince   Santo Domingo

JAMAICA

San Juan
PUERTO RICO *(USA)*

ANTIGUA

DOMINICA

ST LUCIA

BARBADOS

Caribbean Sea

Barranquilla

Caracas   Port of Spain   TRINIDAD & TOBAGO

PANAMA

Panama City

COLOMBIA   VENEZUELA

Orinoco   GUYANA

ATLANTIC OCEAN

to USA, to present

to United Kingdom, 1948-70

to Cuba, Dominican Republic 1880s to1930s to work on sugar plantations

to Central America, 1850s to 1930s to work on infra-structure & fruit plantations

to Netherlands Antilles, 1920s to 1940s to work in oil refinery industry

to Trinidad and Guyana, from1838. Post-emancipation movement away from smaller islands to work on new, larger plantations

Principal migrations from the Caribbean

→ 19th-century post-emancipation movement
→ post-war migration to UK and USA
→ migration to Netherlands Antilles
→ migration to Cuba and Dominican Republic
→ migration to Central America

The black inhabitants of the Caribbean had almost all been brought there as part of a process of involuntary migration (in slave ships). During the 20th century increasing numbers of people from the Caribbean migrated to work elsewhere in the Americas *(map and chart above)*, to sugar plantations in Central America, or to Britain. Some migrants headed for the United States. However, the greatest of all Caribbean migrations was the one to Great Britain between 1948 and 1970: the number of Caribbean-born inhabitants of Britain exceeded 300,000 in 1971. Legislation to restrict immigration coupled with declining job opportunities reduced Caribbean migration to Great Britain, and in the 1980s there was a net return of 27,000 Caribbean migrants to their countries of origin.

Migration into Saudi Arabia *(map below)* was the product of the job opportunities created in Saudi Arabia by the growth of the oil industry. Migrants came mainly from the poor countries of North Africa and the Near East. Such migrants were often Muslims, and hence able to fit into a strictly Islamic culture. They were also generally poor and thus attracted by relatively high wages and the prospect that such earnings might allow them to buy property in their country of origin.

Many inhabitants of the Caribbean were skilled agricultural workers such as cane cutters *(left)*, and they were able to employ these skills in Cuba and Central America. After 1945, however, emigrants were increasingly likely to head for industrialized countries, especially Great Britain, where they rarely found work in agriculture.

American authorities became worried by the influx of immigrants to their country during the early 20th century. For this reason they established increasingly draconian controls over who was allowed to enter. Would-be immigrants disembarking at Ellis Island, such as the ones shown here *(below)*, were subjected to tests to prove that they were healthy, literate and not of subnormal intelligence. Those who failed were deported.

JORDAN   IRAQ

SYRIA

Cairo •

EGYPT   155,100

Tabuk •

Red Sea

Medina •

SUDAN

SOMALIA

SAUDI

Riyadh •

ARABIA

NORTH YEMEN   SOUTH YEMEN

Total: 1,020,000

KUWAIT

IRAN

other Arab countries

other Asian countries

The Gulf

U.A.E.

PAK.

INDIA

OMAN

Arabian Sea

140,000
33,200
24,600
500
93,500
29,700
29,700
10,000
65,000
325,000
8,300
55,600

Migration into Saudi Arabia in the 1980s

→ 8,300
number of migrants

BAGGAGE EXAMINED HERE

REFUGEES

The distinction between refugees fleeing political persecution and migrants seeking to improve their standard of living is hard to make. The poem by Emma Lazarus that accompanied the Statue of Liberty spoke of "huddled masses yearning to breath free", but most of those who entered the United States in the period before 1914 seem to have been primarily concerned with bettering themselves economically. The most obvious refugees during this period were East European Jews who fled from anti-semitism in the tsarist empire and Romania. However, even they sometimes returned to their native land after having saved money working abroad. The Balkan wars and their aftermath marked the beginning of a large-scale refugee problem in Europe. In 1922 and 1923 177,000 Muslim refugees fled into Turkey; at the same stage over a million Greek refugees poured into Greece from western and northern Turkey. The First World War and the Russian Revolution created further refugees, especially in Eastern Europe: by 1921 there were some 800,000 refugees from the Soviet Union alone. In this year the Norwegian explorer Fridtjof Nansen was made League of Nations High Commissioner for Refugees from Russia. The awareness of a specific "refugee problem" was exacerbated by two factors. First, economic conditions no longer permitted the absorption of large numbers of foreign workers that had still seemed possible in many countries before 1914. Second, the increasing emphasis on official identification of nationality through passports and identity cards accentuated the distinction between refugees and citizens.

The political and racial persecutions in Nazi Germany after 1933 created further waves of refugees and European governments responded to this with growing panic. By the late 1930s even the traditionally tolerant French government was beginning to incarcerate political refugees in specially created camps. Refugees came to make up important elements in the anti-Nazi resistance during the Second World War: defeated Spanish Republicans played an important part in the French Resistance.

The Second World War and its aftermath created the high point of the refugee problem in Europe. Millions of Europeans had fled their homes or were liberated from prison camps at a time when their families, their communities or perhaps their entire countries had ceased to exist. However, in the long run post-war Europe did not suffer a refugee crisis comparable to that which had afflicted it before the war. Rapid growth in the Western European economies allowed refugees to be absorbed into employment. Indeed, the millions of ethnic Germans from Eastern Europe who fled to West Germany after 1945 and the million or so European "pieds noirs"

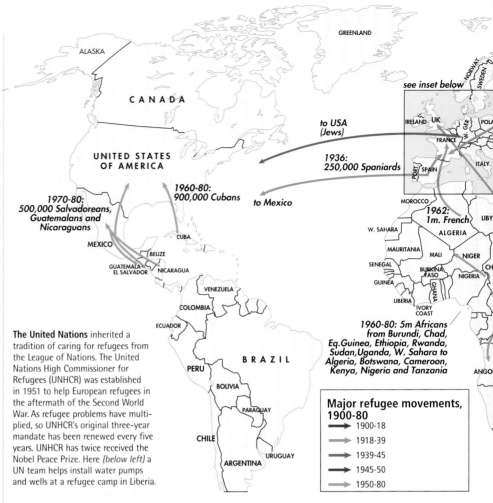

The United Nations inherited a tradition of caring for refugees from the League of Nations. The United Nations High Commissioner for Refugees (UNHCR) was established in 1951 to help European refugees in the aftermath of the Second World War. As refugee problems have multiplied, so UNHCR's original three-year mandate has been renewed every five years. UNHCR has twice received the Nobel Peace Prize. Here (below left) a UN team helps install water pumps and wells at a refugee camp in Liberia.

Major refugee movements, 1900-80
→ 1900-18
→ 1918-39
→ 1939-45
→ 1945-50
→ 1950-80

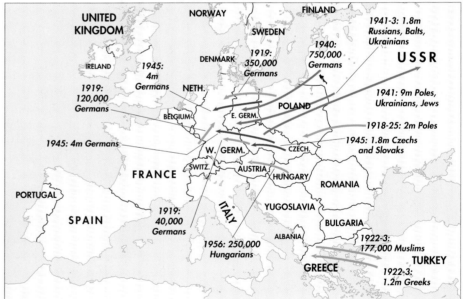

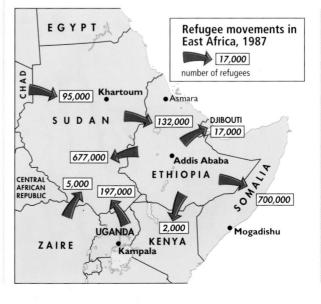

Refugee movements in East Africa, 1987

→ 17,000 number of refugees

The 20th century has seen refugee movements on an unparalleled scale (maps above). The overwhelming majority have been caused by wars. There were massive movements of population after the First and Second World Wars. Continuing unrest since 1945 has sparked further huge movements, the single most striking example being the 15 million Muslims and Hindus who fled their homes following the partition of British India in 1947. Refugees create particularly severe problems in sub-Saharan Africa (map left). Here the countries into which they flee are often too poor to support an influx of population.

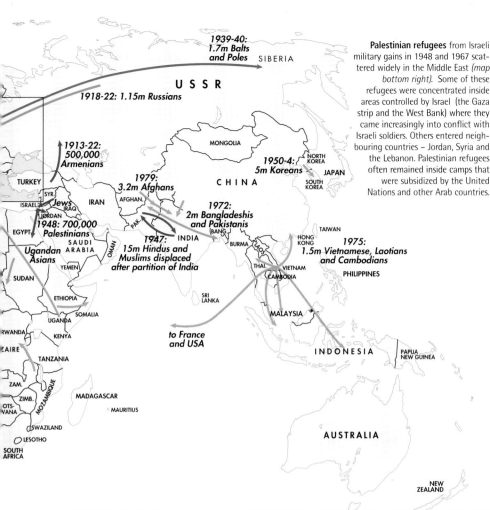

Palestinian refugees from Israeli military gains in 1948 and 1967 scattered widely in the Middle East *(map bottom right)*. Some of these refugees were concentrated inside areas controlled by Israel (the Gaza strip and the West Bank) where they came increasingly into conflict with Israeli soldiers. Others entered neighbouring countries – Jordan, Syria and the Lebanon. Palestinian refugees often remained inside camps that were subsidized by the United Nations and other Arab countries.

Map labels:

1939-40: 1.7m Balts and Poles SIBERIA
1918-22: 1.15m Russians
USSR
1913-22: 500,000 Armenians
1979: 3.2m Afghans
1950-4: 5m Koreans
1972: 2m Bangladeshis and Pakistanis
1947: 15m Hindus and Muslims displaced after partition of India
1948: 700,000 Palestinians
Jews
Ugandan Asians
1975: 1.5m Vietnamese, Laotians and Cambodians
to France and USA

who fled Algeria after it was granted independence in 1962 almost certainly benefited their host economies: they alleviated labour shortages while their desire to rebuild shattered prosperity often made them entrepreneurial and dynamic. West European governments were further helped by the closing of the frontiers between Eastern and Western Europe, which limited the numbers of refugees who were able to seek freedom in the West.

By the 1960s the refugee problem was seen primarily as an extra-European problem. Improved transport and communication forced the wealthy citizens of Europe and America to take notice of problems that occurred in Africa and Asia. The end of colonial control allowed racial and tribal differences, often exacerbated or created by colonialism, to spill over into violent conflict (this drove Asians out of Uganda and Kenya and two million inhabitants out of Rwanda). Finally, the super-powers were increasingly prone to direct their conflicts into proxy wars in the Third World. Some entire nations after 1945 were largely made up of refugees. This was true of Israel (which was founded by European Jews and which continued to take refugees from countries as diverse as Ethiopia and the Soviet Union), Taiwan (a refuge for Chinese nationalist forces), and Pakistan (to where numerous Muslims had fled from India at partition). Increasingly refugees were likely to find themselves trapped in squalid camps or even imprisoned (like the "boat people" who escaped from Vietnam). Refugees may well prove to be the greatest casualties of the end of the Cold War. A world without ideological division leaves no legitimate reason for flight. Unwilling to return, unwanted by their hosts, refugees have come to inhabit a political no-man's-land.

Refugee movements, 1994

country of origin					
Rwanda		**Former Yugoslavia** (other than Bosnia)		**Bosnia-Herzegovina**	
country of refuge					
Zaire	1,000,000	Germany	35,000	Germany	275,000
Tanzania	550,000	Italy	30,000	Croatia	180,000
Burundi	160,000	Hungary	10,000	Austria	55,000
Uganda	5,000	France	8,500	Sweden	50,000
		Czech Republic	50,000	Slovenia	29,000
		Spain	2,500	Turkey	20,000
		Macedonia	1,900	Denmark	18,500
				Switzerland	11,000

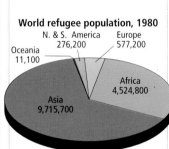

World refugee population, 1980

Oceania 11,100
N. & S. America 276,200
Europe 577,200
Africa 4,524,800
Asia 9,715,700

The chart *(far left)* shows the pattern of refugee settlement in 1994. This has been marked by two disasters. In Africa refugees fled from genocide in Rwanda, while in Europe refugees fled the war in former Yugoslavia. The scale of the first of these crises was illustrated by the fact that the poverty-stricken state of Zaire received a million new inhabitants *(picture below left)*. The crises in both Yugoslavia and Rwanda exposed limitations on the willingness of developed nations to take in refugees and this in a period that saw much of Europe impose increasingly draconian regulations on the movement of such people.

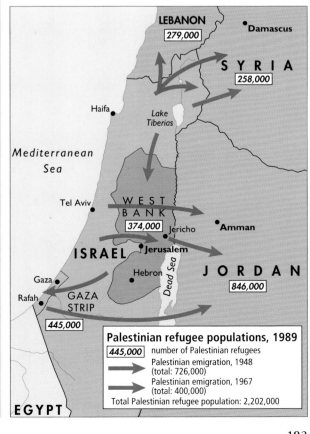

LEBANON 279,000
Damascus
SYRIA 258,000
Haifa
Lake Tiberias
Mediterranean Sea
Tel Aviv
WEST BANK 374,000
Jericho
Amman
ISRAEL
Jerusalem
Hebron
Dead Sea
JORDAN 846,000
Gaza
Rafah
GAZA STRIP
445,000
EGYPT

Palestinian refugee populations, 1989

445,000 number of Palestinian refugees
→ Palestinian emigration, 1948 (total: 726,000)
→ Palestinian emigration, 1967 (total: 400,000)
Total Palestinian refugee population: 2,202,000

Refugees

183

DISEASE AND HEALTH

Health is an issue closely bound up with economic development. Improvements in health and healthcare over the century have depended on improvements in the general level of prosperity and on the breakthroughs in medical science made possible by expensive programmes of advanced research. Throughout the century the level of general health and the prospects of survival have accordingly been higher in the developed regions of the world.

At the beginning of the century there was little international co-operation on health issues, many of which had not yet been properly identified or understood. In 1907 an International Office of Public Hygiene was established in Paris to discuss and advise on public health questions, and in 1919 the League of Nations set up a permanent Health Organization in Geneva. When the UN conference convened in San Francisco in 1945, Brazil proposed the creation of an autonomous international health body, which on 7 April 1948 – now celebrated as World Health Day – became the World Health Organization (WHO). Its function was to monitor world health trends, advise on health care provision and co-ordinate national efforts to promote health and eradicate disease. The main fruit of its early work was the publication in 1969 of International Health Regulations, which member states were supposed to observe.

Achievements outside the developed world were modest by the 1970s, and in 1973 the role of the WHO was strengthened to allow it to act as a full partner in establishing effective healthcare in deprived regions. The organization took the lead in challenging major epidemic diseases. Its most conspicuous success was the eradication of smallpox between 1967 and 1977, when the last recorded case occurred in Somalia.

This success prompted a more grandiose ambition. In 1977 the WHO launched the "Health for All by the Year 2000" programme which aimed to raise the level of primary healthcare globally. The aim of the campaign was not only to eradicate disease but to tackle the basic causes of poor health and hygiene through education, environmental improvements and development economics. The WHO also pledged greater help for the identification and treatment of mental illness, which was estimated to affect more than 50 million people worldwide. In the 1980s this programme was pursued through a comprehensive Programme on Immunization, first launched in 1974 and aimed primarily at the established killers: tuberculosis, measles and polio. The inoculation rate in the developing world quadrupled in ten years. There were remarkable results. In India and Indonesia the rate of the measles/TB inoculation was 0.1 per cent in 1980-2. In 1987-90 it had risen to 86 per cent. The exception to this improvement was in the Soviet

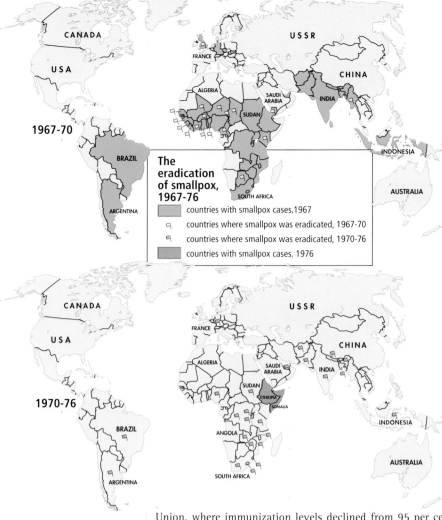

The eradication of smallpox, 1967-76

- countries with smallpox cases, 1967
- countries where smallpox was eradicated, 1967-70
- countries where smallpox was eradicated, 1970-76
- countries with smallpox cases, 1976

In 1967 a world programme was launched to eradicate smallpox, which was endemic in much of Africa and southern Asia. Within a decade the disease was virtually wiped out *(maps above)*. Samples were kept in laboratories in case the disease returned, but they are to be destroyed in 1999.

Union, where immunization levels declined from 95 per cent to 68 per cent over the same period. Simultaneously the 1980s was declared the "International Drinking Water Supply and Sanitation Decade". Within ten years 1.59 billion people in the developing world were provided with safe water – a coverage of 68 per cent compared with 29 per cent in 1980. In 1988 the WHO embarked on a further fight against six major infections, including polio, leprosy and tetanus, which affected over 30 million people. The aim was to eradicate them by the year 2000.

The overall impact of improved healthcare has been to raise life expectancy levels sharply in 40 years. In the developing areas average life expectancy was almost 60 years in 1990, as against 41 years in 1948. In China and East Asia the

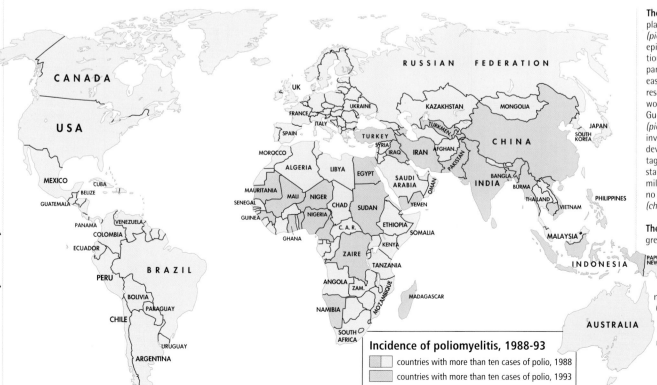

Incidence of poliomyelitis, 1988-93

- countries with more than ten cases of polio, 1988
- countries with more than ten cases of polio, 1993

The aftermath of an outbreak of plague in New Delhi, September 1994 *(picture above right below)*. The epidemic was just one in which infectious diseases revived during the late part of the century. Such events can easily overwhelm primary health care resources. Nonetheless, local care workers, such as this doctor in the Gudular Hills visiting a tribal area *(picture above right top)*, play an invaluable role in disease control. The developing world has the disadvantage of limited numbers of medical staff. In 1990, out of a total of 6.2 million doctors worldwide, Africa had no more than 61,000, for example *(chart above right)*.

The incidence of polio has been greatly reduced by the drive to provide effective immunization in the 1980s *(map left)*. In Bangladesh in 1980 only 0.7 per cent were immunized, but in 1990 the figure was 62 per cent. The general vaccination rate in the developing world was 20 percent in 1982, but reached 84 per cent a decade later.

Access to safe drinking water, 1990

percentage of population with access to safe supply

	below 30%
	30 to 50%
	50 to 70%
	70 to 90%
·	over 90%
	no data

Numbers of health personnel worldwide, 1990 (thousands)				
Regions	Physicians	Dentists	Nurses/Midwives	Pharmacists
Africa	61	7	310	14
Americas	1,139	237	2,140	224
Eastern Mediterranean	175	23	249	23
Europe	2,566	365	4,507	412
South East Asia	442	16	562	225
Western Pacific	1,800	82	1,446	152
All regions	6,183	730	9,214	1,050

figure has risen from 42 to 70. Marked differences in health opportunities between the developed and developing world still remain. The costs of healthcare have risen steeply, and even within developed states there are differences in the levels of provision. Out of the 17 million healthcare personnel worldwide in 1990, 11.5 million were employed in Europe and North America. Health expenditure in developing states in 1988 totalled four per cent of GNP; in the developed economies the figure was 12.6 per cent. In these circumstances "Health for All" by the millennium, in spite of remarkable gains in controlling deadly and debilitating infections, will still fall short of its ambition in what has otherwise been a remarkable century of medical progress.

The World Health Organization has made the provision of safe water to drink one of its top priorities. The proportion in North America and Europe is 99-100 per cent, but in much of Africa, Latin America and Asia waterborne infection is still widespread *(map above)*.

The rate of malaria has declined significantly since 1945 but it is still endemic in large parts of the world *(map left)*. Some 800 million suffer its effects each year. The fight against infectious diseases was revolutionized by the discovery of penicillin, the first antibiotic drug, by Sir Alexander Fleming *(above)*.

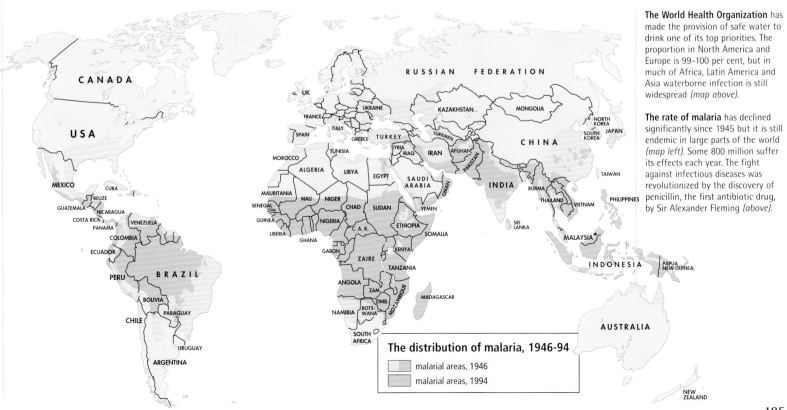

The distribution of malaria, 1946-94

	malarial areas, 1946
	malarial areas, 1994

Disease and Health

185

THE NEW EPIDEMICS

The healthcare revolution which the WHO has led since the 1970s has been involved not only with eradicating long-established diseases but with fighting a crop of new epidemics, some the product of mutations in the stock of viruses and bacteria, some, more dangerously, the consequence of growing immunity to the spectrum of antibiotics used to contain infection.

New microbial strains have been responsible for the re-emergence of cholera and diphtheria. The cholera outbreak in southern India in 1992 spread northwards into most of China and South East Asia. Diphtheria developed in the former Soviet bloc, where immunization programmes and effective disease screening declined with the break-up of the communist state systems. The medicines which had transformed the fight against epidemic diseases after 1945 were faced by a germ pool with rapidly developing resistance in the 1980s and 1990s. The rate of cure for illnesses such as tuberculosis declined; malaria revived, despite extensive public health efforts, because the mosquitoes that carried the disease became resistant to the standard pesticides. The common bacteria that cause intestinal, respiratory or wound infections – streptococci, pneumonococci, enterococci – in some cases became almost entirely immune to antibiotic treatment, and have stimulated an urgent search for an entirely new generation of medicines.

New diseases with exceptionally high death rates and no known cure appeared alongside the resistant strains of bacteria

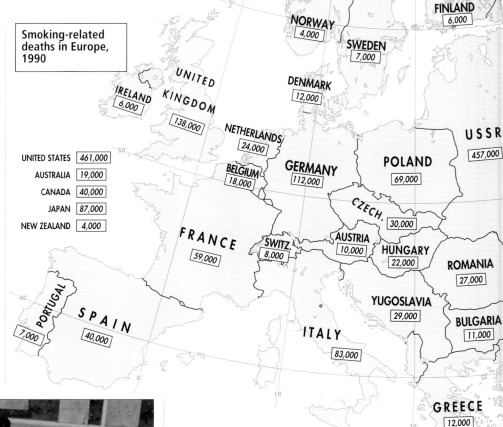

Smoking-related deaths in Europe, 1990

Country	Deaths
UNITED STATES	461,000
AUSTRALIA	19,000
CANADA	40,000
JAPAN	87,000
NEW ZEALAND	4,000

FINLAND 6,000
NORWAY 4,000
SWEDEN 7,000
DENMARK 12,000
IRELAND 6,000
UNITED KINGDOM 138,000
NETHERLANDS 24,000
USSR 457,000
BELGIUM 18,000
GERMANY 112,000
POLAND 69,000
CZECH. 30,000
FRANCE 59,000
SWITZ. 8,000
AUSTRIA 10,000
HUNGARY 22,000
ROMANIA 27,000
PORTUGAL 7,000
SPAIN 40,000
ITALY 83,000
YUGOSLAVIA 29,000
BULGARIA 11,000
GREECE 12,000

Since 1981 AIDS has grown to become a worldwide epidemic. In 1994 an estimated 13 million people were infected with the HIV organism which causes AIDS. It has spread with remarkable speed. Almost unknown in southern Asia in 1987, over 2.5 million people are now infected. AIDS prompted a global campaign of health education, and its growth in Europe and the US (maps far right) has slowed. The children in the Jinja district of Uganda (left) have a lesson in health education. Almost two thirds of HIV infected cases are in sub-Saharan Africa.

Legionnaires' disease
United States, 1976

Cryptosporidiosis
United States, 1976

AIDS
United States, 1981

E. coli O157:H7
United States, 1982

Hepatitis C
United States, 1989

Venezuelan haemorrhagic fever
Venezuela, 1991

New infectious diseases identified since 1976

AIDS United States, 1981	newly identified disease, with country and date first identified

Worldwide cases of turberculosis, 1900-2000

Discovery of first TB drug (1945)

WHO declares global emergency (1993)

millions

5
4
3
2
1
0
1900   1950   2000

Incidence of tuberculosis, 1993

number of cases per 100,000 people

- more than 100
- 25 to 100
- fewer than 25

Though tuberculosis was one of a number of diseases which proved responsive to antibiotic treatment after 1945, it has revived again and now kills three million a year, particularly AIDS victims in Africa whose resistance to infection is seriously impaired (map left). Tuberculosis has also returned to Europe and North America, where it was virtually eliminated in the 1960s.

RABIES CONTROL

The importation of unlicensed animals into Great Britain is prohibited

Severe penalties, including destruction of the animal, may be imposed if an animal is landed without an import licence having first been obtained.

A poster *(above)* in the British port of Ramsgate warns against the spread of rabies. From the 1940s onwards rabies spread west through Europe but was kept out of Britain by strict quarantine law. The incidence in humans is low.

Smoking causes an estimated three million deaths a year, half of them in the developed world *(map left and chart right)*. It is estimated that this figure will rise to ten million in the year 2020. Since 1987 20 European countries have adopted anti-smoking legislation, and in France it is banned in public places.

in the 1980s. The Ebola virus, which first appeared in Zaire in 1977, returned to southern Zaire in 1995, but the rapid response of the local authorities and the WHO restricted the outbreak to just 316 cases, of whom 245 died. Ebola was one of a number of new viruses which cause internal haemorrhaging in humans. Like bovine spongiform encephalopathy (BSE), identified in 1986, which has been linked to the incurable brain disorder Creutzfeldt-Jakob disease in humans, neither the origin nor the behaviour of the disease organisms is well understood.

Of the new viruses by far the most deadly and widely spread was the human immunodeficiency virus (HIV), which reduces the human body's resistance to infection and can lead to the

Lung cancer trends in the US, 1930-90

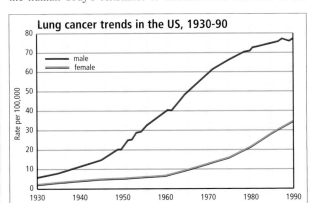

fatal condition of acquired immunodeficiency syndrome (AIDS). HIV was first identified in 1981 on the basis of isolated cases in the 1970s. The virus spread rapidly. In 1990 an estimated five million were infected; in 1991, nine million; by 1996, 24 million. Most of them were in sub-Saharan Africa. The epidemic spread across the United States and Europe in the 1980s, and provoked extensive research and health education programmes which have had the effect of reducing the rate of growth of the disease. In the developed world HIV was closely linked to lifestyle. In the US in 1988, 89 per cent of those infected with AIDS came from the male homosexual or drug-using communities. In the epidemic regions of the developing world, the disease was more socially diverse and, in areas with high population growth, could be passed on to very large numbers of children.

AIDS was not the only disease in the developed world whose spread was closely related to social behaviour. Low levels of death from infectious diseases highlighted other major causes of premature death, such as smoking, alcohol consumption and poor diet. In Russia life expectancy for males actually fell from 65 in 1986 to 59 in 1993, due in large part to a sharp increase in alcohol consumption and a doubling of the rate of homicide. By the year 2020 smoking is expected to kill ten million people annually. The high cost of treatment for such diseases has led to legislation and propaganda in the developed world to encourage healthier lifestyles. Nonetheless, the modern drug-resistant organism is no respecter of prosperity. Epidemic disease has been kept at bay since the 1940s but has by no means been eliminated.

Bovine spongiform encephalopathy
United Kingdom, 1986 animal cases only

Salmonella enteritidis PT4
United Kindom, 1988

Hepatitis D (Delta)
Italy, 1980

Human T-cell lymphotropic virus I
Japan, 1980

Hantaan virus
South Korea, 1977

Brazilian haemorrhagic fever
Brazil, 1994

Vibrio cholerae 0139
India 1992

Ebola haemorrhagic fever
Zaire, 1976

Human and equine morbillivirus
Australia, 1994

Since 1973 30 infectious diseases have been identified, many of which have no known cure, and are difficult to control or prevent *(map above)*. Bovine spongiform encephalopathy (BSE), although a disease of cattle, may have infected humans as a related brain disease, CJD. Efforts to contain it are focused on destruction of infected cattle *(right)*.

Deaths worldwide in 1993 totalled 52 million, 17 million of them caused by major infectious diseases *(chart below)*. There are more than 11 million deaths of children under five, of which nine million are caused by disease, one quarter of which could be checked by effective immunization.

Principal causes of death worldwide

Diseases of the circulatory system 19% 9,676,000

Cancer 12% 6,013,000

External causes 8% 3,996,000

Perinatal and neonatal causes 6% 3,180,000

Chronic lower respiratory diseases 5% 2,888,000

Maternal causes 1% 508,000

Other causes 0.3% 170,000

16% 8,124,000

32% 16,445,000

Other

Acute respiratory infections

Infant diarrhoea

Tuberculosis

Hepatitis B

Measles

Malaria

Whooping Cough

Unknown causes

Infectious and parasitic diseases

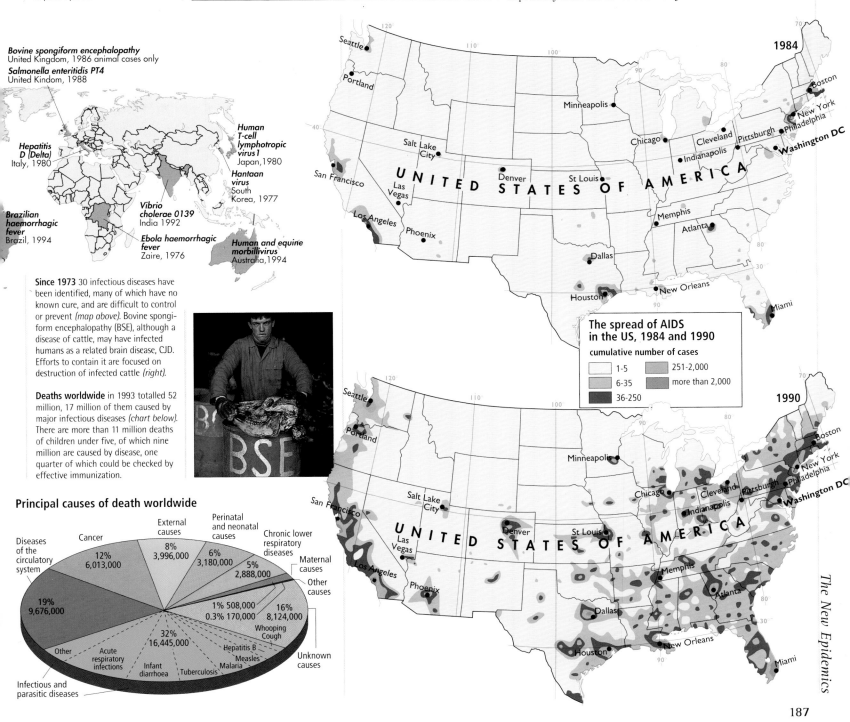

The spread of AIDS in the US, 1984 and 1990

cumulative number of cases

1-5	251-2,000
6-35	more than 2,000
36-250	

1984

1990

TELECOMMUNICATIONS

Communication, like healthcare, has been a central component in the transformation of daily life since 1900. Aviation, motorization and telecommunications all owe their modern development to a cluster of discoveries and inventions made in the late 19th and early 20th centuries during a short period of remarkable scientific endeavour. Telecommunications owed its origins to the pioneering work of two men: Alexander Graham Bell, the inventor of the telephone in 1876; and the Italian inventor Guglielmo Marconi, who brought the first primitive radio transmitter with him to Britain in 1896.

The telephone made rapid strides in the United States, where Bell's discovery was welcomed by a population spread thinly across a vast continent: there were nine million telephones in 1910, more than 50 million in the 1950s. The system was run by a private monopoly, regulated by the federal authorities. In Europe it was the state that took control, to safeguard the systems from any threat to national security. In 1925, at a meeting in Paris, an international regime was established for linking the telephone systems of the European states in order to make possible a continental communications net. Not until 1956 was Europe linked to the US by underwater telephone cable.

The development of radio and, shortly afterwards, television – a word first coined by *Scientific American* in 1907 – owed much to the rapid diffusion of Marconi's technology. The first radios transmitted only Morse code, but on Christmas Day 1906 the American scientist Reginald Fessenden transmitted readings from St Luke's Gospel to startled radio operators along the

Atlantic sea-lanes. The following year Lee de Forest invented the vacuum tube, which made possible the development of modern electronic communications. Almost immediately work began on creating image as well as voice transmission. Radio broadcasting was formally established in the US and Britain shortly after the Great War. By 1928 television images had been developed by the American General Electric Company on a screen no larger than a postcard. In 1936 television broadcasting was officially launched in Britain and in April 1939 Roosevelt inaugurated an American tradition when he became the first in a long line of television politicians.

The Second World War both inhibited and stimulated telecommunications. War-related research produced a remarkable acceleration of the technical threshold, but the development of television and worldwide diffusion of the telephone was postponed for almost a decade. Once restrictions were lifted the growth of the industries was phenomenal. There were 70 million televisions in the United States by 1965, 195 million by 1986, by which time American lifestyles were dominated by the technology. On average television was played for eight hours a day in every American home.

The second wave of development in telecommunications depended on a further set of inventions after 1945. In 1948 the Bell Laboratories in the US developed the transistor, which made possible smaller and more efficient equipment. The same year the first storage computer was invented by scientists at Manchester University in Britain. When silicon was discovered in 1957 to be an effective form of storing and sending electronic information, and the first space satellite, the Soviet Sputnik I, was launched the same year, the scientific basis was laid for an integrated global system of telecommunications. The first purpose-built communications satellite was launched in 1962. At almost exactly the same time the silicon microchip was perfected, which made possible the development of infinitely larger and more sophisticated systems of communication and data-holding. In the 1990s plans were well advanced for a series of up to 800 Low Earth-Orbiting satellites (LEO), which would make possible a single worldwide net for voice telephony, video transmission and multimedia communication. In Japan the "Telecity", an urban utopia based around telecommunications, is on the drawing board.

Advances in telecommunications have been rapid and irreversible. They have contributed to making industry and services more efficient; they have revolutionized the conduct of public affairs – Hitler and Stalin both preferred the telephoned order to the written directive – and they have altered in fundamental ways both the rhythm of life and patterns of social behaviour in societies where the telephone, television and computer are no longer luxuries. The sheer pace of modern life is unthinkable without the electronic web that keeps it in motion.

Average television viewing time in the US, 1987

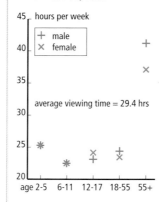

hours per week

+ male
× female

45
40
35
average viewing time = 29.4 hrs
30
25
20

age 2-5   6-11   12-17   18-55   55+

A poster of the Vietnamese communist leader Ho Chi Minh overlooks a telephone box in Hanoi *(bottom)*. With over 500 inhabitants per telephone, Vietnam was one of the least developed states in telecommunications in the 1980s. Telephone technology has been transformed over the century. By the 1990s the small town telephone exchange *(below)* of 1908 had given way to sophisticated automatic exchanges capable of handling thousands of calls a minute.

The telephone and television have transformed communication over the century, though they have been heavily concentrated in the more industrially developed areas of the world *(map right)*. In 1992 there were 153 televisions for every 1,000 people worldwide, but in the developed states the figure was 490, and in the US 797. The developed states had 380 telephones per 1,000, but in sub-Saharan Africa the figure was only ten. Colour television was generally introduced in the late 1960s *(far right above)*. A remarkable technical breakthrough came with the fibre-optic cable developed in the 1970s, which can transmit a much larger number of telephone messages and can carry cable television and computer communications *(far right)*.

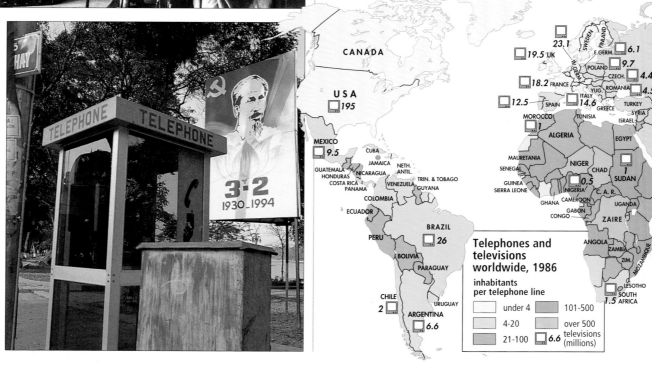

CANADA

USA
195

MEXICO
9.5

CUBA
JAMAICA
HONDURAS
GUATEMALA
NICARAGUA
COSTA RICA
PANAMA
NETH. ANTIL.
TRIN. & TOBAGO
VENEZUELA
GUYANA
COLOMBIA
ECUADOR

BRAZIL
26
PERU
BOLIVIA
PARAGUAY

CHILE
2
URUGUAY
ARGENTINA
6.6

23.1
SWEDEN FINLAND
19.5 UK
E.GERM.   6.1
POLAND   9.7
18.2 FRANCE   CZECH.   4.4
12.5   SPAIN   YUG.   ROMANIA   4.5
ITALY 14.6   GREECE   TURKEY   SYRIA
MOROCCO   TUNISIA   ISRAEL
1   ALGERIA   EGYPT
MAURETANIA
SENEGAL   NIGER   CHAD
GUINEA   NIGERIA   1 SUDAN
SIERRA LEONE   0.5
GHANA   CAMEROON   C.A.R.   UGANDA
CONGO   GABON   ZAIRE

ANGOLA
ZAMBIA
ZIM.
MOZAMBIQUE
LESOTHO
SOUTH AFRICA
1.5

Telephones and televisions worldwide, 1986

inhabitants per telephone line

under 4	101-500
4-20	over 500
21-100	6.6 televisions (millions)

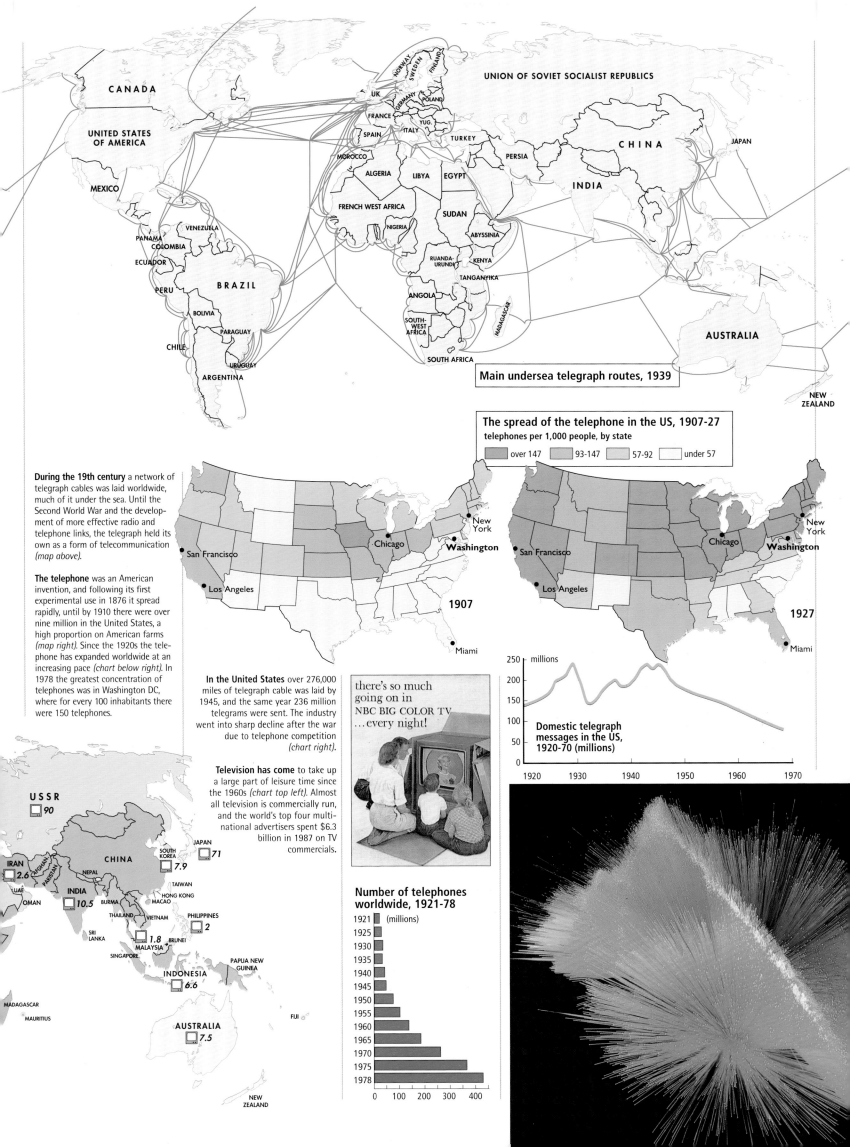

Main undersea telegraph routes, 1939

The spread of the telephone in the US, 1907-27
telephones per 1,000 people, by state

| over 147 | 93-147 | 57-92 | under 57 |

1907

1927

During the 19th century a network of telegraph cables was laid worldwide, much of it under the sea. Until the Second World War and the development of more effective radio and telephone links, the telegraph held its own as a form of telecommunication *(map above)*.

The telephone was an American invention, and following its first experimental use in 1876 it spread rapidly, until by 1910 there were over nine million in the United States, a high proportion on American farms *(map right)*. Since the 1920s the telephone has expanded worldwide at an increasing pace *(chart below right)*. In 1978 the greatest concentration of telephones was in Washington DC, where for every 100 inhabitants there were 150 telephones.

In the United States over 276,000 miles of telegraph cable was laid by 1945, and the same year 236 million telegrams were sent. The industry went into sharp decline after the war due to telephone competition *(chart right)*.

there's so much going on in NBC BIG COLOR TV ...every night!

Television has come to take up a large part of leisure time since the 1960s *(chart top left)*. Almost all television is commercially run, and the world's top four multi-national advertisers spent $6.3 billion in 1987 on TV commercials.

Domestic telegraph messages in the US, 1920-70 (millions)

USSR □ 90
IRAN □ 2.6
INDIA □ 10.5
CHINA
SOUTH KOREA □ 7.9
JAPAN □ 71
PHILIPPINES □ 2
MALAYSIA □ 1.8
INDONESIA □ 6.6
AUSTRALIA □ 7.5

Number of telephones worldwide, 1921-78

Year	(millions)
1921	
1925	
1930	
1935	
1940	
1945	
1950	
1955	
1960	
1965	
1970	
1975	
1978	

0   100   200   300   400

AIR TRANSPORT

Aviation is a true child of the 20th century. The first powered flight was made by Orville and Wilbur Wright at Kill Devil Hills, Kitty Hawk on the North Carolina coast of the US on 17 December 1902, and a sustained flight of 38 minutes was made two years later. The first powered flight without some kind of launch apparatus was made in France in 1908. Within years of the early experiments aircraft were produced in hundreds. In October 1911, during the Italo-Turkish war, the first bombs were dropped from an aircraft, and on the Coronation Day of the British king, George V, in September 1911, the first official air mail service was operated, landing in the grounds of Windsor Castle.

The First World War gave aviation a remarkable boost. By 1918 larger, faster, sturdier aircraft were produced in thousands. Many were converted to start the first scheduled airlines in 1919. A British route was set up between London and Paris to convey officials to the Paris Peace Conference, but the first sustained passenger service was flown between Berlin and Weimar in Germany from February 1919. Air travel was boosted by a series of spectacular long-distance flights. In May 1919 the British aviators John Alcock and Arthur Brown crossed the Atlantic. The first flight from Britain to Australia started on 12 November 1919 and reached Fanny Bay, Darwin, exactly four weeks later. The first round-the-world flight was made in 1924 by two American army officers who took a total of 175 days, although only 15 days were spent in the air.

By the late 1920s a worldwide system of passenger and freight routes was established. In the United States, which later came to dominate world aviation, passenger services were slow to develop and followed the US Post Office mail services, first established across the continent in 1924. By the 1930s a high standard of passenger service was available on airliners of vastly improved performance. The first of the generation of modern airliners was the Boeing 247, an all-metal monoplane capable of 155 miles per hour. In 1930, Boeing's Air Transport service became the first to use air hostesses, who had to be under five feet four inches tall, younger than 25 and weigh no more than 115 pounds. The Boeing 247 was produced the same year as the Douglas DC-2, which cut the transcontinental journey from 27 hours to 13 and became the first American aircraft widely exported to other countries, a pattern repeated down to the 1990s.

The 1930s was the decade when aviation caught the public imagination. Hitler toured Germany in the election campaigns of 1932 in his own aircraft, and the British prime minister, Neville Chamberlain, set the model for modern summitry when in 1938 he flew for the first time to visit Hitler, returning to England to give a press statement on the tarmac at Heston airport. Demand for air travel grew rapidly. In the USSR 14,000 passengers were carried in 1925; in 1940 the number was 369,000. They were

Air travel boomed in the 1980s and 1990s, helped by the progressive deregulation of the industry, which permitted more operators and lower prices. Almost one quarter of all international air traffic flew across the Atlantic in 1993 (map bottom), where increased competition put a severe strain on airport capacity and flight scheduling. The sharp increase in oil prices in the 1970s failed to dent the growth of air transport. The picture (bottom right) shows Changi airport in Singapore, centre of one of the fastest growing air transport networks in the world.

US domestic airports, 1955

- service point for minor domestic airlines
- service point for principal "big 9" domestic airlines
- connection point for all domestic airlines

The air traffic revolution produced a rash of airports large and small across the United States, many of them built with New Deal money (map left). In the mid-1950s the average journey length was 550 miles, but 43 per cent of passengers flew less than 300 miles on the numerous short "feeder" routes which filled in the gaps between the main intercity trunk lines.

Air transport began in the Soviet Union in May 1921 on Lenin's orders. The first state company, Dobrolet, was founded in 1923, but was superseded by Aeroflot in March 1932. Following plans formulated at the Communist Party Congress in 1933, Soviet air routes were greatly expanded in the 1930s (map below left), totalling 146,000 km by 1941.

The Gypsy Moth (left), was one of the most popular British aircraft of the 1930s, in which many pilots began their flying careers. This model dates from 1929 and shows the relaxed, almost casual nature of early flying, without the immense infrastructure necessary to support modern air travel.

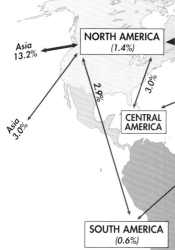

Air routes in the USSR, 1937

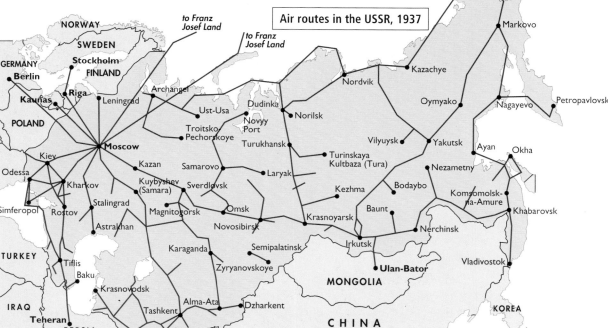

carried in increasing comfort. The first reclining seats, with in-flight tables, and the familiar configuration of aisles, galleys and toilets were installed in the Handley-Page Hannibal in 1931. The technology of modern air transport was in the process of development when war broke out in 1939. The first helicopter and the first experimental jets were developed in Germany before the war. Building on wartime research, the first jet airliner, the British De Havilland Comet, flew in May 1952, but the market was stolen by the Boeing 707, which became a standard aircraft around the world. The new technology of gas turbines, developed in the 1960s, made aircraft more powerful, quieter and cheaper to run. The wide-body design of the 1960s airliners allowed much larger payloads, and the long economic boom fuelled demand for air travel. In 1979 over 700 billion air passenger miles were flown, against a figure of only 74 billion in 1960.

The aviation industries played a central part in maintaining the momentum of growth in the developed economies. At its peak (1989) the American aerospace industry employed 823,000, while the air transport industry employed 540,000 people, 58,000 of them pilots. America's busiest airport in 1992 was O'Hare airport in Chicago, with over 378,000 flights. The expansion of air travel, made possible by increased competition and growing prosperity in the developed world, has led to congested airports and air routes. But air travel has become an essential adjunct to the globalization of the world economy and world politics.

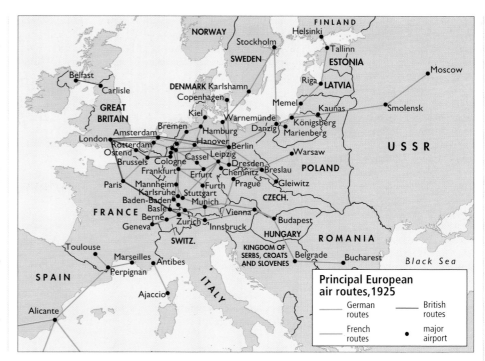

Principal European air routes, 1925

— German routes
— French routes
— British routes
• major airport

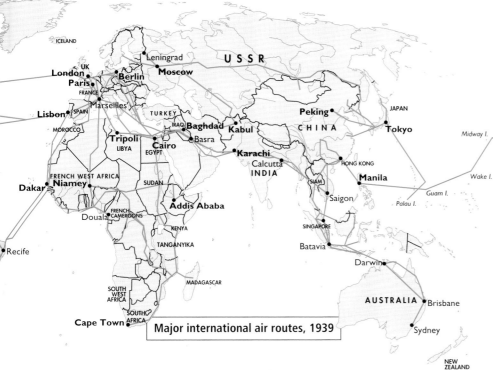

Major international air routes, 1939

Major air routes spread around the globe by 1939 *(map left)*. The short range of most passenger aircraft meant long hours spent refuelling. Oceans could only be crossed with stops at island bases, until the introduction of the Pan-American "Clipper" flying boat in 1939, which could fly the Atlantic non-stop. Early air travel had high levels of comfort *(poster far left)* to compensate for noise and air sickness.

Commercial aviation developed in Europe immediately after the end of the Great War. Despite the ban on military aircraft, Germany took the lead in developing a network of European flights, including the only air route to Moscow *(map above)*. The fastest growth of air travel occurred in the United States, where long distances and higher incomes made it both necessary and feasible *(chart below)*.

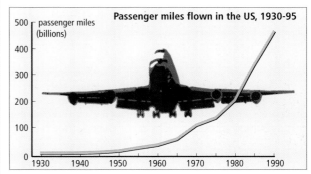

Passenger miles flown in the US, 1930-95

passenger miles (billions)

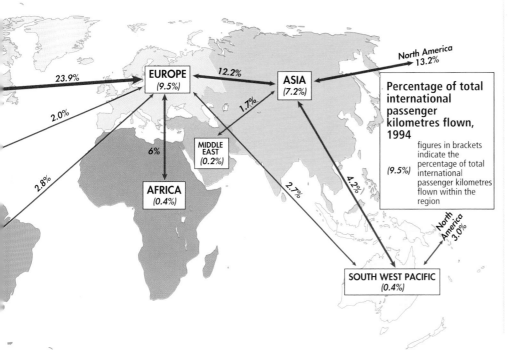

Percentage of total international passenger kilometres flown, 1994

figures in brackets indicate the percentage of total international passenger kilometres flown within the region

MOTORIZATION

The development in the early 1880s of the light vehicle powered by an internal combustion engine, the brainchild of two German engineers, Karl Benz and Gottlieb Daimler, set in motion a remarkable technical and social revolution over the following century. Life in the late 20th century is almost inconceivable without motor vehicles.

The revolution was slow in coming. Even 50 years after the invention, there were only just over three million cars and lorries operating outside the United States. The US had 85 per cent of the world's stock of vehicles, a fact that owed a great deal to the size and wealth of the American market, to the distances in a country stretching across a continent, and to the aggressive modernity of American society. In Europe motor vehicles were associated with the wealthy. They were hand-crafted rather than mass-produced.

Most vehicles were bought at first by business – haulage and taxi firms, delivery and postal services, farms far from their markets, factories attracted to the greater flexibility and efficiency of motor transport. Only in the inter-war years did private customers begin to buy cars as they became both cheaper and more reliable. The expansion of car use depended on higher incomes and good roads. Economic stagnation between the wars confined car ownership largely to the middle classes. But a start was made on worldwide programmes of roadbuilding to replace what were, in the main, crude tracks of earth and stone.

In the United States over 300,000 miles of new road were laid from the 1950s, a vast network of inter-state highways that made up the century's largest single engineering project. In Europe new multi-lane motorways were laid down in Mussolini's Italy and Hitler's Germany, which became models for the modern road networks constructed after 1945. The supply of roads created its own demand. Rail haulage and rail travel were still the main means of transport before the 1950s. But by the 1970s road transport greatly exceeded rail transport in importance in all the major economies and railways went into absolute decline.

After 1945 economic growth took off worldwide. Rising incomes in the Western capitalist economies brought a dramatic increase in demand for motor vehicles. By 1958 there were 119 million in use worldwide; by 1974 the figure was 303 million. The motor industry became a key component in the post-war boom, with high employment and sales and a whole range of indirect effects on economic life: the expansion of oil production; road-building and repair; motor transport. Motorization helped to sustain the pace of economic modernization in the 20th century and gave Western developed economies the means to keep ahead technically and industrially. Western car firms established vast multi-national businesses with branches throughout the developing world.

The application of the motor vehicle transformed traditional society. Peasant agriculture, which had been inefficient and labour intensive, became mechanized and productive. In 1950 there were just five million tractors worldwide; by 1980 the total had risen to 22 million. Mechanization freed rural workers for jobs in the city, while motor vehicles broke down the isolation of even the most distant village. Motor car ownership allowed city dwellers to move to the suburbs and to a further band of commuter villages, bringing town and country together, and creating a more homogeneous society. Motor cars gave a degree of flexibility and choice in shaping living conditions and a social life which had simply been unavailable, except to the very rich or privileged, earlier in the century.

Motorization had a negative side, however. Vehicle production made high demands on the world's mineral and oil resources. The harmful effects of pollution by vehicles in cities – and in 1991 there were a total of 591 million vehicles in the world – have not yet been faced effectively. Roads have scarred the landscape to a much greater extent than railways did in the 19th century. The cityscape is dominated by the needs of the motor vehicle, and in the developed world road fatalities come third behind heart failure and cancer as a cause of adult death. There will come a point when the world vehicle population will face the serious consequences of uncontrolled growth.

The Model T Ford was the first genuinely mass-produced car. Here *(below)* workers inspect the chassis of the Model T. It made Henry Ford a household name in the United States. "Fordism" was soon adopted as the most efficient way to assemble vehicles, with production broken down into thousands of small operations carried out along a moving assembly line. The arrival of the microchip has made possible another revolution in car-making. The robotic workshop *(bottom)* has replaced men by machines.

Canada 16 million

USA 188 million

Mexico 10 million

Brazil 13 million

The pace of worldwide motorization accelerated sharply in the last third of the century. Until then most vehicles were to be found in North America and Europe. Now millions of motor vehicles are spread throughout the developing world. In Latin America there are 25 million; in Africa 13 million. The spread of vehicles has been aided by the growth of multi-national motor companies, particularly American and Japanese, which have set up production in the less developed areas of the world *(map above)*, or cornered the market with cheap imports. The world trade in vehicles is dominated by 14 giant companies.

Evening rush hour in Manila *(opposite page near right)*, where a two-mile drive can take an hour. The motor car is now the chief cause of urban pollution.

The Chinese motor industry was insignificant until 1990 *(chart far right)*. With the help of French, German and Japanese producers, China's motor industry was poised to repeat the remarkable rise of Japanese production *(chart near right)*. By the 1980s, Japan had become the world's largest vehicle producer and exporter. Chinese output in 1990 was just 0.4 per cent of the Japanese figure.

Japanese car production, 1955-90 (thousands)

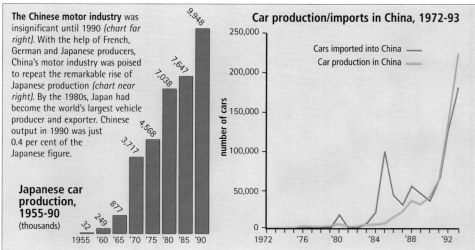

Car production/imports in China, 1972-93

Cars imported into China
Car production in China

number of cars

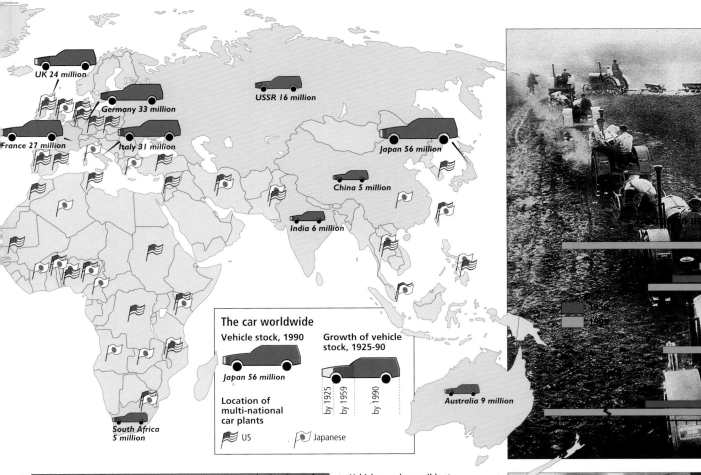

The car worldwide

Vehicle stock, 1990

UK 24 million

Germany 33 million

USSR 16 million

France 27 million

Italy 31 million

Japan 56 million

China 5 million

India 6 million

Japan 56 million

Growth of vehicle stock, 1925-90

by 1925 | by 1959 | by 1990

Location of multi-national car plants

US

Japanese

Australia 9 million

South Africa 5 million

... in use 1950 and 1983

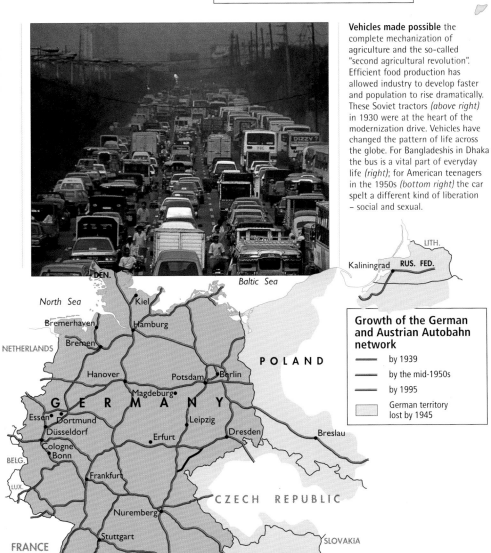

Vehicles made possible the complete mechanization of agriculture and the so-called "second agricultural revolution". Efficient food production has allowed industry to develop faster and population to rise dramatically. These Soviet tractors *(above right)* in 1930 were at the heart of the modernization drive. Vehicles have changed the pattern of life across the globe. For Bangladeshis in Dhaka the bus is a vital part of everyday life *(right)*; for American teenagers in the 1950s *(bottom right)* the car spelt a different kind of liberation – social and sexual.

LITH.

Kaliningrad   RUS. FED.

DEN.

Baltic Sea

Growth of the German and Austrian Autobahn network

—— by 1939

—— by the mid-1950s

—— by 1995

German territory lost by 1945

North Sea

NETHERLANDS

Bremerhaven · Kiel · Hamburg · Bremen

BELG.

Essen · Dortmund · Hanover · Magdeburg · Potsdam · Berlin · Leipzig

Düsseldorf · Erfurt · Dresden · Breslau

Cologne · Bonn

LUX.

Frankfurt

FRANCE

Nuremberg

Stuttgart

SLOVAKIA

Freiburg · Munich · Linz · Vienna

SWITZ. · Salzburg · AUSTRIA · HUNGARY

ITALY · Klagenfurt · SLOVENIA

GERMANY

POLAND

CZECH REPUBLIC

Begun in 1934 as a monument to the Third Reich, the German motorway system pioneered the development of fast multi-lane, purpose-built roads for the age of mass motoring. Much of the system has been built since 1945 *(map left)*.

Motorization

MASS LEISURE

The poor have always had free time. It was estimated that a labourer at the turn of the century was lucky if he worked for one day in four. But the poverty that resulted from this ensured that leisure and consumption were mutually exclusive. During the 20th century this began to change. The campaign for the eight-hour day, and then for the 40-hour week, gave workers more time in which to enjoy themselves, while rising prosperity meant that commercialized leisure was no longer beyond them.

The impact of the new culture of mass leisure was first felt in sport. The first organized sports in the 19th century – horse racing, sailing, rowing – were the activities of the well-to-do. But at the turn of the century new mass sports emerged across Europe and the United States. An international sports federation was set up for cycling in 1892, for football in 1904, for ice hockey and swimming in 1908. Over the course of the new century 44 sports were organized internationally, as the principle of competition came to replace the idea of simple recreation. European elites saw mass sport as a means to create a healthier and less dissident working class. In Russia before 1914, English factory owners introduced football to their Russian workforce to instil a sense of corporate loyalty. Sporting competitions were provided for frontline soldiers in the First World War to mitigate the pressures of trench warfare and to avert mutiny.

Nothing symbolized the new age of mass sport so much as the revival in 1896 of the classical Olympic Games. The first

The map (bottom) shows the sites for the Olympic and Commonwealth Games organized in modern times. The location of the Olympics sometimes had political implications (notably in Berlin in 1936 and in Moscow in 1980). However, increasingly cities have competed for the right to stage Olympic Games simply because of the commercial benefits that are to be reaped.

The Swedish javelin champion Eric Lemmings (bottom left), who won Olympic medals in 1904, 1908 and 1912, represented elite sport whose pursuit required considerable means. The boys practising football in the back streets of Rio de Janeiro (below left) show how sport has attracted a mass audience and how it provides a small number of the poor with an escape from their background.

Games, held in Greece, were modest in scope. Only 13 countries and 295 athletes (all of them male) took part. By 1936, when the Games were held in Hitler's Berlin, there were 49 states represented and almost 4,000 athletes (both men and women). By this stage sport had become a potent symbol of national rivalries. At the Berlin Games Hitler wanted to demonstrate the superior athleticism of his master race. Although Germany won 33 events, and came second in 26 more, the coveted 100 metres gold medal went to the American black athlete Jesse Owens, and the real endurance test – the marathon – was won by the Korean Kitei Son. Since 1936 the Olympics have mixed idealism with politics. In the 1960s South Africa and Rhodesia were banned; the 1972 Games at Munich were marred by the murder of 11 Israeli team members by Palestinian terrorists; and the United States boycotted the 1980 Games in Moscow because of the Soviet invasion of Afghanistan.

194

The universal appeal of sport was matched by the universal spread of the most successful form of mass leisure – the cinema. The first film shown to a paying public was in 1895. The film industry grew rapidly throughout the Western world, but its home was the United States. American films were exported worldwide and stimulated imitators everywhere. Film entertainment was cheap and the technology easily mastered. The high point of the American film industry came after the Second World War, when the rest of the film-making world was recovering from the war. In 1949 the US exported 48 million feet of film footage; by 1960 the figure was only 29 million. In the post-war years large film industries developed in India and China, where cinemas could reach a mass audience still too poor to have television widely available. In 1991 India made over 900 films and the Pacific Rim over 1,100, whereas the US produced 345 and Britain, once one of the world's largest film-producing countries, only 54, many of them with American money.

Cinema contributed in many ways to the establishment of a more global culture. American habits of speech or dress had a world audience. Films were made with the idea of export in mind, which meant producing them with universal appeal and clear themes. Cinema was also ideally suited to the art of political persuasion. The rapid spread of mass cinema in the Soviet Union in the 1920s and in China in the 1950s was a direct product of those regimes's use of films as propaganda. In the 1930s mobile cinemas toured the villages of the USSR, showing peasants the wonders of the new socialist system. In Nazi Germany films were used to project the values of a racist and militarist regime, though Hitler's favourite film remained the American *Gone with the Wind*. Since 1945 cinema has given ground to television and video, and a range of other leisure pursuits made possible by the prosperity of the long boom. Sport and film established the contours of the mass leisure industry, which is now one of the world's largest employers and biggest earners.

The map *(below)* shows the distribution of world cinema in 1991. Cinemas remain numerous in rich countries, but are being replaced by the home entertainment of television, video and home-computing. By contrast cinema is becoming a more important form of entertainment in some relatively underdeveloped countries, where young populations are seeking cheap entertainment. It is especially important in those countries, such as India, which have a strong indigenous film industry *(poster and chart below right)*.

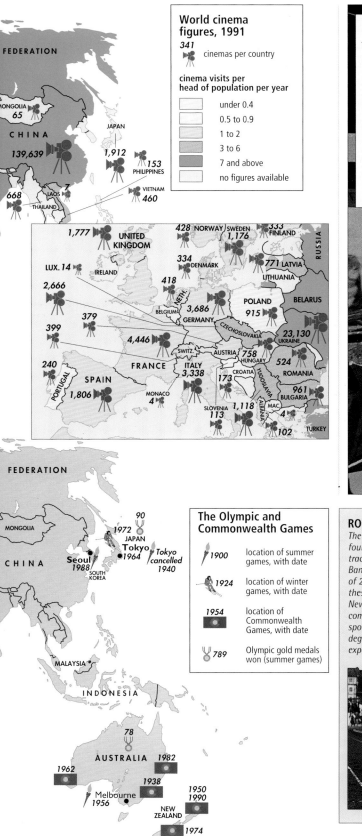

World cinema figures, 1991

341
🎥 cinemas per country

cinema visits per
head of population per year

	under 0.4
	0.5 to 0.9
	1 to 2
	3 to 6
	7 and above
	no figures available

World film production, 1970-91

No. of feature films produced

■ 1970
■ 1991

India, USA, Hong Kong, Japan, Thailand, France, Philippines, China, Korea, Italy, Pakistan, Brazil

The Olympic and Commonwealth Games

1900	location of summer games, with date
1924	location of winter games, with date
1954	location of Commonwealth Games, with date
789	Olympic gold medals won (summer games)

ROGER BANNISTER

The first man to run a mile in under four minutes (at Oxford's Iffley Road track on 6 May 1954), Roger Bannister exemplifies two features of 20th-century sport. The first of these is the search for heroes. Newspapers and broadcasting companies subjected successful sportsmen to an ever increasing degree of attention. Athletes were expected to exemplify particular

values as well as to display physical ability. Bannister was one of the last men to be associated with the values of amateurism that had grown up around English public schools and universities - values celebrated by the film Chariots of Fire. *He was a professional doctor, and it was characteristic that he made his record-breaking run at an Oxford University track. After this*

date athletes were increasingly likely to be dedicated - to the point of professionalism - and to come from social groups for whom sporting achievement might be the only means of social mobility. The values stressed by men like Carl Lewis and Linford Christie were ones of aggressive competition rather than gentlemanly amateurism.
The second value exemplified by Bannister's run was simply the obsession with records. Accurate time-keeping, measurement and good communication meant that athletic competitions between particular individuals at particular times yielded in importance to more universal attempts to beat the clock. Certain times and distances acquired particular significance in the eyes of public opinion. The athletes who have now cut some 15 seconds from Bannister's original record have not achieved the same fame.

TOURISM

The revolution in communications had a profound effect on travel for its own sake. The first decade of the 20th century saw the high point of tourism for the wealthy. Railways and steam ships meant that most cities were accessible to anyone with sufficient money. The *Guides Michelins*, published from 1900, gave advice to motorists on locations worth visiting. Members of smart society might spend the winter on the French Riviera, then go to Switzerland, where the British were beginning to develop skiing and climbing resorts; or they might arm themselves with the Baedeker guide and head for Venice or Athens.

During the inter-war period, and especially in the 1930s, tourism became more democratic. The French Popular Front Government introduced paid holidays in 1936, and by 1938 11 million British workers also enjoyed this right. Seaside resorts flourished in Britain, and throughout Europe youth hostels encouraged the young to pursue healthy outdoor recreations. The Nazi regime in Germany used holidays as an instrument of propaganda, providing subsidized trips for workers.

However, the great development of mass tourism came during and after the 1960s. Air travel became cheaper; in 1957 for the first time the number of people crossing the Atlantic by plane exceeded the number who did so by boat. The package tour (providing accommodation as well as travel) became increasingly common: by 1970 two million package holidays by air were sold in Britain, while only half a million independent holidays were taken by air. The result of these changes, and of growing prosperity, was that the numbers taking foreign holidays increased greatly. In 1950 300,000 North Americans visited Central America and the Caribbean; by 1970 this figure had increased to seven million per year. The number of, mainly north European, tourists visiting Spain and Italy exceeded 50 million by the late 1980s. Tourism also changed in nature during this period. Previous tourists had sought either to improve their minds with inspections of monuments and art or to improve their bodies with bracing walks and fresh sea air. Tourists on package holidays often sought nothing but sunshine and beaches. An industry grew

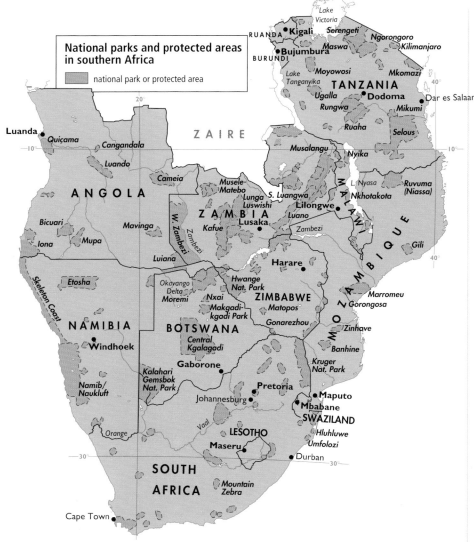

National parks and protected areas in southern Africa

national park or protected area

One feature of the late 20th century has been the attempt to market tourist destinations on the basis that they reveal "unspoiled" nature. African safaris have become fashionable forms of recreation for Western tourists *(left)* and states across Africa have established game reserves partly for this purpose *(map above)*.

The growth of tourism in the 20th century has meant a dramatic change in the nature of those who enjoy such recreation as well as in the quality of the destinations. During the early part of the century tourism was an activity that only the wealthy could afford. Figures from fashionable society might spend the entire winter season in the south of France *(left)*. Now much larger numbers of people travel longer distances over shorter periods. This has meant that tourist destinations tend to lose their particular character. The beach in Hawaii *(right)* seems almost the same as a beach anywhere in the world.

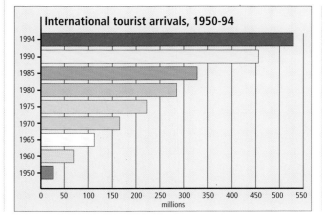

International tourist arrivals, 1950-94

(bar chart showing years 1950, 1960, 1965, 1970, 1975, 1980, 1985, 1990, 1994 with scale 0 to 550 millions)

Tourism has expanded so greatly over the 20th century that many countries depend on the money received from tourists *(map right)*. The number of tourists travelling per year increased by a factor of 20 between 1950 and 1994 *(map above right and chart left)*. This was a shift that reflected growing prosperity, increased ease of travel (particularly by air) and a culture that stressed the search for new experience. However, tourism left many parts of the world untouched. Few inhabitants of sub-Saharan Africa, for example, travelled unless compelled to do so by famine, war or the search for work.

up to cater for them. Whole areas of southern Spain, Greece and even Thailand were specifically developed to receive tourists. The tourism industry could no longer rely on taking people to see things that already existed: "theme parks", such as Disneyland, were developed to provide entertainment.

The growth of mass commercial tourism generated hostility. Many see the industry as the epitome of passive consumerism and blame it for exploiting underdeveloped countries or for ruining the environment of key "heritage sites". In 1968 rioting students in Paris broke the windows of the Club Med holiday company. At the same time a parallel industry grew up that appealed precisely because it seemed to offer an alternative to mass commercialized tourism. This alternative industry stressed the search for new "unspoilt" destinations and encouraged travellers to think of themselves as independent adventurers. A whole industry of "eco-tourism" developed in the 1980s and 1990s dedicated to "sustainable" tourism and sometimes involving tourists in environmental project work.

Mass tourism has had a dramatic impact, both on those countries receiving tourists and on their home countries. The economic, and perhaps even political, transformation of a country such as Spain during the last 30 years owes much to receipts from the tourist trade. The other side of this coin is that the ubiquity of "national" dishes such as paella and spaghetti owes much to the experiences of tourists. Most significantly, tourism has had a political impact. This is particularly true of Eastern Europe. The gradual easing of travel restrictions allowed citizens of Warsaw Pact countries to witness for themselves the prosperity of the West and of the more liberal Eastern bloc countries at first hand. Anyone observing the floods of travellers heading west in the aftermath of the collapse of the Berlin Wall in 1989 might well have assumed that the "right to be a tourist" was the most fundamental demand of those who had brought down communism. Conversely, authoritarian regimes, such as those in China and Myanmar, have used an opening to tourism to legitimize themselves in the eyes of the world community. If tourists can turn a blind eye to political oppression, their governments may well be prepared to do the same.

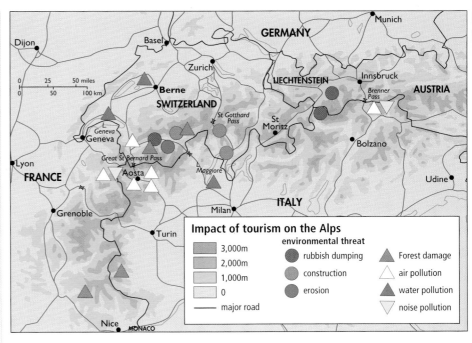

Impact of tourism on the Alps

	environmental threat	
3,000m	rubbish dumping	Forest damage
2,000m	construction	air pollution
1,000m	erosion	water pollution
0		noise pollution
major road		

Tourism has become a big business in the course of the 20th century. Travel agents have marketed exotic locations to wealthy clients *(poster right)* and tourism now plays a major part in many national economies. The economic impact of tourism is best measured in terms of the proportion of total national earnings that come from the industry. A rich country such as the US may regard tourism as just one of a variety of sources of income. Comparatively poor countries, such as Thailand in the 1980s *(chart below right)* or Spain in the 1960s *(chart bottom right)*, may be changed by increasing numbers of tourists.

The Alps *(map above)* have attracted tourists throughout the 20th century (particularly skiers, who now make up two thirds of those visiting the mountains). However, tourism carries dangers as well as benefits. Visits by excessive numbers of people may damage the delicately balanced mountain environment: seven million cars cross the Alps each year and the acid rain that such vehicles help to produce has now destroyed 60 per cent of trees in the area. There is a risk that such damage will undermine the very attractions that bring tourists to the area in the first place.

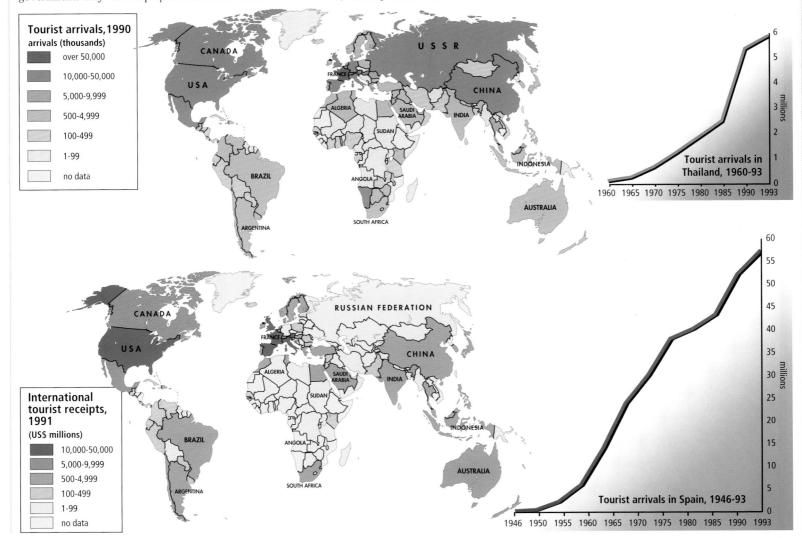

Tourist arrivals, 1990
arrivals (thousands)
- over 50,000
- 10,000-50,000
- 5,000-9,999
- 500-4,999
- 100-499
- 1-99
- no data

Tourist arrivals in Thailand, 1960-93

International tourist receipts, 1991
(US$ millions)
- 10,000-50,000
- 5,000-9,999
- 500-4,999
- 100-499
- 1-99
- no data

Tourist arrivals in Spain, 1946-93

Tourism

197

EDUCATION

Few aspects of the revolutionary century have touched more people than the spread of educational opportunity. Although there remain almost one billion illiterates worldwide, there has been a remarkable growth in the numbers receiving full-time education at every level in the second half of the century.

Early in the century most of the world's population was illiterate or nearly so, and formal long-term education for whole populations was confined to Europe and areas of European settlement. Even here most children left school after receiving only basic instruction. The numbers going on to secondary and then to higher education were tiny. In colonial regions education was provided by the imperial power, most of it by Christian organizations who saw education, as they did in Europe or America, as a means of moral instruction.

Not until the Second World War did the idea of education as a right, valuable in itself, become more generally accepted. The League of Nations established a Committee on Intellectual Co-operation in 1922, chaired by the French philosopher Henri Bergson, but its work was centred on Europe, where the provision of higher levels of instruction made substantial progress before 1939, at least for boys. A new agenda emerged from the war. In 1942 the British minister of education, Richard Butler, called a conference of Allied education ministers at which education was pronounced a basic human right to be promoted for its own sake. Butler's group laid the foundation for what, in November 1945, became the United Nations Educational, Scientific and Cultural Organization (UNESCO), whose first director-general was the British scientist Julian Huxley.

In 1947 UNESCO published the report, *Fundamental Education*, which laid the grounds for the post-war campaign against illiteracy and educational discrimination. Nine years later the UN adopted the Free and Compulsory Education Project, which laid down the principle that everyone was entitled to education regardless of race, sex or religion for a minimum period of six years. The project was piloted in Latin America, where it made substantial progress. In Karachi in 1960 the UN endorsed a global programme for the provision

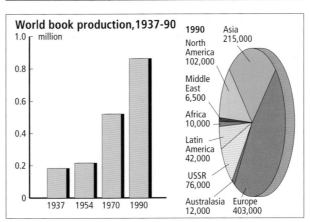

An open-air primary school in the Sudan *(right)*. World aid programmes have targeted sub-Saharan Africa as a priority for educational reform. Japanese children at the Tsukuba Science City school *(left)* have the advantages of one of the most comprehensive and highly funded educational systems in the world.

The output of books and newspapers has grown rapidly with improvements in education and rising prosperity *(charts below left and right)*. Newspaper readership has expanded rapidly in Latin America and Asia since the mid-1950s, but has remained almost static in North America for 40 years.

World book production, 1937-90

1990	
Asia	215,000
North America	102,000
Middle East	6,500
Africa	10,000
Latin America	42,000
USSR	76,000
Australasia	12,000
Europe	403,000

The Turkish leader Kemal Atatürk attending a professorial discourse at Istanbul University *(below left)*. His drive to modernize Turkey in the 1920s helped higher education but left large parts of the rural population with little or no formal schooling.

Great strides have been made in the provision of primary education over the last 30 years *(map left)*, though in parts of the world attendance, particularly for girls, is still low. The total for the world in 1995-6 was 83.6 per cent of all boys and 75.5 per cent of all girls. Improvements in education in the developing world have come about despite low levels of funding for schools *(chart below)*.

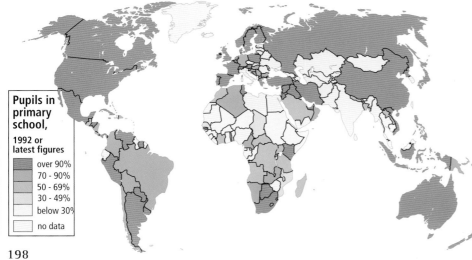

Pupils in primary school,

1992 or latest figures

- over 90%
- 70 - 90%
- 50 - 69%
- 30 - 49%
- below 30%
- no data

Expenditure per school pupil (in US$)	
East Asia/Australasia	76
South Asia	104
Latin America	287
Sub-Saharan Africa	58
Arab States	262
Developed world	2,419

of primary education, and the following year produced the first comprehensive survey of global illiteracy. The survey showed that two fifths of the world's adult population was illiterate, and that in some states almost the entire female population could neither read nor write. Since there was widespread agreement that education was a central explanation for differing levels of success in economic development – a fact highlighted by the attention lavished on education in the high-growth Pacific Rim – the UN established the fight against illiteracy as the central educational ambition.

The literacy drive had mixed results. In 1970 one third of adults were still illiterate and the absolute number was growing rather than falling. The most striking gains were made only after 1980, when international funding rose sharply and economic success was no longer confined to the wealthy north. The aim of the UN in International Literacy Year (1990) was to eradicate illiteracy by the year 2000. particularly in what it called the "Least Developed Economies", where fewer than 50 per cent of adults were literate. In some respects the gap between the developed and developing world has narrowed in the last ten years. Technology and information is easily transferred between regions: in 1992 1.2 million students studied abroad, mainly in Europe and the US. The numbers enrolled in secondary and tertiary education have risen dramatically in the developing regions, and those countries' expenditure on research and development, though still lagging significantly behind levels in Europe and the United States, has broken the near monopoly the latter enjoyed until the 1980s.

Where the gap still matters is in educational expenditure. Between 1980 and 1992 world spending on education rose from $526 billion to $1,196 billion. But by 1992 the developed world accounted for $927 billion of this, and the Least Developed Economies for just $4 billion. Expenditure per pupil in the developed world was $2,419 in 1990, in sub-Saharan Africa it was $58 and in East Asia $76. Spending on this scale has helped to maintain the knowledge gap between north and south and the gap in economic achievement. The focus for the future is no longer on the problem of illiteracy but on other skills and opportunities which literacy makes possible.

Daily newspaper circulation, 1956-92

Circulation (millions)	1956	1975	1992
Africa	2	3	5
North America	65	66	66
Latin America	9	23	40
Asia	51	127	203
Middle East	1	3	9
Europe	120	221	287
Australasia	5	6	6

Enrolment in formal education, 1970-90

numbers of pupils (millions)	1970	1990
primary education	313	490
	122	111
secondary education	81	210
	80	90
tertiary education	7	29
	21	35

developing countries ▢ developed countries

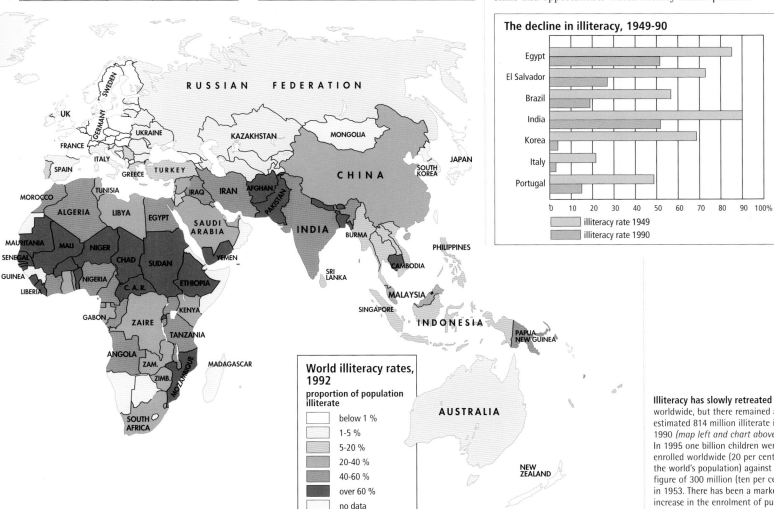

World illiteracy rates, 1992

proportion of population illiterate

- below 1 %
- 1-5 %
- 5-20 %
- 20-40 %
- 40-60 %
- over 60 %
- no data

The decline in illiteracy, 1949-90

Egypt, El Salvador, Brazil, India, Korea, Italy, Portugal

0 10 20 30 40 50 60 70 80 90 100%

▢ illiteracy rate 1949
▢ illiteracy rate 1990

Illiteracy has slowly retreated worldwide, but there remained an estimated 814 million illiterate in 1990 *(map left and chart above).* In 1995 one billion children were enrolled worldwide (20 per cent of the world's population) against a figure of 300 million (ten per cent) in 1953. There has been a marked increase in the enrolment of pupils in secondary and higher education.

Education

RELIGION

In 1900 it seemed reasonable to assume that religion was losing the power that it had previously exercised. Liberal, educated opinion in much of the world had moved significantly against the primacy of religion, especially its traditional influence over the conduct of public affairs. Religion seemed arrayed against powerful forces of progress and rationality. The competition between progressive forces and the Catholic Church was particularly acute. The declaration of Papal Infallibility (1870) seemed to have pitted the Church against tolerance and scepticism. This struggle acquired dimensions of gender and social class with the most enthusiastically religious seen as uneducated and backward.

The struggle between religious and secular power manifested itself across the globe. It took place in France, where Church and state were separated in 1905, and in Spain in 1932, where disestablishment of the Catholic Church helped fuel the crisis that led to civil war. It was seen in Turkey, where Atatürk secularized the state during the 1920s, and in the Middle East since the 1940s.

However, the 20th century has not ended with a secularized world. In some countries religion continues to have a formal place alongside the state, despite falling church attendances. In Britain the Anglican Church – though attracting little devotion from its members – remains formally at the centre of public life: the head of state is also head of the church. In the United States, religious attendance remains high despite the fact that there is no formal role for religion alongside the state. In India politics now revolve around religious issues as the Hindu nationalist BJP party overtakes the more secular Congress group. Most importantly, Muslim fundamentalism has gained a presence in many parts of the world, provoking fears in the secularized world that the division between Islam and the West may prove as damaging as that in the Cold War between communism and capitalism.

Religion has remained important for several reasons. First of all, the very upheavals of the 20th century often created a new need for religious structures. Those migrating to cities, or even to new countries, often clung to their religion as something to provide them with a sense of belonging. Catholic priests have done much to organize Irish, Polish and Italian communities in North America; radical Islam has found a ready audience among the uprooted second generation Muslim immigrants of Great Britain and France. Secondly, religion has often blended into broader secular structures. In areas such as Brittany and Ireland, nationalism has been seen, in part, as a conflict between local Catholicism and the domination of a Protestant or secular state. In Eastern Europe religious and ethnic divisions coincided and reinforced each other. Poles flocked to Catholic churches because these seemed to symbolize resistance to communist rule, even though statistics on matters like abortion suggest that few Poles accept all the teachings of a conservative Polish Pope; the secularized Muslims of Bosnia were forced to accept help from Iranian Islamic militants in order to defend themselves against their Serbian Orthodox neighbours.

Religion also ties in with nationalism in many parts of the Third World. The most prominent leader of Indian nationalism,

Relative size of Christian denominations, 1985

Protestants 19% 278m
- others 48m
- Pentecostal 24m
- United 65m
- Lutheran 45m
- Methodist 24m
- Baptist 35m
- Reformed 39m

independent indigenous 6% 95m

Orthodox 12% 170m

Roman Catholics 59% 872m

others 1% 16m

Anglicans 3% 51m

GREENLAND

CANADA

USA

Protestant

Catholic

● Waco (Branch Davidians, 1993)

Catholic

BRAZIL

Catholic

Jonestown (People's Temple, 1978)

North Atlantic Ocean

South Atlantic Ocean

Northern Ireland, from 1969

Irish Free State (Catholic), Northern Ireland (Protestant) from 1922

near Lausanne (Solar Temple, 1994)

Rif Kabils 1921-6

The map *(above)* shows changes in the influence of world religions across the 20th century. These changes have sometimes been the result of conversion. However, they have also been linked to migration (taking Islam into Western Europe); extermination (driving Judaism out of Central Europe); different birth rates (favouring Muslims over Christians and Catholics over Protestants); and politically motivated campaigns (seeking to eliminate religious practice in communist countries).

World religious affiliations, 1920-95

- Christianity
- Islam
- no religion
- Hinduism
- Buddhism
- Judaism

(y-axis: millions, 0 to 1,800; x-axis: 1920 1930 1940 1950 1960 1970 1980 1990 1995)

CULTS

In much of the world organized religion has lost power during the 20th century. However, this decline has been accompanied by the rise of small groups of particularly fervent believers. In some respects such groups can be seen as an extension of the Reformation. Like the early Protestant church, they emphasize a return to "original" values – sometimes based on a particular reading of the Bible. Often such movements revolve around a single charismatic leader who claims to be in direct communication with God. Cults appeal to the marginalized and alienated and have particular success in countries that are undergoing rapid social change or some other form of disruption.

Sometimes such cults reject the values of state and society and those of established religion. The power of those cults has become a source of concern for many. Parents sometimes sought to have children "deprogrammed", and courts were asked to rule on the extent to which such

individuals were able to exercise free will. Two very different but equally striking manifestations of cult power were seen in 1995: in Switzerland the Solar Temple Cult organized a mass suicide, while in Seoul the Unification Church (the "Moonies") organized a mass wedding (picture below).

The graph *(above)* reflects the growth of world religions since 1920. It shows that Christians, Hindus and Muslims have increased as a percentage of world population, while Buddhists have tended to fall back. (The statistics reflect varying population growth rates in areas associated with particular religions, as well as the ability to inspire or maintain faith.) Within each religion there are variations between denominations or sects. Roman Catholics make up 59 per cent of all the world's Christians, while the Anglican Church, which still commands a degree of cultural and political influence, accounts for only three per cent of Christians.

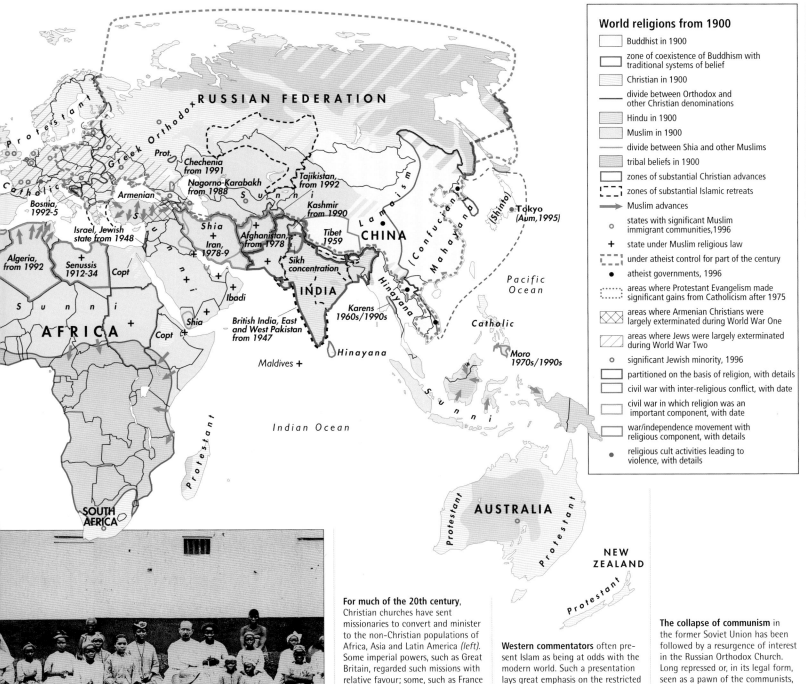

World religions from 1900

- Buddhist in 1900
- zone of coexistence of Buddhism with traditional systems of belief
- Christian in 1900
- divide between Orthodox and other Christian denominations
- Hindu in 1900
- Muslim in 1900
- divide between Shia and other Muslims
- tribal beliefs in 1900
- zones of substantial Christian advances
- zones of substantial Islamic retreats
- Muslim advances
- states with significant Muslim immigrant communities, 1996
- state under Muslim religious law
- under atheist control for part of the century
- atheist governments, 1996
- areas where Protestant Evangelism made significant gains from Catholicism after 1975
- areas where Armenian Christians were largely exterminated during World War One
- areas where Jews were largely exterminated during World War Two
- significant Jewish minority, 1996
- partitioned on the basis of religion, with details
- civil war with inter-religious conflict, with date
- civil war in which religion was an important component, with date
- war/independence movement with religious component, with details
- religious cult activities leading to violence, with details

For much of the 20th century, Christian churches have sent missionaries to convert and minister to the non-Christian populations of Africa, Asia and Latin America *(left)*. Some imperial powers, such as Great Britain, regarded such missions with relative favour; some, such as France sought to restrain missionary activity. Missionaries often provided education and healthcare for the populations of the countries to which they were sent. They sometimes also helped exacerbate cultural and religious divisions that have lasted to the present day.

Western commentators often present Islam as being at odds with the modern world. Such a presentation lays great emphasis on the restricted position of women. However, even highly Islamic states (such as Iran and Pakistan) live in a world of technological modernity. The photograph *(below left)* shows the apparent paradox of two girls in traditional Muslim dress being given a modern scientific education.

The collapse of communism in the former Soviet Union has been followed by a resurgence of interest in the Russian Orthodox Church. Long repressed or, in its legal form, seen as a pawn of the communists, since 1991 the Church has been able to tap a deep ground-swell of Russian nationalism and uncertainty about Russia's economic and political future. Congregations have burgeoned and many new converts have come forward for baptism *(below far left)*.

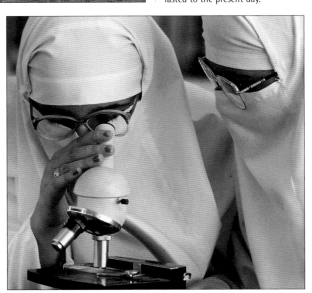

Gandhi, sought to exploit a complicated blend of Hindu traditionalism, anti-caste modernization and nationalism. His own subsequent assassination, by a Hindu extremist, and the importance religion assumed in underwriting political conflict in India suggests that he did not succeed in controlling the forces that he unleashed.

Party politics was another area that proved inseparable from religion. Christian Democrat parties became the dominant force in Italy and Germany after 1945. The influence of religion was less explicit, though perhaps even more effective, in the politics of the United States. Here a powerful religious lobby made up of Catholics and born-again Protestants endorsed candidates who would defend school prayers and attack the right to abortion. A century that had begun with many assuming that the diffusion of Darwin's ideas would undermine religious belief ended with a large group of people in an industrial super-power insisting that children should be taught a strict biblical view of the creation of the world. Secularism is in retreat. At the century's end there are many more religious devotees than there were in 1900.

Religion

201

THE ENVIRONMENT: POLLUTION

The idea that the environment is something to be cherished is comparatively new. Until recently, human industry only affected small parts of the world. Peasants who endured heat, barren soil, mosquitoes and crop disease would have been surprised to be told that nature was a "resource" to be conserved. The growth of industrial cities in the 19th century began to create an awareness of the problems of pollution. As a result, the middle classes sought escape by moving to country summer houses or to the pseudo-rural atmosphere of the suburbs. Cities were seen as unhealthy, while the technological developments that made them so also made the countryside more accessible and more tolerable. With the growth of private motor cars, rural living seemed like an escape rather than an imprisonment. The decline of rural industries made the countryside in developed countries seem more unspoilt than it had done for centuries.

The rapid economic growth of the period after the Second World War saw the human impact on the environment at its most dramatic. Factories produced emissions that lowered air quality and sometimes produced "acid rain", which poisoned forests, rivers and lakes. Environmental damage by industry was particularly bad in Eastern Europe. In the late 1980s, Czechoslovakia, with a population of 18 million, produced sulphur dioxide emissions that were twice as great as those of West Germany, with a population of 60 million. Sometimes

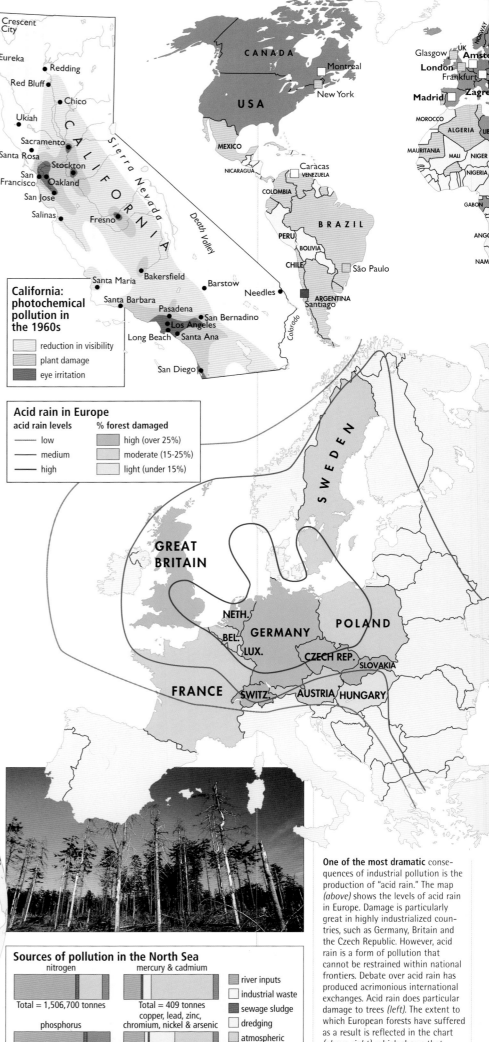

California: photochemical pollution in the 1960s

- reduction in visibility
- plant damage
- eye irritation

Acid rain in Europe

acid rain levels	% forest damaged
low	high (over 25%)
medium	moderate (15-25%)
high	light (under 15%)

A worker in protective clothing (above) at the Chernobyl site in the Ukraine after the explosion of a nuclear reactor in 1986. Engineers showed enormous courage in their attempts to contain the damage caused by this explosion and many died as a result of radiation doses they received. Considerable numbers of people in the Ukraine continue to be affected by the fall-out. The numbers of cancers caused in the rest of the world is hard to compute, although in the US radiation from nuclear fall-out is a relatively modest proportion of that from natural and medical sources *(chart far right)*.

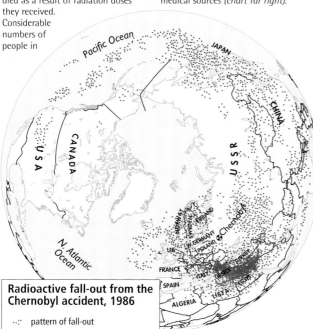

Radioactive fall-out from the Chernobyl accident, 1986

- ⫶ pattern of fall-out

Sources of pollution in the North Sea

nitrogen	mercury & cadmium	
Total = 1,506,700 tonnes	Total = 409 tonnes	
phosphorus	copper, lead, zinc, chromium, nickel & arsenic	
Total = 104 tonnes	Total = 40,557 tonnes	

- river inputs
- industrial waste
- sewage sludge
- dredging
- atmospheric
- direct discharge

One of the most dramatic consequences of industrial pollution is the production of "acid rain." The map *(above)* shows the levels of acid rain in Europe. Damage is particularly great in highly industrialized countries, such as Germany, Britain and the Czech Republic. However, acid rain is a form of pollution that cannot be restrained within national frontiers. Debate over acid rain has produced acrimonious international exchanges. Acid rain does particular damage to trees *(left)*. The extent to which European forests have suffered as a result is reflected in the chart *(above right)*, which shows that Britain has suffered worst (with over 60 per cent of her forests damaged).

The Revolutionary Century: Themes

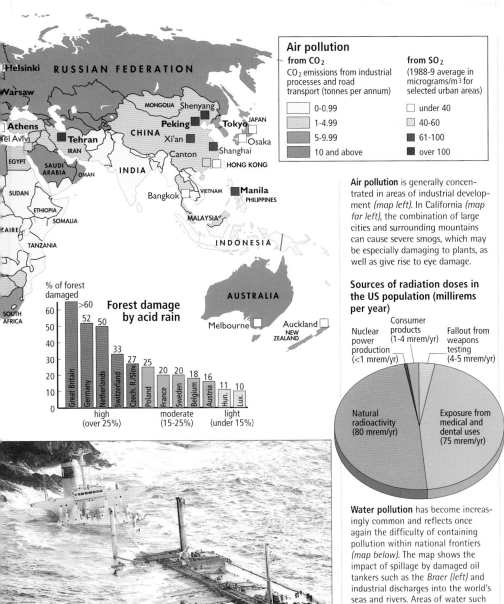

Air pollution

from CO₂
CO₂ emissions from industrial processes and road transport (tonnes per annum)

	0-0.99
	1-4.99
	5-9.99
	10 and above

from SO₂
(1988-9 average in micrograms/m³ for selected urban areas)

	under 40
	40-60
	61-100
	over 100

Air pollution is generally concentrated in areas of industrial development *(map left)*. In California *(map far left)*, the combination of large cities and surrounding mountains can cause severe smogs, which may be especially damaging to plants, as well as give rise to eye damage.

Forest damage by acid rain

% of forest damaged

Great Britain	>60	high (over 25%)
Germany	52	
Netherlands	50	
Switzerland	33	
Czech. R./Slov.	27	moderate (15-25%)
Poland	25	
France	20	
Sweden	20	
Belgium	18	
Austria	16	light (under 15%)
Hun.	11	
Lux.	10	

Sources of radiation doses in the US population (millirems per year)

- Natural radioactivity (80 mrem/yr)
- Exposure from medical and dental uses (75 mrem/yr)
- Nuclear power production (<1 mrem/yr)
- Consumer products (1-4 mrem/yr)
- Fallout from weapons testing (4-5 mrem/yr)

Water pollution has become increasingly common and reflects once again the difficulty of containing pollution within national frontiers *(map below)*. The map shows the impact of spillage by damaged oil tankers such as the *Braer (left)* and industrial discharges into the world's seas and rivers. Areas of water such as the North Sea, which are bounded by industrial regions, are particularly vulnerable to pollution. The table *(below left)* illustrates the quantities of industrial and other waste being dumped in the North Sea.

Western companies moved production to countries where environmental regulations were less onerous.

Private motor cars were the second great source of environmental damage. Car usage increased vastly, especially in North America. Motor cars caused direct damage (through exhaust emission) and ensured that ever larger areas of the country were taken up by new roads. Developments in agriculture and forestry also caused great environmental damage. Forests throughout the world were cut down to provide timber. The use of pesticides and fertilizers to increase production in agriculture meant that increasing quantities of harmful nitrates entered the soil and rivers. Environmental damage even became a weapon of war during the late 20th century: American forces used defoliants to strip away the jungle that provided cover for communist forces in the Vietnam war. In the Gulf War of 1991 Saddam Hussein turned to "environmental terrorism" by setting fire to Kuwaiti oil wells.

Concern for the environment became an increasing focus for political activity during the late 20th century. At first this concern came mainly from the political right. Conservatives lauded the countryside as the repository of real virtue and contrasted it with the degeneracy and political radicalism of the cities. The French conservative Charles Maurras wrote an article on the damage done by oil refineries in the south of France. "Peasant" often became a political label under which conservatives chose to present themselves: even urban conservatives took up the rhetoric of ruralism and in the 1940s "peasant" candidates won parliamentary elections in western Paris and gained places on the Budapest municipal council. By contrast, the left usually identified itself with technological progress, urbanization and the need to expand the industrial working class. The upheavals emerging from the student demonstrations of 1968 changed all this. Young radicals who rejected the perspectives of the old left began to doubt whether unlimited economic growth was a possible or desirable aim. New "Green" parties were established and achieved particular success in West Germany, where they gained over five per cent of the vote and 27 seats in the 1983 elections. Greens argued that conventional politics was distorted by an emphasis on short-term material benefits. They advocated direct action against projects that were seen to be environmentally damaging. That a concerted international effort may finally be under way to counteract the damaging effects of pollution was demonstrated by the "Earth Summit" held in Rio de Janeiro in 1992. An unprecedented number of government leaders attended and committed themselves to action. A symbolic gesture perhaps, but in the field of pollution control even such gestures have been infrequent.

Ocean and river pollution

oil slicks and tar balls
- high occurrence
- low occurrence

oceanic pollution
- frequent and severe
- partial and intermittent
- major oil tanker disaster
- oil rig blowout
- natural seepage

river pollution
- severe
- background

Exxon Valdez 1989
Burmah Agate 1979
Atlantic Empress 1979
Braer 1993
Sea Empress 1995
Amoco Cadiz 1978
Aegean Sea 1992
Ellen Conway 1976
Independenta 1979
Asaimi 1983
Tadotsu 1978
Yuyo Maru No.10 1974
Castillo de Beliver 1983

203

The Environment: Pollution

THE ENVIRONMENT: CLIMATIC CHANGE

A curious shift in power has taken place over the past century. In 1900 most people in the world were still utterly dependent on the climate. Drought, frost, hail storms or floods could destroy homes or ensure crop failure and starvation. National cuisines were tailored to suit climate: most inhabitants of north European countries never saw bananas, oranges or even coffee. Armies could not move until the beginning of the "campaigning season" (spring). Peasants lived by the weather. It determined their prosperity and rhythm of work: frantic labour to get in a good harvest could be followed by months of inactivity during the winter. Even in a comparatively developed country, such as France, Alpine villages might be cut off by snow for part of the year. Even the most privileged sectors of society had to go to some lengths to escape from the impact of climatic change. Kings moved to summer palaces; the British administrators of India retreated from Delhi to the hill station at Simla during the hot season. George Orwell's novel *Burmese Days* describes the obsessive interest taken by the British community in a Burmese town at the arrival of fresh consignments of ice at the local club.

Dependence on climate can still be seen. Countries like Bangladesh are highly vulnerable to floods and typhoons. Droughts can still produce starvation in Africa and hail storms can depress connoisseurs of Bordeaux wine. However, in general, industrialized countries have become less and less dependent on climate. Few people in Western countries would even bother to notice the link between the weather and the prices that they pay for vegetables in supermarkets. Air conditioning allows investment bankers in Saigon to ensure that their offices are slightly cooler than those of their colleagues in London and has contributed much to the new prosperity of certain cities in the American South.

However, the very technology that has allowed Western countries to gain control over climate has also begun to have wider and more dangerous implications. Industrial activity emits a variety of gases (nitrous oxide, methane and carbon dioxide) that create a so-called "greenhouse effect" by surrounding the earth and trapping heat. The build-up of gases might normally be alleviated by the fact that trees consume carbon dioxide and emit oxygen, but the destruction of the tropical rainforests means that there are now fewer trees. The fact that many felled trees are then burnt, allied to the uses to which the cleared land is put, further exacerbates the greenhouse effect. The result is that the overall temperature of the world is slowly rising. It is estimated that the global

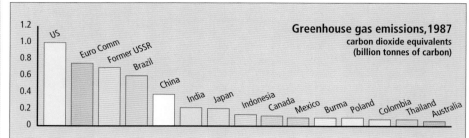

Flooding is the oldest environmental problem. Floods are a hazard for inhabitants of underdeveloped countries who live near large rivers in order to obtain access to fertile land, but who lack the resources to protect themselves against sudden increases in the water level. The map *(right)* shows the countries in the world that are prone to flooding. Bangladesh is particularly vulnerable to highly destructive floods *(left)*.

Greenhouse gas emissions, 1987
carbon dioxide equivalents
(billion tonnes of carbon)

Many areas of the world are severely affected by droughts *(map bottom)*. Drought can be explained in terms of physical geography – it is caused by lack of rainfall where the climate is dry and variable. The human impact of drought cannot be separated from the social and economic nature of the countries in which it occurs. Lack of rainfall in southern California is unlikely to have the same consequences as it would have in Ethiopia.

Predictions for sea level rise by 2100
— high
— best estimate
— low

CENTRAL AMERICA *between 33% and 50% destroyed*

COLOMBIAN CHOCÓ

ECUADOR

WESTERN AMAZONIA

BRAZIL *10% of forest destroyed*

SOUTH AMERICA

WEST AFRICA *forest almost completely gone*

SOUTH CAMEROON

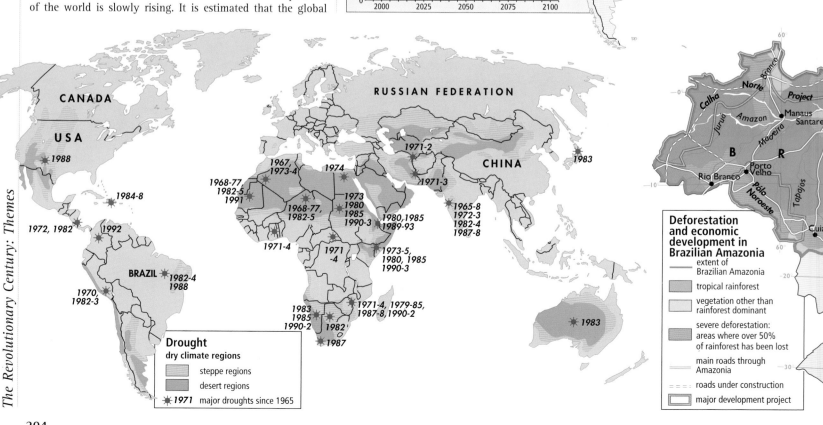

Drought
dry climate regions
- steppe regions
- desert regions
- *1971* major droughts since 1965

Deforestation and economic development in Brazilian Amazonia
— extent of Brazilian Amazonia
- tropical rainforest
- vegetation other than rainforest dominant
- severe deforestation: areas where over 50% of rainforest has been lost
— main roads through Amazonia
- - - roads under construction
- major development project

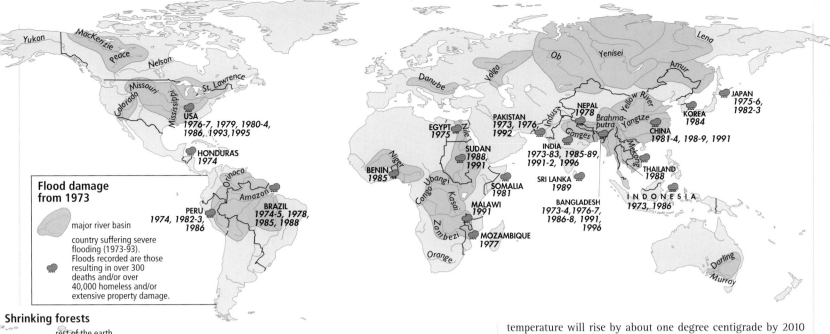

Flood damage from 1973

major river basin

country suffering severe flooding (1973-93). Floods recorded are those resulting in over 300 deaths and/or over 40,000 homeless and/or extensive property damage.

USA *1976-7, 1979, 1980-4, 1986, 1993,1995*

HONDURAS *1974*

PERU *1974-5, 1982-3, 1986*

BRAZIL *1974-5, 1978, 1985, 1988*

EGYPT *1975*

BENIN *1985*

SUDAN *1988, 1991*

SOMALIA *1981*

MALAWI *1991*

MOZAMBIQUE *1977*

PAKISTAN *1973, 1976, 1992*

NEPAL *1978*

INDIA *1973-83, 1985-89, 1991-2, 1996*

SRI LANKA *1989*

BANGLADESH *1973-4,1976-7, 1986-8, 1991, 1996*

THAILAND *1988*

INDONESIA *1973, 1986*

CHINA *1981-4, 198-9, 1991*

JAPAN *1975-6, 1982-3*

KOREA *1984*

Shrinking forests

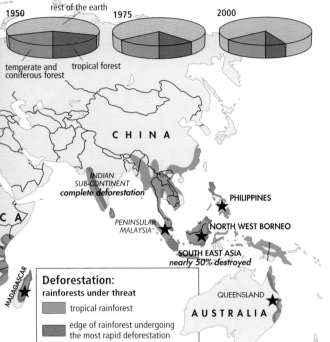

1950   1975   2000

rest of the earth

temperate and coniferous forest   tropical forest

CHINA

INDIAN SUB-CONTINENT *complete deforestation*

PENINSULAR MALAYSIA

PHILIPPINES

NORTH WEST BORNEO

SOUTH EAST ASIA *nearly 50% destroyed*

MADAGASCAR

QUEENSLAND

AUSTRALIA

Deforestation: rainforests under threat

tropical rainforest

edge of rainforest undergoing the most rapid deforestation

★ threatened area with large concentrations of endemic species

The destruction of tropical rainforests *(map and chart left)*, which take a long time to grow and which are part of a delicately balanced eco-system, has been pushed forward by logging for timber and the desire to gain access to resources such as minerals that are located in the forests. Destruction has been particularly severe around the Amazon (which accounts for about a third of the world's rainforest), especially in Brazil, a rapidly developing country which is keen to exploit the economic possibilities of the Amazon *(map below left)*.

Global warming produced by the emission of gases into the atmosphere may be having a worldwide effect *(map below)*. The result of these emissions *(chart above left)* is the "greenhouse effect", as heat is trapped by a layer of gases. One consequence is that parts of the two polar ice caps are breaking up. The melting of the ice caps is contributing to a worldwide rise in sea level. The rise in sea level by 2010 will cause grave problems for some low-lying coastal regions *(chart far left)*.

temperature will rise by about one degree centigrade by 2010 and that it may rise by another two degrees by the end of the next century. This increase is likely, in the long term, to have a variety of effects. It will change agricultural productivity in certain areas and it will melt the polar ice caps, so causing floods and reducing the habitable surface of the globe. Not all the effects of global warming will be negative: most Ethiopians would be happy to receive more rain and many English people would like a climate in which it was easy to cultivate grapes. But generally it is assumed that the effects of global warming are to be feared. Countries have discussed measures that may alleviate the problem and the European Union has agreed to prevent further increases in the emission of carbon dioxide after the year 2000, but there are good reasons to doubt that such good resolutions will be implemented. The most dramatic consequences of global warming are likely to be borne by the impoverished agricultural economies of the south, while emissions of greenhouse gases currently come mainly from the rich countries of the northern hemisphere (which also have the resources to protect themselves from the consequences of global warming). Newly developing countries see what pressure there is for a reduction in greenhouse emissions as hypocritical: denying them the means which the rich world had earlier employed for increasing its industrial base. Under these circumstances it seems hard to believe that those most responsible for global warming will have much interest in curtailing it.

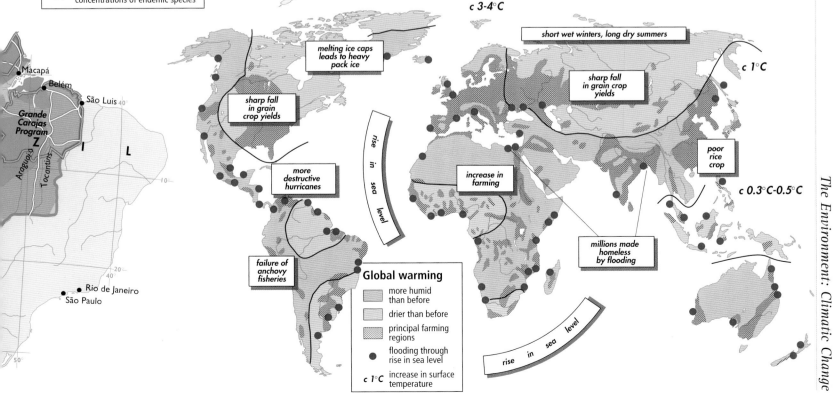

Macapá

Belém

São Luís

Grande Carajás Program

Araguaia

Tocantins

Rio de Janeiro

São Paulo

c 3-4°C

melting ice caps leads to heavy pack ice

sharp fall in grain crop yields

more destructive hurricanes

failure of anchovy fisheries

rise in sea level

short wet winters, long dry summers

c 1°C

sharp fall in grain crop yields

increase in farming

poor rice crop

c 0.3°C-0.5°C

millions made homeless by flooding

rise in sea level

Global warming

more humid than before

drier than before

principal farming regions

● flooding through rise in sea level

c 1°C increase in surface temperature

ENERGY RESOURCES

The 19th-century world was created by coal. Coal powered steam engines and drove factory machinery, railway trains and ocean liners. Coal remained important in the 20th century and underlay many of the most dramatic social and international conflicts of the age. In Britain the coal strikes of 1926, 1973 and 1984 reflected the changing balance of forces between capital and labour; on the European continent, the Franco-German conflict that lasted from 1870 to 1945 was in part a conflict about access to coal reserves in Alsace-Lorraine and the Ruhr. The end of this conflict was symbolized by the creation of the European Coal and Steel Community in 1950 and its subsequent metamorphosis into the European Union (*see* pages 114-7).

However, the characteristic technologies of the 20th century were driven by a new form of energy: oil. Petrol refined from oil fuelled motor cars and aeroplanes. Oil reserves became associated with political and social power. The two super-powers of the mid-century – the United States and the Soviet Union – were both major producers of oil, and the latter used oil supplies to assert its hegemony over Eastern Europe. However, oil reserves on their own, without military and industrial strength, did not bring power. This was shown in 1973, when the Organization of Petroleum Exporting Countries (OPEC) raised world prices of crude oil by a factor of four. The result was a downturn in the world economy and much apocalyptic talk about the danger that Arab power posed to the West. In reality, however, it soon became clear that Arabs could only use their new wealth to buy property and investments in the West, which thus made their own interests inseparable from those of the Western economies. Oil prices stabilized quickly and the OPEC cash surplus dropped from $65 billion dollars in 1974 to $10 billion in 1978. In 1991, Saudi Arabia and Kuwait were obliged to recognize their dependence on the Western industrialized nations when they accepted the protection of a United Nations coalition against Iraqi invasion.

The other characteristic energy source of the 20th century was electricity. Electricity might be generated by coal or oil but it was much easier to control and distribute than these original sources. Electricity powered the radios, televisions, computers and simple electric lights that made the lives of most people in the late 20th century almost unimaginably different from those of their grandparents.

Coal and oil were both used for purposes other than the generation of electricity, but nuclear fission, which began to be used in the 1960s, had no civil use outside power generation. Nuclear reactors seemed to offer the solution to many problems. They were "clean" (because they did not generate carbon emissions in the way that fossil fuels did), they promised to free industrialized nations from dependence on

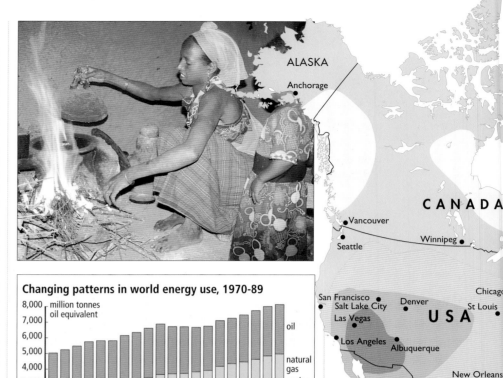

Changing patterns in world energy use, 1970-89

Global distribution of energy use varies widely *(chart below)*. Developed countries draw a higher proportion of their total energy needs from oil, gas and nuclear power. Less developed countries are much more dependent on "biomass" fuels such as wood, peat or dung, which are often used for simple tasks *(above)*.

The chart *(right)* shows the changing sources of energy used in the US since 1900. Oil became the dominant energy source in the late 1940s and its use grew sharply until the 1970s. The increase in the oil price of 1973 stimulated an interest in "alternative" sources of energy, of which the most promising seemed to be solar energy *(below left)*. The map *(above right)* shows the potential for the use of solar energy in the US.

World energy use, 1987

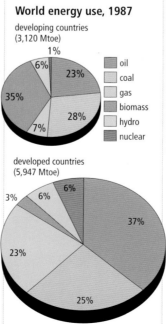

developing countries (3,120 Mtoe)

developed countries (5,947 Mtoe)

- oil
- coal
- gas
- biomass
- hydro
- nuclear

1 Mtoe = million tonnes oil equivalent

oil-producing states and they provided a means to reduce the power that trade unions had acquired in coal mines. However, nuclear reactors were so costly to construct and to keep safe after their active life was over that it was hard to calculate the long-term cost of the energy that they produced. More importantly, the "cleanliness" of nuclear power in the short term was balanced by its catastrophic capacity to do environmental damage when nuclear reactors failed to function properly.

Concern about environmental damage, and the prospect that some energy sources might become exhausted, encouraged many to look for "renewable" sources of energy: sun, wind, water and tides. However, these sources themselves proved problematic. Dams interfered with the natural course of rivers and flooded large areas. Even "wind farms" excited the ire of those who wished windswept areas to preserve their natural charm. In fact there is evidence that emphasis on energy production had distracted planners from interesting themselves in its consumption. Some studies have suggested that per capita consumption of energy in the Czech Republic is 30 to 50 per cent higher than that of more developed Western nations, while highly developed countries - such as those of Scandinavia - are notably efficient in their energy use. The simplest solution to the problems of energy seems to be to use less of it.

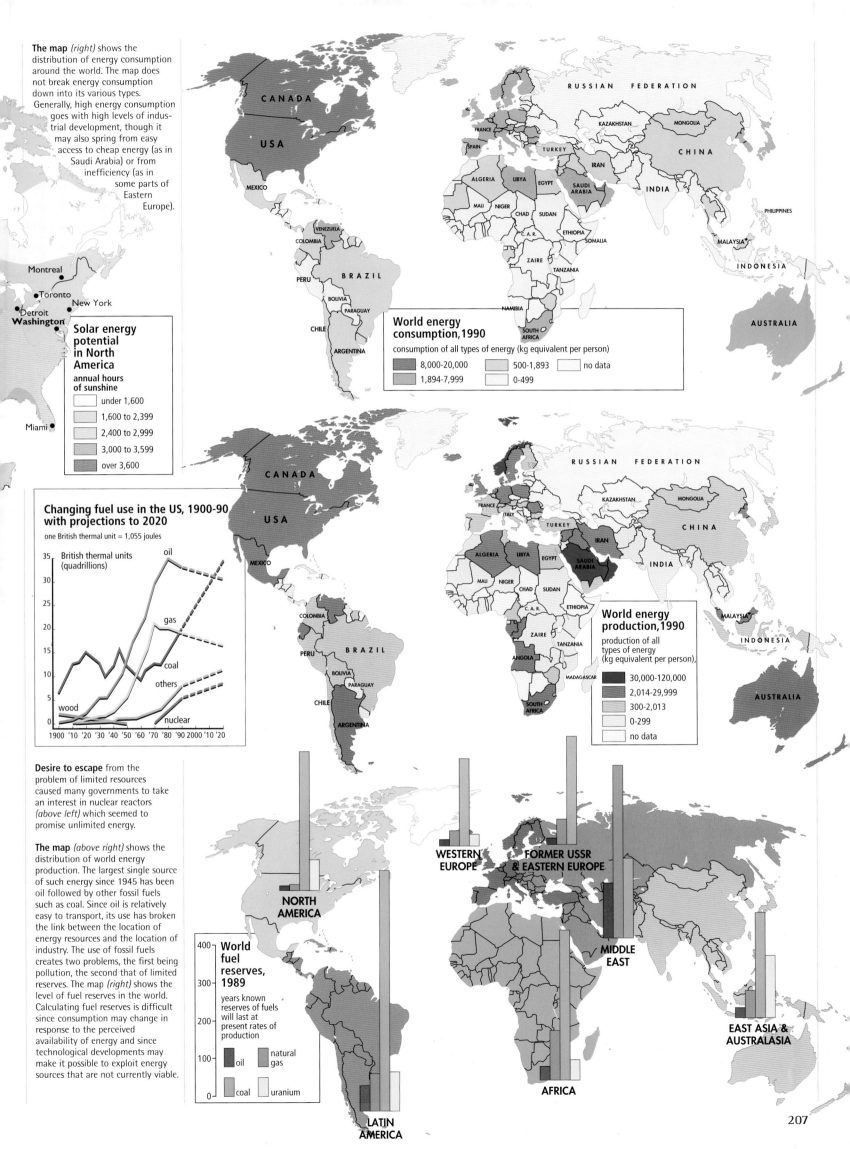

The map *(right)* shows the distribution of energy consumption around the world. The map does not break energy consumption down into its various types. Generally, high energy consumption goes with high levels of industrial development, though it may also spring from easy access to cheap energy (as in Saudi Arabia) or from inefficiency (as in some parts of Eastern Europe).

Solar energy potential in North America

annual hours of sunshine

- under 1,600
- 1,600 to 2,399
- 2,400 to 2,999
- 3,000 to 3,599
- over 3,600

Changing fuel use in the US, 1900-90 with projections to 2020

one British thermal unit = 1,055 joules

British thermal units (quadrillions)

Desire to escape from the problem of limited resources caused many governments to take an interest in nuclear reactors *(above left)* which seemed to promise unlimited energy.

The map *(above right)* shows the distribution of world energy production. The largest single source of such energy since 1945 has been oil followed by other fossil fuels such as coal. Since oil is relatively easy to transport, its use has broken the link between the location of energy resources and the location of industry. The use of fossil fuels creates two problems, the first being pollution, the second that of limited reserves. The map *(right)* shows the level of fuel reserves in the world. Calculating fuel reserves is difficult since consumption may change in response to the perceived availability of energy and since technological developments may make it possible to exploit energy sources that are not currently viable.

World energy consumption, 1990

consumption of all types of energy (kg equivalent per person)

- 8,000-20,000
- 1,894-7,999
- 500-1,893
- 0-499
- no data

World energy production, 1990

production of all types of energy (kg equivalent per person),

- 30,000-120,000
- 2,014-29,999
- 300-2,013
- 0-299
- no data

World fuel reserves, 1989

years known reserves of fuels will last at present rates of production

- oil
- coal
- natural gas
- uranium

207

TERRORISM

Terrorism is a dangerous word. All governments like to label their violent opponents as "terrorists": the word implies a small, isolated and irrational group willing to cause immense suffering to innocent civilians in pursuit of its aims. It also implies that such groups can be lumped together regardless of their aims or the context in which they operate. Changes in political circumstances may make yesterday's terrorist seem like today's freedom fighter. "Terrorists" have often been young men who subsequently went on to enjoy respectable careers. One of the seven men who carried out the single most important act of terrorism in history (the assassination of Archduke Franz Ferdinand, which sparked the First World War) ended his days as director of the Institute of Historical Research in Belgrade. Independence movements in countries such as Cyprus and Kenya were once labelled terrorist by their opponents. Israel, some of whose founders bombed the King David Hotel in Jerusalem, is now seen as the state that confronts the most persistent terrorist problem. In Algeria the Front de Libération Nationale (FLN) once employed terrorist tactics against the French rulers of the country, but has now become a government party that campaigns against the "terrorism" of Islamic militants in its own domain.

In general terms terrorism in the 20th century has been characterized by an awareness that its perpetrators cannot hope to gain outright military victory. Their actions are designed to provide "propaganda by the deed" to draw attention to their grievances and possibly to provoke their opponents into counter-productive repression. Terrorism has been rendered effective by two things. The first is technology, especially in the form of portable high explosives, which has allowed small numbers of determined people to do great damage. The second is the publicity provided first by the press, and later by radio and television.

Assassinations by anarchist and nationalist groups in France, Russia and the Balkans during the years leading to the First World War spread the idea of terrorism. During the inter-war period there were also some spectacular terrorist attacks, such as the murder of the Yugoslav king and French foreign minister in 1934 in Marseilles. However, repressive regimes, such as those of Hitler, Stalin and, in the early years of his rule, General Franco, were not very susceptible to terrorist threats. Such regimes were impervious to the loss of civilian life and willing to repress terrorism with extreme ferocity.

After the Second World War terrorism began to be practised by many in the Middle East and, particularly, by Palestinians who wished to attack the state of Israel. Supporters of the Palestinian Liberation Organization launched a spectacular

The Provisional IRA's campaign of violence, which aimed at establishing a united Ireland, began in the late 1960s. In 1996, after a ceasefire during which they had hoped to negotiate with the British government, the IRA resumed their bombing campaign on the British mainland. The photograph *(above)* shows a bus destroyed by a bomb that exploded prematurely in central London shortly after this resumption of violence.

The map *(right)* shows the location of terrorist assaults on state power in the 20th century. Successful Marxist revolutions involving actions that some defined as terrorism occurred in parts of Africa and Central America during the 1970s. These attacks involved large bands of well-organized troops with

foreign support that in some ways resembled more the military/political seizures of power in Eastern Europe after 1945 than the isolated and marginalized groups of young people who are associated with terrorism in much of Western Europe. Marxist organizations were also active in developed countries (France, Germany, Italy) as well as in parts of the developing world such as Peru. There were considerable differences between groups in Western Europe (usually small, urban and middle class) and those in the developing world (often enjoying a large base of popular support in the countryside). Alongside Marxist assaults on state power were those of nationalist movements, characterized in Europe by groups such as ETA and the IRA.

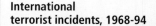

International terrorist incidents, 1968-94

year	kidnap	hijack	bombing	hostage barricade	assassination and shooting
1994	56	16	121	2	61
1993	41	20	120	10	48
1992	28	10	141	3	45
1991	24	13	155	2	52
1990	22	13	155	2	52
1989	34	10	175	5	46
1988	38	7	225	8	57
1987	52	5	181	6	51
1986	53	5	210	4	36
1985	71	15	231	4	71
1984	43	22	203	–	63
1983	35	34	190	1	39
1982	31	31	249	8	59
1981	20	24	169	6	46
1980	17	23	121	15	51
1979	16	12	139	13	43
1978	24	10	127	13	33
1977	23	20	157	4	28
1976	25	8	183	4	46
1975	35	3	160	12	25
1974	17	7	179	11	46
1973	31	10	93	10	19
1972	10	31	121	3	12
1971	12	16	90	1	8
1970	28	51	88	2	14
1969	4	72	62	–	5
1968	–	28	73	–	8

THE RED BRIGADES

The Red Brigades were supported by young left-wing intellectuals in north Italian cities. They drew their inspiration from the resistance to German occupation in 1943-5 and from Latin American guerrilla movements of the 1960s and 70s. The Brigades were also influenced by the student and labour protests of Italy's "hot autumn" (1969). The ideology of the movement was diverse: some were Maoists, others rebels from the mainstream Communist Party. The Brigades argued that terrorist attacks would expose "the latent violence of the state" and accelerate the coming of civil war between oppressors and the oppressed. In 1972, the kidnapping of a manager at the Sit Siemens factory lasted for just 20 minutes. In 1974 the kidnapping of the Genoese judge Mario Sossi lasted 25 days but ended with his release after the government refused to release political prisoners. In 1978 the kidnapping of the minister of the interior, Aldo Moro, lasted for 54 days and finished with his murder. Terrorist violence increased after this date (30 people were killed in 1980 alone). However, police action and the use of "repentant" terrorists to provide information eventually ensured that the few hundred activists who provided the backbone of the Red Brigades were all either dead, despairing or in prison.

The chart *(above)* illustrates various types of terrorist attack. The use of such tactics is governed by the nature of the state that terrorists confront. Kidnapping is particularly appropriate to capitalist societies. Airline hijackings are appropriate to Middle Eastern terrorists who wish to bring their cause to the attention of an international community that would take less notice of events happening within Lebanon or Iran. Many of the attacks depend on cultures where fighters place little value on their own lives. Much of the fear of Islamic fundamentalist terrorism *(above)* comes from the perception that "suicide bombers" will not be deterred by risks to their own safety.

NORTH AMERICA

114 78 126 129 131 52 21 20 11

LATIN AMERICA

200 160 222 147 195 228 307 402 188

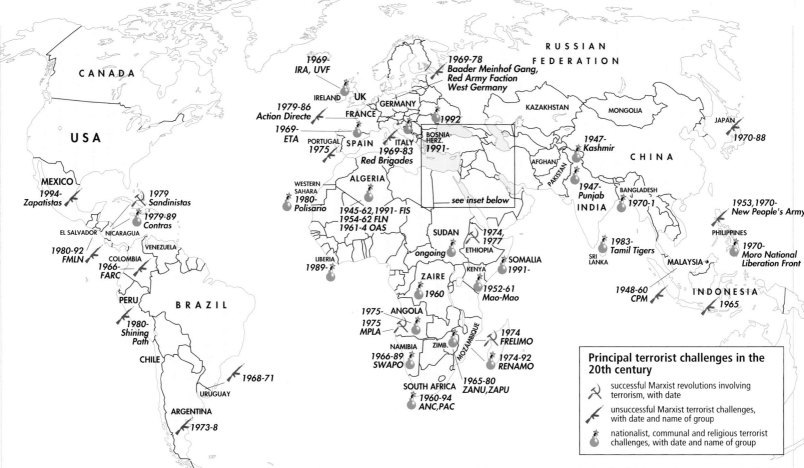

CANADA

USA

MEXICO
1994-
Zapatistas

EL SALVADOR
1980-92
FMLN

NICARAGUA
1979
Sandinistas
1979-89
Contras

VENEZUELA

COLOMBIA
1966-
FARC

PERU
1980-
Shining Path

BRAZIL

CHILE

URUGUAY
1968-71

ARGENTINA
1973-8

IRELAND
UK
1969-
IRA, UVF

FRANCE
1979-86
Action Directe

GERMANY
1969-78
Baader Meinhof Gang, Red Army Faction
West Germany

RUSSIAN FEDERATION

1992

BOSNIA-HERZ.
1991-

see inset below

SPAIN
1969-
ETA

PORTUGAL
1975

ITALY
1969-83
Red Brigades

WESTERN SAHARA
1980-
Polisario

ALGERIA
1945-62, 1991- FIS
1954-62 FLN
1961-4 OAS

LIBERIA
1989-

SUDAN
ongoing

ETHIOPIA
1974, 1977

SOMALIA
1991-

ZAIRE
1960

KENYA
1952-61
Mao-Mao

ANGOLA
1975
1975
MPLA

NAMIBIA
1966-89
SWAPO

ZIMB.

MOZAMBIQUE
1974
FRELIMO
1974-92
RENAMO

1965-80
ZANU, ZAPU

SOUTH AFRICA
1960-94
ANC, PAC

KAZAKHSTAN

MONGOLIA

JAPAN
1970-88

AFGHAN.
PAKISTAN
1947-
Kashmir

CHINA

1947-
Punjab
INDIA

BANGLADESH
1970-1

SRI LANKA
1983-
Tamil Tigers

1953, 1970-
New People's Army

PHILIPPINES
1970-
Moro National Liberation Front

MALAYSIA
1948-60
CPM

INDONESIA
1965

Principal terrorist challenges in the 20th century

- successful Marxist revolutions involving terrorism, with date
- unsuccessful Marxist terrorist challenges, with date and name of group
- nationalist, communal and religious terrorist challenges, with date and name of group

1994- CHECHENIA

GEORGIA

GREECE
1948-9 ELAS
1973-
November 17

TURKEY
1968-
Dev Sol

1984- PKK

ARMENIA

AZERBAIJAN
1992

1991-

1976-
PUK

CYPRUS
1951-60, 1971-4
EOKA EOKA B

SYRIA
LEBANON

ISRAEL

1982-
Hezbollah

IRAQ

IRAN

LIBYA
EGYPT
1992-
Gamaat Islamiya, Islamic Jihad

1937-48 Irgun Zvai Leumi, Stern Gang
1956-94 PLO
1987- Hamas, Islamic Jihad

SAUDI ARABIA

series of attacks during the 1970s and 1980s and gained publicity with a new tactic of hijacking international aircraft. Such attacks excited interest from middle class students in developed countries who made up the core of movements such as the French Action Directe, the West German Red Army Faction and the Italian Red Brigades. In 1978 the Red Brigades kidnapped and murdered the Italian minister of the interior, Aldo Moro. These terrorist groups were characterized by a desire to challenge the social order and a sympathy for liberation struggles in the Third World. However, in the long term, such groups were defeated. Left-wing terrorism was undermined by the gradual spread of disillusion with ideologies that had seemed so fashionable in the 1960s, as many young people retreated from political activism altogether.

The terrorism that survived during the 1980s was increasingly

The map *(above)* shows the locations of terrorist attacks between 1968 and 1994. The figures are deceptive. Attacks are more likely to be reported and recorded in Western Europe or North America than in underdeveloped countries where forces of order may be reluctant to acknowledge the scale of a problem or where victims of terrorism may be unable to report their status. It could be argued that the most effective terrorists are those who are able to hold sway over an area without open displays of violence.

Marxist and nationalist terrorists often saw the US as the supreme representative of the capitalism and imperialism that they wished to attack. However, the heartlands of the US remained remarkably free of terrorist violence. The most spectacular terrorist attack in recent times was that in Oklahoma City on 20 April 1995 *(below right)*. Paradoxically the attack appears to have been the work not of those foreign terrorists who hated the US but of a right-wing militia whose members believed themselves to be defending American values.

linked to various forms of nationalism. Obvious examples were the activities of the Provisional IRA in Northern Ireland and ETA in the Spanish Basque country. Palestinian nationalism also continued to be linked to terrorism, even as the Palestine Liberation Organization groped its way towards a peaceful settlement with Israel, and this nationalism became increasingly associated with Islamic fundamentalism, much of it sponsored by Iran. Groups such as Hamas and Hezbollah have taken the ground vacated by the PLO in its avowed shift to more peaceful operations. The US, which had long escaped terrorist action on its own soil, also began to suffer attacks from both new and old forms of terrorism. The bombing of the World Trade Centre in 1994 by Islamic fundamentalists and the bomb attack on Oklahoma City, which seemed linked to extreme right-wing nationalism, proved no country was immune from terrorist attack.

EUROPE
89 180 299 285 500 476 329 366 349

ASIA
360 334 295 240 203 167 100

MIDDLE EAST
82 103 154 225 167

29 41 42 51 40 63

SUB-SAHARAN AFRICA
9 11 29 32 34 39 57 38 98

International terrorist incidents by region, 1968-94

Period	Incidents
1968-70	9
1971-73	11
1974-76	29
1977-79	32
1980-82	34
1983-85	39
1986-88	57
1989-91	38
1992-94	98

number of incidents in region, by three-year period

DRUGS

In 1988 106 states signed a United Nations Drugs Convention, the most comprehensive programme yet devised for combating the traffic in illicit drugs. By the late 1980s the value of the drugs trade was put at $300 billion, half of it from Latin America. By the mid-1990s the trade was estimated to be worth $500 billion, making it the single most lucrative business sector in the world.

There was nothing new about the drugs trade, but its scale and the geography of drug use and abuse has changed over the century. In 1900 narcotic drugs were widely produced and consumed in much of Asia. Their medicinal properties were already known. The trade in morphine (derived from the opium poppy) and in cocaine was legal where it was used for medical and scientific purposes – and large quantities were used in the First World War to dull the pain of horrific wounds. Drugs were also traded illegally to supply addicts, most of whom were to be found in the Far East and southern Asia. It was this traffic that first led to international attempts to control the production and movement of drugs, and to see the whole issue as a "drugs problem".

The first international effort to curb the flow of drugs came in 1912 with a convention signed at The Hague by states keen to eradicate a trade they equated with vice and crime. In 1914 the Harrison Act in the United States made hard drugs illegal – and thereby pushed the whole trade into the hands of criminal syndicates. Further conventions came in 1925 and 1936 as the League of Nations struggled to find ways to enforce control. Production of opium was regulated in British India, but flourished in Iran and China, where in the 1930s an estimated 80 per cent of all opium was produced and ten per cent of the population – 40 million people – were regular smokers of the drug.

The pattern of drug abuse and drug-related crime was affected by political events and social changes, particularly large-scale urbanization, which threw up an endless stream of vulnerable new customers. The key political event was the Chinese revolution in 1949. The communist regime began a tough programme of eradication of drugs, which was virtually

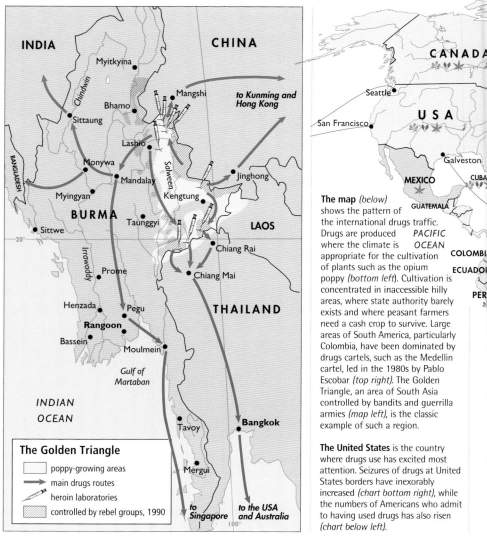

The Golden Triangle

☐	poppy-growing areas
→	main drugs routes
⚷	heroin laboratories
▨	controlled by rebel groups, 1990

The map (below) shows the pattern of the international drugs traffic. Drugs are produced where the climate is appropriate for the cultivation of plants such as the opium poppy (bottom left). Cultivation is concentrated in inaccessible hilly areas, where state authority barely exists and where peasant farmers need a cash crop to survive. Large areas of South America, particularly Colombia, have been dominated by drugs cartels, such as the Medellín cartel, led in the 1980s by Pablo Escobar (top right). The Golden Triangle, an area of South Asia controlled by bandits and guerrilla armies (map left), is the classic example of such a region.

The United States is the country where drugs use has excited most attention. Seizures of drugs at United States borders have inexorably increased (chart bottom right), while the numbers of Americans who admit to having used drugs has also risen (chart below left).

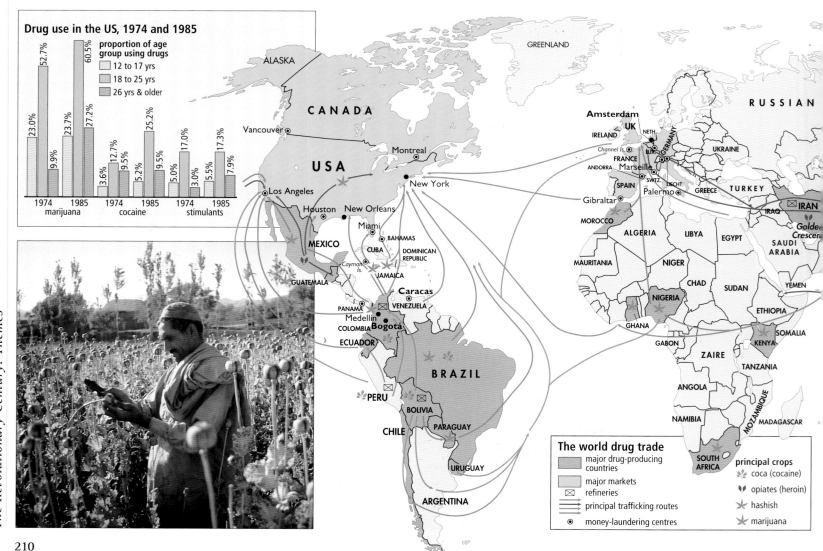

Drug use in the US, 1974 and 1985

proportion of age group using drugs
- ☐ 12 to 17 yrs
- ☐ 18 to 25 yrs
- ☐ 26 yrs & older

marijuana
- 1974: 23.0%, 52.7%, 9.9%
- 1985: 27.2%, 60.5%, 17.0%

cocaine
- 1974: 3.6%, 12.7%, 5.2%
- 1985: 9.5%, 25.2%, 9.5%

stimulants
- 1974: 5.0%, 3.0%, 5.5%
- 1985: 17.3%, 7.9%

The world drug trade

▨	major drug-producing countries
☐	major markets
⊠	refineries
≡	principal trafficking routes
⊙	money-laundering centres

principal crops
- 🌿 coca (cocaine)
- ❦ opiates (heroin)
- ✳ hashish
- ✴ marijuana

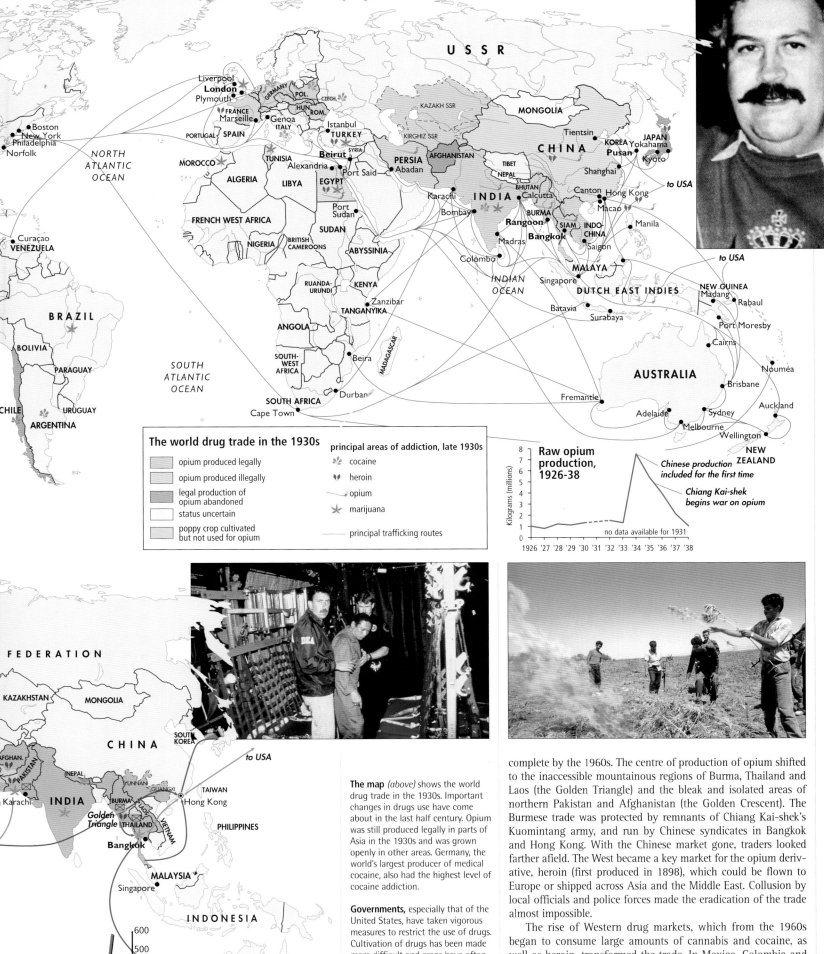

The world drug trade in the 1930s

- opium produced legally
- opium produced illegally
- legal production of opium abandoned
- status uncertain
- poppy crop cultivated but not used for opium

principal areas of addiction, late 1930s

- cocaine
- heroin
- opium
- marijuana
- principal trafficking routes

Raw opium production, 1926-38

Chinese production included for the first time

Chiang Kai-shek begins war on opium

no data available for 1931

Kilograms (millions): 8 7 6 5 4 3 2 1

1926 '27 '28 '29 '30 '31 '32 '33 '34 '35 '36 '37 '38

Narcotics seizures at US borders, 1970-87

US$ (millions): 600 500 400 300 200 100 0

1970 1975 1980 1985 1990

Golden Triangle

The map *(above)* shows the world drug trade in the 1930s. Important changes in drugs use have come about in the last half century. Opium was still produced legally in parts of Asia in the 1930s and was grown openly in other areas. Germany, the world's largest producer of medical cocaine, also had the highest level of cocaine addiction.

Governments, especially that of the United States, have taken vigorous measures to restrict the use of drugs. Cultivation of drugs has been made more difficult and crops have often been destroyed in operations such as that in Lebanon *(above right)*. The US has also pressured governments in Latin America to take action against prominent drugs barons. In the case of Manuel Noriega, the president of Panama who was suspected of facilitating the cocaine trade, the US government invaded a whole country in order to bring about a single arrest *(above)*.

complete by the 1960s. The centre of production of opium shifted to the inaccessible mountainous regions of Burma, Thailand and Laos (the Golden Triangle) and the bleak and isolated areas of northern Pakistan and Afghanistan (the Golden Crescent). The Burmese trade was protected by remnants of Chiang Kai-shek's Kuomintang army, and run by Chinese syndicates in Bangkok and Hong Kong. With the Chinese market gone, traders looked farther afield. The West became a key market for the opium derivative, heroin (first produced in 1898), which could be flown to Europe or shipped across Asia and the Middle East. Collusion by local officials and police forces made the eradication of the trade almost impossible.

The rise of Western drug markets, which from the 1960s began to consume large amounts of cannabis and cocaine, as well as heroin, transformed the trade. In Mexico, Colombia and Bolivia, new drugs cartels emerged to supply cocaine to North America. The drug barons wielded enormous power, dominating large parts of central South America. The vast profits from the trade were laundered through offshore banking houses, and ended up in the international financial system as "legal" funds for investment. The $80 million available to the three UN agencies fighting the drugs trade was eclipsed in the 1990s by the vast wealth of the drug traders, able to live beyond the law, in a global underworld with its own power-brokers and its own rules.

Drugs

SPACE EXPLORATION

The idea of space exploration has haunted the 20th-century imagination. At the beginning of the century, H. G. Wells fantasized about the possibility. By its end a whole genre of films and television programmes (*Star Trek*, *Alien*, *2001 A Space Odyssey*) had grown up around the theme of space travel. Early progress in space flight was necessarily slowed by technological constraints. The basic theory of space flight was laid down by the Russian Konstantin Tsiolkovsky in 1903; in 1925 the *Verein für Raumschiffahrt* (Society for Spaceship Travel) was founded in Germany; in 1926 the American Robert Goddard built and flew the first liquid-propelled rocket.

Space travel has been intimately tied up with military developments. The most effective early rockets were the German V2 missiles that were developed during World War Two by Wernher von Braun, who subsequently helped design the mechanisms which propelled the first US satellite, Explorer 1, in 1958 and the Saturn V rocket that took the Apollo spacecraft to the moon in the late 1960s. It was the USSR, however, which launched Sputnik, the first artificial earth satellite, in October 1957.

It was always understood that space travel had military applications. Observation satellites transformed the possibility of intelligence gathering and allowed the Russians and Americans to understand each other's military potential with a new degree of precision. Satellites also lay at the heart of Ronald Reagan's "Strategic Defence Initiative" – a project to destroy incoming Soviet missiles with lasers fired from space, which seems to have been taken seriously by Soviet leaders if not by many scientists in the West.

Space travel also became connected with a more subtle struggle between the US and the Soviet Union. The Soviet success in launching the first satellite to orbit the earth in 1957 and, in 1961, the first manned space flight shocked the Americans. These developments encouraged the US to invest greater resources in its space programme and provoked President Kennedy to pledge that Americans would be the first to land on the moon (an aim achieved in 1969). Even after the hostilities of the Cold War began to subside, the United States and the USSR continued to compete in space and even their moments of co-operation (symbolized by astronauts shaking hands in the space stations) were self-consciously tied to the development of international relations.

But space travel was also linked to the 20th century's need for heroes. Spacecraft never needed to be manned. The first satellites were little more than radio transmitters, while the first living creatures to be sent into space were a dog and a monkey. However, after the Soviet Yuri Gagarin became the first man to fly in space (in 1961), an obsession developed with putting human beings in space. The early astronauts became international heroes and Neil Armstrong's words "one

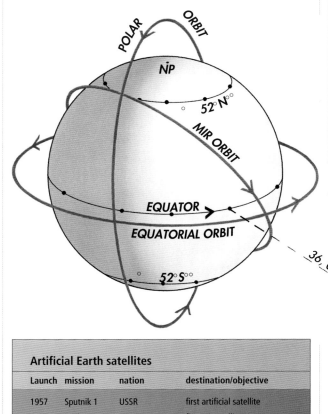

Artificial Earth satellites

Launch	mission	nation	destination/objective
1957	Sputnik 1	USSR	first artificial satellite
1958	Explorer 1	USA	first US satellite
1960	TIROS 1	USA	first weather satellite
1961	Vostok 1	USSR	Yuri Gagarin: first man in space
1962	Mercury	USA	John Glenn: first American orbits the earth
1963	Syncom 2	USA	first 24-hr synchronous orbit
1963	Vostok 3	USSR	Valentina Tereshkova: first woman in space
1971	Salyut 1	USSR	first space station
1975	Soyuz/Apollo	USSR/USA	docking and meeting in space
1977	Meteosat 1	ESA	weather satellite in 24-hr orbit
1986	Mir	USA/Russia/ESA	first permanently occupied station
1990	Hubble S.T.	USA/ESA	large space telescope in orbit
1995	Shuttle/Mir	USA/Russia	first of many dockings
1996	SOHO	ESA	continuous solar observation
1998	ISS	USA/Russia/ESA	international space station

The Hubble Space Telescope, launched in 1990, has provided astonishing pictures of deep space with a clarity previously not possible. Here (*below left*), embryonic stars emerge from a pillar of dense molecular hydrogen and dust in the Eagle Nebula.

The pictures and charts (*above and above right*) illustrate the distances that spacecraft have been propelled. A variety of orbits (*above*) around the earth have been achieved by artificial satellites since the launch of the Russian Sputnik in 1957. A 24-hour synchronous orbit (first achieved by Syncom 2 in 1963) requires a distance of 36,000 km from the earth. The orbit of the Russian space station Mir passes over all points between 52° N and 52° S. The distances travelled by space probes designed to investigate other planets in the solar system are vastly greater (*above right*). Travel to the moon involved a journey of 400,000 kilometres. The Voyager 1 and 2 probes had to travel thousands of millions of kilometres in order to reach Jupiter, Saturn, Uranus and Neptune. It took four hours for pictures of Neptune taken by Voyager 2 to reach the earth.

small step for [a] man, one giant leap for mankind" (spoken as he stepped onto the moon) became among the most widely quoted of the century.

During the 1970s and 1980s heavier and more sophisticated satellites were put into orbit and probes sent to more distant planets. No fewer than five craft were sent to observe Halley's comet. Flights by the US space shuttles to launch and service satellites have become almost routine. The European Space Agency (ESA), China and Japan now launch satellites and ESA and Japan build planetary probes.

The two most important consequences of space exploration were probably ones that had nothing to do with the drama of manned flight. The first concerned international communications. Satellites made it much easier for broadcasters to beam their programmes around the world and much more difficult for governments to control these broadcasts. The second consequence of space exploration was that scientists acquired a far more detailed knowledge of the universe. Surveys have been made of all the planets from Mercury to Neptune, and of many of their satellites. Probes were sent to other planets in the solar system and devices positioned in space, such as the Hubble Telescope, were able to pick up signals that came from far away in both space and time – even to acquire the first pictures of the Big Bang that created the universe.

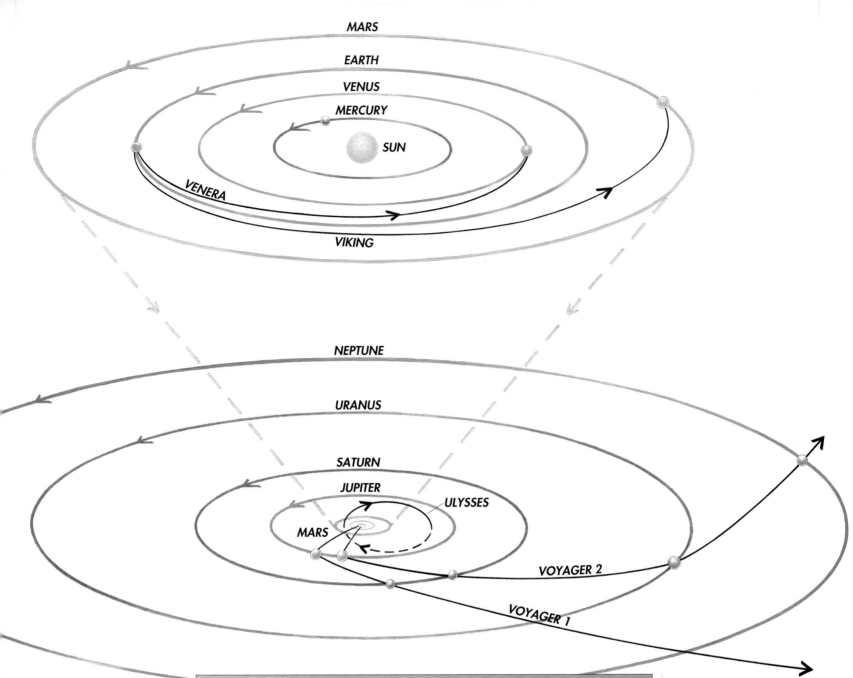

MARS

EARTH

VENUS

MERCURY

SUN

VENERA

VIKING

NEPTUNE

URANUS

SATURN

JUPITER

ULYSSES

MARS

VOYAGER 2

VOYAGER 1

Probes to the moon and planets

launch date	mission	nation	destination/objective
1959	Luna 3	USSR	first pictures of far side of moon
1966/68	Surveyor 1-7	USA	landing on moon: Apollo sites
1968	Apollo 8	USA	first men round the moon
1969	Apollo 11	USA	first manned moon landing
1970	Venera 7	USSR	first pictures from surface of Venus
1973	Mariner 10	USA	only probe to Mercury to date
1976	Viking 1&2	USA	first pictures from surface of Mars
1977	Voyager 1&2	USA	Jupiter, Saturn, Uranus, Neptune
1989	Magellan	USA	detailed radar mapping of Venus
1989	Galileo	USA	orbiter and probe to Jupiter
1990	Ulysses	USA/ESA	probe over sun's poles

Space travel provided a means by which the US and the Soviet Union could express their rivalry without actual violence. However, the two countries sometimes sought to use space as a platform for ostentatious gestures of international solidarity and co-operation. A spectacular example of this was provided in 1996 when the American space shuttle Atlantis linked up with the Russian space station Mir while in orbit around the earth *(top left)*.

Yuri Gagarin *(below left)* was the first man to fly in space when his Vostok 1 craft completed one revolution around the earth on 12 April 1961. Valentina Tereshkova *(below right)* became the first woman to fly in space when her Vostok 6 craft completed 48 revolutions around the earth between 16 and 19 June 1963. Cosmonauts – and astronauts – became national heroes and symbols of their nation's technological advance.

In 1972 Apollo 17 took Eugene Cernan and Harrison Schmitt to the moon *(left)*. They spent 72 hours on the planet and made three expeditions in the Lunar Roving Vehicle. After this, the United States abandoned its moon-landing programme and devoted its energies to sending unmanned probes to more distant destinations and building a space station to be placed in permanent orbit around the earth. Eugene Cernan thus became the last man to stand on the moon.

BIBLIOGRAPHY

Aldcroft, D., *From Versailles to Wall Street* (London, 1977)

Allen, D., Long, N., *Coming to Terms: Indo-China, the United States and the War* (Bolder, Co., 1991)

Allen, G.C., *A Short Economic History of Modern Japan* (London, 1972)

Ambrose, S., *Rise to Globalism: American Foreign Policy since 1918* (London, 1971)

Anslinger, H.J., Tompkins, W.F., *The Traffic in Narcotics* (New York, 1953)

Armstrong, P., Glyn, A., Harrison, J., *Capitalism since World War Two* (London, 1984)

Armytage, W.H., *A Social History of Engineering* (London, 1961)

Ashworth, W., *A Short History of the International Economy since 1850* (London, 1975)

Badger, A.J., *The New Deal: The Depression Years 1933-1940* (London, 1989)

Bale, J., *Sports Geography* (London, 1989)

Banks, A., *A Military Atlas of the First World War* (London, 1989)

Barber, J., Harrison, M., *The Soviet Home Front 1941-1945* (London, 1991)

Barnhart, M., *Japan Prepares for Total War* (New York, 1987)

Barnouw, E., *Tube of Plenty: The Evolution of American Television* (2nd ed., New York, 1990)

Berghahn, V., *Modern Germany: Society, Economy and Politics in the Twentieth Century* (Cambridge, 1982)

Bernard, P., Dubief, H., *The Decline of the Third Republic 1914-1938* (Cambridge, 1985)

Bernstein, M.A., *The Great Depression* (Cambridge, 1987)

Bhatia, B.M., *Famines in India 1860-1965* (New Delhi, 1967)

Biddis, M., *The Age of the Masses: Ideas and Society in Europe since 1870* (London, 1977)

Blanning, T.C.W., *The Oxford Illustrated History of Modern Europe* (Oxford, 1996)

Bragadin, M., *The Italian Navy in World War Two* (Annapolis, 1957)

Brandt, K., *Management of Agriculture and Food in the German Occupied and Other Areas of Fortress Europe* (Stanford, 1953)

Brenan, G., *The Spanish Labyrinth* (Cambridge, 1964)

Breuilly, J., *Nationalism and the State* (2nd ed. Manchester, 1993)

Brogan, H., *The Longman History of the United States* (London, 1985)

Brown, C., Mooney, P., *Cold War to Détente 1945-80* (London, 1981)

Brown, K., Pearce, D.W. (eds), *The Cause of Tropical Deforestation* (London, 1994)

Bullock, A., *Hitler and Stalin: Parallel Lives* (London, 1991)

Byron, M., *Post-War Caribbean Migration to Britain: The Unfinished Cycle* (Aldershot, 1994)

Cariedes, C.N., *Elections in Chile: The Road toward Democratization* (Boulder, Co., 1991)

Carr, W., *A History of Germany 1815-1990* (4th ed., London, 1991)

Caute, D., *Sixty-Eight: The Year of the Barricades* (London, 1988)

Cheston, T.S., Loeffke, B., *Aspects of Soviet Policy toward Latin America* (New York, 198?)

Cipolla, C., *The Economic History of World Population* (6th ed., London, 1975)

Clutterbuck, R., *Terrorism in an Unstable World* (London, 1994)

Collins, J.M., *American and Soviet Military Trends since the Cuban Missile Crisis* (Washington D.C., 1978)

Cook, C., Paxton, J., *European Political Facts 1918-1973* (London, 1975)

Crnobrnja, M., *The Yugoslav Drama* (London, 1994)

Crowder, M. (ed.), *The Cambridge History of Africa: Vol. 8 c.1940 to c.1975* (Cambridge, 1984)

Cunningham, W.P., Saigo, B.W., *Environmental Science: A Global Concern* (New York, 1992)

Dallek, R., *Franklin D. Roosevelt and American Foreign Policy 1932-1945* (Oxford, 1979)

Dallin, D., Nicholaevsky, O., *Forced Labour in Soviet Russia* (London, 1947)

Danchev, A. (ed.), *Fin de Siècle: The Meaning of the Twentieth Century* (London, 1995)

Davenport, T.R.H., *South Africa: A Modern History* (3rd ed., London, 1987)

Davies, R.E.G., *A History of the World's Airlines* (London, 1964)

Davies, R.W., Harrison, M., Wheatcroft, S.E. (eds), *The Economic Transformation of the Soviet Union 1913-1945* (Cambridge, 1994)

Dear, I.C.B., Foot, M.R.D. (eds), *The Oxford Companion to the Second World War* (Oxford, 1995)

Deist, W. et al, *Germany and the Second World War: Vol. 1* (Oxford, 1990), *Vol. 2* (Oxford, 1991)

Dollinger, H., *Der Erste Weltkrieg* (Munich, 1965)

Dülffer, J., *Nazi Germany 1933-1945* (London, 1996)

Dulles, J.W.F., *Vargas of Brazil: A Political Biography* (Austin, Tex., 1967)

Eatwell, R., *Fascism: A History* (London, 1995)

Eichengreen, B., *Golden Fetters* (Oxford, 1992)

Erickson, J., Dilks, D. (eds), *Barbarossa: The Axis and the Allies* (Edinburgh, 1994)

Erickson, J., *The Road to Berlin: Stalin's War with Germany* (London, 1983)

Erickson, J., *The Road to Stalingrad* (London, 1975)

Fage, J.D., *A History of Africa* (3rd ed., London, 1995)

Fairbank, J.K., Lui, K-C., *The Cambridge History of China 1800-1911* (Cambridge, 19??)

Fairbank, J.K., Reischauer, E.O., *China: Tradition and Transformation* (London, 1979)

Faulkner, H.U., *American Economic History* (9th ed., New York, 1976)

Fearon, P., *The Origins and Nature of the Great Slump, 1928-1932* (London, 1979)

Fearon, P., *War, Prosperity and Depression in the US Economy 1917-1945* (London, 1986)

Feldman, G., *The Great Disorder: Politics, Economics and Society in the German Inflation 1914-1924* (New York, 1993)

Fejtö, F., *A History of the People's Democracies* (London, 1974)

Ferro, M., *The Great War 1914-1918* (London, 1973)

Ferro, M., *October 1917: A Social History of the Russian Revolution* (London, 1980)

Foreman-Peck, J., *A History of the World Economy: International Economic Relations since 1850* (2nd ed., London, 1995)

Freedman, L., *Britain and Nuclear Weapons* (London, 1980)

Freeman, M., *Atlas of Nazi Germany* (London, 1987)

Fukatake, T., *The Japanese Social Structure: 20th Century Development* (Tokyo, 1982)

Galbraith, J.K., *The Great Crash, 1929* (London, 1961)

Geary, R., *European Labour Protest 1848-1939* (London, 1981)

Getty, J.A., *The Origins of the Great Purges: The Soviet Communist Party Reconsidered* (Cambridge, 1985)

Glantz, D., *The Military Strategy of the Soviet Union: A History* (London, 1992)

Gould, P., *The Slow Plague: Geography of the Aids Pandemic* (Oxford, 1993)

Gray, J., *Rebellion and Revolution: China from the 1800s to 1980* (Oxford, 1990)

Grenville, J.A.S., *The Major International Treaties 1914-1973* (London, 1974)

Guback, T.H., *The International Film Industry* (Bloomington, Ind., 1969)

Guttmann, A., *The Olympics: A History of the Modern Games* (Chicago, 1992)

Haigh, R.W., Gerbner, G., Byrne, R.B., *Communications in the Twenty-First Century* (New York, 1981)

Halliday, J., Cumings, B., *Korea: The Unknown War* (London, 1988)

Hannah, L., *The Rise of the Corporate Economy* (2nd ed., London, 1983)

Harries, M., Harries, S., *Soldiers of the Sun: The Rise and Fall of the Imperial Japanese Army* (London, 1991)

Harrison, M., *Soviet Planning in Peace and War 1938-1945* (Cambridge, 1985)

Hart, J.M., *Revolutionary Mexico: The Coming and Process of the Mexican Revolution* (Berkeley, 1987)

Herbert, U., *Fremdarbeiter* (Bonn, 1985)

Higashi, C., *Internationalisation of the Japanese Economy* (Boston, MA, 1987)

Hilberg, R., *Perpetrators, Victims, Bystanders: The Jewish Catastrophe 1933-1945* (London, 1993)

Hogan, M.J., *The Marshall Plan: America, Britain and the Reconstruction of Western Europe 1947-1952* (Cambridge, 1987)

Holland, R.F., *European Decolonization* (London, 1978)

Hopfinger, K., *Beyond Expectation: The Volkswagen Story* (London, 1954)

Horsley, W., Buckley, R., *Nippon: Japan since 1945* (London, 1990)

Hosking, G., *A History of the Soviet Union* (London, 1985)

Hoyt, E., *Japan's War: The Great Pacific Conflict* (London, 1986)

Hsü, I.C.Y., *The Rise of Modern China* (4th ed., Oxford, 1990)

Jelavich, B., *History of the Balkans: The Twentieth Century* (Cambridge, 1983)

Joll, J., *Europe since 1870* (London, 1980)

Jones, H.R., *Population Geography* (2nd ed., London, 1990)

Kaelbe, H., *A Social History of Western Europe 1880-1980* (Dublin, 1989)

Kahn, H. *The Emerging Japanese Superstate* (London, 1971)

Karlsson, M., Sturesson, L. (eds), *The World's Largest Machine: Global Communications and the Human Condition* (Stockholm, 1995)

Karnow, S., *Vietnam: A History* (London, 1983)

Kater, M., *The Nazi Party: A Social Profile of Members and Leaders 1919-1945* (Oxford, 1983)

Kemp, A., *The Maginot Line: Myth and Reality* (London, 1981)

Kennedy, P., *The Rise and Fall of the Great Powers* (London, 1988)

Keylor, W.R., *The Twentieth Century World* (Oxford, 1992)

Kim, Byoung-Lo, *Two Koreas in Development* (London, 1992)

Kindleberger, C., *The World in Depression 1929-39* (London, 1973)

Knox, M., *Mussolini Unleashed: 1939-1941* (Cambridge, 1983)

Kolb, E., *The Weimar Republic* (London, 1988)

LaFeber, W., *The American Age* (2nd ed., New York, 1994)

Langhorne, R., *The Collapse of the Concert of Europe: International Politics 1890-1914* (London, 1981)

Lavigne, M., *The Socialist Economies of the Soviet Union and Eastern Europe* (London, 1974)

Le Heron, R., Park, S.O., *The Asian Pacific Rim and Globalization* (Alershot, UK, 1995)

Lehman, D., *Democracy and Development in Latin America: Economics, Politics, and Religion in the Postwar Period* (Cambridge, 1990)

Leifer, M., *ASEAN and the Security of South East Asia* (New York, 1988)

Leighton, R.M., Coakley, R.W., *Global Logistics and Strategy* (2 vols., Washington, 1955-1968)

Lewin, R., *The American Magic: Codes, Ciphers and the Defeat of Japan* (London, 1982)

Lieberman, S., *The Growth of European Mixed Economies 1945-1970* (New York, 1977)

Lintner, B., *Cross-Border Drugs Trade in the Golden Triangle* (Durham, UK, 1991)

Loth, W., *The Division of the World 1941-1955* (London, 1988)

Luebbert, G., *Liberalism, Fascism and Social Democracy: Social Class and the Political Origin of Regimes in Inter-War Europe* (Oxford, 1991)

Macdonald, H., *Aeroflot: Soviet Air Transport since 1923* (London, 1975)

Mack Smith, D., *Mussolini's Roman Empire* (London, 1976)

Maddison, A., *Economic Growth in the West* (London, 1964)

Maddison, A., *Monitoring the World Economy 1870-1992* (Paris, 1995)

Maier, C., *Recasting Bourgeois Europe: Stabilization in France, Germany and Italy in the Decade after World War One* (Princeton, N.J., 1975)

Malefakis, E., *Agrarian Reform and Peasant Revolution in Spain* (Yale, N.J., 1970)

Marcks, S., *The Illusion of Peace: International Relations in Europe 1918-1933* (London, 1979)

Martin, L., *Arms and Strategy. The World Power Structure Today* (New York, 1973)

Mathews, J.D. et al (eds), *The Earth as Transformed by Human Action* (Cambridge, 1990)

Mattson, C., Mattson, M.T., *Contemporary Atlas of the United States* (New York, 1990)

Mayer, A.J., *Politics and Diplomacy of Peacemaking: Containment and Counter-Revolution at Versailles* (London, 1967)

Mayne, R., *Postwar: The Dawn of Today's Europe* (London, 1984)

McCauley, M., *The Russian Revolution and the Soviet State 1917-1921* (London 1980)

McCauley, M., *The Soviet Union since 1917* (London, 1981)

McDonald, S.B., Zagaris, B. (eds), *International Handbook on Drug Control* (Westport, Conn., 1992)

McManners, J. (ed.), *The Oxford Illustrated History of Christianity* (Oxford, 1990)

Meadows, D.H. et al, *The Limits to Growth: A Report for the Club of Rome's Project on the Predicament of Mankind* (London, 1974)

Miller, G.T., *Living in the Environment: Principles, Connections and Solutions* (8th ed., Belmont, CA., 1994)

Milward, A.S., *War, Economy and Society 1939-1945* (London, 1987)

Morishima, M., *Why Has Japan Succeeded? Western Technology and the Japanese Ethos* (Cambridge, 1982)

Morrow, J.H., *The Great War in the Air: Military Aviation from 1909 to 1921* (Washington D.C., 1993)

Mühlberger, D. (ed.), *The Social Bases of European Fascist Movements* (London, 1987)

Myers, R.H., Peattie, M.R., *The Japanese Colonial Empire 1895-1945* (Princeton, N.J., 1984)

Nachmai, A., *International Intervention in the Greek Civil War* (New York)

Nicholas, L., *The Rape of Europe: The Fate of Europe's Treasures in the Third Reich and the Second World War* (London, 1994)

Noakes, J. (ed.), *The Civilian in War* (Exeter, 1992)

Northedge, F.S., *The League of Nations: Its Life and Times* (Leicester, 1986)

Northedge, F.S., *The Troubled Giant: Britain among the Great Powers 1916-1939* (London, 1966)

Nove, A., *An Economic History of the USSR* (2nd ed., London, 1989)

O'Ballance, E., *Civil War in Bosnia 1992-94* (London, 1995)

O'Ballance, E., *The Greek Civil War 1944-49* (London, 1966)

Oshima, H.T., *Strategic Processes in Monsoon Asia's Economic Development* (Baltimore, 1993)

Ovendale, R., *The Middle East since 1914* (London, 1992)

Overy, R.J., Wheatcroft, A., *The Road to War* (London, 1989)

Overy, R.J., *Why the Allies Won* (London, 1995)

Parker, R.A.C., *Struggle for Survival: The History of the Second World War* (Oxford, 1989)

Pearson, R., *National Minorities in Eastern Europe 1848-1945* (London, 1983)

Perkin, H., *The Age of the Automobile* (London, 1976)

Pickering, K., Owen, L.A., *An Introduction to Global Environment Issues* (London, 1994)

Pool, I. de S., *The Social Impact of the Telephone* (Cambridge, Mass., 1977)

Porter, A.N., *European Imperialism* (London, 1994)

Porter, B., *The Lion's Share: A Short History of British Imperialism 1850-1970* (London, 1975)

Porter, R., Teich, M. (eds), *Fin de Siècle and its Legacy* (Cambridge, 1990)

Pratt, J.W., *America and World Leadership 1900-1921* (New York, 1967)

Preston, P., *The Coming of the Spanish Civil War* (London, 1978)

Preston, P., *The Triumph of Democracy in Spain* (London, 1986)

Rae, J.J., *Climb to Greatness: The Story of the American Aircraft Industry* (Cambridge, Mass., 1968)

Rae, J.J., *The Road and Car in American Life* (Cambridge, Mass., 1971)

Reynolds, D., Kimball, W., Chubarian, A.O., *Allies at War: The Soviet, American and British Experience 1939-1945* (New York, 1994)

Reynolds, D., Dimbleby, D., *An Ocean Apart: The Relationship between Britain and America in the Twentieth Century* (London, 1988)

Rhodes, R., *The Making of the Atomic Bomb* (New York, 1986)

Rich, N., *Hitler's War Aims: The Establishment of the New Order* (London, 1974)

Roberts, A.D. (ed.), *The Cambridge History of Africa: Vol. 7, 1905-1940* (Cambridge, 1986)

Roberts, J., *Europe 1885-1945* (2nd ed., London, 1994)

Rogger, H., *Russia in the Age of Modernization and Revolution 1881-1917* (London, 1983)

Roskill, S.W., *The War at Sea 1939-1945* (4 vols., London, 1954-61)

Rowe, D. (ed.), *International Drug Trafficking* (Chicago, 1988)

Saint-Etienne, C., *The Great Depression: 1929-1938: Lessons for the 1980s* (Stanford, CA., 1984)

Sapolsky, H.M. et al (eds), *The Telecommunications Revolution* (London, 1992)

Schuker, S., *American 'Reparations' to Germany 1919-1933: Implications for the Third World Debt Crisis* (Princeton, N.J., 1988)

Segal, A., Chalk, P., Shields, J.G., *An Atlas of International Migration* (New York, 1994)

Shaw, S.J., Shaw, E.K., *History of the Ottoman Empire and Modern Turkey* (2 vols., Cambridge, 1977)

Sheehan, N., *A Bright Shining Lie* (London, 1990)

Sherry, M., *The Rise of American Air Power* (New Haven, 1987)

Sherry, M., *In the Shadow of War: The United States since the 1930s* (New Haven, 1995)

Silverman, D.P., *Reconstructing Europe after the Great War* (New York, 1982)

Simons, T.W., *Eastern Europe in the Post-War World* (New York, 1991)

Skidmore, T.E., Smith, P.E., *Modern Latin America* (2nd ed., Oxford, 1989)

Sluglett, P., Farouk-Sluglett, M. (eds), *The Times Guide to the Middle East* (London, 1995)

Smith, D.D., *Communications via Satellite* (Boston, Mass., 1976)

Soon, L.T., Suryadinata, *Moving into the Pacific Century: The Changing Regional Order in the Asia-Pacific* (Singapore, 1988)

Spence, J., *The Search for Modern China* (New York, 1990)

Spence, C.P., Navaratnam, V. (eds), *Drug Abuse in East Asia* (Kuala Lumpur, 1988)

Staudohar, P.D., Mangan, J.A. (eds), *The Business of Professional Sport* (Chicago, 1991)

Steven, R., *Japan's New Imperialism* (London, 1990)

Stevenson, D., *The First World War and International Politics* (Oxford, 1988)

Stockholm International Peace Research Institute, *The Arms Race and Arms Control* (London, 1982)

Stoetzer, O.C., *The Organisation of American States* (Westport, Conn., 1993)

Stolfi, R., *Hitler's Panzers East: World War Two Reinterpreted* (Norman, Okl., 1991)

Taheri, A., *The Spirit of Allah: Khomeini and the Islamic Revolution* (London, 1985)

Takenaka, H., *Contemporary Japanese Economy and Economic Policy* (Ann Arbor, Mi., 1991)

Tan, J.L.H., *Regional Economic Integration in the Asia Pacific* (Singapore, 1993)

Taneja, N.K., *Airlines in Transition* (Lexington, Mass., 1981)

Terraine, J., *Business in Great Waters: The U-Boat Wars 1916-1945* (London, 1989)

Thayer, G., *The War Business: The International Trade in Armaments* (New York, 1969)

Thomas, A. et al, *Third World Atlas* (2nd ed., Milton Keynes, 1994)

Thompson, R., *Exporting Entertainment: America in the World Film Market* (London, 1985)

Thorne, C., *Allies of a Kind: The United States, Britain and the War against Japan, 1941-1945* (Oxford, 1978)

Tietze, W. et al (eds), *Geographie Deutschlands: Bundesrepublik Deutschland, Staat-Natur-Wirtschaft* (Stuttgart, 1990)

Tordoff, W., *Government and Politics in Africa* (London, 1993)

UNESCO, *World Education Report 1995* (Oxford, 1995)

Urban, M., *War in Afghanistan* (2nd ed., London, 1990)

Urwin, D.W., *Western Europe since 1945* (3rd ed., London, 1981)

Valderrama, F., *A History of UNESCO* (Paris, 1995)

Vatter, H.G., *The US Economy in World War Two* (New York, 1985)

Vinen, R., *France, 1934-1970* (London, 1996)

Watt, D.C., *How War Came* (London, 1989)

Webster, C., Frankland, N., *The Strategic Air Offensive against Germany* (4 vols., London, 1961)

Wee, H. van der, *Prosperity and Upheaval: The World Economy 1945-1980* (London, 1987)

Wegs, J.R., *Europe since 1945: A Concise History* (London, 1984)

Weiss, T., Forsythe, D.P., Coate, R.A., *The United Nations and Changing World Politics* (Oxford, 1994)

Wilkins, M., *The Maturing Multinational Enterprise: American Business Interests Abroad from 1914 to 1970* (New York, 1974)

Williamson, E., *The Penguin History of Latin America* (London, 1972)

Wood, D.M., Yesilada, B.A., *The Emerging European Union* (London, 1996)

World Health Organisation, *Health in Europe* (Copenhagen, 1994)

World Health Organisation, *World Health Report* (Geneva, 1996)

Wright, G., *The Ordeal of Total War 1939-1945* (New York, 1968)

Yergin, D., *The Prize: The Epic Quest for Oil, Money and Power* (New York, 1991)

Yeung, Y-M., *Pacific Asia in the 21st Century: Geographical and Developmental Prospects* (Beijing, 1993)

Yoder, A., *The Evolution of the United Nations* (New York, 1989)

Young, M., *The Vietnam Wars 1945-1990* (London, 1991)

Young, R.J., *In Command of France: French Foreign Policy and Military Planning 1933-1940* (Cambridge, Mass., 1978)

PICTURE ACKNOWLEDGEMENTS

The following abbreviations have been used: b, bottom; c, centre; l, left; r, right; t, top

Advertising Archives: 126cr; 189c; 190cr

AKG London: 19tc; 20-1c; 35tr; 43bl; 49tr; 86-7; 102tr; 155

Associated Press: 131tl; 156; 157c; 198cl; 200bc; 208bc

Bundesarchiv, Koblenz: 115tl

e.t.archive: 18; 20; 22; 23tl and cr; 26br; 30tr; 31bl; 32t; 35cr; 36; 46c; 53br; 56bl; 65tr; 67tc; 73bc; 79c; 83; 89; 89bc; 90-1; 92br; 93c; 94bc and tr; 95tr; 97tl; 99br; 100; 129tr; 185cr; 192c

Black Star: 193br

Thomas Cook Archive: 196bl; 197c

Corbis/Range: 19b; 46cr; 49cr; 52cr; 53c; 63cr; 64cr; 65bl; 103tl; 110; 111cr; 112bl; 114-5; 115r; 121br; 126bl; 130; 181bc; 188cl

James Davis Travel Photography: 158

Collection of Mr and Mrs Barney Ebsworth Foundation, St Louis: 62br

The Environmental Picture Library: 177; 178; 203

Sally and Richard Greenhill: 74-5bc

Robert Harding Picture Library: 206tc

Hulton-Getty: 20-1tc; 28bl and c; 32br; 34; 47br; 48br; 51br; 74c; 74-5c and tc; 86c; 108bl; 126tl; 127; 128; 132cr; 133; 134bl; 155bl

Robert Hunt Library: 30r and c; 33t and bl

Hutchison Library: 157tl; 160bc; 176bl; 179bl; 181br; 182bl; 186tl; 188bc; 196cl; 199; 202cl; 204; 210

Imperial War Museum: 69tl; 79cr; 89tr; 92cr; 97tc; 98cr; 99tl; 105-6

David King Collection: 38-9; 41tl and br; 43lc; 58; 60cl; 61; 73br; 79br; 85tc; 86tc; 108c; 111bl; 122; 123c and tl; 129br; 193br

London Transport Museum: 65cr

Magnum: 111br; 112cr; 117tl and cr; 118c; 119; 125; 129cr; 132bl; 135tl and br; 140br; 147bl; 149c and br; 162tr and bc; 163; 168cl; 201bc

Musée d'Histoire Contemporaine, Paris: 77tr

Novosti Photo Library: 40; 41tr; 84cr; 87cr

Peter Newark's Western Americana: 63br; 65cl

Photo Assist Inc: 91br; 102-3c

Popperfoto: 23tr and bl; 24; 25; 26t; 26-7; 27; 28t; 29; 36-7; 37; 43tr; 48-9; 50; 51l; 52tc; 53br; 54; 57br; 59; 60bl; 66; 67cl; 68; 69r; 70; 71; 72; 72-3; 77bl and cr; 82; 84br; 87; 94br; 96cr; 101; 102-3bc; 114bl; 120cr; 129bl; 173cr; 180bl; 194; 195br; 201c

Powerstock: 173cl; 187tr; 191br; 206

Quadrant Picture Library: 191cl

Rex Features: 158cr; 198tc; 211

Roger-Viollet, Paris: 77br

Science Photo Library: 189br; 192cr; 212; 213

Still Pictures: 186; 187bl; 193

Sygma: 103tc; 112cl; 116b; 117br; 118cl; 121bc; 123br; 124; 134c; 139bl; 140tr; 141br; 142; 146; 148tc; 149bl; 150; 151; 152; 153; 160tr; 164; 165; 166; 167; 168bl; 169; 170; 171; 183bl; 184cl; 200-1; 202bc; 208; 209

Toyota: 159

TRH Pictures: 91c and tr; 93br; 141bl

Trip/Helene Rogers: 176tc; 195cr
Trip/F. Torrance: 170cr

Ullstein Bilderdienst: 62-3

US Naval Historical Center: 91cl

Zefa: 196cr

GLOSSARY

This glossary is intended to provide supplementary information about some of the individuals, peoples, events, treaties and movements treated in the text. It is not a general encyclopedia of the 20th century and only names mentioned in the atlas proper are included. Subjects such as Hitler and Lenin, who are covered in the panels elsewhere in the book, are cross-referenced to the panels. Names printed in bold within entries have their own glossary entries.

ADENAUER, KONRAD (1876-1967) West German leader. A lawyer by profession, he became mayor of Cologne in 1917 and held the post until 1933. He was active in politics in the 1920s, representing the Catholic Centre Party, but was forced out of public life by the Nazi takeover of power in 1933. After the Second World War, he returned to politics and helped to set up the Christian Democratic Union, a new centre-right party. He became West Germany's first federal chancellor in 1949. His 14 years in office saw a remarkable economic recovery and Germany's re-acceptance into the Western world: not only did the country join NATO, but Adenauer cultivated especially good relations with **de Gaulle** of France. He also obtained recognition for West Germany from the USSR, although some believed that this reduced the likelihood of German reunification. He remained politically active after his replacement by Ludwig Erhard in 1963.

AFRICA, PARTITION OF In the latter part of the 19th century, much of the continent was divided into European-ruled colonies. Britain, Belgium, France, Germany, Italy and Portugal all participated in this "Scramble for Africa". Although the Congress of Berlin (1884-5) had sought to accommodate rival claims and had provided for free trade areas, the partition increased tension among the European powers in the years leading up to the First World War. Many of the boundaries resulting from the partition exist to this day.

AFRICAN NATIONAL CONGRESS Set up in 1912, the ANC was originally a small group of black South Africans who sought to improve the position of blacks through peaceful methods. Failure to moderate white minority rule led to growing militancy after the Second World War, shown by the formation of the Youth League. The new tough line proved popular and led to co-operation with India's Congress nationalist movement. In 1948, the whites-only National Party gained power and firmly enforced white rule: unrest led to the ANC's suppression in 1960. The ANC led an underground campaign until 1990, when it was legalized amid moves to bring about full democracy in South Africa.

AFRICAN UNITY, ORGANIZATION OF The OAU was set up in 1963 by 32 countries at a conference in Addis Ababa. It was intended to encourage co-operation and to support independence movements in those areas still under colonial rule. Most African states have since joined it. Heads of government meet regularly, and there is a council of ministers, as well as various committees. Though a valuable forum for discussion, the OAU has proved unable to resolve differences between states and to tackle effectively the many problems facing the continent.

ALEXANDER, HAROLD (1891-1969) British field marshal. He fought in the First World War and against the **Bolsheviks** in the Russian Civil War. In 1940 he was a British Expeditionary Force commander, and was involved in the Dunkirk evacuation. In August 1942, after leading operations in Burma, he was appointed commander-in-chief (Middle East) and presided over Allied successes in the Mediterranean, capturing Rome in June 1944. He continued to advance northwards through Italy until the German surrender in May 1945. After the war, he was governor general of Canada and minister of defence (1952-4).

ANTI-COMINTERN PACT Agreement between Germany and Japan, reached in November 1936, providing for co-operation against world communism. Italy associated herself with the Pact a year later. The Pact encouraged Japan to launch operations in China in 1937 and increased the Soviet Union's fear of encirclement. The **Nazi-Soviet Pact** of August 1939 angered Japan, but Germany, Italy and Japan renewed their association by the Tripartite Pact of September 1940. The Anti-Comintern Pact was re-affirmed after the German invasion of the USSR in June 1941.

AQUINO, CORAZON (1933-) President of the Philippines, 1986-92. In 1956 she married the opposition leader, Benino Aquino, and later supported his campaign against the increasingly corrupt regime of Ferdinand Marcos. When her husband was assassinated in Manila in 1983 (allegedly on government orders), she took over the opposition movement. In 1986 Marcos fled to the US and Aquino was established as his replacement. Her years in office were difficult, with persistent economic problems and various natural disasters. She also had to contend with a series of coups, put down largely with US help.

ARAB LEAGUE Association of Arab states, set up in 1944-5. The League was designed to promote Arab solidarity, to weaken any remaining colonial regimes and to oppose the **Zionist** movement in Palestine. Most states, plus the Palestine Liberation Organization (PLO), had joined by 1979, when Egypt was suspended for making a separate peace agreement with Israel and the League's headquarters were transferred from Egypt to Tunisia. Egypt was readmitted in 1987, but the League was weakened by divisions over the Gulf War crisis of 1990-1.

ARAFAT, YASSER (1929-) Palestinian leader. Involved in anti-Israeli activity after the establishment of the State of Israel in 1948, he emerged as leader of the Palestine Liberation Organization (PLO) after 1964. He became more moderate over the years, hoping to build up worldwide support, and in 1988 he recognized Israel, to the fury of many of his followers. This led to peace talks with Israel (1993-4). Despite hostility among many Arabs to his continuing accommodation with Israel, Arafat remained the most prominent of the Palestinian leaders.

ARMSTRONG, NEIL (1930-) US astronaut and the first man to walk on the moon. After service as a pilot in the US navy, Armstrong was a civilian test pilot on the X-15 rocket plane. He was accepted as an astronaut by NASA in 1962. He made his first space flight on Gemini 8 in 1966, successfully overcoming severe spinning to land the spacecraft undamaged. He was assigned command of Apollo 11 in early 1969 and, with astronaut Buzz Aldrin, landed on the moon on 20 July 1969, in the process confirming US dominance over the USSR in space flight. Armstrong left NASA in 1971 and became a professor of engineering at the University of Cincinnati. Despite the inevitable fame attaching to the first man to walk on the moon, Armstrong has consistently shunned the spotlight.

ATLANTIC CHARTER A joint statement by US president **Roosevelt** and British prime minister **Churchill** in August 1941. The leaders condemned illegal seizures of territory and declared their support for self-determination, free trade and disarmament. The Charter gained in significance with the US entry into the Second World War in December 1941.

ATTLEE, CLEMENT (1883-1967) British prime minister. After studies at Oxford, social work in the East End led him to join the Labour movement. After serving in the First World War, he became Labour MP for Limehouse in 1922. He held office in the first Labour government of 1924, but refused to co-operate with MacDonald's National government in 1931 and became leader of the Labour opposition in 1935. He was a member of **Churchill**'s war cabinet and led Labour to an overwhelming electoral victory in July 1945, representing Britain at the **Potsdam Conference** soon afterwards. He supported the creation of NATO and the granting of independence to India and Pakistan. At home, his government introduced the National Health Service and a far-reaching programme of nationalization, which was effectively to set the tone of Britain's economic life for the next 30 or more years. Internal divisions weakened his government, which was voted out of office in 1951, but he led the Opposition until his retirement in 1955.

AUSCHWITZ Nazi Germany's most notorious extermination centre, sited in Galicia near the German-Polish border. There were three camps at Auschwitz, built between 1940 and 1942: one to house political prisoners; one for slave labourers; and one for victims of the Nazis' "Final Solution". All the camps were operated by the SS under the command of Rudolf Hoess. Jews were brought to Auschwitz from all over Europe. When each consignment arrived, there was a medical inspection. Some were selected as slave-labourers, the rest were executed. A small number were also used in medical experiments. Estimates of the death toll vary greatly, but it appears that between one and two million prisoners may have perished. Auschwitz has since become the pre-eminent symbol of the Holocaust.

AYUB KHAN, MOHAMMED (1907-74) Pakistani president. After military training in Britain, he joined the British Indian Army and served in the Second World War. In 1951 he was appointed commander-in-chief of Pakistan's new army, and three years later he became minister of defence. In 1958, having declared martial law, he took power. He was then elected president in 1960, in which capacity his rule became increasingly autocratic. Growing economic difficulties by the late 1960s led to his resignation. He was succeeded by General Yahya Khan.

BADOGLIO, PIETRO (1871-1956) Italian soldier. He served in the Abyssinian campaign of 1896 and was a general in the First World War, playing an important part in the Italian recovery after the defeat at **Caporetto** in 1917. He was made a field marshal in 1925 and governor general of Libya from 1929 to 1933. He completed Italy's conquest of Abyssinia in 1936. Opposed to Italy's entry into the Second World War on the grounds that the country's armed forces were unprepared, he resigned in December 1941 after Italy's disastrous Greek campaign. He began to plan **Mussolini**'s downfall and replaced him in July 1943 with the collusion of King Victor Emmanuel III. Shortly afterwards he concluded peace with the Allies and joined them against the Germans. After the liberation of Rome in June 1944, he retired from public life.

BATTLE OF BRITAIN See page 83

BECK, JOZEF (1894-1944) Polish statesman. He supported the fight for Polish independence before 1914, associating himself with **Pilsudski**. After the creation of the new Polish state in 1919, he was ambassador in several European capitals before rejoining the military and assisting Pilsudski's return to power in 1926. He became foreign minister in 1932 and tried to avoid dependence on either Nazi Germany or Soviet Russia, although he was increasingly prepared to side with Hitler, notably during the Munich Crisis of September 1938. While this policy temporarily preserved Poland's independence, it left her vulnerable to a Russo-German partition, which duly occurred in September 1939.

BELL, ALEXANDER GRAHAM (1847-1922) Inventor of the telephone. Born in Scotland into a family specializing in work for the deaf, he was taken to Canada in 1870. In 1873 he became professor of vocal physiology at Boston University. His attempts to create new methods of communication led to his production of a telephone apparatus in 1876. The invention led to complicated legal actions over patents. In 1885 he set up his own research centre, continuing his work until his death.

BENEŠ, EDUARD (1884-1948) Czech statesman. After a successful university career, he became a leading figure in the Czech independence movement during the First World War. He became Czechoslovakia's foreign minister (1918-35), favouring good relations with France, the Soviet Union and the "Little Entente" of Eastern European states. He became president in 1935, but resigned after the Munich Crisis of September 1938, in which Britain and France colluded in the award of Czech territory to Germany, Hungary and Poland. He led a Czech government in exile during the Second World War. Though he returned to Czechoslovakia as president (1946-8), he resigned after the Soviet-engineered communist takeover.

BEN-GURION, DAVID (1886-1973) Israeli prime minister. A Polish Jew by origin, he arrived in Palestine in 1906 and joined the **Zionist** movement, fighting against the Turks in the First World War. He became secretary general of the labour movement in 1921 and was chairman of the Jewish Agency from 1935 to 1948. He became the first prime minister of Israel when the country was formed in 1948, remaining in office for most of the period to 1963. His influence on the new state was profound. At the same time, despite his pacific nature, he proved an inspiring war leader. He remained a father figure to the nation after his retirement.

BERIA, LAVRENTI (1899-1953) Soviet police chief. He joined the **Bolsheviks** in 1917 and became a high-ranking communist in his native Georgia. **Stalin** (a fellow Georgian) appointed him head of the NKVD (secret police) in December 1938. He presided over the final stages of the Great Terror and organized police operations in Eastern Europe after the Second World War. He was feared by his rivals, but owed his position to Stalin's support. After Stalin's death in March 1953, Beria was soon arrested and executed.

BERLIN WALL Heavily fortified barrier built in 1961 by the East German government around the Western zones of Berlin. The Berlin Wall was the single most potent and visible symbol of the Cold War. As levels of emigration from East to West Berlin increased in the 1950s and early 1960s, the East German government determined to construct what it called the Anti-Fascist Exclusion Wall. Though its stated purpose was to keep West Berliners out of East Berlin, in reality the Wall was a crude but effective means of preventing East Germans from escaping to the West. The opening of the Wall in late 1989 after the collapse of the communist government of East Germany was a vivid symbol of the end of the Cold War.

BLITZ, THE Name given to the bombing of British cities by the German air force from September 1940 to May 1941. London was frequently bombed (over a million homes were damaged), and most other large cities were also attacked. Coventry suffered an especially intensive raid, with most of the city centre destroyed. Despite the devastation caused, the Blitz did not destroy British morale, and the policy of attacking cities rather than military targets enabled the RAF to recover from losses sustained earlier in the war. From mid-1941 the Blitz subsided as Germany turned to other fronts, but in 1944-5 further damaged was inflicted by the V1 flying bombs and V2 rockets.

BOER WAR (1899-1902) Conflict between British forces and established European settlers (Boers) for control of South Africa, especially the prosperous Transvaal. After early Boer successes, resulting in the sieges of Kimberley and Mafeking, the British counter-attacked under Roberts and Kitchener, lifting the sieges, capturing Pretoria and gaining a fragile military control, though this was always threatened by the Boers' guerrilla tactics. After the conclusion of the peace treaty of Vereeniging, the Boers obtained self-rule (1907), while promising to defend British rights in South Africa.

BOLSHEVIKS Members of the Russian Social Democratic Labour Party, as opposed to the **Mensheviks**. Their leader, Lenin, wanted the RSDLP to become a small grouping of professional revolutionaries who would lead the working classes to final victory. Menshevik opposition caused the Bolsheviks to form their own party in 1912. During the revolution of 1917, they were the most hardline socialist group, committed to the overthrow of the "compromisers" of the provisional government after Lenin's return from exile in April. Their rising in July was put down, but worsening conditions and popular demands for peace enabled them to take power in November. They established a Soviet government and banned opposition parties, changing their own name to Communist in 1918.

BOTHA, P. W. (1916-) South African politician. He entered parliament in 1948, representing the National Party, and became a government minister in the 1960s. He was minister of defence for 12 years from 1966, favouring intervention in neighbouring countries. A strong supporter of apartheid, he replaced Vorster as prime minister in 1978. He made concessions to some non-white groups, thus causing tension inside his party, though he resisted moves towards black enfranchisement. He became president in 1984. A stroke in 1988 led to his retirement in the following year.

BRANDT, WILLY (1913-1992) West German leader. A socialist, he fled Germany in 1933 to escape Nazi persecution and lived in Norway and, after 1940, in Sweden. He returned to Germany in 1945 and joined the democratic socialist movement, becoming an SPD deputy in West Germany in 1949. As mayor of West Berlin between 1957 and 1966, he defended the city's interests stubbornly in the face of often intense Soviet demands. In 1969 he became federal chancellor. In addition to continuing the economic policies which had already made West Germany one of the world's most prosperous countries, he worked hard to normalize relations with East Germany, with whom diplomatic relations were established in 1972. Claims that he had been compromised by the East German government led to his resignation in 1974.

BRAUN, WERNHER VON (1912-1977) German (later US) rocket scientist.

After graduating in 1932, he became the head of a German rocket science research team. The group, supported by the military, developed a number of effective rockets, notably the V-2, which was used against British targets during the Second World War. In 1945 Braun and his colleagues surrendered to the Americans, rather than the Russians, and went to the US to continue research into rocketry. Having established – though never with complete conviction – that he was no supporter of the Nazis, in 1955 Braun became a US citizen. As chief of NASA's rocket research, he played a key role in the US space programme. He oversaw the development and building of every major US rocket type, including the Saturn V, the most powerful rocket ever to be launched successfully, and without which America's triumphant moon landings would have been impossible.

BREST-LITOVSK, TREATY OF Peace agreement signed in March 1918 between Russia and the Central Powers. Russia recognized the independence of Poland, Finland, Georgia, the Baltic States and the Ukraine and agreed to pay a large indemnity. The treaty was declared void under the Armistice of 1918.

BREZHNEV, LEONID (1906-82) Soviet leader. A Ukrainian, he became a communist party official in the 1920s and served in a political capacity during the Second World War. He rose rapidly in the party after the war and entered the Politburo in 1957. He replaced **Khrushchev** as first secretary in October 1964, effectively becoming Soviet leader, although he did not formally assume the presidency until 1977. Brezhnev rapidly developed a reputation as a dour hardliner determined to maintain the Soviet system. He ordered tanks into Prague in August 1968 to crush the burgeoning democracy movement there. In 1979 he ordered the Soviet invasion of Afghanistan. Any improvements in East-West relations earlier in the decade were effectively rendered void by the invasion. Despite failing health and a marked decline in the Soviet Union's already poor economic performance toward the end of his rule, he maintained a tenacious hold on office until his death.

BUCHAREST, TREATY OF Settlement made in August 1913 after the Second Balkan War, in which Bulgaria was defeated by a combination of states, including her former anti-Turkish allies. Bulgaria was obliged to accept territorial losses, especially to Greece, Romania and Serbia. The possibility of regaining territory at Serbia's expense encouraged the Bulgarians to join the Central Powers after the outbreak of the First World War.

BUKHARIN, NIKOLAI (1888-1938) Leading Bolshevik. After the communist takeover of Russia in 1917, he opposed Lenin over the question of peace with Germany, favouring a revolutionary war, but subsequently became a strong supporter of Lenin's **New Economic Policy** (NEP). During the power struggle following Lenin's death in 1924, he initially co-operated with **Stalin** against **Trotsky**, but Stalin's policy of rural collectivization drove him into opposition. Bukharin was expelled from the Politburo in 1929. He was tried and shot in 1938.

BULGE, BATTLE OF THE German Second World War offensive. By late 1944, with the Western Allies advancing towards Germany, Hitler hoped to regain the initiative by launching a counter-attack through the weakly defended Ardennes (the scene of the German breakthrough in 1940) towards Antwerp. The attack opened on 16 December, catching the Allies almost entirely off guard. The Germans advanced rapidly, thus causing the "bulge" in the Allied line. However, the Allies regrouped and, aided by their air squadrons, forced the Germans back. The Germans ultimately gained nothing, and the cost in men and material weakened their defences against future attacks from both east and west.

BUSH, GEORGE (1924-) US president. He served in the navy in the Second World War, and was educated at Yale before entering the oil business. In 1966 he joined the House of Representatives and became the US ambassador to the UN in 1970. From 1980 to 1988 he served as **Ronald Reagan**'s vice-president, becoming president in 1989 after his defeat of the Democrat Michael Dukakis. His four years in office saw an unusual concentration on foreign affairs: US troops intervened in Panama in 1989-90, and he led the UN coalition which drove the Iraqis from Kuwait in 1991. Bush was also able to announce the end of the Cold War in January 1991 as the communist system in Eastern Europe collapsed. Many felt that he had neglected domestic affairs, however, and he was defeated by Bill Clinton in the 1992 presidential election.

BUTHELEZI, CHIEF MANGOSUTHU (1928-) South African Zulu leader. He was originally a member of the **African National Congress** and became head of the Buthelezi tribe in 1957. In the 1970s he became chief minister of Kwa Zulu province and re-founded the Inkatha Freedom Party, opposing apartheid but refusing to co-operate with the ANC's armed struggle. This led to violence between ANC and Inkatha supporters in the 1980s. In 1994 Inkatha accepted the results of multi-racial elections, and Buthelezi joined the Mandela government. However, ANC-Inkatha rivalry continued to generate violence, undermining the new order in South Africa.

BUTLER, RICHARD A. (1902-82) British politician. Educated at Cambridge, he entered parliament as a Conservative in 1929. He held government office throughout the 1930s, while in the Second World War he was president of the Board of Education, introducing the 1944 Education Act, which largely determined the thrust of British education for more than a decade. He was a key minister in the Conservative governments of 1951-64, and gained a wide following for his generally moderate views. "Butskellism", named after Butler and the then leader of the Labour Party, Hugh Gaitskill, was a recognized short-hand for the left-of-centre politics which characterized the period and which he championed so effectively. Widely tipped for the leadership of the Conservative Party after the resignation of Anthony Eden in 1956, he allowed himself to be outwitted by Harold Macmillan. He was no more successful in 1963, when Alec Home became leader. He remained a much-admired

figure after his retirement from active politics in 1964.

CADORNA, LUIGI (1850-1928) Italian general in the First World War, chief of staff from July 1914. After Italy's entry into the war in May 1915, he organized several offensives against Austro-Hungarian forces, but these failed due to unimaginative tactics. He blamed an indifferent home front for Italy's difficulties, but failure to secure direct military aid from Italy's allies and the disastrous defeat at **Caporetto** (October 1917) led to his dismissal in November.

CAPORETTO, BATTLE OF On 24 October 1917, Austro-Hungarian forces, supported by German troops, launched a major offensive along the Italian front. They achieved surprise and advanced rapidly to the River Piave, inflicting heavy casualties on the Italians. Anglo-French reinforcements enabled the Italians to recover and halt the offensive. The battle ended two years of stalemate on this front and represented a major, if shortlived, breakthrough for the Central Powers.

CARLOS, JUAN (1938-) King of Spain from 1975. Franco long regarded him as a likely successor, designating him future king in 1969. After Franco's death, Juan Carlos played an important part in restoring democracy in Spain, authorizing a new constitution in 1978. His clear commitment to democracy – and his moral authority – were highlighted by his courageous stand against the leaders of the military coup in 1981. His position thus reinforced, he has proved a highly popular and successful constitutional monarch, and his reign has seen Spain's international rehabilitation and entry into the European Community.

CASABLANCA CONFERENCE Meeting of Allied war leaders, held in January 1943 and attended by **Roosevelt**, **Churchill** and pro-Allied French leaders such as **de Gaulle**. The leaders agreed to create a "Second Front" in Europe to relieve Russia: but while Churchill and the British wanted to invade Italy from Africa, many US generals called for a direct invasion of France. It was eventually agreed to concentrate on Italy for the time being. Plans were also made for offensives against Japanese forces in Asia. Meanwhile, Roosevelt declared that the Allies would only accept the "unconditional surrender" of the Axis powers, a view echoed by his colleagues.

CASTRO, FIDEL (1926-) Cuban revolutionary leader. After legal training, he led a failed uprising against the Batista regime in 1953 and was later exiled. He returned to Cuba in December 1958 and took over the government, having announced the Cuban Revolution. His socialist measures alienated the US, but the US-sponsored invasion of Cuba in 1961 – the "Bay of Pigs" – both failed miserably and enhanced Castro's standing. He was often obliged to submit to Soviet influence, notably during the 1962 Cuban missile crisis. He became president in 1976, but the decline of communism in Europe from the late 1980s left him increasingly isolated.

CEAUÇESCU, NICOLAE (1918-89) Romanian communist leader. He joined the Romanian Communist

Party in 1936 and helped to build it up before and during the Second World War, coming to prominence after 1945. He joined the Party Central Committee in 1952 and succeeded Gheorgiou-Dej as general secretary in 1965. He became president two years later. Ceauçescu persistently proclaimed his independence from Soviet influence – in the process gaining a degree of respectability in the West. However, his regime became progressively more corrupt and autocratic. He was overthrown by a popular revolution in 1989. After a brief army trial, he and his wife (a prominent figure in his regime) were executed, amid general rejoicing.

CHAMBERLAIN, NEVILLE (1869-1940) British prime minister. After several years in local politics (including a spell as mayor of Birmingham), he entered parliament in 1918. He was minister of health for much of the 1920s. From 1931, his administrative ability allied to solid political sense saw him become chancellor, a post he held until 1937, when he became prime minister. As prime minister, his most immediate problem was the growing threat of Nazism in Germany. Though he implemented a programme of re-armament in the event that war would prove unavoidable, he was a firm champion of what came to be called "appeasement" – the accommodation of Hitler's demands in the belief that the German Nazi leader would not ultimately risk another European war. With France, Britain duly allowed the Nazi takeover of the Sudetenland in Czechoslovakia, a deal agreed in September 1938 in Munich. Chamberlain's assertion that the deal had brought "peace for our time" was shown to be false six months later when Nazi Germany marched into Prague. The Nazi invasion of Poland in September 1939 duly forced Chamberlain into the war he had striven so hard to avoid. Never convincing as a war leader, he resigned in May 1940 and was succeeded by **Churchill.**

CHARTER 77 Name given to a declaration made in January 1977 by a group of reformist Czechs who called for greater political freedom in their country. The Soviet invasion of Czechoslovakia in 1968 had led to increased repression, and the Charter was essentially a challenge to the Czech government to honour its 1976 promises to respect human rights. The Charter helped to stimulate opposition to Czechoslovakia's communist regime. With the fall of the country's communist government in 1989, the first president of the new Czechoslovakia was **Václav Havel**, a member of the Charter 77 group.

CHIANG KAI-SHEK (1887-1975) Chinese nationalist leader. After the death of Sun Yat-Sen in 1925, he assumed the leadership of the nationalist movement and used its army (largely organized by himself) to gain control of much of the country. He was highly thought of in the West, but occasional conflicts with the communists and the Japanese occupation of Manchuria in 1931 weakened his authority. He enjoyed US support during the Second World War, but his armies failed to play a decisive part in the war against Japan. Renewed conflict with the communists after the war led to his fall in 1949, but he continued to lead a nationalist government in exile on the island of Formosa (Taiwan) until his death.

CHUIKOV, VASILY (1900-82) Soviet general. In the army from 1919, his communist sympathies enabled him to rise rapidly. He was a commander during the Russo-Finnish war of 1939-40 and took over the defence of Stalingrad in 1942-3. His dogged tactics resulted in heavy casualties but tied down the German forces in the city and made possible the encircling Soviet counter-attack. He remained a prominent commander during the Russian advance westwards and received the German surrender after the battle for Berlin in 1945. He later headed the Soviet occupation forces in Germany.

CHURCHILL, SIR WINSTON (1874-1965) British statesman and prime minister. He became an MP in 1900 and represented both the Liberal and Conservative Parties over the next 64 years. He entered the cabinet in 1906 and became first lord of the admiralty in 1911, but was discredited by the failure of the Gallipoli expedition (which he had advocated) in 1915. He was the Conservative chancellor from 1924 to 1929, returning Britain to the Gold Standard and vigorously opposing the General Strike. During the 1930s, he opposed the "appeasement" of Hitler and **Mussolini** and concessions to Indian nationalism. In 1940 he succeeded **Chamberlain** as prime minister after the disaster in Norway. He was a dynamic war leader, who powerfully symbolized the nation's will to fight on. He consulted regularly with Britain's allies and foresaw post-war Soviet policy in Eastern Europe. Opposition to post-war reforms led to his electoral defeat in 1945, but he later returned for a final four years in power (1951-5).

CLEMENCEAU, GEORGES (1841-1929) French statesman. He entered politics in 1870 and became a prominent Radical, his fierceness in debate earning him the nickname "the tiger". He became prime minister in October 1906 and remained in office for three years, favouring good relations with Britain and firm handling of industrial disputes. He was not originally part of France's wartime government, but was recalled as prime minister in November 1917, strongly opposing a negotiated settlement with Germany. He rallied France after the German spring offensive of 1918, and insisted on harsh terms at the peace conference after the Allied victory, although nationalists criticized him for failing to ensure the Rhineland's separation from Germany. He retired in 1920.

COMINTERN The Communist International, set up by the Soviet government in 1919 as a forum for European communists. When revolutions failed to occur elsewhere in Europe, the Russian leaders used the Comintern to take control of the foreign communist parties. The foreign parties were required to modify their policies in accordance with Soviet wishes, and were instructed not to co-operate with moderate socialist parties against fascist movements. The Comintern was eventually dissolved in 1943 by **Stalin**, who wished to reassure his wartime allies about Russian intentions.

DALADIER, EDOUARD (1884-1970) French politician and a Radical deputy from 1919 to 1940. He became minister for colonies in 1924 and was war minister for much of the 1930s (1932-4, 1936-40). He also had three spells as prime minister, notably after April 1938. He was associated with "appeasement" and signed the Munich Agreement in September, but he was not convinced that long-term peace was possible and favoured rapid re-armament. During the first months of the war, his leadership style was criticized, and Reynaud replaced him in March 1940. He was imprisoned by the Germans for much of the war, later returning as a Radical deputy from 1946 to 1958.

DE HAVILLAND, SIR GEOFFREY (1882-1965) British aircraft designer. He built and flew his first airplane in 1910, afterwards becoming an important figure in the developing British aviation industry. During the First World War he was responsible for the design of war planes. He set up his own company in 1920: the success of the small Moth plane helped bring it to the public's attention. During the war, the company also produced the Mosquito, which was regularly used in night operations. De Havilland continued his research after 1945, receiving the Order of Merit in 1962.

DELORS, JACQUES (1925-) French politician. He joined the Socialist Party in 1973, and was **Mitterrand**'s finance minister from 1981 to 1984, showing willingness to compromise socialist ideals and introduce unpopular spending cuts. In 1985 he became president of the European Commission. He strongly supported moves towards closer European integration, putting forward the Delors Plan for monetary union and presiding over the introduction of the Single Market in January 1993. He also obtained increased powers for the Commission. He was succeeded by Jacques Santer, but decided not to stand as the Socialist candidate in the 1995 French presidential elections.

DENG XIAOPING *See* page 158

DOLLFUSS, ENGELBERT (1892-1934) Austrian leader. After studying theology, he served with distinction in the First World War. He became minister of agriculture in 1931 and chancellor in May 1932, representing the Christian Socialist Party. He suspended parliament in March 1933 and came into violent conflict with the socialist movement in February 1934, ordering the bombardment of workers' housing areas. His foreign policy was based on friendship with Hungary and Italy, and his plans for a new constitution were strongly influenced by Mussolini's fascist system. However, this failed to satisfy Austrian Nazis, whose attempts to bring about union with Hitler's Germany led to Dollfuss's assassination in July 1934. He was succeeded by Schuschnigg.

DÖNITZ, KARL (1891-1980) German naval commander and briefly head of state. In the First World War he served as a submarine officer. After the succession of Hitler he supervised the construction of a new U-boat fleet. He was appointed commander of submarine forces in 1936, head of the German navy in 1943 and head of the northern military and civil command in 1945. Named in Hitler's political testament as the next president of the Reich, he assumed control of the government for a few days after Hitler's suicide in May 1945. He was sentenced to ten years' imprisonment as a Nazi war criminal in 1946 and was released in 1956.

DULLES, JOHN FOSTER (1888-1959) US statesman. He attended the Paris Peace Conference in 1919 and remained active in international diplomacy between the wars. He was involved in the setting up of the United Nations after the Second World War and helped to organize the peace treaty with Japan in 1951. He became secretary of state in January 1953 and adopted a firmly anti-Soviet policy, arguing that aggression should be met by "massive retaliation", the only effective deterrent. He also took an interest in the Middle East, condemning the Anglo-French Suez action of 1956. Dogged by ill health, he resigned shortly before his death.

EISENHOWER, DWIGHT D. (1890-1969) US general and president. A graduate of military school, he was **MacArthur**'s assistant in the Philippines from 1935 to 1940. He commanded the Allied invasion of French North Africa in November 1942: his successes in the Mediterranean in 1942-3 led to his appointment as supreme commander of Allied forces in Western Europe. He oversaw the Normandy landings of June 1944 and the advance into Germany, although his cautious tactics were opposed by many of his subordinates. He was chief of staff (1945-8) before being elected president in 1952. He was a moderate leader, opposing the excesses of anti-communist zealots and introducing some social reforms, and he was comfortably re-elected in 1956, retiring from public life when his term expired in 1961.

EL ALAMEIN, BATTLE OF (23 October -4 November 1942) Decisive engagement of the "desert war". The British-led Allied forces enjoyed overwhelming superiority in men and equipment over a German-led Axis army weakened by supply problems. The Allies attacked after a massive artillery bombardment. Initial progress was slow, and casualties heavy, but eventually the Germans were forced to retreat. The battle was followed by a steady Allied advance along the North African coast, although Axis forces frustrated Allied attempts to encircle them.

ENIGMA Cipher machine, used in radio transmissions, developed in Germany in the 1920s. British, Polish and French intelligence networks made initial attempts to decipher Enigma messages before the Second World War. Although the Germans regularly changed the codes, some messages were deciphered by the Allies throughout the war. These breakthroughs were crucial to Allied successes in naval wars in the Mediterranean and Atlantic, and enabled them to keep informed of German preparations before the Normandy landings of June 1944.

EUROPE, COUNCIL OF The Council was established in May 1949 to encourage co-operation in various spheres between members, and to uphold democracy in Europe. Most European democracies joined at once, with others (including Austria and West Germany) becoming members soon afterwards. The Council consists of various assemblies and committees. It has no real power, but has presided over a number of European agreements, notably those relating to human rights.

EUROPEAN ECONOMIC CO-OPERATION, ORGANIZATION FOR In 1948, most Western European states formed the OEEC, along with the US and Canada. Their immediate objective was to implement the Marshall Plan for European economic recovery. The OEEC became a symbol of post-war European co-operation. It was renamed the Organization for Economic Co-Operation and Development (OECD) in 1960. The OECD, based in Paris, continues to monitor the economic well-being of members and works for the improvement of trade relations. It also assists the economic progress of developing countries.

EUROPEAN MONETARY UNION, *See* page 153

FALANGE Spanish fascist movement, set up in 1933 by José Antonio Primo de Rivera, son of the 1920s dictator. Despite miserable performances in elections, it attracted the support of frustrated right-wingers from early 1936 as the Popular Front took office. Falange supported the nationalist revolt of July 1936 but soon lost independence, especially after José Antonio's execution by republicans in November. In April 1937, General Franco obliged Falange to join with the Carlists. Under his regime (1939-75), Falange continued to exist but occupied a clearly subservient position.

FERDINAND, FRANZ (1863-1914) Austrian archduke and heir to the Habsburg throne when the Emperor Francis Joseph's son died in 1889. An independent-minded man, he incurred the emperor's displeasure in 1900 by marrying a Czech countess, and his sympathetic attitude to the empire's Slavic minorities added to this unpopularity among the Austro-Hungarian ruling classes. On 28 June 1914 he and his wife were shot dead by a Serbian terrorist in Sarajevo, in revenge for Austria's annexation of Bosnia. The resulting Austrian ultimatum to Serbia led directly to the outbreak of the First World War.

FLEMING, SIR ALEXANDER (1881-1955) British scientist, born in Scotland. He was working as a medical researcher in London when, in 1928, he discovered that penicillin had remarkable anti-bacterial powers. Further work, with H. Florey and E. Chain, resulted in increased awareness of penicillin's uses. Mass production began during the Second World War, and penicillin remains a much-used antibiotic. In 1945 Fleming and his two collaborators received the Nobel Prize.

FOCH, FERDINAND (1851-1929) French general, in the army from 1870, notably as a teacher and strategic thinker. He commanded an army at the Marne in 1914 and was appointed chief of staff by Pétain in May 1917. After the German spring offensive of 1918, he was appointed supreme Allied commander and led the counter-attacks which brought about the final Allied victory in November. He became a marshal of France but felt that the peace terms imposed on Germany at Versailles did not guarantee French security.

FOURTEEN POINTS Proposals put forward by the US president Woodrow **Wilson** in January 1918, designed to form the basis of a peace settlement. Particularly important Points advocated self-determination for Europe's peoples; the ending of "secret diplomacy"; disarmament; freedom of the seas; and the creation of a League of Nations to settle international disputes. Germany and her allies, who had requested armistices on condition that the final settlement be based on the Points, felt betrayed by the peace treaties, which seemed to violate the Points by forcing Germans to live under foreign rule and banning union *(Anschluss)* between Germany and Austria.

FRANCO, FRANCISCO *See* page 73

FRANK, HANS (1900-46) German Nazi politician. He was the party's legal specialist before 1939, and was instrumental in the "Nazification" of the state apparatus. After the defeat of Poland in September 1939, he was chosen as ruler of the areas not allocated to Germany or Russia. His regime, based at Cracow, soon became notorious, with Jews and other "subversives" (such as nationalists and intellectuals) being summarily executed, while the remaining Poles were brutally exploited. He was charged with war crimes at Nuremberg and executed.

FUJIMARO, PRINCE KONOE (1891-1945) Japanese politician. An aristocrat by birth, he sat in the House of Peers before becoming prime minister (1937-9). He favoured Japanese expansion in Asia, regardless of US opposition, and defended his policy of war against China by claiming that Japan was establishing a "New Order" in East Asia. He became prime minister again in 1940-1 and strengthened Japan's links with the Axis powers, although he attempted to avoid war with the US by vainly suggesting a summit meeting with President **Roosevelt**. With Japan's defeat in 1945, he killed himself to avoid a war crimes trial.

GADDHAFI, MUAMMAR (1942-) Libyan leader. After military training, he participated in the 1969 military coup and became chairman of the Revolutionary Command Council and head of the Libyan armed forces. He used his new powers to remove foreigners and Anglo-American military bases. He also favoured a strictly Islamic state. He came into conflict with various neighbouring countries, especially Chad. The US accused his regime of complicity in various acts of international terrorism, and launched bombing raids on Libya during the 1980s. Libya has since been relatively inactive in international affairs.

GAGARIN, YURI (1934-68) Soviet cosmonaut. He learned flying in the 1950s, qualifying from the Air Force Training School in 1957. From 1959 he was trained as an cosmonaut and in April 1961 became the first man to journey in space, travelling around Earth in the Vostok spacecraft. He was immediately hailed as a Soviet hero. He became a specialist in space travel, although he never repeated his historic flight. After his death in an air crash in 1968, his home town of Gzhatsk was renamed Gagarin in his honour.

GANDHI, INDIRA (1917-84) Indian prime minister. The daughter of **Jawaharlal Nehru**, she joined the **Indian National Congress** in 1939 and was the party's president from 1959 to 1960. She entered the council in 1964, becoming prime minister in 1966. Her popularity was increased by the Indo-Pakistani war of 1971, but allegations about her party's dubious electoral methods led to her fall in 1977. She returned to power after the 1979 elections, but was unable to control internal conflict. She was killed by a Sikh in 1984 and was succeeded by her son Rajiv Gandhi.

GANDHI, MOHANDAS See page 56

GAULLE, CHARLES DE (1890-1970) French president. He served in the First World War and became a prominent figure in the army between the wars, calling for rapid modernization. After the fall of France in 1940, he fled to Britain and set up the Free French movement to continue resistance to Nazi Germany at home and abroad. He returned to liberated Paris in triumph in August 1944. He was provisional president until January 1946, resigning after arguments with political parties. The collapse of the Fourth Republic in 1958 led to de Gaulle's return, as head of an emergency government. In December, he became president of the Fifth Republic, having been granted wide-ranging powers by the electorate. He ended the war in Algeria despite opposition from right-wing extremists, surviving several assassination attempts. He pursued a vigorously independent foreign policy, withdrawing French forces from NATO in 1966, but the conservatism of his regime led to widespread rioting and strikes in 1968. He resigned after a referendum defeat in 1969 and died a year later.

GEORGE V (1865-1936) British king-emperor. After serving in the Royal Navy, he came to the throne in 1910 and led Britain during the First World War. He tacitly supported General **Haig** (a personal friend) in several of the latter's disputes with the prime minister, **Lloyd George**. He was an exceptionally popular monarch, widely respected for his sense of duty, and he began the tradition of royal radio broadcasts to the nation and the empire. His silver jubilee in 1935 was a great popular success, but persistent ill health led to his death less than a year later.

GLASNOST Russian word (meaning "openness") applied to policies followed by the Soviet leader Mikhail **Gorbachev** during the mid-1980s. Gorbachev intended to encourage freer debate about the Soviet system: to facilitate this, information was made more readily available through the media. At the same time the regime became somewhat less obsessively secret, though by this point the habit was so deeply ingrained as to be all but unbreakable.

GOEBBELS, PAUL JOSEPH (1897-1945) German Nazi leader. He escaped national service during the First World War because of a club foot. His early career as a journalist was undistinguished but his rise

within the Nazi party was meteoric, and in 1926 he was appointed district party leader in Berlin. He was instrumental in bringing Hitler to power by utilizing all modern methods of propaganda. Goebbels was elected to the Reichstag in 1928 and when Hitler became chancellor in 1933 was appointed propaganda minister. In this capacity he had absolute control over the radio, press, cinema and theatre, which he manipulated with utter cynicism. As an orator, he was second only to Hitler, whom he served with unswerving devotion. He remained with Hitler in the Berlin bunker and, as the city fell to the advancing Soviets, killed himself and his family.

GOERING, HERMANN (1893-1946) German air ace of the First World War and a prominent Nazi from 1933. An early confidant of Hitler, he was elected to the Reichstag in 1928 and became its president in 1931. In 1933, on coming to power, Hitler made him air minister, in which capacity Goering created the *Luftwaffe*. An advocate of air power as a decisive factor in war, Goering encouraged his commanders to develop revolutionary techniques of tactical air support based on battlefield dive-bombers. From 1937 to 1942 he was virtual dictator of the German economy and in 1939 was designated Hitler's successor. In the following year he was given the unique rank of *Reichsmarschall*. His great popularity in Germany waned, however, after Germany's lack of success in the Battle of Britain and, contrary to his many bombastic predictions, Allied air forces increasingly bombed Germany. Further promises concerning the relief of beleaguered German forces at Stalingrad in 1942 caused his fall from favour with Hitler. Goering began to withdraw from public life but remained head of the *Luftwaffe* to the last, his poor strategic judgement inhibiting the air force from making the most of the great increase in aircraft production in 1944. Goering surrendered to US troops in 1945 and, despite defending himself with brilliance at the Nuremberg trials, was convicted and sentenced to death. Two hours before he was due to be hanged he committed suicide by swallowing poison.

GOLD STANDARD See pages 20-1

GORBACHEV, MIKHAIL (1931-) Soviet leader. He joined the Communist Party in 1952, and was elected to the Supreme Soviet in 1970. He took his place in the Politburo in 1980, serving as secretary for agriculture. In 1985, after Chernenko's death, he became party general secretary and effective head of state. The Gorbachev years saw considerable liberalization, as shown by the programmes of **glasnost** and **perestroika**, which encouraged reform of the Soviet economic and political structure. Relations with the West improved (arms limitation agreements were reached with the US), while Soviet troops withdrew from Afghanistan. Gorbachev's acceptance of German reunification and the collapse of the **Warsaw Pact** in 1989-90 led to unrest at home. He survived a coup attempt in August 1991 but was swiftly eclipsed by Boris **Yeltsin**. In December 1991, as the USSR disintegrated, he resigned as Soviet president.

GRAZIANI, RODOLFO (1882-1955) Italian general. In the army from the late 19th century, he fought in the First World War and in the Abyssinian war (1935-6). In 1940 he became commander of Italian forces in North Africa but was heavily defeated by the British. Dismissal followed in February 1941. After the Italian surrender in September 1943, he supported **Mussolini**'s German-sponsored republic in northern Italy, becoming defence minister. He was jailed for treason in 1950 but released soon afterwards, spending his last years in right-wing politics.

GREAT EAST ASIA CO-PROSPERITY SPHERE Term used by the Japanese to describe the empire established by them in East Asia from the 1930s. They claimed to be establishing a "New Order" in Asia in 1938 after attacks on China, arguing that Asia must be united against communism and Western colonialism. The concept of a co-prosperity sphere enjoyed some support, especially in territories (such as Burma and the Philippines) given independence by the Japanese, but the "sphere" was in general little more than a cover for Japanese exploitation of the conquered regions. It collapsed in 1945 with the defeat of Japan and attempts to restore colonial rule afterwards were mostly unsuccessful.

HAIG, DOUGLAS (1861-1928) British field marshal. He enjoyed a successful military career before the First World War, including a spell at the War Office (1906-9). He became a British Expeditionary Force commander in 1914 before succeeding Sir John French as commander-in-chief of British forces in France in December 1915. He was criticized for employing costly "attrition" tactics, especially by the prime minister, **Lloyd George**, but managed to contain the great German offensive of March 1918 and subsequently helped the new Allied supreme commander, **Foch**, to bring about final victory in November. After the war, he became president of the British Legion and was a leading figure in the veterans' movement.

HARRIS, ARTHUR T. (1892-1984) British air force commander. He became a pilot in the First World War and held several RAF posts during the inter-war period. He became commander-in-chief of Bomber Command in February 1942 and demanded enough aircraft to begin a sustained campaign of area bombing, which attempted to destroy civilian morale as well as damaging German industry. This policy was symbolized by the "1,000 bomber" raid on Cologne in May 1942, and culminated in the attack on Dresden in February 1945. "Bomber" Harris's tactics have been widely regarded as inefficient (since he failed to make full use of precision-bombing techniques) and morally repugnant. Nonetheless, his contention that bombing non-military targets was more effective than costly strikes against military installations had influential supporters and was endorsed by Albert Speer. He was dismissed shortly after the war and retired to South Africa.

HAVEL, VÁCLAV (1936-) Czech dramatist and president. He was interested in drama from his youth and wrote several plays during the 1960s, but his work was banned

after the 1968 Prague Spring. He was involved in the **Charter 77** declaration of 1977, and was subsequently imprisoned. By 1989 he had become a symbol of the reform movement. He was elected president of Czechoslovakia in 1990 after the collapse of the communist regime, but in 1993 he was obliged to accept the division of the country.

HIMMLER, HEINRICH (1900-45) German Nazi leader, appointed head of the SS in 1929. On Hitler's assumption of power in 1933, Himmler was made chief of police for Munich and then for Bavaria. After Hitler eliminated Ernst Roehm in June 1934 (the "Night of the Long Knives"), Himmler's SS became the dominant police arm of the Nazi state. From this powerful position, he terrorized not only the wartime occupied states of Europe, but also the German people. Himmler was a hugely gifted administrator but a racial fanatic and without compunction sent millions (mainly Jews) to concentration camps and extermination centres. Himmler suppressed the conspiracy against Hitler in 1944. As Germany's approaching collapse became clear in 1945, he attempted for his own safety to negotiate a separate peace with the Western Allies. Hitler thereupon expelled him from the Nazi Party and stripped him of authority. He was captured by the British in May 1945 and committed suicide.

HINDENBURG, PAUL VON (1847-1934) German general and president. He retired from the army in 1911 but was recalled after the outbreak of the First World War and inflicted several heavy defeats on the Russians before assuming overall military control as chief of staff in August 1916. Together with **Ludendorff**, he began to control German domestic policy as well, notably through the "Hindenburg Programme", which was designed to raise German armament production. He left the army in 1919 when the government decided to accept the Versailles Treaty. In 1925, he was elected president of the republic, holding the post until his death. He reluctantly appointed Hitler as chancellor in January 1933, and his death in August 1934 enabled Hitler to become head of state.

HIROHITO (1901-89) Emperor of Japan. He became regent in 1921 and emperor in 1926. Despite supposedly supreme powers, he showed little interest in politics before the Second World War, allowing military and political leaders to adopt increasingly aggressive policies. He insisted, however, on surrendering to the Allies in August 1945 to prevent further destruction. He was spared a war crimes trial, and the monarchy was preserved by the post-war Japanese constitution, although Hirohito was obliged to give up his divine status and agree not to interfere in political matters. He took little further part in public life and devoted much of his time to marine biology.

HITLER, ADOLF See page 71

HO CHI MINH (1890-1969) Vietnamese politician. From 1918 he was involved in the worldwide communist movement, spending time in France, the Soviet Union and China, while unsuccessfully attempt-

ing to spread communism in South East Asia. During the Second World War, he took command of communist forces in Vietnam, and set up a Democratic Republic after the Japanese defeat. When the French sought to regain control of Vietnam, he called for resistance, and victory at Dien Bien Phu (1954) enabled him to restore his communist regime in North Vietnam. He then attempted to assist communist forces in South Vietnam, but this caused conflict with the US, which continued until after his death.

HONECKER, ERICH (1912-94) East German communist leader. He joined the communist movement at an early age and was active in anti-Nazi resistance. He became an influential figure in the new German Democratic Republic, entering the Politburo in 1958 and becoming party leader in 1971. He maintained a rigid autocracy in the GDR, although relations with West Germany improved. He was powerless to prevent the collapse of communism and German reunification in 1989-90. He was held responsible for the deaths of those who had tried to escape from his regime, but failing health prevented criminal proceedings.

HORTHY, ADMIRAL (1868-1957) Hungarian regent. He served in the Austro-Hungarian navy in the First World War, becoming commander-in-chief. In 1920 he was made regent, refusing to support the Habsburg claimant to the empty throne. His regime was conservative and undemocratic, although the press retained its freedom. His foreign policy was designed to recover lands lost after 1918, and Hungary gained territory from Czechoslovakia and Yugoslavia through co-operation with Hitler. Horthy was suspicious of Nazi Germany, but joined the German invasion of Russia in June 1941. As the Axis powers fell back, he tried to arrange peace with the Allies but was imprisoned by the Germans. After the war, he lived in Portugal.

HORTON, MAX (1883-1951) British naval commander. After serving in the submarine force during the First World War, he commanded the Reserve Fleet before the Second World War and took over as flag officer, submarines, in January 1940. He was made an admiral in 1941. His grasp of submarine warfare led to his appointment as commander-in-chief, Western Approaches in November 1942. He gradually overcame the threat posed by German U-Boats, employing groups of specially-prepared escort vessels. He retired after the war.

HU YAOBANG (1915-89) Chinese communist politician. From a peasant background, he joined the communist movement at an early age and took part in the Long March of 1934-5. He came to prominence after the communist takeover in 1949, introducing youth schemes, but was criticized during the Cultural Revolution in the mid-1960s. He was rehabilitated in the 1970s and, thanks to patronage from Deng Xiaoping, became effective party leader in 1981. His policies of liberalization, in both the economic and political spheres, aroused condemnation from many of his colleagues. He was dismissed in 1987. His death two years later helped cause the student unrest which was crushed at Tiananmen Square.

HULL, CORDELL (1871-1955) US statesman. After a career in law, he sat in the Senate as a Democrat from 1931 to 1933 before becoming secretary of state under **Roosevelt.** He was not a prominent international figure before the war, largely because of the US's isolationist foreign policy, although he did improve relations with Latin American countries. During the Second World War he was involved in the planning for the United Nations. He retired because of ill health in 1944, receiving the Nobel Peace Prize the following year.

HUSSEIN (1935-) King of Jordan. Educated in Britain, he came to the throne in 1952 and attempted to follow a moderate course in Middle Eastern affairs. Jordan was, however, involved in the 1967 Arab defeat by Israel. The country then became a base for Palestinian guerrillas. Hussein's regime removed them in the early 1970s, despite condemnation by many Arab states. He abandoned Jordanian claims to the West Bank in 1988. During the Gulf War crisis of 1990-1, he was criticized for adopting a conciliatory attitude towards Iraq, though his country's heavy economic dependence on Iraq effectively left him little or no choice in the matter.

HUSSEIN, SADDAM (1937-) Iraqi dictator. He joined the Ba'th Socialist Party in 1957 and was often in exile before taking part in the 1968 revolution. In 1969 he became vice-president of the Revolutionary Command Council. In 1979 he succeeded Bakr as president and established personal control over the Ba'th movement. An eight-year war with Iran began in 1980. Saddam was tacitly supported by the West, as he seemed a bulwark against Islamic fundamentalism. In 1990, however, his invasion of Kuwait aroused worldwide condemnation of what came to be seen as an increasingly brutal and repressive dictatorship. Contrary to widespread predictions that Iraq's crushing defeat by UN coalition forces would quickly result in Saddam's overthrow, he tightened his grip on the country and ruthlessly stamped out dissent. Continuing economic sanctions and economic disarray have done little to prise loose his tenacious grip on power.

INDIAN CONGRESS MOVEMENT The Indian National Congress was founded in 1885 as a nationalist grouping. At first, the movement was comparatively moderate in its attempts to obtain concessions from the British, but from the 1920s a more uncompromising position was adopted by leaders such as Gandhi and **Nehru,** who organized campaigns of civil disobedience and called for complete Indian independence. The movement's determination, coupled with Britain's refusal to allow India greater autonomy during the Second World War, increased its popular support. After independence (1947), the Congress Party became the most powerful force in Indian politics.

IRISH REPUBLICAN ARMY Irish nationalist terror organization. The IRA was established in 1919 to oppose British rule in Ireland. After being banned by the Irish government in Dublin, it remained a fairly insignificant force until the late 1960s. At that time, the "Provisional" IRA was formed. The Provisionals began a sustained campaign against the Protestant majority in Northern Ireland, launching bomb attacks there and on the British mainland. A ceasefire in 1994 suggested that a breakthrough might be at hand. But the IRA's resumption of terrorist tactics in March 1996, coupled with its refusal to decommission any of its weapons, suggests that peace in Northern Ireland remains as elusive a prospect as it has ever been.

JANG QING (1914-91) Chinese politician and wife of Mao Tse-tung. After a short career as an actress, she fled the Japanese invasion of Shanghai in 1937, becoming a communist. She married Mao in 1939. After Mao's takeover of China in 1949, she took particular interest in cultural affairs. Her political views were consistently extreme and she was a firm supporter of the Cultural Revolution. She joined the Politburo in 1969. After Mao's death in 1976, political opponents managed to have her and three others (the "Gang of Four") expelled from the party. She was tried and sentenced to death in 1980. The sentence was subsequently changed to life imprisonment.

JENKINS, ROY (1920-) British politician. A Labour MP from 1948, he became a minister in 1964 and was chancellor (1967-70) and home secretary (1974-6). He then served as European Commission president from 1977 to 1981. Returning to British politics, he found the Labour party now too left-wing for his liking. He helped to found the Social Democratic Party and became its leader. He was replaced by David Owen in 1983, but continued to support the SDP-Liberal Alliance. He lost his parliamentary seat in 1987 and was given a peerage.

JIANG ZEMIN (1926-) Chinese communist politician. He was trained as an engineer and became an economic adviser to the Chinese ambassador in Moscow in the 1950s. After working many years for state industries, he joined the Central Committee in 1962 and the Politburo in 1987. In June 1989, when party leader Zhao Ziyang was dismissed after the Tiananmen Square massacre, Jiang succeeded him. As party leader Jiang has allowed some economic liberalization but has rigorously maintained the Party's political control.

JUTLAND, BATTLE OF (31 May 1916) Naval battle of the First World War and the only major engagement between the British and German fleets, caused by the German admiral Scheer's attempt to launch a surprise attack on British bases and so lessen Britain's naval superiority. The British intercepted the German fleet and eventually forced it to withdraw late in the day and return to port. Although the British fleet suffered heavier losses, the battle in fact confirmed its overall dominance at sea, and the Germans did not seriously challenge it again. The Germans subsequently employed submarine warfare to try to cut British supply lines.

KAKUEI, TANAKA (1918-) Japanese politician. After setting up a construction business, he was elected to the House of Representatives in 1947, representing the Liberal Democratic Party. He became minister of finance (1962-5) and later minister of trade and industry (1971-2). He was prime minister from 1972 to 1974, when he was forced to resign because of the Lockheed bribery scandal. He remained politically active but was entangled in prolonged legal conflicts, having been found guilty of corruption in 1983.

KAMMHUBER LINE German defence system against Allied bombers during the Second World War. Dividing the important air zones into a series of overlapping "boxes", the Germans used radar and searchlights to direct fighter pilots when enemy bombers entered their box. The system managed to limit the effectiveness of Allied bombing until July 1943, when the British defeated it and launched a heavy raid on Hamburg. The Germans subsequently employed less restrictive tactics.

KARAMANLIS, CONSTANTINE (1907-) Greek politician. After a legal career, he entered parliament in 1935. After the war he became a prominent figure, helping to reorganize the democratic right wing. He became minister of public works in 1952 and prime minister in 1955. His years in power saw unsuccessful Greek attempts to join the European Common Market and continued unrest in Cyprus. He was voted out of office in 1963, and was forced out of the country by the military coup of 1967. When Greek democracy was restored in 1974, he returned as prime minister. In 1980 he became president of the republic. He returned to this post in 1990.

KEMAL, MUSTAPHA (1880-1938) Turkish leader. After military training, he came to prominence during the Balkan Wars of 1912-13, and was a successful commander during the First World War, distinguishing himself at Gallipoli in 1915. Turkey's final defeat led to substantial territorial losses which he resisted by force until 1922, driving the Greeks out of Anatolia and threatening Constantinople. The Lausanne Treaty revised the post-war settlement and Kemal set up a Turkish Republic in October 1923, retaining power until 1938. He was determined to replace the antiquated Ottoman empire with a modernized, Western-style state. He favoured industrialization, abolished laws discriminating against women and encouraged Turkish nationalism, as opposed to religious fervour. He named himself "Atatürk", or "Father of the Turks".

KENNEDY, JOHN F. (1917-63) US president. Educated at Harvard, he served with distinction in the Second World War and joined Congress as a Democrat in 1947. In 1952 he became a senator. After participating in the Senate's foreign relations committee, in 1961 he became America's youngest president. He introduced various social reforms at home, while his image abroad was improved by his firmness during the **Berlin Wall** (1961) and Cuban missile (1962) crises. He was assassinated in Dallas in November 1963. Lee Harvey Oswald was generally blamed, but alternative theories abound. Kennedy's youthful dynamism and his tragic end have probably led many to overestimate the achievements of his presidency.

KERENSKY, ALEXANDER (1881-1970) Russian politician. After legal training, he entered parliament in 1912 and became a strong critic of tsarist autocracy. After the tsar's abdication in March 1917, he became minister of justice in the Provisional Government. His energy and ability enabled him to become war minister in May and prime minister in July. He was a moderate socialist, but failure to introduce popular industrial and agricultural reforms left him increasingly isolated. His insistence on continuing the costly war against Germany reduced his popularity still further. He survived an attempted coup by General Kornilov in September, but was removed from office by the **Bolshevik** revolution in November. When his attempt to recover power failed, he left Russia and spent the rest of his life in exile.

KESSELRING, ALBERT (1885-1960) German general. A veteran of the First World War, he later joined the air force and successfully commanded an air fleet in the early stages of the Second World War. In December 1941 he was transferred to the Mediterranean theatre and assisted **Rommel** in North Africa. In 1943-4 he led German forces in Italy, skillfully holding up the advance of the numerically superior Allied forces. Hitler gave him command on the Western Front in March 1945, but by this point the defeat of Germany was all but guaranteed. He was sentenced to life imprisonment for complicity in war crimes and released in 1952.

KEYNES, JOHN MAYNARD (1883-1946) British political economist. Born and educated in Cambridge, he was part of the British team at the 1919 Paris Peace Conference and afterwards published a devastating attack on the settlement, arguing that reparations would harm the European economy. During the inter-war period he developed economic theories, rejecting orthodox thinking and advocating state intervention in the economy to deal with the unemployment problem. His publication in 1936 of the *General Theory* led to the spread of Keynesian thinking. He was a government economic adviser during the Second World War and attended the Bretton Woods conference in 1944, calling for the introduction of a World Bank. His influence was felt especially in the 1950s and 1960s but have since declined in importance.

KGB Soviet security service. The KGB was set up in 1953 to replace the Stalinist NKVD. It was designed to protect the Soviet regime (by monitoring potential dissidents) and to handle intelligence work. The KGB became highly influential, acquiring a sinister reputation abroad. The Soviet leader Andropov (1982-4) had previously been KGB chief for many years. The KGB became less secretive after the collapse of the Soviet Union in 1990-1.

KHOMEINI, AYATOLLAH RUHOLLAH *See* page 167

KHRUSHCHEV, NIKITA (1894-1971) Soviet premier. Brought up in the Ukraine, he became a senior communist party official in the 1930s. After the Second World War, he led the Ukrainian soviet republic and oversaw recovery, afterwards being chosen by **Stalin** to organize agricultural programmes. After Stalin's death in 1953, he became the party's first secretary and shared power for a time with Bulganin. After launching a fierce attack on the Stalinist regime in January 1956, he became prime minister in March 1958. His period in power was marked by occasional tension with the US and China, while at home he adopted a policy of gradual liberalization. He was forced from office in October 1964 and withdrew into private life.

KIEL MUTINY At the end of October 1918, with the German army close to defeat, several admirals of the High Seas Fleet decided (without authorization from the government) to launch a last desperate attack on the British Grand Fleet, believing that this would salvage national honour. Many of the sailors at the Kiel base, already angered by poor conditions and rigid discipline, mutinied and formed Soviet-style councils, which the increasingly impotent authorities could not suppress. The mutiny was an important part of the revolutionary movement which spread throughout Germany at the end of the war and helped force the Kaiser to flee the country.

KIM IL SUNG *See* page 131

KING, MARTIN LUTHER *See* page 119

KISSINGER, HENRY (1923-) US statesman. Of German-Jewish origin, he arrived in the US in 1938 and became an academic. He was an adviser to **Nixon** during his successful presidential campaign of 1968, afterwards becoming national security adviser. He made efforts to improve relations with the Soviet Union and China, while his attempts to mediate in Vietnam won him the Nobel Peace Prize in 1973. At the same time his willingness to make concessions to communist states aroused criticism at home. He became secretary of state in September 1973 and retained this position under President Ford after Nixon's resignation. He retired from politics in 1977.

KOHL, HELMUT (1930-) German statesman. He joined the Christian Democratic Union in his youth, becoming the party's deputy chairman in 1969. He was the party's candidate for the chancellorship in 1976 but was defeated. The collapse of Schmidt's centre-left coalition in 1982 enabled Kohl to recover, and he led the CDU to victory in the elections which followed. As chancellor, Kohl adopted a cautious economic policy and reduced government spending. His foreign policy involved support for closer European integration. The CDU's disappointing showing in the 1987 election forced him to rely increasingly on coalition partners. After the collapse of the **Berlin Wall** in November 1989, Kohl was able to bring about German reunification during 1990. This success led to victory in "all-German" elections soon afterwards. Reunification caused serious economic difficulties, however, and Kohl's austerity measures created discontent, but he was still in power in mid-1996.

KUN, BELA (1886-1937) Hungarian revolutionary leader. He led the communist insurgents who overthrew the Karolyi regime in 1918. On becoming premier in 1919, he attempted to reorganize the country on Soviet principles, but was forced into exile four months later.

LATERAN PACTS Agreements between **Mussolini** and the papacy, concluded in February 1929, which ended the long church-state conflict in Italy. The state agreed to recognize the Vatican and to pay it compensation for the territories seized in 1870. The church was also granted some influence in the educational field. In return, the Vatican recognized the Italian state. Mussolini was willing to make concessions because the Pacts increased his popularity and forestalled further criticisms by the papacy of the fascist state.

LEE KUAN YEW (1923-) Singaporean politician. After studying law in Britain, he returned to Singapore in 1951 and entered politics. In 1954 he set up the democratic socialist People's Action Party. In 1959, having secured autonomy from Britain, he became the country's first prime minister. Under Lee's autocratic and decisive leadership, Singapore enjoyed rapid development, becoming one of the foremost of the region's "Tiger Economies". Lee retired in 1993.

LEMAY, CURTIS (1906-89) American air force general, who developed the techniques of strategic bombing in the Second World War. He commanded the US 8th Air Force (in Europe) from 1942 and his "pattern bombing" tactics proved brutally effective over Germany. In 1944 he was transferred to the Far Eastern sphere, where his B-29 squadrons inflicted severe damage on Japan. He became chief of the strategic air command in 1948 and later chief of the US air staff.

LEND-LEASE Term applied to arrangements between the US and Allied nations during the Second World War. In March 1941, President **Roosevelt** was officially allowed by Congress to sell, lend or give arms to countries whose survival was considered important to US security. This allowed Roosevelt (who wanted the US to become the "arsenal of democracy") to compromise US neutrality, though isolationist sentiment continued to prevent entry into the war. Lend-Lease continued after the US joined the war, and goods worth over $43 billion were eventually transferred, mostly to Britain, before 1945.

LENIN, VLADIMIR ILYICH See page 41

LIN PIAO (1908-71) Chinese general. After military training, he began to associate with the communists against **Chiang Kai-shek's** nationalist forces and was involved in the communist Long March of 1934-5. He was instrumental in bringing about the communist victory in the civil war, capturing Mukden in October 1948. He became minister of defence in 1959 and was a fervent supporter of the Cultural Revolution. In 1969 he was recognized as successor to Mao. He was killed in an air crash while fleeing to the USSR, following the discovery of his plans to overthrow Mao.

LLOYD GEORGE, DAVID (1863-1945) British prime minister. Brought up in Wales, he entered Parliament in 1890 as a Liberal and became president of the board of trade in 1905, afterwards spending seven years as chan-

cellor (1908-15). He introduced a number of social reforms (notably the Pensions Act of 1908) despite strong opposition from the House of Lords. From May 1915 he was an energetic minister of munitions, succeeding Asquith as prime minister in December 1916. He proved a vigorous war leader, coming into conflict with military leaders who opposed his plans to weaken Germany by attacking her eastern allies. At the Paris Peace Conference, he generally encouraged a lenient attitude towards Germany, believing that a healthy Germany was necessary for overall European recovery. He remained prime minister until 1922, when controversy over his handling of the Greco-Turkish war caused his resignation. He remained politically active until his death but was unable to regain high office.

LONDON, TREATY OF Secret agreement ensuring Italy's entry into the First World War on the Allied side, signed 26 April 1915. The Allies promised to supply the Italian war effort and to grant Italy reparations and substantial territorial gains (mostly in Central and south east Europe) in the event of victory. Italy duly entered the war within a month, but in 1917 the new **Bolshevik** government in Russia published the treaty's terms, much to Allied embarrassment. At the post-war peace negotiations, the emphasis on national self-determination meant that the Italians did not receive many of the promised territories. The resulting resentment in Italy increased support for nationalist parties.

LUDENDORFF, ERICH (1865-1938) German general. After a series of successes in both east and west in the early stages of the First World War, he became chief quartermaster general in August 1916 and, with **Hindenburg**, effective joint ruler of Germany. He favoured unrestricted submarine warfare and the imposition of harsh peace terms on defeated Russia in 1918. He organized the German spring offensive of 1918, but after the eventual German defeat he left the army. After the war he was a leading figure in extreme right-wing politics, working with the Nazis in the mid-1920s, but his political influence soon declined.

MACARTHUR, DOUGLAS (1880-1964) US general. After serving in the First World War, he became a noted army commander and was chief of staff from 1930 to 1935. He then retired and became military adviser in the Philippines. He was recalled in 1941 to command US (and later Allied) forces in the Pacific. After early failures, he developed the successful "island-hopping" approach towards Japan and took the Japanese surrender in 1945. He subsequently headed the occupation forces, and was highly active politically. He led UN forces in the Korean War from 1950 to 1951, but was dismissed by President **Truman** for publicly opposing official policy. After returning home, he became involved in Republican politics.

MAJOR, JOHN (1943-) British prime minister. A banker by profession, he became a Conservative MP in 1976. He enjoyed a rapid rise in the late 1980s, becoming foreign secretary in 1989 and chancellor shortly afterwards. After Mrs **Thatcher**'s resigna-

tion in November 1990, he defeated Michael Heseltine and Douglas Hurd in the party leadership election and became prime minister. He unexpectedly won the general election of 1992, despite economic problems. Criticism of his leadership style, especially from right-wing MPs, led to an unsuccessful leadership challenge by John Redwood in 1995.

MANDELA, NELSON See page 165.

MANHATTAN PROJECT Name given to the US-led Allied programme to produce an atomic bomb during the Second World War. The project was officially set up in 1942, by which time US and exiled European scientists had already made considerable progress. From August 1943, Britain also co-operated in the programme. The research was intensive and fruitful, partly because of the contributions made by German scientists. Three bombs were constructed with the use of uranium and plutonium. The first was successfully exploded in New Mexico in July 1945. The others were dropped on Japan in August and brought an immediate end to the war.

MANSTEIN, ERICH VON (1887-1973) Outstanding German general and strategist. In 1939, he developed a plan for the invasion of France, involving an armoured attack through the Ardennes to divide the Allied armies. The plan, approved by Hitler and implemented in May and June 1940, was spectacularly successful. He later led armies on the Eastern Front, achieving particular success in the south, where his forces took Sebastopol in the Crimea. He prevented a German collapse after the surrender at Stalingrad in 1943, but his willingness to undertake tactical retreats angered Hitler, who dismissed him in April 1944.

MAO TSE-TUNG See page 75

MARCONI, GUGLIELMO (1874-1937) Italian inventor. In 1896, after work on electro-magnetism, he sent the first wireless message by electro-magnetic waves. During the following years he refined his methods, sending messages across the Channel and in 1901 across the Atlantic. In recognition of this breakthrough, Marconi received the Nobel Prize for Physics. He spent his remaining years developing and marketing the techniques of wireless telegraphy.

MARNE, BATTLE OF THE By the end of August 1914, the German invasion of France seemed likely to be successful. General von Kluck, the commander of the German "right wing", was poised to encircle Paris, as specified by the Schlieffen Plan. However, confusion over the position of the Anglo-French armies and the diversion of German troops to meet a Russian offensive in the east caused the German high command to hesitate. The offensive ground to a halt and the French, some of their troops driven to the front in fleets of taxis, counter-attacked along the River Marne to remove the threat to Paris. The Germans were forced to retreat to the River Aisne, but avoided a complete defeat. The battle ended German hopes of a quick victory in the west and led rapidly to the stalemate of the Western Front.

MARSHALL, GEORGE C. (1880-1959) US general. He served in the First World War and was a senior army planner before 1939, when he became army chief of staff. He oversaw the rapid development of the US army during the Second World War and was the main Anglo-American strategist, arguing that the European war should take priority. After the war he retired from the army and became an international figure, but his attempts to end the Chinese civil war were unsuccessful. He became secretary of state in January 1947 and proposed the Marshall Plan for European economic recovery, as well as taking part in the setting-up of NATO. He was awarded the Nobel Peace Prize in 1953.

MARX, KARL (1818-83) The founder of modern communism. Born in Germany of Jewish descent, he studied law before joining the European socialist movement. With his closest associate, Engels, he published the *Communist Manifesto* in 1848. Marx believed that the class struggle was the dominant feature of world history and called upon the workers of all countries to unite and overthrow the oppressive capitalist system. The "dictatorship of the proletariat" would be followed by the "withering away" of the state and the achievement of a classless, communist society. Towards the end of his life, he lived in London, where he wrote *Das Kapital* and attempted vainly to promote unity in the world-wide socialist movement.

McCARTHY, JOSEPH (1908-57) US senator. He became a circuit judge in 1939 and afterwards served in the Second World War. He became a senator in 1946, and soon embarked on the anti-communist campaign with which his name is associated. Most of his "red-baiting" accusations were unproven, notably his claim that the state department had been infiltrated by communists, but widespread fear of communism enabled him to head a Senate-investigating committee from 1952. His increasingly wild allegations eventually aroused hostility, and he was formally condemned by the Senate in December 1954, at which point his career was effectively over.

MEIN KAMPF A book written by Hitler during his imprisonment (1924-5) after his attempted coup in 1923. He described his early career (the title means "my struggle") and set out his world view. He argued that racial struggles determined history and that the Germans, as "Aryans", must conquer living space (*Lebensraum*) if they were to flourish and defeat the threat posed by inferior races (Jews and Slavs). The territory would be found in Eastern Europe, and this meant an anti-Russian foreign policy. German foreign policy also had to isolate France, the power most opposed to German expansion, through alliances with Britain and Italy. Since 1945, historians have continued to debate the extent to which the principles of *Mein Kampf* determined Hitler's policies during his 12 years as German dictator.

MEIR, GOLDA (1898-1978) Israeli prime minister. Of Russian-Jewish background, she was taken to the US in 1907 and left for Palestine in 1921. She worked on a kibbutz and was involved in left-wing politics, joining the Mapai Labour Party. After

the establishment of Israel in 1948, she became ambassador in Moscow and then minister of labour. From 1956 to 1965 she was foreign minister. In 1969 she became head of a coalition government. She survived the Yom Kippur War of 1973 but failed to hold her cabinet together, resigning suddenly in 1974. She was replaced by Yitzhak Rabin.

MENGISTU, HAILE MARIAM (1941-) Ethiopian leader. Trained as a soldier, he took part in the 1974 military coup and became a member of the Provisional Military Administrative Council. In 1977 he seized control of the country and imposed his authority with harsh measures against political opponents. His hopes for a successful socialist state were not fulfilled, party because of government corruption. Ethiopia was weakened by famines and droughts throughout the 1980s and by wars in Eritrea and with Somalia. Mengistu was forced into exile in 1991.

MENSHEVIKS The non-Leninist members of the Russian Social Democratic Labour Party. The Mensheviks often worked with the **Bolsheviks** before 1917, but wished to build up a mass workers' movement, while Lenin preferred a well-organized party of professional revolutionaries. This division was highlighted during the 1905 revolution, when the Mensheviks took a more moderate stance. The revolution of 1917 exposed their internal divisions: while most Mensheviks agreed that Russia was not prepared for socialism, only right-wingers were prepared to take part in the Provisional Government and continue the war. When the Bolsheviks seized power in November, the Mensheviks were too disorganized to provide effective opposition and were swiftly suppressed.

MEXICAN REVOLUTION See page 54

MILOŠEVIĆ, SLOBODAN (1941-) Serbian president. He joined the communist movement in his youth, and from 1966 held various Yugoslav government posts. He became head of the Serbian Communist League in 1984 and Serbian president four years later. In 1991-2 he attempted to prevent the break-up of Yugoslavia by vigorously opposing the aspirations of Croatia, Slovenia and Bosnia-Herzegovina. Claims that he had sponsored rebel Serb armies in Bosnia and Croatia led to UN sanctions against Serbia. As hopes for a Greater Serbia faded, he became more moderate, and in late 1995 he took part in the negotiations which led to a temporary peace agreement in the former Yugoslavia.

MITTERRAND, FRANÇOIS (1916-96) French president. He was active in the resistance during the Second World War and became a deputy in 1946, holding various offices in the Fourth Republic. He was critical of de Gaulle's Fifth Republic (set up in 1958), and unsuccessfully challenged the general in the 1965 presidential elections. He tried to unify the French left, achieving success with the setting-up of the Parti Socialiste in 1971. He displaced President Giscard d'Estaing in 1981 and introduced a policy of nationalization, but economic problems soon forced him to adopt austerity measures and abandon many socialist principles. He was obliged to co-operate with a

Gaullist prime minister, Chirac, after 1986. Mitterrand defeated Chirac in the 1988 presidential election, thus restoring socialist pre-eminence. Despite ill health and troubles with his party, he managed to remain in office until 1995, when he retired and was replaced by Chirac.

MONTGOMERY, BERNARD (1887-1976) British field marshal. After military training, he served in the First World War and was a commander during the battle for France in 1940. In August 1942 he took over command of the 8th Army in North Africa and enjoyed a series of successes in the Mediterranean theatre, winning the battle of **El Alamein** and advancing into Italy. He became famous for his ability to inspire his troops. He commanded the land forces during the Allied invasion of Western Europe after June 1944, although his plan for a rapid advance into Germany was rejected by **Eisenhower**. In May 1945 he received the German surrender. He later became deputy commander of NATO forces in Europe (1951-8) before his retirement.

MORGENTHAU PLAN In September 1944, the US treasury secretary, Henry Morgenthau, put forward a plan to strip Germany of its industrial capacity and transform it into a pastoral country after the war. At first, Allied leaders seemed sympathetic to the plan, but it was soon exposed as impractical and was swiftly dropped. In an attempt to strengthen the resolve of the German people, Nazi propagandists seized on the plan as proof of Allied vindictiveness.

MAU-MAU REBELLION Nationalist revolt in Kenya in the 1950s. In 1944, dissatisfaction among Africans had led to the setting-up of the Kenya African Union. The failure of the KAU to obtain concessions from the colonial authorities led to the so-called Mau-Mau movement. Attacks on Europeans and "disloyal" Africans caused the authorities to declare a state of emergency in 1952. A confused and often brutal struggle followed, causing concern in Britain. In 1959 it was agreed to start negotiations on Kenyan independence, which was achieved in 1963.

MUGABE, ROBERT (1924-) Zimbabwean leader. Educated in South Africa and Britain, he was an important figure in the Zimbabwean nationalist movement in the early 1960s. He was imprisoned, eventually being released in 1975. With Joshua Nkomo, he then led the Patriotic Front against Ian Smith. When elections were held in 1980, he became prime minister. He was increasingly estranged from Nkomo, and this led to simmering conflict throughout the 1980s. He became president in 1987. His political career has borne witness to his fervent, if pragmatic, socialism and his strong opposition to South Africa's apartheid regime.

MUJAHEDDIN Name given to the groups of Islamic freedom fighters, particularly in Afghanistan after the Soviet invasion in 1979. They proclaimed a *jihad* (holy war) and received aid from Islamic states and the US. Their tenacity and knowledge of the terrain made it impossible for the Russians to subdue them, and the Soviets duly withdrew in 1989. After that, however, the divisions among the Mujaheddin became apparent, and their failure to form a generally acceptable government rapidly led to civil war.

MUSSOLINI, BENITO (1883-1945) Italian fascist leader. He served in the First World War and subsequently entered right-wing politics, setting up a fascist group in March 1919. He demanded strong government to deal with the communist threat and the territorial expansion denied Italy at the Paris Peace Conference. He was appointed prime minister in October 1922, eventually turning Italy into a one-party state in 1928. He maintained his popularity through public works schemes and friendly relations with the papacy. His foreign policy was relatively moderate until 1935, when he invaded Abyssinia. Condemnation by the Western democracies led him to side increasingly with Hitler. He joined the Second World War on the German side in June 1940, but military failures, notably in Greece, forced him to accept a clearly subordinate position. He was dismissed by the king in July 1943 and imprisoned as Italy made peace with the Allies. He was rescued by German troops and installed as ruler of the German-occupied areas of Italy, but had little independence. He was captured and shot by partisans in April 1945.

NAGY, IMRE (1896-1958) Hungarian leader. A notable communist in the inter-war period, he became minister of agriculture in the new Hungarian regime after the Second World War. He became prime minister in July 1953, shortly after **Stalin**'s death, and adopted a policy of liberalization. This led to his deposition in 1955, but he returned to power in 1956 and tried to withdraw Hungary from the Soviet-dominated **Warsaw Pact**. Soviet forces arrived in November to crush the uprising. Nagy was arrested, and later executed. He was rehabilitated in 1989.

NASSER, GAMAL ABDEL (1918-70) Egyptian leader. After military training, he became an ardent Egyptian nationalist and was involved in the military takeover in July 1952. He became prime minister in 1954 and president soon after. He survived the Suez crisis of 1956, when an Anglo-French-Israeli invasion was condemned by the UN. Nasser introduced socialist economic polices at home, while his attempts to encourage Arab co-operation made him a symbol of the pan-Arab movement, despite the failure of his United Arab Republic – an abortive federation set up with Syria in 1958 in the name of Arab unity which ended ignominiously in 1963. Having made increasingly belligerent threats against Israel from the mid-1960s, in 1967 Israel launched a pre-emptive strike against Egypt. Israel's crushing victory greatly weakened Nasser's authority. He died in office and was succeeded by **Sadat**.

NAZI-SOVIET PACT (23 August 1939) Agreement signed by Ribbentrop and Molotov, the foreign ministers of Nazi Germany and Soviet Russia. The two powers signed an official non-aggression pact, and also secretly agreed to divide up much of Eastern Europe into spheres of influence. The Pact ensured that Russia would not have to face a united attack by the "capitalist powers" and allowed

Stalin time to prepare for a possible German attack later. It also allowed Germany to invade Poland without fear of Soviet intervention. Russo-German co-operation continued until June 1941, when Germany invaded the Soviet Union.

NEHRU, JAWAHARLAL (1889-1964) Indian leader. After education in England, he joined the Congress nationalist movement, becoming Congress president in 1929 and remaining a prominent figure despite several periods of imprisonment. He was instrumental in arranging the transfer to independence in 1946-7, afterwards becoming prime minister and foreign minister. He favoured industrialization and gradual liberalization of Indian society, but encountered strong opposition from conservatives. He adopted a neutral world policy, while attempting to solve more local problems (especially over Kashmir) by peaceful methods. He remained in office until his death.

NEW DEAL Term applied to policies of US president **Roosevelt** during the mid-1930s. While seeking election in 1932, he advocated government intervention in the economy to ameliorate the problems caused by the Great Depression. Once elected, he concentrated on recovery from 1933-4, adopting more ambitious social policies from 1935, symbolized by the Social Security Act and measures to improve industrial relations. The new approach, involving high government spending, aroused criticism from liberal economists but brought relief to those impoverished by the Depression, and ensured Roosevelt's re-election in 1936.

NEW ECONOMIC POLICY Policy introduced by the Soviet government to replace **War Communism** and announced by Lenin in March 1921. The NEP was above all designed to increase agricultural production by abolishing the food-requisitioning system and introducing a more moderate tax in kind. The government also abandoned its ownership of many enterprises, while retaining control of heavy industry. These policies led to a limited revival of capitalism, and traders ("Nepmen") were allowed to prosper. The NEP enabled Russia to recover slowly from the upheavals of the revolution and civil war, but some **Bolsheviks** saw it as a betrayal of socialist ideals.

NICHOLAS II (1868-1918) Russian tsar (1894-1917). He was a resolutely autocratic ruler who, together with his wife, firmly opposed moves towards a constitutional monarchy. After the humiliating defeat by Japan in 1904-5 and the revolution of 1905, he reluctantly agreed to the creation of a parliament (duma), but imposed strict limits on its powers. The outbreak of the First World War temporarily strengthened his authority, but a series of disastrous defeats (especially after his takeover of supreme command in September 1915) and failure to make political concessions to a war-weary population caused unrest, culminating in the revolution of February 1917. He abdicated on 15 March and was subsequently arrested by the Provisional Government and exiled to the Urals. As civil war spread after the **Bolshevik** takeover in November 1917, he and his family were executed by Bolshevik agents, probably in July 1918.

NIMITZ, CHESTER (1885-1966) US admiral. After serving in the submarine force, he became a rear admiral in 1939 and commander of the US Pacific Fleet after Pearl Harbor (December 1941). His victory at Midway in June 1942, assisted by superior intelligence work, began a gradual American advance in the Pacific. He often disagreed with the army commander, General **MacArthur**, but was made a fleet admiral in December 1944 and helped to oversee the Japanese surrender in August 1945. A highly respected commander, he later became a UN mediator.

NIXON, RICHARD (1913-94) US president. A lawyer, he was elected to Congress in 1946, and soon proved a firm anti-communist. He became vice-president in 1953 but was defeated in the presidential election of 1960 by **Kennedy**. He was the Republican candidate again in 1968 and defeated Herbert Humphrey. He reversed many of his predecessors' social reforms and twice devalued the dollar. After making a final effort in Vietnam, Nixon began to withdraw US forces and accepted a ceasefire in 1973. He improved relations with the USSR through bilateral agreements, and he recognized China's communist government, visiting the country in 1972. He was comfortably re-elected in 1972, but resigned in August 1974 following the Watergate scandal, and withdrew from public life.

NKRUMAH, KWAME (1909-72) Ghanaian leader. He was educated in the US and Britain, and was involved in the **Pan-African Congress** in Manchester in 1945. He founded a nationalist party in the Gold Coast (as Ghana was then known) in 1949 despite a period of imprisonment. He secured independence for Ghana from Britain in 1957. Nkrumah became president in 1960. He was a keen supporter of African independence movements, but his popularity was reduced by economic problems. He was forced from office by a military coup in 1966, dying in exile in Romania.

NORIEGA, MANUEL (1939-) Panamanian general. Originally trained by the CIA, he was the head of Panama's armed forces from 1983. By 1989 he was effectively ruler of the country. The US accused him of involvement in espionage and international drug-dealing, and sent a task force to depose him. Noriega gave himself up in January 1990 and was taken to the US to face charges. He was replaced as leader by the more pro-American Endara.

NOVOTNÝ, ANTONÍN (1904-75) Czech communist leader. He joined the party in 1921 and came to prominence after the Second World War, during which he had spent four years in a concentration camp. He became first secretary of the party in September 1953, and president four years later. Novotný was opposed to reforms and his refusal to relax economic controls led to a severe recession in the 1960s. Novotný began to make some concessions, but was replaced as first secretary by Dubček in January 1968. He left the presidency two months later.

NUREMBERG LAWS The first important racial laws of Nazi Germany, announced at the party rally in 1935.

One law restricted citizenship to those of German descent, while the others banned, among other things, marriage and extra-marital relations between Germans and Jews. Other laws were subsequently issued to deal with the status of "part-Jews". The laws paved the way for the exclusion of Jews from German society.

NUREMBERG TRIALS From October 1945 to September 1946, some of the political and military leaders of Nazi Germany were tried by an international tribunal assembled by the victorious Allies. The defendants were charged with conspiracy to wage wars of aggression, war crimes and crimes against humanity. Many claimed that they could not be held personally responsible for the crimes of the Nazi regime. Ten, including Ribbentrop, were executed. **Goering**, the most prominent of the condemned men, managed to commit suicide before the executions. Some, such as Hess and Speer, were given long prison terms. A few were acquitted.

PAHLAVI, SHAH MUHAMMAD REZA (1918-80). Shah of Iran from 1942. Coming to power after the deposition of his father, he hoped to encourage economic development in Iran, but his maintenance of a despotic system of government prevented him from enjoying popular support. Economic problems and resentment of the luxurious lifestyle of the ruling class led to growing discontent in the 1970s. This strengthened the position of fundamentalists, who opposed the regime because of its perceived failure to uphold the Islamic faith. The shah left Iran in January 1979. The Islamic revolution which followed prevented his return. He died in exile in Egypt soon afterwards.

PAN-AFRICAN CONGRESS South African political movement. The PAC was formed in 1959 by disaffected **African National Congress** members, led by Robert Sobukwe, who considered the ANC too moderate. The PAC's hardline tactics led to an increase of civil strife. The Sharpeville massacre of 1960 followed a PAC-sponsored demonstration. In that year the Congress was banned along with the ANC, and some leaders were imprisoned while others fled to neighbouring countries, encouraging a continuation of the armed struggle inside South Africa. As the country moved away from apartheid in the 1990s, the PAC was re-legalized and showed willingness to co-operate with the ANC.

PAN-SLAVISM A movement advocating greater unity among Slavic peoples, which had some influence in Eastern Europe in the late 19th and early 20th centuries. The leading figures in the movement were mostly Russians, such as Aksakov, who called for Russian "liberation" of Slavs oppressed by their Germanic or Turkish rulers. The movement had religious overtones and the Serbs (who, like the Russians, were Orthodox) attracted particular sympathy. Its aims also coincided with those of Russian nationalists who dreamed of capturing Constantinople. Russia's pro-Serb policy in the Balkans before the First World War was strongly supported by Pan-Slavists.

PASHA, ENVER (1881-1922) Turkish soldier and politician. He came to prominence in 1908 as an important member of the Young Turk movement of ambitious army officers. After a spell in Berlin as military attaché, he became an enthusiastic supporter of close relations with Germany. As war minister, he was instrumental in bringing Turkey into the First World War on the German side in 1914, and enlisted the aid of German military advisers. After Turkey's surrender in 1918, he joined the resistance to Soviet forces in Turkestan and is thought to have died in action.

PATTON, GEORGE (1885-1945) US general. He served in Mexico in 1916. In the First World War, in 1918, he took part in the first US tank action. He was a prominent commander in the Second World War, although his single-mindedness sometimes led to friction with his superiors. In North Africa and Sicily, and later in France, he proved a skillful user of tank units. His troops occupied Bavaria, where he became military governor after the war. He was killed in a road accident.

PERESTROIKA Russian term meaning "reconstruction". Perestroika was introduced after 1985 by the Soviet leader Mikhail **Gorbachev**. The policy involved the loosening of state economic control and moves towards greater democracy across the Soviet bloc. However, despite the vigour with which the policy was championed by Gorbachev, it had little impact on what by then was the almost terminal state of the Soviet economy and succeeded mainly in alienating almost all strands of political opinion. While old-style communists decried perestroika as an abandonment of traditional communism, those in favour of reform became impatient at what they saw as the slowness with which it was introduced. With the collapse of communism in Eastern Europe from 1989, hard-line elements within the Soviet Communist Party staged a coup against Gorbachev in August 1991. Despite its failure, the coup hastened Gorbachev's fall from power in December the same year.

PERÓN, GENERAL JUAN (1895-1974) Argentinian soldier and politician. After serving as a military attaché in Rome and Berlin, where his thinking was influenced by fascism, he participated in the military coup of 1943 and became minister of labour and social security. His social policies attracted wide support and he was elected president in February 1946, advocating strongly nationalistic policies. At first, his attempts to make Argentina economically independent had some success, but economic decline in the 1950s, and conflicts with the Church over his proposed social reforms, reduced his popularity. He was forced into exile by an armed rising in 1955 but returned in 1973 and was again elected president, dying the following year.

PÉTAIN, PHILIPPE (1856-1951) French general and head of state. He was a senior commander in the First World War, leading the heroic defence of Verdun in 1916. After mutinies in the French army in May 1917, he was made commander-in-chief and restored morale, ensuring that France would play its part in the

Allied victories in 1918. He was made a marshal after the war and led French forces in Morocco during the inter-war period. After the fall of France in June 1940, he became prime minister and negotiated an armistice with the Germans. He became the head of state in the unoccupied zone of France (Vichy France), but increasingly found himself obliged to collaborate with the Germans. After the liberation, he was tried for treason and sentenced to life imprisonment.

PILSUDSKI, JOSEF (1867-1935) Polish leader. He was a vigorous opponent of Russian rule in Poland before the First World War, and set up a nationalist force with the help of Austria. He co-operated with the Central Powers until mid-1917, when his refusal to accept a German-dominated Polish state led to his imprisonment. After the war, he quickly became the leader of the new Poland, waging an indecisive war against Bolshevik Russia but resigning in 1923 after disagreements with political parties. He took power again in 1926, but ill health and continuing internal conflict weakened his autocratic regime. He remained hostile to Russia, and feared in his last years that Nazi Germany would also threaten Poland's independence.

PINOCHET, GENERAL AUGUSTO (1915-) Former Chilean head of state. He rose to prominence when appointed commander of the Santiago zone by Chile's Marxist president, Allende, in 1972. He succeeded General Carlo Prats as commander of the army and emerged after a violent coup as head of the ruling military junta the following year. He assumed sole leadership in 1974 but stood down as head of state in 1989 following a plebiscite in favour of democratic elections.

PKK The Worker's Party of Kurdistan, set up in the late 1970s by Abdullah Öcalan. An advocate of armed struggle against the governments ruling the area claimed as the Kurdish homeland, the PKK provoked violent reaction from Turkey in the 1990s, and was condemned by other Kurdish organizations.

POL POT (1927-) Cambodian communist leader. From a peasant background, he was involved in anti-French activity during the Second World War and joined the Cambodian communist party in 1946. He led the Khmer Rouge (communist guerrillas) from the 1960s and seized power in 1975, whereupon he attempted to create a self-sufficient socialist state. The methods he employed in his attempt to reduce Cambodia to "year zero" were brutal by even the most extreme standards of 20th-century totalitarianism. Though precise figures may never be known, the death toll is believed to number two million, one fifth of the population of Cambodia. Vietnamese military intervention brought him down in 1979, but he remained a Khmer Rouge leader during a renewed guerrilla campaign.

PORTSMOUTH, TREATY OF (5 September 1905) Settlement following Japan's victory over Russia in the war of 1904-5. Russia surrendered southern Sakhalin and her rights in southern Manchuria, notably the lease of Port Arthur. Japan was also

confirmed as the dominant power in Korea. However, Russia refused to pay an indemnity, and many Japanese felt that the terms were unduly lenient. The treaty reduced Russo-Japanese rivalry and diverted Japan's attention southward, paving the way for later Sino-Japanese conflicts.

POTSDAM CONFERENCE Last conference of Allied Second World War leaders, held in July 1945. **Stalin** was joined by **Truman** (**Roosevelt** had died in April) and **Churchill**, who was replaced in mid-conference by **Attlee**, the new British prime minister. It was agreed that the four occupying powers should retain control over their respective zones of occupation in Germany and Austria. The Western Allies also accepted Stalin's suggestions for the boundaries of Poland. At the same time, Stalin agreed to enter the war against Japan at once. The conference effectively accepted Russian dominance of Eastern Europe. However, the failure to convene a final peace conference paved the way for the later division of Germany.

PRINCIP, GAVRILO (1895-1918) Serbian nationalist. A member of the secret Serbian nationalist society known as the Black Hand, he assassinated Archduke Franz Ferdinand and his wife at Sarajevo in 1914. He died in an Austrian prison of tuberculosis.

RABIN, YITZHAK See page 169

RAEDER, ERICH (1876-1960) German admiral, commander-in-chief of the navy from 1935. Although his build-up of a powerful fleet was incomplete when the Second World War broke out, he encouraged the attack on Norway in 1940. He then adopted a cautious approach after the fall of France, hoping to weaken Britain by cutting her supply lines. He considered the attack on Russia in 1941 reckless. His disagreements with Hitler led to his resignation in January 1943. He was jailed for ten years at the Nuremberg trials.

RAHMAN, MUJIBUR (1920-75) Bangladeshi leader. From a landowning family, he studied law before helping to set up the Amori League, calling for an independent East Pakistan ("Bangladesh"). The Pakistani government resisted these demands, but the League's electoral success in 1970 enabled Mujibur to become prime minister of Bangladesh in 1972. His policies were strongly socialist, and by 1975 he had effectively become a dictator. He was overthrown by a military coup and executed along with his family.

RASPUTIN, GRIGORI (c. 1871-1916) Russian mystic. He claimed to have healing powers, and gained the favour of the tsarina, who was convinced that he could alleviate the sufferings of Alexis, heir to the throne and a haemophiliac. During the First World War, he became increasingly influential at court, taking advantage of the tsar's absences at the Front. His growing hold over the tsarina and his scandalous private life made him highly unpopular and discredited the monarchy. He was murdered in December 1916 by a group of aristocrats led by Prince Yusupov.

REAGAN, RONALD (1911-) US president. Formerly a radio announcer and film star, he went into Republican politics in the 1960s and was governor of California from 1967 to 1975. After several attempts, he gained the Republican nomination for the 1980 presidential election, in which he defeated the incumbent, Jimmy Carter. Though a fierce champion of economic liberalism and a noted opponent of the Soviet Union ("the evil empire"), he was never less than pragmatic and presided over a huge increase in government spending. The economic boom this brought about, coupled with general approval for his anti-Soviet stance, ensured him wide popularity which his genial personality only served to reinforce. Interventions in Nicaragua and Grenada boosted his standing and America's renewed self-confidence in equal measure. He was re-elected in 1984. During his second term, there were growing economic problems (the stock market collapsed in October 1987), but relations with Moscow improved as the Soviet regime became less autocratic. Before the expiry of his term in 1989, the "Iran-Contra" arms scandal damaged the reputation of his administration.

RED BRIGADES, THE See page 208

RHEE, SYNGMAN (1875-1965) Korean politician. A supporter of Korean independence movements from an early age, he formed a government in exile in 1919. He spent the Second World War in the US and returned to Korea after 1945, hoping to create a unified and non-communist state. He was president of South Korea from 1948 (the division of the country was confirmed by the Korean War of 1950-3) and remained in power until 1960, when economic discontent and criticism of his regime led to his overthrow. He remained in exile in Hawaii until his death.

RIVERA, GENERAL PRIMO DE (1870-1930) Spanish leader. After a long period of military service, including a spell as captain general of Barcelona, he seized power in September 1923, promising to restore law and order after bitter industrial unrest. He soon established an autocracy, but was unable to solve the country's fundamental socio-economic problems. Attempts were made to modernize industrial relations and win over the working classes, but these alienated the industrialists and army leaders who had originally supported him. He was increasingly isolated, resigning in January 1930 as the economic depression struck.

ROME, TREATY OF Agreement, signed on 25 March 1957, which set up the European Economic Community, or "Common Market", forerunner of today's European Union (EU). In June 1955, the six members of the European Coal and Steel Community (France, West Germany, Italy, Holland, Belgium and Luxembourg) agreed to set up such a community and asked the Belgium statesman, **Spaak**, to put forward specific proposals. As set up by the Treaty of Rome, the EEC was a customs union, with common external tariffs. Member countries would trade freely, with restrictions on the movement of labour and capital to be abolished. The EEC duly came into existence in January 1958.

ROMMEL, ERWIN (1891-1944) German soldier. He served with distinction in the First World War and became a commander of armoured units in the Second World War. In May 1940, his troops broke through the Allied lines on the Meuse and advanced deeply into northern France, thus paving the way for the German victory in the West. In 1941 he was sent to North Africa as commander of the Afrika Korps. Here he achieved notable successes, driving the British to the borders of Egypt and threatening Alexandria and Cairo. Allied superiority in men and material were to prove irresistible, however, and defeat at **El Alamein** in November 1942 presaged the start of a long retreat. Posted to Europe, he was placed in charge of Germany's forces in northern France. The success of the Allied D-Day landings in June 1944 dented his reputation, though in truth the German forces were forced onto the back foot from the very start of the campaign. After being implicated in anti-Hitler conspiracies, he committed suicide in October.

ROOSEVELT, FRANKLIN D. (1882-1945) US president. He sat in the New York State Senate as a Democrat (1911-13) before becoming assistant secretary to the navy (1913-20). Having failed to become vice-president in 1920, his political career seemed over when he was crippled by polio in 1921. In the face of his disability, he became governor of New York in 1928 and president four years later. He sought to combat economic depression with the **New Deal**, which involved high government spending. The resulting recovery, partial though it was, coupled with his consistently winning personality, won him immense popularity and ensured his re-elections in 1936 and, for an unprecedented third term, in 1940. He encouraged support for Britain against Nazi Germany, but an isolationist public only accepted entry into the Second World War when the Japanese attacked Pearl Harbor in December 1941. He was influential in Allied strategic debates, favouring concentration on the European war. He was re-elected once more in 1944, but died in April 1945, shortly before the final Allied victory.

ROSENBERG, ALFRED (1893-1946) Nazi theorist. Born in Estonia, he came to Germany after the Russian Revolution and joined the Nazi movement, becoming editor of the party newspaper. He helped to develop the theory of an Aryan master-race and was especially hostile towards Christianity and Judaism. His influence on Nazi ideology was considerable, but he never became politically powerful, despite being appointed minister for the occupied East in 1941. He was executed after the war by the Allies for war crimes.

RUSSO-FINNISH WAR After the Nazi-Soviet Pact and the outbreak of the Second World War, the Soviet Union embarked on a policy of expansion in Eastern Europe, hoping to strengthen itself against any German attack. In October 1939 Moscow demanded strategically important territory from Finland. By 30 November, with no agreement, the Russians attacked. Despite a series of humiliating military defeats at the hands of the Finns, in the end

Russia's huge numerical superiority proved decisive. Finland accepted the Soviet demands in March 1940. The war demonstrated the damage done to the Red Army by the purges of the 1930s and encouraged Germany to attack Russia in 1941 - with the help of the Finns.

SADAT, ANWAR (1918-81) Egyptian leader. After military training, he was involved in the 1952 coup and became an influential supporter of President **Nasser**, succeeding him in 1970. In alliance with Syria, he launched the unsuccessful invasion of Israel in 1973 (the Yom Kippur War). By 1977 Sadat had decided that continued war with Israel was not only expensive but futile, and he began a drive to normalize relations, addressing the Israeli parliament and, in March 1979 under US auspices, signing the Camp David Accord with Israeli prime minister, Menachem Begin. In revenge for his "betrayal" of the Arab cause, he was assassinated by fundamentalists in October 1981.

SALAZAR, ANTONIO DE (1889-1970) Portuguese dictator. After studying economics, he was appointed minister of finance in 1928, in which capacity he ensured that Portugal avoided the worst effects of the Depression. He became prime minister in 1932 and established a right-wing dictatorship. At first, his sound financial policies were popular, as socio-economic conditions improved. However, modernization was not undertaken and Portugal, neutral in the Second World War, languished after the war, becoming one of the poorest countries in Europe. The country was also involved in costly and futile colonial wars, as Salazar resisted moves to grant independence to its African colonies. A stroke ended his career in 1968.

SAN FRANCISCO, TREATY OF Peace treaty between Japan and most of her Second World War enemies, signed on 8 September 1951. Japan agreed to recognize Korean independence and to abandon claims to various Pacific territories, effectively giving the US a free hand in these areas. The signing of the treaty symbolized Japan's return to international respectability, with the post-war Allied occupation ending soon afterwards. With the USSR and the rival Chinese states not party to the treaty, Japan's position as an ally of the US was duly confirmed.

SAVIMBI, JONAS (1934-) Angolan political leader. An active supporter of the campaign for Angolan independence from Portugal, he formed the UNITA nationalist movement in the 1960s. After independence in 1975, UNITA fought the rival MPLA and FNLA groups for mastery in the country. Savimbi's opposition to communism enabled him to gain the support of South Africa and the US, in the process ensuring that he became a highly controversial figure in post-colonial Africa. In 1991 he accepted a ceasefire, but conflict between MPLA and UNITA supporters continued.

SISULU, WALTER (1922-95) South African politician. He joined the **African National Congress** in 1940, becoming secretary general in 1949. Associated with the hardline Youth League, he continued agitation after the banning of the ANC in 1960. In 1964 he was sentenced to life imprisonment. Along with **Nelson Mandela**, he became a symbol of the African nationalist movement. After his release in 1989, he helped to reorganize the ANC. He later took part in negotiations for the transition to full democracy in South Africa.

SOUTH EAST ASIA TREATY ORGANIZATION (SEATO) Based at Bangkok, SEATO was set up by the Manila Treaty (September 1954). The signatories were the US, Britain, France, Australia, New Zealand, Pakistan, Thailand and the Philippines. The objective was to protect members from internal or external threats, above all from what the treaty described as "communist aggression". Differences between the members (France and Pakistan withdrew in the early 1970s), and the refusal of many Asian countries to join, ensured that SEATO was never truly effective. Its failure to develop a clear policy on Vietnam helped cause its final demise in 1977.

SPAAK, PAUL-HENRI (1899-1972) Belgian statesman. After legal training, he entered parliament as a socialist in 1932 and joined the cabinet in 1935. He held the posts of foreign minister and prime minister before 1940, supporting Belgium's return to neutrality after 1936. He participated in the government in exile during the Second World War, and had three more spells as foreign minister in the 20 years after the war. He favoured the development of the alliance with Holland and Luxembourg ("Benelux") and was a firm supporter of moves towards closer European integration. He was instrumental in the setting up of the European Economic Community in 1957. He also recognized the importance of the North Atlantic alliance, and served as NATO secretary general from 1957 to 1961.

STALIN (1879-1953) Born Joseph Vissarionovich Dzhugashvili. Son of a Georgian shoemaker, he trained for the priesthood but was expelled in 1899 after becoming a Marxist. In 1917 he became Peoples' Commissar for Nationalities in the Soviet government and general secretary of the Communist Party of the Soviet Union from 1922 until his death. He eliminated all rivals after the death of Lenin in 1924; and promoted an intensive industrialization, the forced collectivization of agriculture (in which millions died) and the development of a police state. A series of show trials in the late 1930s further strengthened his grip on power though also decimated the leadership of the Red Army. He signed a non-aggression pact with Nazi Germany in 1939, resulting in the Soviet occupation of eastern Poland and Finland. Having refused to believe that Germany would invade the USSR in 1941, in the event **Stalin** proved a commanding war leader, rarely interfering in the decisions of his generals. He oversaw a massive increase in Soviet influence worldwide after the war, especially in Eastern Europe, which was firmly placed within the Soviet sphere of influence.

STRATEGIC DEFENCE INITIATIVE (SDI) Plan put forward by US President **Reagan** in 1983 to improve the US defences against nuclear attack. According to the "star wars" theory, enemy missiles would be intercepted and destroyed in space. SDI was soon criticized as expensive and unworkable while others claimed that even if it could be made to work it would destroy the system of deterrence which had prevented nuclear war since 1949. The initiative also caused friction with Moscow. As its extreme technical complexity became clear, it was effectively shelved after Reagan's retirement in 1989.

STRESA CONFERENCE Anti-Nazi doctrine propounded after a meeting between Italian, French and British leaders in April 1935 to consider Hitler's announcement that Germany would no longer obey the Versailles limits on her armed forces. The three powers condemned Hitler's action and confirmed their support for Austrian independence and the 1925 Locarno Treaties. The "Stresa Front" was weakened when Britain signed a Naval Agreement with Germany in June, and collapsed when Italy invaded Abyssinia in October.

STRESEMANN, GUSTAV (1878-1929) German statesman. As a National Liberal, he sat in the Reichstag from 1907 until the end of the First World War, frequently calling for an expansionist foreign policy. After the war, he set up a new right-wing party and was largely responsible for German foreign policy from 1923 to 1929. He remained a keen nationalist but believed that co-operation with Germany's former enemies would be the most effective way to achieve revision of the Versailles Treaty. He was the architect of the 1925 Locarno Treaties, which recognized Germany's "Versailles frontiers" in the west, and he presided over Germany's entry into the League of Nations in 1926. His successes caused a reduction in Germany's reparations bill and the early evacuation of the Rhineland by Allied occupation forces, which was agreed upon shortly before his death.

SUKARNO, ACHMED (1901-70) Indonesian leader. In 1927 he set up a nationalist movement, calling on the Dutch authorities to grant the Indonesian people greater autonomy. He co-operated with occupying Japanese forces during the Second World War. After the Japanese defeat in August 1945, he proclaimed an Indonesian republic, with himself as leader. The Dutch finally recognized his regime in 1950. Sukarno was a notable figure in the non-aligned world, as shown by his hosting of the Bandung Conference in 1955. Persistent economic depression and domestic unrest led to Sukarno's eclipse in 1967, although he officially remained president until March 1968.

SUN YAT-SEN *See* page 23

TANNENBERG, BATTLE OF First World War battle between Germany and Russia in which Russia suffered a crushing defeat. The Russians began an advance into East Prussia shortly after the outbreak of the war in August 1914. The Germans initially fell back, but poor communications between the Russian commanders, Rennenkampf and Samsonov, enabled the Germans to encircle Samsonov's army at the end of August. The Russians were unable to break the German ring and German artillery inflicted enormous casualties, Samsonov himself committing suicide. The battle decisively halted the Russian advance and contributed to the growing popularity in Germany of the commanders **Hindenburg** and **Ludendorff**.

TARANTO, BATTLE OF Second World War naval engagement between Britain and Italy. By late 1940 the British were determined to cripple the Italian fleet, and on 11 November they sent Swordfish bombers (based on an aircraft carrier) to attack it at Taranto in south east Italy. One battleship was sunk and two others badly damaged, at the cost of only two aircraft. The victory gave Britain clear naval superiority in the Mediterranean and improved the chances of Allied success in North Africa. The effectiveness of this surprise attack also demonstrated the growing importance of naval air power and may have encouraged the Japanese to take similar action against the US fleet at Pearl Harbor in 1941.

TAYLOR, FREDERICK (1856-1915) American engineer and management scientist. His promising academic career was hindered by poor eyesight, but he recovered to obtain a degree in mechanical engineering, becoming a successful chief engineer at the Midvale Steel Company in 1884. He resigned soon afterwards to develop theories of industrial management based on his experiences. His main argument was that efficiency would be greatly increased if employees avoided unnecessary movements while working. As a lecturer and consultant, he encouraged the spread of "scientific management". The Taylor system proved influential in the age of mass production, but often attracted hostility from workers' movements.

TEHERAN CONFERENCE (28 November-1 December 1943) First meeting of the "Big Three" Allied leaders of the Second World War: **Roosevelt**, **Churchill** and **Stalin.** It was agreed that the Western Allies would ease the pressure on the Soviet Union by launching an invasion of France in mid-1944. In return, Stalin agreed to join the war against Japan once Germany had been defeated. The future of Germany (and its border with Poland) was discussed, as was the situation in the Mediterranean and Middle East. A notable aspect of the conference was the understanding built up between Roosevelt and Stalin.

THANT, U (1909-74) Burmese statesman. Originally a teacher, he joined the Burmese government in the late 1940s and represented his country at the United Nations. In 1961 he succeeded Hammarskjöld as UN secretary general, his position being confirmed in the following year after various successes, notably during the Cuban Missile crisis of 1962. During his ten years in office he presided over UN mediation efforts in Cyprus, the Congo and the Middle East. He retired in 1971 and died in New York.

THATCHER, MARGARET (1925-) British prime minister. A Conservative MP from 1959, she served in Edward Heath's 1970-4 government. In 1975 she replaced Heath as party leader. In 1979 the Conservatives were re-elected, with Thatcher becoming Britain's first woman prime minister. Combining a fierce and instinctive Conservatism with a dominating personality, she stamped herself indelibly on the 1980s. "Thatcherism", though defined in very different ways by most people, became one of the key doctrines of the period. Privatization of state-owned industries, reductions in union power and state spending, and vehement opposition to totalitarianism in any guise were the hallmarks of her leadership. Though opponents accused her of being being in thrall to the anti-Soviet policies of the **Reagan** administration in the US and of encouraging a culture of greed, supporters championed her forthright belief in freedom and democracy and her regeneration of the British economy. She was re-elected in 1983 and 1987, partly because of divisions within opposition parties. Increasing resentment of her and of what many decried as an ever more regal manner climaxed in 1990 when, in November, she was forced to resign the leadership of the Conservative Party. Having been succeeded by the more moderate John **Major**, she retired from parliament in 1992, taking a seat in the House of Lords.

TITO (1892-1980) Yugoslav head of state. Born of Croatian origin as Josip Broz, he served in the First World War for the Central Powers and later fought for the Red Army in the Russian civil war. He became leader of the Yugoslav communist movement after the war, despite periods of imprisonment, and later organized effective resistance to German occupation forces in the Second World War. He led the new Yugoslav state from November 1945, and his independent policy led **Stalin** to cut ties in June 1948. Tito's refusal to accept Moscow's authority (shown by his condemnation of Soviet repression in Hungary and Czechoslovakia) enabled Yugoslavia to become prominent in the association of non-aligned nations. After his death, the old rivalries between the various ethnic groups inside Yugoslavia resurfaced in sharper form.

TOJO, HIDEKI (1884-1948) Japanese general and war leader. In the army from his youth, he was involved in the occupation of Manchuria in the 1930s. His keen advocacy of expansionist policies enabled him to enter the government in 1938, and he strongly supported the war against China. He became war minister in July 1940 and encouraged the conclusion of the Tripartite Pact with Germany and Italy. In October 1941 he replaced Prince Konoe as prime minister and presided over the Japanese successes after Pearl Harbor. He suppressed opposition at home and favoured exploitation of the conquered territories, though he attempted to win over many East Asian peoples with talk of a "co-prosperity sphere". He resigned in July 1944 as the Japanese forces retreated and was executed by the Allies as a war criminal.

TROTSKY, LEON (1879-1940) Russian revolutionary. He joined the socialist movement in the late 1890s and soon became a prominent figure, despite periods of exile. His political allegiance was uncertain until he finally sided with the **Bolsheviks** during the 1917 Revolution, in which he played a key role in their seizure of power in November. He became commissar for foreign affairs, leading the peace negotiations with Germany. He then led the Red Army to victory in the civil war. His overbearing personality and belief in "worldwide revolution" (as opposed to **Stalin**'s plan for "socialism in one country") made him unpopular in the party. At the same time, Stalin, ever conscious of threats to his leadership, was determined to dispose of his rival. Trotsky was accordingly exiled in 1929. He continued to criticize Stalin's government from abroad, and was murdered by a Stalinist agent in Mexico.

TRUMAN DOCTRINE In March 1947, US president **Truman** declared that the US would assist democratic states threatened by internal or external force. This "Truman Doctrine", sparked in large measure by the communist uprising in Greece, was clearly directed against the Soviet Union and underlined the US's decisive move away from the isolationist policies of the 1920s and '30s. As the Cold War developed, the assertion of the Truman Doctrine was followed by the US-sponsored Marshall Plan for European economic recovery and by the creation of NATO in 1949-50.

TRUMAN, HARRY S. (1884-1972) US president. After serving in the First World War, he was a county judge and sat in the US Senate from 1934. He unexpectedly became vice-president in January 1945 and succeeded **Roosevelt** upon the latter's death in April. He participated in the July 1945 **Potsdam Conference** and ordered the atom bomb attacks on Japan in August. As the Cold War developed, he put forward the **Truman Doctrine** (March 1947), which promised American support for those threatened by communism. He also oversaw the setting-up of NATO in 1949-50, having been re-elected in 1948. His domestic policies included concessions to the black population, shown by the de-segregation of the army in 1948, though other reforms were held up by Congress. He demonstrated his independence, however, by sacking General **MacArthur** in 1951 after disagreements over the Korean War. He retired into private life in 1952.

TUDJMAN, FRANJO (1922-) Croatian president. He fought with **Tito**'s partisans in the Second World War, later becoming a Yugoslav army general and then a professional historian. In 1990, he became president of Croatia. His proclamation of Croatian independence from Yugoslavia in the following year led to an invasion by the Serb-led Yugoslav federal army. Large areas of Croatia were lost to rebel Serbs, but international recognition of Croatia and the diversion of Serb attention to Bosnia strengthened Tudjman's position. In 1995 the Croatian army defeated the Croatian Serbs and restored the country's territorial integrity. By 1996 it appeared that Tudjman had succeeded in establishing a strong Croatian state.

VARGAS, GETÚLIO (1883-1954) Brazilian leader. After a period as finance minister (1926-8), he led a successful uprising in November 1930, thereafter ruling the country as a dictator until 1945, except for three years of semi-constitutional rule (1934-7). He was elected president in 1950, but popular discontent led to his suicide four years later.

VENIZELOS, ELEUTHERIOS (1864-1936) Greek prime minister. In 1896, he participated in the revolt to free his native Crete from Turkish rule and was later instrumental in bringing about Crete's union with Greece. After entering Greek politics, he became prime minister in 1910, presiding over Greece's successes in the Balkan Wars of 1912-13. He sought to lead Greece into the First World War on the Allied side, and allowed an Anglo-French force to base itself at Salonika. When King Constantine I (who opposed war with Germany) dismissed him, he went into open opposition and in 1917 forced Constantine to abdicate. His vigorous support for the Allies enabled Greece to make substantial territorial gains at the post-war peace negotiations. He was prime minister again for brief periods in the inter-war years, but was forced into exile in 1935 after an unsuccessful rising.

WALDHEIM, KURT (1918-) Austrian politician. He studied in Vienna before serving on the Eastern Front during the Second World War. He entered the Austrian foreign service after 1945 and held various posts before becoming foreign minister in 1962. From 1971 to 1981 he was the UN secretary general. Tensions between the super-powers reduced the UN's effectiveness during the period. After returning to Austria he campaigned for the presidency amid accusations that he had been involved in Nazi atrocities in Eastern Europe during the war. He was nonetheless elected president in 1986 and has remained an important if controversial figure in Austrian politics.

WALESA, LECH (1943-) Polish politician. A shipyard worker by profession, he became politically active in the 1970s, calling for a relaxation of the communist system. His Solidarity union movement, supported by the Church, was outlawed in 1981 when the government proclaimed martial law. Walesa himself was arrested, but was released soon afterwards. With the collapse of Soviet communism in Eastern Europe in 1989, Walesa was widely recognized as the country's natural leader and became president the following year. However, despite his international standing, increasing economic difficulties undermined his position in Poland. He was defeated as president in the elections of 1995.

WALL STREET CRASH *See* page 62

WAR COMMUNISM Policy adopted by the **Bolsheviks** during the Russian civil war. Lenin insisted that the Russians must give the war absolute priority and take control of every sector of the economy. War Communism involved large-scale nationalization, seizure of food from peasants and state control of distribution and exchange. These measures caused revolts by industrial workers and peasants. At the same time, a thriving black market developed. The policy was replaced by the **New Economic Policy** (NEP) in 1921. War Communism may have enabled the Bolsheviks to supply the Red Army and win the civil war but it imposed enormous strains on the Russian economy and reduced the Soviet regime's popularity.

WARSAW PACT Alliance of communist East European states, formed in May 1955 as a response to the establishment of NATO. The agreement obliged the signatories to defend one another and to accept a unified military command. The pact also enabled the USSR to retain control of Eastern Europe and was used to justify intervention in Hungary (1956) and in Czechoslovakia (1968). As the communist regimes collapsed in 1989-90, the Warsaw Pact rapidly disintegrated. It was abolished officially in June 1991.

WASHINGTON CONFERENCE (1921-2) Conference called to deal with naval disarmament and Far Eastern problems, with nine powers participating. A Five-Power Treaty banned the construction of heavy capital ships for ten years and fixed the proportional naval strengths of the US, Britain, Japan, France and Italy according to a ratio of 5:5:3:1.75:1.75 respectively. Britain, France, Japan and the US also agreed to co-operate against aggression in the Far East. All nine powers agreed to respect the integrity of China. The Conference underlined the growing power of Japan in the Far East.

WEST WALL After the German remilitarization of the Rhineland in March 1936, Hitler ordered the construction of a line of defensive fortifications along Germany's western frontier. This West Wall, or "Siegfried Line", was the equivalent of France's Maginot Line and was similarly equipped with a formidable array of pill-boxes and anti-tank defences. By 1945, however, the strength of the Allied forces was sufficient to overrun it.

WILHELM II (1859-1941) German Kaiser (1888-1918). His reign began controversially when, in 1890, he dismissed the long-serving chancellor, Bismarck, whose cautious foreign policy he opposed. Wilhelm's policies of naval and colonial expansion, together with strident public statements, caused concern across Europe. He was determined to extend Austro-German power in Central Europe, even at the risk of war, and he encouraged Austria to take a tough stance towards Serbia after the assassination of the Archduke **Franz Ferdinand** and his wife in Sarajevo in June 1914. During the First World War, he tolerated the unofficial takeover of power by the military commanders **Hindenburg** and **Ludendorff**. In 1918, the German defeat and resulting turmoil forced him to flee to Holland, were he remained until his death, protected by the Dutch government from a war crimes trail.

WILSON, WOODROW (1856-1924) US president. Generally thought an outstanding chief executive of the United States, his two terms in office (1913-21) covered the First World War and the Paris peace conference, at which he was the dominating figure. His influence on the re-shaped Europe after the war, not least as a result of his **Fourteen Points**, was substantial, though many of Wilson's ideological convictions in favour of nationalism were subverted by the desire of the other victorious Allied leaders that Germany should be made to pay reparations for the war. Nonetheless, though a controversial figure during his first years in office, Wilson won temporary fame as the world's foremost leader after the war, though his power and influence later declined. The US Senate repudiated the League of Nations, which he had ardently advocated, while ill health sapped his capacity to govern.

WITTE, SERGEI (1849-1915) Russian politician, finance minister from 1892. He believed that Russia needed to industrialize rapidly to maintain its Great Power status, and that central planning was needed for this. He particularly encouraged the construction of new railways (notably the Trans-Siberian), hoping that this would extend Russian influence in the Far East without the need for war. He considered that his policy had been vindicated by Russia's defeat in the 1904-5 war with Japan. He negotiated peace terms with the Japanese and became Russia's first prime minister in November 1905. He was admired in the West as an able modernizer, but was forced to leave office after less than six months.

WRIGHT, ORVILLE (1871-1948) American inventor. With his brother, Wilbur (1867-1912), he began experimental work on aviation in the late 19th century, making the first powered airplane flight in Kitty Hawk, North Carolina, in 1903. The event was greeted by public apathy. By 1905, the Wright brothers had managed a flight lasting half an hour. Well before Wilbur's death, the immense significance of their breakthrough had become abundantly clear.

YALTA CONFERENCE Meeting of Allied war leaders (**Roosevelt**, **Stalin** and **Churchill**) in February 1945. With the Axis powers clearly close to defeat, positive decisions could be made about the post-war order. Plans for the treatment of Germany were developed, with France admitted as one of the major Allied victors and accordingly granted zones of occupation in Germany and Austria on the same lines as those assumed by the US, the USSR and Britain. A declaration stressed that democracy should be encouraged in the liberated countries of Europe, and it was agreed that Poland should cede territory to the USSR and gain some from Germany. However, the exact frontiers and the composition of the new Polish government remained uncertain. The leaders also discussed the form of the new United Nations. Most of the Yalta agreements were unclear and differences of interpretation contributed to the rapid collapse of the wartime alliance.

YAMAMOTO, ISOROKU (1884-1943) Japanese naval commander, admiral from 1927. During the inter-war period, he stressed the importance of naval air power. He became deputy navy minister in 1937 and naval commander-in-chief in August 1939. Although he doubted Japan's ability to defeat the US and Britain in a long war, he organized the attack on Pearl Harbor in December 1941. After several further successes in the Pacific, Japanese forces under Yamamoto's command were defeated by the US at Midway in June 1942. The battle proved a decisive turning point in the Pacific war, and Japan found herself increasingly on the defensive thereafter. Yamamoto remained commander until April 1943, when his airplane was shot down by Allied forces over the Solomon Islands.

YELTSIN, BORIS (1931-) Russian president. He joined the Soviet Communist Party in 1961 and its central committee in 1981. After **Gorbachev**'s takeover in 1985, he became party leader in Moscow, where he introduced a programme of rapid reform. However, his outspoken attacks on hardliners led to his demotion. In May 1990 he became president of the Russian Republic and his criticism of Gorbachev became more open. In August 1991, he played a key role in defeating the hardline coup mounted against against Gorbachev. The prominence he gained led swiftly to his elevation to the leadership of the new Russian Federation after the collapse of the USSR at the end of 1991. He survived another coup attempt in 1994, but growing lawlessness and economic hardships have since weakened his administration while his standing, especially in the West, has been undermined by persistent reports of alcoholism. He stood for re-election in 1996.

ZHOU ENLAI (1898-1976) Chinese communist politician. He was associated with Mao and the Chinese communists from his student days, and in 1927 he was arrested in Shanghai by **Chiang Kai-shek**'s nationalist forces. After escaping, he continued his agitation, and was a prominent communist leader during the Second World War. When the People's Republic of China was established in 1949, he became prime minister and foreign minister. In foreign affairs, he helped to end wars in Korea and Indo-China. His diplomacy greatly enhanced the new China's international standing, as shown by US President **Nixon**'s visit in 1972. At home he encouraged party unity, notably during the Cultural Revolution. A much-respected figure, he remained premier until his death.

ZHUKOV, GEORGI (1896-1974) Soviet general. From a peasant background, he served in the First World War and later in the Red Army, showing his ability by defeating the Japanese in Mongolia in 1939. He led the defence of Leningrad against the Germans in 1941 and, in December, organized the successful counter-attack before Moscow. He enjoyed further successes at Stalingrad and Kursk in 1942-3 before planning the final attack on Berlin in 1945. He became minister of defence in 1955 but was retired soon afterwards. He is generally considered the outstanding Soviet commander of the Second World War, although his victories were usually costly.

ZIONISM Movement advocating the return of the Jewish people to their original home in Palestine. Growing persecution of European Jews, especially in Russia, led to demands for a Jewish state in the early 20th century. The British Balfour Declaration of 1917 seemed to encourage this. In the inter-war period, the World Zionist Organization (led by Chaim Weizmann) encouraged Jews to move to Palestine. The Zionists achieved success with the establishment of the State of Israel in 1948, to the dismay of Palestine's subjugated Arab population. Zionism subsequently became one of a number of all-purpose targets for revolutionary movements around the world.

INDEX

*page numbers in italics
refer to pictures*

motorization, 192-3
Nazi-Soviet conflict, 84-7
new economic super-power, 154-5
partition of, 108-9
persecution of Jews, 70-1, 100-1, 182
post-war settlement, 46-9, 52
refugee problems, 182-3
reparation, 108-9
scorched earth policy, 84
Second World War, 78-9, 82-5, 92-4
States of the New Order, 70-1, 84-5, 94
unemployment, 64-5
unification with East Germany, 146
withdrawal of army, 47
Yugoslav civil war, 152
Ghana, 53, 132, 133
Gibraltar, 94
Giscard d'Estaing, Valéry, 117
glasnost, 148
global
economy, 20-1, 54-5, 62-3, 172-3
markets, 173
warming, 204-5
Goddard, Robert, 212
Golan Heights, 137
gold, standard, 20-1
Gold Coast, see Ghana
Golden Triangle, 210
Gomorrah Operation, 98
Gonzolo, Chairman, 142
Gorbachev, Mikhail, 112, 146, 147-9
Göring, Hermann, 83
Gotha bombers, 33
Graziani, Marshal, 95
Great Britain, see Britain
Great Leap Forward, 128-9
Great War, see First World War
Greece, 26, 76
American intervention, 121
civil war, 108-9
the costs of war, 36-7
First World War, 32
Ottoman rule, 24
post-war settlement, 46, 50
refugee problems, 182
Second World War, 84, 94, 104
tourism, 196
greenhouse effect, 204-5
Greenland, 92
Grenada, 112, 121
Grove, Marmaduke, 56

Grozny, 149
Guadalcanal, 90
Guam, 54, 89
Guatemala, 121
Gudular Hills, tribal area doctor, 185
Guernica, bombing of, 73
Guevara, "Che", 142
Guides Michelins, 196
Guinea-Bissau, 138
Gulf of Tongking, 130
Gulf War, 166-7
Gush Emunim, 169
Guzmán, Abimael, 142
gypsies, persecution of, 100-1
Gypsy Moth, 190

H

Habyarimana, Juvenal, 138
Habsburg empire, 18, 26, 32, 42-3, 46
The Hague, drugs convention, 210
Haifa, arrival of Jewish settlers, 135
Haig, Field Marshal Douglas, 30
Haiti, 56
Halley's comet, 212
Hama (city), 166
Hamas, 168-9, 209
Hamburg, 98
Hammarskjöld, Dag, 171
Handley-Page bomber, 33
Hannibal, 191
Hanoi, 130
Harris, Air Marshal Arthur, 98
Harrison Act, 210
hashish, 210-11
Hatoyama, Ichiro, 156
Havel, Václav, 147
Hawaii, 54, 89, 104
Hawley Smoot tariff, 62
Hay-Pauncefote Treaty, 54
health, disease and, 184-5
healthcare revolution, 186-7
Hebron, 168
helicopter, first, 191
Heligoland Bight, 34
Helsinki Accords, 113
Herceg-Bosna, 151
heroin, 210-11
Herzegovina, 26
Hezbollah 166, 168-9, 209
Himmler, Heinrich, 70, 100-1
Hindenburg, Field Marshal Paul von, 30, 38, 70
Hindus, 140-1
Hirohito, Emperor, 66
Hiroshima, 91
Hitler
Adolf, 188
appointed chancellor, 64
democracy and dictatorship, 76, 77
enters political limelight, 49
and Italy, 69
Jewish persecution, 70-1, 100-1
overture to Second World War, 78-9
Second World War,

82-7, 91-101
States of the New Order, 70-1
tour of Paris, 82
Hitler Youth, 71
HIV, 187
Ho Chi Minh, 53, 130
Hohenzollern empire, 43
holidays, political use of, 196-7
Holocaust, 100-1
Honecker, Erich, 146
Hong Kong, 160-1, 211
Chinese sovereignty, 158
emigration from, 159
Japanese attack, 89
lease granted to Britain, 22
Special Economic Zones, 158
Hood, HMS, 92
Horn of Africa
Africa after independence, 139
civil wars, 112, 165
Italian territories, 68
Horthy, Admiral, 46, 76
Horton, Admiral Max, 93
Hsin-hai revolution, 23
Hu Yaobang, 159
Hua Guofeng, 129, 158
Hubble Space Telescope, 212
Hughes, Charles Evans, 54
Hull, Cordell, 170
Hungary
collapse of communism, 146-7
communism in, 58, 110, 124-5
democracy and dictatorship, 76
Jewish persecution, 101
Munich Agreement, 68
and Nazi-Soviet conflict, 87
post-war settlement, 46
Husky, Operation, 94
Hussein, King, 137, 168
Hussein, Saddam, 166, 167, 203
Hutu, 138
Huxley, Julian, 198
Hyderabad, 140

I

I-ho-ch'üan, see Boxer Rising
Ibáñez, Carlos, 56
IBM, 121
Iceland, 92
immunization, 184-7
India
cholera outbreak, 186
cinema industry, 195
Congress party, 52
the costs of war, 36-7
drug trade, 210
economic modernization, 141
Five Year Plans, 140
immunization programmes, 184-5
nationalism, 104
partial self government, 52-3
partition, 132, 140
Quit India campaign, 52, 132
religion and secular power, 200-1
Indian Ocean, 90-1
Indo-China, see also Vietnam
Japanese occupation, 88-9
Manchu dynasty, 22-3
Indonesia, 160-1
Cold War in Asia, 130
immunization programmes, 184-5
industrialization, 20-1, 31
basis of, 20
revolution in Asia, 160-1
Inkatha Freedom Party, 164
inoculation, 184-5
Institute of Historical Research, 208
Inter-American Development Bank, 121
Inter-American Reciprocal Assistance Treaty, 121
International Bank for Reconstruction and Development, 114
International Futures Exchange, London, 172
International Health Regulations, 184
International Labour Office, 50
International Office of Public Hygiene, 184

N

V

W

Y

Z